Birkhäuser Advances in Infectious Diseases

For further volumes:
http://www.springer.com/series/5444

Rino Rappuoli · Giuseppe Del Giudice
Editors

Influenza Vaccines for the Future

Second Edition

 Springer

Editors
Dr. Rino Rappuoli
Novartis Vaccines & Diagnostics
S.r.l.
Via Fiorentina 1
53100 Siena
Italy

Dr. Giuseppe Del Giudice
Novartis Vaccines & Diagnostics
S.r.l.
Via Fiorentina 1
53100 Siena
Italy

ISBN 978-3-0346-0278-5   e-ISBN 978-3-0346-0279-2
DOI 10.1007/978-3-0346-0279-2

Library of Congress Control Number: 2010938614

Cover illustration: Fig. 2 from E. C. Settembre, P. R. Dormitzer, R. Rappuoli, H1N1: Can a pandemic cycle be broken? Sci. Transl. Med. 2, 24ps14 (2010). Image: Adapted by C. Bickel/Science Translational Medicine. Reprinted with permission from AAAS.

Cover design: deblik, Berlin

Printed on acid-free paper

Preface

The pandemic caused by the 2009 A/H1N1 influenza virus has changed the manner in which the world will respond to pandemics in the future and will have an important place in history. Why is a relatively mild pandemic so important that it will leave a mark on history? The fact is that this event has represented a test of the global pandemic preparedness and has highlighted weaknesses and strengths of the health protection system worldwide. The best strategy to protect mankind against future pandemics is by vaccination. Thanks to the H5N1 avian influenza, during the past 10 years our ability to control a pandemic has improved considerably. Nevertheless, the 2009 A/H1N1 influenza pandemic has demonstrated the many weaknesses of the current pandemic preparedness plans. These weaknesses would have been fatal had this pandemic resulted in the global spread of a more lethal influenza strain. It can be said that this pandemic has provided a unique opportunity, a "fire drill", to identify the deficiencies that must be urgently addressed to develop a better and more efficient plan for the next pandemics of the twenty-first century.

This second edition of "Influenza Vaccines of the Future" intends to provide the grounds for developing such plans. The major points to be addressed for our future preparedness plans include prediction of pandemics (viral evolution and epidemiology), the features of the immune response to the virus, the development of safe and effective vaccination strategies (including quick reaction by the productive infrastructures), planning for vaccine distribution and coverage of populations at risk, and the major need of global awareness and communication.

In this perspective, the first chapters cover the latest information on the complex biology of the influenza virus and of its epidemiology in different areas of the world, to come to the evolution of the H1N1 pandemic viruses and to the features of the 2009 H1N1 pandemic. This information is instrumental to the understanding of human immunity to influenza and to the consequent development of vaccines. Several chapters are dedicated to the latest studies in searching for new vaccine antigens and effective adjuvants, in setting up predictive *in vitro* and *in vivo* models, in identifying relevant correlates of protection, in tackling possible side effects, in developing novel methodologies for vaccine production, in designing new

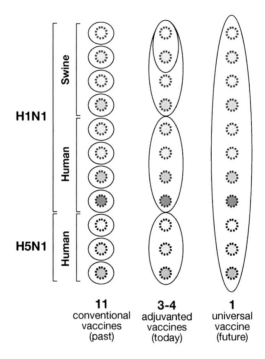

Fig. 1 The development of influenza vaccines

approaches to prophylaxis and treatment. The path of progress of influenza vaccines is summarized in the Fig. 1. Traditionally, we have used a different vaccine for every single virus variant. However, today we can protect against a subgroup of strains using oil-in-water adjuvants that induce an immune response able to cover the diversity of closely related viruses. Hopefully, in future, universal vaccines will be available which may be the final solution to pandemic and seasonal influenza.

The last chapters are dedicated to more perspective considerations, including the economic and social impact and costs of pandemic influenza, and the strategies for implementing global preparedness to the future threats.

The 2009 A/H1N1 influenza pandemic has confirmed that once a pandemic begins, the time to react is limited. The only way to address and control a pandemic is to be prepared. The response to the first influenza pandemic of the twenty-first century benefited from the extensive preparation for an avian influenza pandemic and the mild nature of the 2009 A/H1N1 swine influenza virus. However, the pandemic demonstrated the limited ability to predict influenza pandemics, to anticipate levels of cross-protection, and to deliver vaccines in a timely manner, particularly to low-income countries. The lessons learned from the 2009 H1N1 pandemic are of paramount importance to develop more effective preparations against future pandemics. We must exploit such information straight away. And get ready.

Acknowledgment

The Editors would like to thank Diana Boraschi for her professional support in the coordination of this volume preparation. Her endless monitoring has made possible the realization of this project that brings to colleagues and students the best knowledge for future vaccines for global infections.

Siena   Rino Rappuoli
August 2010   Giuseppe Del Giudice

Contents

Contributors

W. Abdullah Brooks Head Infectious Diseases Unit, International Centre for Diarrhoeal Disease Research, Bangladesh (ICDDR,B), Mohakhali, GPO Box 128, Dhaka 1000, Bangladesh, abrooks@jhsph.edu; Johns Hopkins Bloomberg School of Public Health, 615 North Wolfe Street, Suite E8132, Baltimore, MD 21205, USA

Barbara Capecchi Novartis Vaccines and Diagnostics, Via Fiorentina 1, 53100 Siena, Italy

Robert B. Couch Department of Molecular Virology and Microbiology, Baylor College of Medicine, One Baylor Plaza, MS: BCM280, Houston, TX 77030, USA, ikirk@bcm.edu

Peter C. Doherty Department of Microbiology and Immunology, The University of Melbourne, Parkville, VIC 3010, Australia; Department of Immunology, St Jude Children's Research Hospital, Memphis, TN 38105, USA

Philip R. Dormitzer Novartis Vaccines and Diagnostics, 350 Massachusetts Avenue, Cambridge, MA 02139, USA, philip.dormitzer@novartis.com

Kathryn M. Edwards Department of Pediatrics, Vanderbilt Vaccine Research Program, Vanderbilt University School of Medicine, Nashville, TN 37232, USA

Robert H. Friesen Crucell Holland BV, Archimedesweg 4-6, 2333 CN Leiden, The Netherlands

Anthony Gilbert London Bioscience Innovation Centre, Retroscreen Virology Ltd, 2 Royal College Street, London NW1 ONH, UK

Giuseppe Del Giudice Research Center, Novartis Vaccines and Diagnostics, Via Fiorentina 1, 53100 Siena, Italy, giuseppe.del_giudice@novartis.com

Jaap Goudsmit Crucell Holland BV, Archimedesweg 4-6, 2333 CN, Leiden, The Netherlands, jaap.goudsmit@crucell.com

Benjamin Greenbaum The Simons Center for Systems Biology, Institute for Advanced Study, Princeton, NJ, USA

Harry Greenberg Departments of Medicine and Microbiology and Immunology, Stanford University School of Medicine, Stanford, CA, USA; Veterans Affairs Palo Alto Health Care System, Palo Alto, CA, USA

Carole Guillonneau Department of Microbiology and Immunology, The University of Melbourne, Parkville, VIC 3010, Australia

Katja Hoschler Health Protection Agency, Specialist and Reference Microbiology Division, ERNVL, Influenza Unit, Centre for Infections, 61 Colindale Avenue, London, UK

George Kemble MedImmune, Mountain View, CA, USA, kembleg@medimmune.com

Hossein Khiabanian Department of Biomedical Informatics, Center for Computational Biology and Bioinformatics, Columbia University College of Physicians and Surgeons, New York, NY, USA

Wouter Koudstaal Crucell Holland BV, Archimedesweg 4-6, 2333 CN Leiden, The Netherlands

Robert Lambkin-Williams London Bioscience Innovation Centre, Retroscreen Virology Ltd, 2 Royal College Street, London, NW1 ONH, UK

Arnold Levine The Simons Center for Systems Biology, Institute for Advanced Study, Princeton, NJ, USA

Romina Libster INFANT Fundacion, Buenos Aires, 1406, Argentina, romina.p.libster@Vanderbilt.Edu; Department of Pediatrics, Vanderbilt Vaccine Research Program, Vanderbilt University School of Medicine, Nashville, TN 37232, USA

Catherine J. Luke Laboratory of Infectious Diseases, National Institute of Allergy and Infectious Diseases, National Institutes of Health, Bethesda, MD 20892, USA, cluke@niaid.nih.gov

Cyrus Maher School of Public Health, University of California, Berkeley, CA 94720-7360, USA

Mark A. Miller Fogarty International Center, National Institutes of Health, Bethesda, MD, USA

Justine D. Mintern Department of Microbiology and Immunology, The University of Melbourne, Parkville, VIC 3010, Australia, mintern@wehi.edu.au

Emanuele Montomoli Department of Physiopathology, Experimental Medicine and Public Health, Laboratory of Molecular Epidemiology, University of Siena, Via Aldo Moro 3, 53100 Siena, Italy, Montomoli@unisi.it

Samira Mubareka Department of Microbiology, Mount Sinai School of Medicine, One Gustave L. Levy Place, PO Box 1124, New York, NY 10029, USA, samira.mubareka@mssm.edu; Department of Microbiology and Division of Infectious Diseases, Sunnybrook Health Sciences Centre and Research Institute, 2075 Bayview Avenue, Suite B 103, Toronto, ON, Canada M4N 3M5; Department of Laboratory Medicine, University of Toronto, Toronto, ON, Canada

Derek T. O'Hagan Novartis Vaccines and Diagnostic, 350 Massachussetts Avenue, Cambridge, MA 02139, USA, derek.ohagan@novartis.com

John Oxford London Bioscience Innovation Centre, Retroscreen Virology Ltd, 2 Royal College Street, London NW1 ONH, UK, j.oxford@retroscreen.com

Peter Palese Department of Microbiology, Mount Sinai School of Medicine, One Gustave L. Levy Place, PO Box 1124, New York, NY 10029, USA, peter.palese@mssm.edu

Michael Perdue Department of Human and Health Services (HHS), Biomedical Advanced Research and Development Authority (BARDA), 330 Independence Avenue, SW Rm G640, Washington, DC 20201, USA

Raul Rabadan Department of Biomedical Informatics, Center for Computational Biology and Bioinformatics, Columbia University College of Physicians and Surgeons, New York, NY, USA, rabadan@dbmi.columbia.edu

Rino Rappuoli Research Center, Novartis Vaccines and Diagnostics, Via Fiorentina 1, 53100 Siena, Italy, rino.rappuoli@novartis.com

Steven Reed IDRI, 1124 Columbia Street, Seattle, WA 98104, USA

Alan Shaw VaxInnate, 3 Cedar Brook Drive, Suite #1, Cranbury, NJ 08512, USA, Alan.Shaw@vaxinnate.com

Lone Simonsen George Washington University School of Public Health and Health Services, Washington, DC, USA

Marc Steinhoff Global Health Center, Cincinnati Children's Hospital Medical Center, 3333 Burnet Avenue, ML 2048, Cincinnati, OH 45229, USA, mark.steinhoff@gmail.com

Kanta Subbarao Laboratory of Infectious Diseases, National Institute of Allergy and Infectious Diseases, National Institutes of Health, Bethesda, MD 20892, USA

Robert J. Taylor SAGE Analytica, LLC, Bethesda, MD, USA, mark.steinhoff@gmail.com

Vladimir Trifonov Department of Biomedical Informatics, Center for Computational Biology and Bioinformatics, Columbia University College of Physicians and Surgeons, New York, NY, USA

Theodore Tsai Novartis Vaccines and Diagnostic, 350 Massachussetts Avenue, Cambridge, MA 02139, USA

Stephen J. Turner Department of Microbiology and Immunology, The University of Melbourne, Parkville, VIC 3010, Australia

Fons G. UytdeHaag Crucell Holland BV, Archimedesweg 4-6, 2333 CN Leiden, The Netherlands

Cécile Viboud Fogarty International Center, National Institutes of Health, Bethesda, MD, USA

Julia A. Walsh School of Public Health, University of California, Berkeley, CA 94720-7360, USA, jwalsh@berkeley.edu

Richard J. Webby Department of Infectious Diseases, St. Jude Children's Research Hospital, 262 Danny Thomas Place, Memphis, TN 38105, USA

Robert G. Webster Department of Human and Health Services (HHS), Biomedical Advanced Research and Development Authority (BARDA), 330 Independence Avenue, SW Rm G640, Washington, DC 20201, USA; Department of Infectious Diseases, St. Jude Children's Research Hospital, 262 Danny Thomas Place, Memphis, TN 38105, USA, Robert.Webster@STJUDE.org

Part I
Evolution and Epidemiology

Influenza Virus: The Biology of a Changing Virus

Samira Mubareka and Peter Palese

Abstract Influenza viruses are members of the family *Orthomyxoviridae* and include influenza virus types A, B, and C. This introduction provides an overview of influenza virus classification, structure, and life cycle. We also include a brief review of the clinical manifestations of influenza and the molecular determinants for virulence. The genetic diversity of influenza A viruses and their capability to successfully infect an array of hosts, including avian and mammalian species, are highlighted in a discussion about host range and evolution. The importance of viral receptor-binding hemagglutinins and host sialic acid distribution in species-restricted binding of viruses is underscored. Finally, recent advances in our understanding of the seasonality and transmission of influenza viruses are described, and their importance for the control of the spread of these viruses is discussed.

1 Introduction

Influenza has had significant historical impact and continues to pose a considerable threat to public health. Since the transmission of H5N1 avian influenza from birds to humans in 1997, virologists and public health officials alike anticipated global human spread of this virus. More recently, however, pandemic spread of a novel

S. Mubareka
Department of Microbiology and Department of Medicine, Division of Infectious Diseases, Sunnybrook Health Sciences Centre and Research Institute, 2075 Bayview Avenue, Suite B 103, Toronto, M4N 3M5 ON, Canada
Department of Laboratory Medicine, University of Toronto, Toronto, ON, Canada
e-mail: samira.mubareka@sunnybrook.ca

P. Palese (✉)
Department of Microbiology and Department of Medicine, Division of Infectious Diseases, Sunnybrook Health Sciences Centre and Research Institute, 2075 Bayview Avenue, Suite B 103, Toronto, M4N 3M5 ON, Canada
e-mail: peter.palese@mssm.edu

G. Del Giudice and R. Rappuoli (eds.), *Influenza Vaccines for the Future*, 2nd edition, Birkhäuser Advances in Infectious Diseases,
DOI 10.1007/978-3-0346-0279-2_1, © Springer Basel AG 2011

H1N1 influenza virus arose from an unpredicted source; precursors of the pandemic influenza A (H1N1) 2009 virus have been circulating among pigs for over a decade [1, 2]. Additional reassortment events have led to the current pandemic influenza A (H1N1) 2009 virus. Features observed in past pandemics, including atypical seasonality and shifting of the burden of disease to younger populations, are evident during the influenza pandemic of 2009.

Our understanding of the biology of influenza virus and its effect on the host has advanced considerably in recent decades. Recent events in influenza virus research have contributed to this progress [3]. These include the development of plasmid-based reverse genetics systems [4, 5], the generation of the 1918 pandemic H1N1 influenza virus [6], improved access to biosafety level 3 facilities, the establishment of international influenza virus sequence databases, and bioinformatics [7, 8]. Advances have also led to the production of FDA-approved antivirals for influenza, and a heightened understanding of host–virus interactions resulted in the exploration of novel therapies including immunodulatory approaches [9]. New vaccine technologies such as the use of live-attenuated vaccines [10–13] and the development of novel vaccine production methods, including cell culture-based approaches, are the benefits of scientific progress. Continued acceleration of influenza virus research has direct implications for the development of improved vaccines, infection control, and clinical management during pandemic and interpandemic periods.

2   Overview and Classification

Influenza viruses are members of the family *Orthomyxoviridae* and include influenza virus types A, B, and C. Influenza viruses possess seven (influenza C) or eight (influenza A and B) genome segments composed of negative sense single-stranded RNA. These types differ in various aspects, the most important of which include antigenicity, host range, pathogenicity, transmission, and seasonality. Standard nomenclature for human influenza viruses includes type, geographic location of isolation, isolate number, and year of isolation. For example, an influenza A virus isolated in Panama in 1999 would be referred to as A/Panama/2002/1999. Subtypes of influenza A viruses are described by hemagglutinin (HA) and neuraminidase (NA) designations. To date, 16 HA and 9 NA subtypes have been described.

Influenza A viruses are mostly responsible for seasonal epidemics, global pandemics, and the burden of disease attributable to influenza. Clinical disease includes systemic and respiratory manifestations, and rarely may be complicated by central nervous system involvement, toxic shock, or multiorgan system failure [14, 15]. Circulating strains of influenza A viruses are targets for annual vaccination to mitigate morbidity and mortality imparted by these viruses. In addition to infecting humans, influenza A viruses circulate in other mammals, including swine and horses. Waterfowl harbor several lineages of influenza A viruses and serve as a reservoir. Transmission among wild and domestic fowl and mammalian species is an important characteristic of influenza A, enabling viral reassortment and emergence of novel subtypes in susceptible human populations.

In contrast, influenza B virus has a restricted host range, circulating only in humans, although the virus has been isolated in seals [16]. Influenza B virus demonstrates seasonality and is responsible for human disease, although the clinical manifestations are generally less severe compared with influenza A virus-associated illness. Nonetheless, rare cases of encephalitis and septic shock have been described in children [17, 18]. At present, the two major lineages are represented by influenza B/Victoria/2/1987 and B/Yamagata/16/1988 viruses [19]. Re-emergence of the Victoria lineage after a decade of absence was associated with an outbreak during the 2001–2002 influenza season, affecting healthy but immunologically naive children [20]. Influenza B virus is included in inactivated and live-attenuated annual influenza vaccines.

Unlike influenza A and B, influenza C virus lacks neuraminidase and codes for a single-surface hemagglutinin–esterase–fusion (HEF) glycoprotein. This virus does not demonstrate marked seasonality and is not included in the annual influenza vaccine, although it has been responsible for occasional outbreaks, predominantly in children [21]. Illness in humans is generally mild and consists of an upper respiratory tract infection. Influenza C has also been isolated in swine, raising the possibility that this species may serve as a reservoir [22].

3   Structure and Genomic Organization

Influenza viruses are enveloped, deriving the lipid bilayer from the host cell membrane during the process of budding. Viral particles are pleomorphic in nature and may be spherical or filamentous, ranging in size from 100 to over 300 nm [3]. Spikes consisting of HA and NA project from the surface of virions at a ratio of roughly 4:1 in influenza A viruses (Fig. 1) [3]. The viral envelope is also associated with the matrix (M2) protein which forms a tetrameric ion channel.

The polymerase proteins PB1, PB2, and PA, the nucleoprotein (NP), and the virion RNA comprise the ribonucleoprotein (RNP) complex. This complex is present in the core of virions, which also includes the nuclear export and nonstructural protein (NEP/NS1). Influenza virus genes, gene products, and primary functions are summarized in Table 1.

4   Influenza Virus Life Cycle

4.1   Attachment, Entry, and Nuclear Import

In humans, influenza viruses are transmitted by the respiratory route. Host cellular receptors consist of oligosaccharides residing on the surface of respiratory

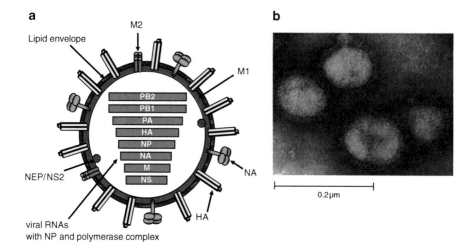

Fig. 1 Schematic structure and electron micrograph of influenza virus A. (**a**) The viral envelop anchors the HA and NA glycoproteins and M2 protein and is derived from the host cell during the process of budding. M1 lies beneath the viral envelope. NEP/NS1 and the core of the virion are contained within. The core consists of eight segments of viral RNA associated with the polymerase complex (PB2, PB1, and PA) and NP. Adapted from [1] and kindly provided by M.L. Shaw. (**b**) Negatively stained electron micrograph of mouse-adapted influenza A WSN/33. Glycoprotein spikes are visible on the surface of the virion. Kindly provided by M.L. Shaw

Table 1 Influenza A genes and primary functions of their encoded proteins

Genome segment[a]	Length in nucleotides	Encoded proteins	Protein size in amino acids	Function
1	2341	PB2	759	Polymerase subunit, mRNA cap recognition
2	2341	PB1	757	Polymerase subunit, endonuclease activity, RNA elongation
		PB1-F2[b]	87	Proapoptotic activity
3	2233	PA	716	Polymerase subunit, protease activity, assembly of polymerase complex
4	1778	HA	550	Surface glycoprotein, receptor binding, fusion activity, major viral antigen
5	1565	NP	498	RNA binding activity, required for replication, regulates RNA nuclear import
6	1413	NA	454	Surface glycoprotein with neuraminidase activity, virus release
7	1027	M1	252	Matrix protein, interacts with vRNPs and glycoproteins, regulates RNA nuclear export, viral budding
		M2[c]	97	Integral membrane protein, ion channel activity, uncoating, virus assembly
8	890	NS1	230	Interferon antagonist activity, regulates host gene expression
		NEP/ NS2[c]	121	Nuclear export of RNA

[a]Influenza A/Puerto Rico/8/1934

[b]Encoded by an alternate open reading frame

[c]Translated from an alternatively spliced transcript

epithelial cells. Specificity of binding is imparted by the linkage of the penultimate galactose (Gal) to *N*-acetylsialic acid (SA). α2,6 linkage (SAα2,6Gal) is distributed in the human respiratory tract and is associated with binding to human influenza virus HA. In contrast, avian hosts including waterfowl and domestic poultry harbor sialic acid with α2,3 linkage (SAα2,3Gal) which is distributed in the gastrointestinal tract, reflecting the fecal-oral mode of transmission of avian influenza strains in these species [23]. Specificity of viral HA binding is imparted by the receptor-binding pocket on the surface of the HA molecule (Fig. 2). The HA is a rod-shaped trimer anchored in the virion's envelope and contains three

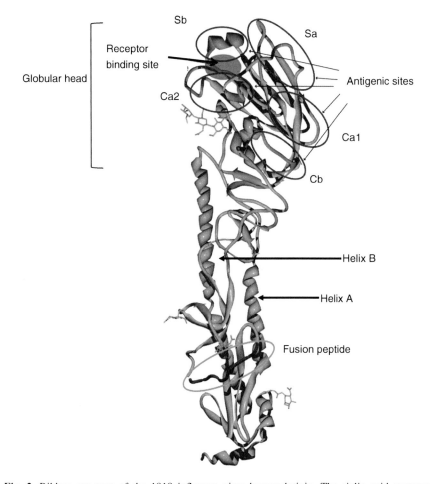

Fig. 2 Ribbon structure of the 1918 influenza virus hemagglutinin. The sialic acid receptor-binding site and the five antigenic sites are located on the globular head. This structure also possesses a cleavage site where HA is cleaved into HA1 and HA2 for fusion of viral and endosomal membranes and subsequent uncoating. Adapted from [1] and kindly provided by J. Stevens and I. Wilson

primary ligand-binding sites on a globular head [24, 25]. Specificity of binding has been linked to certain amino acid residues in the HA receptor-binding domain. In H3 subtypes, amino acid 226 is one such residue, where the presence of leucine allows binding of SAα2,6Gal, whereas the presence of glutamine at this position permits binding of SAα2,3Gal. Amino acid changes in the HA of other subtypes, such as H1 viruses (including the H1N1 virus responsible for the 1918 pandemic), have been associated with adaptations in receptor-binding specificity, translating into a switch in host specificity with disastrous consequences [26, 27]. Specifically, changes at amino acid position 225 impart the ability of A/New York/1/18 to bind both avian and human host influenza virus receptors [26]. Strains of the 2009 pandemic H1N1 influenza viruses retain amino acids (aspartic acids) at positions 190 and 225 of the HA consistent with human sialic acid receptor-binding specificity, although conflicting data exist regarding binding specificity for these viruses. One approach utilizing carbohydrate microarrays suggests that dual (human and avian) sialic acid receptor binding occurs [28]; data obtained using a different approach, namely biotinylated α2,3- and α2,6-sialylated glycans, suggest currently circulating pandemic viruses preferentially bind human sialic acid receptors with α2-6 linkage [29]. The importance of these amino acid residues to respiratory droplet transmission has recently been described using the ferret transmission model. H1N1 viruses containing aspartic acids at residues 190 and 225 were capable of aerosol transmission. This contrasted with H1N1 viruses with glutamic acid and glycine at residues 190 and 225, respectively (consistent with avian sialic acid receptor-binding specificity), which did not transmit through the air [30]. Furthermore, other changes in the HA (and NA) of an avian H9N2 after adaptation in the ferret conferred a more efficient respiratory transmission phenotype [31].

Several possible pathways for the entry of influenza viruses into host cells have been postulated and recently reviewed [32]. Endocytosis is a multistep process consisting of surface receptor-mediated binding, internalization, and intracellular trafficking. Clathrin-mediated and clathrin-independent internalization via caveolae and caveolae-independent endocytosis have been demonstrated [33, 34]. An initial acidification step in early endosomes is followed by trafficking to low-pH late endosomes, a process mediated by members of the Rab host protein family. Fusion of influenza virus to the endosome is triggered by low pH conditions and mediated by the fusion peptide of HA2 after cleavage of HA, creating a pore in the endosome through fusion of viral and endosomal membranes (Fig. 3) [3].

Subsequent steps in the uncoating process involve the influenza virus tetrameric M2 protein, which is involved in the release of RNP into the host cell cytoplasm through ion channel activity [35, 36]. Viral RNA (vRNA) synthesis occurs in the nucleus, and viral RNPs must therefore be imported. This process is primarily mediated by viral NP, which coats viral RNA and possesses nuclear localization signals (NLSs), including an unconventional NLS which binds host karyopherin-α and is essential for energy-dependent RNP nuclear import [37, 38].

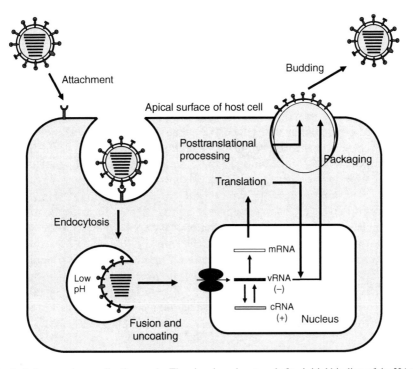

Fig. 3 Influenza virus replication cycle. The virus is endocytosed after initial binding of the HA to host cell sialic acid receptors. Acidification of the cleaved HA facilitates approximation of viral and endosomal membranes and release of RNP. Transcription follows importation of RNPs into the nucleus. Assembly occurs at the apical surface of the host cell where budding and release occur. Adapted from [1] and kindly provided by M.L. Shaw. See text for detail

4.2   Transcription, Replication, and Nuclear Export

Viral RNA serves as a template for the production of messenger RNA (mRNA) and subsequent transcription, as well as for the generation of complementary RNA (cRNA), which is positive sense and functions as a template for the generation of more vRNA (viral replication). RNA segments are coated by NP through nonspecific interactions between the arginine-rich positively charged NP and the negatively charged RNA phosphate backbone [3]. The viral polymerase complex consists of tightly associated PB1, PB2, and PA and associates with NP-coated RNA without disrupting this interaction [39]. PB1 is an endonuclease involved in both replication and transcription and binds the promoter region of RNA segments [40]. It functions as an RNA-dependent RNA polymerase and catalyzes RNA chain elongation. Interaction with PA is required for this function and viral replication [41]. PB2 binds both NP and PB1 via separate binding sites [42]. Initiation of transcription is reliant on PB2, which binds the cap on host pre-mRNA, and this cap serves as a primer for transcription [43, 44]. In addition, interactions between PB2 and host proteins may be species specific and potentially plays a role in restricting host range

[45]. PA is a component of the polymerase heterotrimer, is cotransported into the nucleus with PB1, and is thus important in the formation of this complex [46, 47].

Synthesis of mRNA begins with a host cell 5′-capped primer, generated by host cell RNA polymerase II and obtained from host pre-mRNA [44]. Transcription is thus initiated and synthesis on the template occurs in a 3′ to 5′ direction. A polyadenylation signal consisting of 5–7 uridines at the 5′ end of vRNA prematurely terminates transcription after inducing stuttering of the viral polymerase [48–50]. The generation of NP and NS1 tends to occur earlier after infection compared with the generation of surface glycoprotein and M1 mRNAs [3]. Mechanisms for the regulation of gene expression remain evasive, although NP has been implicated in the control of gene expression [51].

Viral replication requires the synthesis of vRNA, which is primer independent and occurs through a cRNA intermediate. Nascent cRNA is therefore not capped or polyadenylated upon termination. The notion that cRNA synthesis is initiated after a switch from mRNA synthesis has been challenged [52].

RNP complexes subsequently associate with M1 at its C-terminal domain, and aggregation of this complex leads to inhibition of transcription [53]. M1 also interacts with NEP at its C-terminal domain [38, 54]. NEP, in turn, associates with host nuclear export receptor Crm1 via the NEP N-terminal domain [54], thus orchestrating the export of viral RNP from the nucleus.

4.3 Viral Assembly, Budding, and Release

Posttranslational modification of the HA consists of glycosylation in the Golgi apparatus [55]. Along with viral RNP, protein components of the virion are coordinately trafficked to the apical surface of the host cell for assembly into progeny virus.

Two models for the packaging of viral RNA segments exist and include the random incorporation [56, 57] and the selective incorporation models [58, 59]. The latter implies that each RNA segment possesses a packaging signal, resulting in virions with exactly eight segments. Putative packaging signals in coding regions of polymerase genes, spike glycoprotein genes, and the NS gene have been identified [58, 60–63].

Viral assembly is coordinated by the M1 protein, which associates with the cytoplasmic tails of the viral glycoproteins [19, 64, 65], as well as RNP and NEP, as described above. Lipid rafts navigate viral membrane glycoproteins to the apical surface of the host cell [66, 67]. In addition, there is evidence that targeting of NP and polymerase proteins to the apical surface also involves lipid rafts [68].

Genomic packaging and viral assembly occurs at the apical membrane and is associated with accumulation of M1 and the formation of lipid rafts. The M1 protein has also been implicated in viral morphology [69, 70]. Because the HA binds cell surface sialic acid receptors, virions must be released. The NA functions as a sialidase and cleaves sialic acids from the host cell and viral glycoproteins to minimize viral aggregation at the cell surface [71]. Balance between the HA and NA is thus required for optimal receptor binding and destruction [64, 72]. In addition to its receptor-destroying activity, NA is a viral spike glycoprotein and important surface antigen [73].

5   Evolution

Among the influenza virus types, influenza A demonstrates the most genetic diversity and is capable of successfully infecting an array of hosts, including avian and mammalian species. Influenza A viruses exhibit an evolutionary pattern, which is complex and consists of antigenic drift and shift. Drift occurs on an annual basis and has been attributed to low fidelity of the RNA polymerase and subsequent selection from immune pressure exerted by the host [74]. This results in antigenic diversity of the hemagglutinin and neuraminidase glycoproteins and is one of the major challenges to vaccine production, requiring annual changes to vaccine components. The HA1 domain contains several epitopes and is the most dynamic as a consequence, demonstrating clusters of antigenic variance over time [75]. Antigenic shift results after a viral reassortment event where exchange of one or more of the viral segments with that of another strain may result in a novel serotype, potentially diversifying the host range of the virus. It is in this setting that pandemic strains have emerged in immunologically naïve populations in the past, including the H2N2 (with new HA, NA, and PB1 segment) subtype in 1957 and the H3N2 influenza virus (with new HA and PB1 segments) which caused a pandemic in 1968 (Fig. 4).

Since 1997, several avian influenza viruses, including H5N1, H7N2, H7N3, H7N7, H9N2, and H10N7 subtypes, have infected humans [76], though limited evidence for person to person spread exists [77, 78]. Lack of transmission among humans remains a barrier to pandemic spread of these viruses. The H5N1 subtype isolated from avian species has undergone genetic reassortment, and several genotypes exist. Genotypes Z and V are largely responsible for outbreaks of highly pathogenic influenza viruses (HPAI) in domestic birds in Southeast Asia beginning in 2003 [77]. H5N1 viruses may also be divided into clades based on the genomic analysis of the HA genes, and clade 2 is further divided into subclades; up to ten clades have been identified in avian species, four of which have infected humans [79, 80]. Less than 1% divergence from avian isolates has been reported in viruses isolated from humans in Asia [7].

The pandemic influenza A (H1N1) 2009 virus has been described as a "triple reassortant" of swine, human, and avian influenza viruses; the H1 gene from this virus has been circulating among swine for decades, with limited drift compared with genes of H1 viruses that have been circulating in humans, and is thus antigenically different from seasonal human H1N1 viruses. The pandemic influenza A (H1N1) 2009 virus is composed of six segments from the triple reassortant, including a human PB1 segment, classical swine-origin HA, NP, and NS, and avian-origin PB2 and PA segments that have been circulating in swine since approximately 1998.

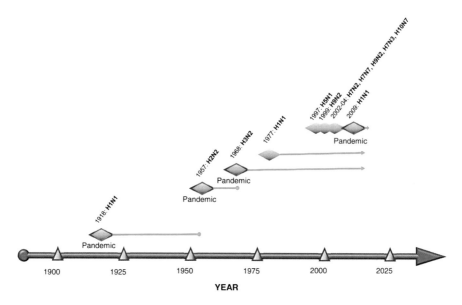

Fig. 4 Influenza A virus subtypes in humans. Three pandemics occurred during the twentieth century, including the "Spanish" influenza pandemic of 1918, the "Asian" pandemic in 1958, and the "Hong Kong" pandemic in 1968. H1N1 viruses re-emerged in 1977 and continue to circulate in the human population, along with the H3N2 subtype. In addition, H5N1 viruses have been reported to infect humans throughout Asia and Africa. Several other avian viruses have also recently caused sporadic infection in humans. A swine-origin influenza virus (pandemic influenza A H1N1 2009 virus) emerged during the spring of 2009 and spread globally, inciting the World Health Organization to declare a pandemic in June of 2009. Adapted from [68]

The NA and M segments originate from a Eurasian lineage of swine influenza viruses [1, 2, 81] (Fig. 5).

In order to tackle the challenge of understanding the evolution of influenza virus, large-scale collaborative efforts such as the Influenza Genome Sequencing Project have been undertaken. The presence of several cocirculating clades in the human population has been described, accounting for reassortment. This can result in limited vaccine effectiveness, as seen with A/Fujian/411/2002-like virus during the 2003–2004 season [8]. Genetic evolution appears to be a relatively gradual process; however, antigenic changes in the HA1 domain tend to cluster [75]. Ongoing changes of the H3 hemagglutinin in the human population result from selective pressure exerted by the host immune system. In contrast, the H3 lineage in birds has remained relatively stable [82]. The rate of change of the H3 subtype is greater when compared with H1 viruses and influenza B, with estimated nucleotide changes per site per year of 0.0037 for H3, 0.0018 for H1, and 0.0013 for influenza B [83]. As greater numbers of influenza virus genome sequences become available and we gain insight into antigenic patterns of change, this knowledge may be applied to annual vaccine development. Prediction of future influenza sequences could lead to more timely development of effective vaccines [84] though modeling methods have yet to be validated.

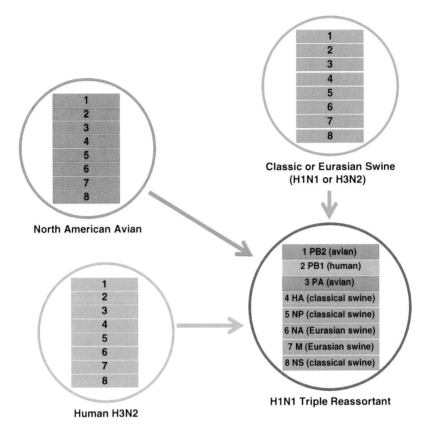

Fig. 5 Origins of pandemic influenza A H1N1 2009 virus. Swine (classical), human, and avian influenza viruses reassorted in North America in 1998 to produce an H3N2 virus which circulated in swine. Further reassortment with a Eurasian lineage of swine influenza virus resulted in the current pandemic influenza virus which has spread globally in humans

6 Host Range

Influenza A virus is a zoonotic pathogen capable of infecting birds (waterfowl and chickens), swine, horses, felines, and other species. Host range restriction of different types of influenza viruses is observed. Species-restricted binding of viruses is mediated by different types of receptor-binding hemaglutinins [85–89]. The distribution of different types of SA linkages has recently been elucidated in humans though the type of cell infected (ciliated vs. nonciliated) is under debate [90, 91]. SA with α2,6Gal linkage predominates on epithelial cells of the upper airway, including nasal mucosa, sinuses, bronchi, and bronchioles [92]. In human tracheobronchial epithelial (HTBE) cells, oligosaccharides with SA with α2,6Gal linkage predominate on nonciliated epithelial cells [91] although these oligosaccharides have been described on ciliated and goblet cells in the human airway [93].

Lower airways contain SA with mostly α2,3Gal linkage, in addition to SA with α2,6Gal linkage [92, 94].

Host restriction is not absolute, and human infections with avian influenza viruses (including H5N1, H9N2, and H7N7 viruses) have been extensively described [95–100]. H5N1 binds type II pneumocytes and macrophages of the lower respiratory tract in humans [92, 94, 101]. H5N1 infection of ciliated cells in HTBE cell culture with limited cell-to-cell spread [90] and of human nasopharyngeal, adenoid, and tonsillar ex vivo cell cultures has been shown [102]. Binding of H5N1 viruses to saccharides terminating in α2,6Gal SA linkage has been achieved by mutating HA amino acid residues at positions 182 and 192, suggesting potential for adaptation to the human host [103].

Differences in influenza virus receptors among avian species have been described and are reflected in differential binding of different types of avian influenza viruses. Although chicken and duck influenza viruses preferentially bind α2,3Gal-linked SA, viruses from chickens had greater affinity for SA where the third sugar moiety was a β(1-4)GlcNAc-containing synthetic sialylglycopolymer. Duck viruses preferred β(1-3)GalNAc sugar moieties in the third position [104]. Distribution of influenza virus receptors reflects the sites of replication. In chickens and waterfowl, SA with α2,3Gal linkage is found in the upper respiratory tract and intestines. Some species demonstrate the ability to support replication of both human and avian influenza viruses. The respiratory tract and intestines of quail contain both α2,3Gal- and α2,6Gal-linked terminal sialic acids [105]. In swine, oligosaccharides with both types of linkages may be found and suggest this species serves as a mixing vessel where human, avian, and swine influenza viruses can reassort [106, 107].

7   Clinical Manifestations, Pathogenesis, and Virulence

7.1   *Clinical Manifestations*

Uncomplicated influenza in humans is an upper respiratory tract infection characterized by cough, headache, malaise, and fever (influenza-like illness). These symptoms are nonspecific and are not predictive of influenza virus infection, particularly in individuals <60 years old [108]. Pulmonary and extrapulmonary complications may arise. The latter consist of central nervous system involvement (encephalitis, acute necrotizing encephalopathy, Reye's syndrome, and myelitis) [14], myositis/rhabdomyositis [109], myocarditis [109, 110], increased cardiovascular events [111], disseminated intravascular coagulation [109], and toxic and septic shock (bacterial and nonbacterial) [15, 18, 109]. Pulmonary complications include primary viral pneumonia, secondary bacterial pneumonia (see below), and exacerbation of chronic lung disease [109, 112]. Acute lung injury (ALI)/acute respiratory distress syndrome (ARDS), multiorgan failure, profound lymphopenia, and hemophagocytosis have been associated with H5N1 infection and carry high mortality rates [15, 95, 113–115].

Bacterial pneumonia following influenza virus infection is a well-recognized complication of influenza since the pandemic of 1918 [116]. More recently, pediatric deaths have been attributed to copathogenesis between influenza virus and *Staphylococcus aureus*, accounting for 34% of pediatric deaths reported to the CDC during the 2006–2007 influenza season [117]. In one case series, 43% of coinfected cases involved methicillin-resistant *S. aureus*, thus contributing to management challenges for these patients. Coinfection was also associated with a worse prognosis compared with influenza virus or *S. aureus* infection alone [118].

To date, secondary bacterial lower respiratory tract infection has not been a dominant feature in adults during the current 2009 pandemic but has been described in children [119]. Severe pandemic 2009 influenza has been predominantly associated with viral pneumonitis and subsequent ALI, particularly in pregnant women in their third trimester [120] and indigeous people including Aborigines in Australia [121], Maoris and Pacific Islanders in New Zealand [122], and First Nations People in Canada [123].

7.2 Pathogenesis

Few human histopathological studies of uncomplicated influenza exist. Pathological findings from postmortem examination of 47 fatal pediatric influenza A cases included major airway congestion (90%), inflammation (73%), and necrosis (50%) [112]. Lower airway pathology included hyaline membranes (67%), interstitial cellular infiltrates (67%), and diffuse alveolar damage (DAD). Secondary pneumonia, intraalveolar hemorrhage, and viral pneumonitis were noted in a quarter of cases [112]. Fulminant DAD with acute alveolar hemorrhage and necrosis followed by paucicellular fibrosis and hyaline membrane formation is observed in H5N1-infected human lungs [124]. Extrapulmonary pathology includes reactive hemophagocytosis in the hilar lymph nodes, bone marrow, liver, and spleen [125]; white matter demyelination [124] and cerebral necrosis [101]; and acute tubular necrosis of the kidneys [113]. Despite the presence of diarrhea and H5N1 virus replication in the gastrointestinal tract of humans, no pathological lesions have been described in the bowel [101, 114]. Immune dysregulation has been implicated in the pathogenesis of ARDS and reactive hemophagocytosis. Elevated levels of neutrophil, monocyte, and macrophage chemoattractants (IL-8, IP-10, MIG, and MCP-1) and proinflammatory cytokines (IL-10, IL-6, and IFN-γ) are observed in H5N1-infected humans [95]. In addition, increased levels of IL-2 (in a human case) [113] and RANTES (in primary human alveolar and bronchial epithelial cells) [126] have also been reported. Contribution of proinflammatory mediators to lung pathology has also been demonstrated using Toll-like receptor 3 knockout mice infected with mouse-adapted WSN influenza A virus. These mice demonstrated enhanced survival despite higher virus replication and lower levels of RANTES, IL-6, and IL-12p40/p70 compared with wild-type mice [127].

Likewise, host response has been implicated in the copathogenesis of bacterial pneumonia post-influenza virus infection. Specifically, sensitization by type I interferons [128], induction of IL-10 [129], and upregulation of interferon-α [130] have been linked to secondary bacterial pneumonia after influenza virus infection. Viral determinants for copathogenesis have also been elucidated and include PB1-F2 and viral neuraminidase [131, 132].

7.3 Virological Determinants of Virulence

The HA, PA, PB1, PB2, PB1-F2, NA, and NS1 gene products have been implicated in virulence. Virulence determinants have been explored using the reverse genetic system for influenza viruses and mammalian (ferret and mouse) models for influenza virus pathogenicity.

The polymerase gene complex, consisting of PA, PB1, and PB2 genes, is involved in replication and transcriptional activity. A single-gene reassortant containing the PB2 from A/Hong Kong/483/97 (H5N1, which is fatal in mice) in the background of A/Hong Kong/486/97 (H5N1, causing mild respiratory infection in mice) demonstrated a lethal phenotype in this animal model [133]. In addition, reassortants containing polymerase complex genes from A/chicken/Vietnam/C58/04 (H5N1), a nonlethal virus, in the background of A/Vietnam/1203/04 (H5N1) influenza virus isolated from a fatal human case were attenuated in an animal model [134]. When a single point mutation K627E in the PB2 gene was generated in A/Vietnam/1203/04 [134] and in A/Hong Kong/483/97 [133], virulence was reduced in mice, although in other studies this substitution did not reduce virulence substantially [135]. The molecular mechanism(s) responsible for virulence have yet to be completely elucidated. Enhanced replication of viruses retaining a lysine at position 627 in PB2 at the lower temperatures of the upper respiratory tract (33°C) [136] may be responsible for robust transmission in mammals [137]. This theory is supported by recent work demonstrating that replacement of the lysine at position 627 with glutamic acid (avian consensus sequence) abrogates aerosol transmission of a 1918 influenza A virus [30]. Currently circulating strains of pandemic H1N1 influenza virus have a glutamine in PB2 at position 627. This may account for reduced efficiency of aerosol transmission of this virus in ferrets, compared with a seasonal H1N1 virus [29].

PB1-F2 is the gene product arising from a second reading frame of the PB1 gene and has been implicated in immune cell apoptosis through the VDAC1 and ANT3 mitochondrial pathways [138]. Knockout of PB1-F2 did not alter viral replication, but enhanced clearance of the virus and reduced lethality in mice was demonstrated, suggesting that PB1-F2 may play a role in viral pathogenesis [139]. Enhanced pathogenicity was observed in mice infected with recombinant influenza virus containing the PB1-F2 gene from a highly pathogenic H5N1 virus isolated from a fatal human case in Hong Kong in 1997 [139]. Currently circulating strains of the pandemic influenza A H1N1 2009 virus do not express PB1-F2.

Evasion of the host immune response is a key virulence determinant, permitting viruses to establish sustainable infection. The innate immune system is the first line of host defense, and the influenza virus possesses the ability to interfere with this response. Type I interferons (IFN-α/β) are central to establishing an antiviral state in host cells. Interferon antagonism has been primarily attributed to the NS1 protein of influenza virus, which plays a multifunctional role in preventing the activation of IFN transcription factors (for review, see [140, 141]).

The effect of avian influenza virus NS1 on IFN production has also been explored. A/goose/Guangdong/1/96 virus with an NS1 that differs by one amino acid from A/goose/Guangdong/2/96 at position 149 is lethal in chickens and antagonizes IFNα/β [142]. In addition, the C-terminus of the NS1 protein contains a PDZ ligand domain, capable of binding PDZ protein interaction domains of host proteins, thus potentially disrupting host cellular pathways. Viruses causing pathogenic infection in humans between 1997 and 2003 contained avian motifs at the NS1 PDZ ligand-binding site. These and the motif found in the 1918 influenza virus NS1 had stronger binding affinities to PDZ domains of human cellular proteins compared with low pathogenicity influenza viruses [143].

Neurovirulence has been associated with glycosylation of the NA glycoprotein [144]. The HA glycoprotein has also been associated with virulence. Although cleavability of the HA gene has been primarily implicated in pathogenicity in chickens, lethality has also been demonstrated in mice. Basic amino acids at the HA cleavage site are determinants for HA cleavage and HA2 fusion activity [145]. Enhanced cleavage of the HA by ubiquitous host proteases is made possible by the presence of a polybasic cleavage site, contributing to the virulence of highly pathogenic avian influenza viruses [146, 147]. Replacement of the polybasic cleavage site in a high pathogenicity H5N1 virus from Hong Kong (HK483) with an amino acid sequence typical of low pathogenicity viruses resulted in attenuation [133]. Pandemic influenza A H1N1 2009 virus strains do not appear to have the polybasic cleavage site.

Virulence determinants for the pandemic 2009 H1N1 virus are currently investigation. Data obtained from mammalian models early in the course of the spread of this virus indicate that compared with a seasonal H1N1 influenza virus, strains of the pandemic virus replicate more efficiently in the lower respiratory tract, and are stronger inducers of proinflammatory mediators, and induce bronchopneumonia [148].

8   Seasonality and Transmission

Influenza A and B viruses exhibit marked seasonality, and this pattern dictates the annual vaccination schedule. Several theories with respect to the mechanism(s) responsible for this seasonal pattern have been proposed (for review, see [149]).

Year-round human influenza virus activity in equatorial regions may be a reservoir for annual outbreaks in the northern and southern hemispheres. As research progresses in this area, factors determining seasonality may be exploited for the control of the spread of influenza virus [150].

Transmission of influenza virus among humans is poorly understood and the mode(s) of spread are currently under debate [151, 152]. It is widely accepted that influenza virus is transmitted by the respiratory route in humans, though the contribution of small particle aerosols relative to large respiratory droplets is unknown. In addition, the role of fomites is questionable. Until recently, ferrets have served as the only animal model for the study of influenza virus transmission. A novel mammalian model using the guinea pig has recently been developed to overcome the limitations of the ferret model. Guinea pigs are highly susceptible to infection with an unadapted human H3N3 (A/Panama/2002/1999, or Pan99) influenza virus, with a 50% infectious dose of 5 PFU, and this virus grows to high titers in the upper respiratory tract and to moderate titers in the lungs. Transmission of Pan99 by direct contact and aerosol in this system is 100% (Fig. 6) [153]; however, transmission efficiency may vary among influenza virus subtypes [154]. Environmental factors such as temperature and relative humidity also appear to play a substantial role [155, 156]. Control of influenza virus spread during interpandemic and pandemic periods through vaccination [157] and physical means will be paramount to abrogating person-to-person transmission and is crucial where viruses are resistant to currently available antivirals.

9 Perspectives

Effective and timely vaccine development depends on in-depth understanding of influenza virus biology. Although recent advances have been made, ongoing research will be required to fulfill this goal. Identification and characterization of

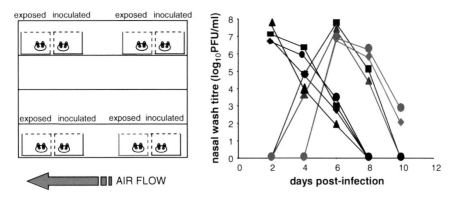

Fig. 6 Close range transmission of human influenza A among guinea pigs. Inoculated animals placed in proximity to uninoculated animals (without direct contact) spread Pan99 to all exposed animals. Adapted from [109]

the molecular signatures required for transmission will be of utmost importance to preventing further influenza virus pandemics. Globalization of H1N1 infection in humans requires parallel efforts on behalf of virologists in conjunction with epidemiologists and other members of the public health community to translate the growing body of knowledge into means by which influenza spread can be controlled.

Acknowledgments The work completed in this laboratory was partially supported by the W.M. Keck Foundation, National Institutes of Health grants P01 AI158113, the Northeast Biodefense Center U54 AI057158, the Center for Investigating Viral Immunity and Antagonism (CIVIA) U19 AI62623. S.M. is grateful for the Ruth L. Kirschstein Physician Scientist Research Training in Pathogenesis of Viral Diseases Award (5T32AI007623-07) and support from Sunnybrook Health Sciences Center, Toronto, ON, Canada.

References

1. Garten RJ, Davis CT, Russell CA, Shu B, Lindstrom S, Balish A, Sessions WM, Xu X, Skepner E, Deyde V et al (2009) Antigenic and genetic characteristics of swine-origin 2009 A(H1N1) influenza viruses circulating in humans. Science 325:197–201
2. Shinde V, Bridges CB, Uyeki TM, Shu B, Balish A, Xu X, Lindstrom S, Gubareva LV, Deyde V, Garten RJ et al (2009) Triple-reassortant swine influenza A (H1) in humans in the United States, 2005–2009. N Engl J Med 360:2616–2625
3. Shaw ML, Palese P (2007) Orthomyxoviridae: the viruses and their replication. In: Knipe DM, Howley PM (eds) Fields virology, 5th edn. Lippincott Williams & Wilkins, Philadelphia, pp 1647–1689
4. Fodor E, Devenish L, Engelhardt OG, Palese P, Brownlee GG, Garcia-Sastre A (1999) Rescue of influenza A virus from recombinant DNA. J Virol 73:9679–9682
5. Neumann G, Watanabe T, Ito H, Watanabe S, Goto H, Gao P, Hughes M, Perez DR, Donis R, Hoffmann E et al (1999) Generation of influenza A viruses entirely from cloned cDNAs. Proc Natl Acad Sci USA 96:9345–9350
6. Tumpey TM, Basler CF, Aguilar PV, Zeng H, Solorzano A, Swayne DE, Cox NJ, Katz JM, Taubenberger JK, Palese P et al (2005) Characterization of the reconstructed 1918 Spanish influenza pandemic virus. Science 310:77–80
7. World Health Organization Global Influenza Program Surveillance Network (2005) Evolution of H5N1 avian influenza viruses in Asia. Emerg Infect Dis 11:1515–1521
8. Ghedin E, Sengamalay NA, Shumway M, Zaborsky J, Feldblyum T, Subbu V, Spiro DJ, Sitz J, Koo H, Bolotov P et al (2005) Large-scale sequencing of human influenza reveals the dynamic nature of viral genome evolution. Nature 437:1162–1166
9. Kugel D, Kochs G, Obojes K, Roth J, Kobinger GP, Kobasa D, Haller O, Staeheli P, von Messling V (2009) Intranasal administration of alpha interferon reduces seasonal influenza A virus morbidity in ferrets. J Virol 83:3843–3851
10. Steel J, Lowen AC, Pena L, Angel M, Solorzano A, Albrecht R, Perez DR, Garcia-Sastre A, Palese P (2009) Live attenuated influenza viruses containing NS1 truncations as vaccine candidates against H5N1 highly pathogenic avian influenza. J Virol 83:1742–1753
11. Hai R, Martinez-Sobrido L, Fraser KA, Ayllon J, Garcia-Sastre A, Palese P (2008) Influenza B virus NS1-truncated mutants: live-attenuated vaccine approach. J Virol 82:10580–10590
12. Suguitan AL Jr, McAuliffe J, Mills KL, Jin H, Duke G, Lu B, Luke CJ, Murphy B, Swayne DE, Kemble G et al (2006) Live, attenuated influenza A H5N1 candidate vaccines provide broad cross-protection in mice and ferrets. PLoS Med 3:e360
13. Murphy BR, Coelingh K (2002) Principles underlying the development and use of live attenuated cold-adapted influenza A and B virus vaccines. Viral Immunol 15:295–323

14. Studahl M (2003) Influenza virus and CNS manifestations. J Clin Virol 28:225–232
15. Sion ML, Hatzitolios AI, Toulis EN, Mikoudi KD, Ziakas GN (2001) Toxic shock syndrome complicating influenza A infection: a two-case report with one case of bacteremia and endocarditis. Intensive Care Med 27:443
16. Osterhaus AD, Rimmelzwaan GF, Martina BE, Bestebroer TM, Fouchier RA (2000) Influenza B virus in seals. Science 288:1051–1053
17. Newland JG, Romero JR, Varman M, Drake C, Holst A, Safranek T, Subbarao K (2003) Encephalitis associated with influenza B virus infection in 2 children and a review of the literature. Clin Infect Dis 36:e87–e95
18. Jaimovich DG, Kumar A, Shabino CL, Formoli R (1992) Influenza B virus infection associated with non-bacterial septic shock-like illness. J Infect 25:311–315
19. Chen JM, Guo YJ, Wu KY, Guo JF, Wang M, Dong J, Zhang Y, Li Z, Shu YL (2007) Exploration of the emergence of the Victoria lineage of influenza B virus. Arch Virol 152: 415–422
20. Hite LK, Glezen WP, Demmler GJ, Munoz FM (2007) Medically attended pediatric influenza during the resurgence of the Victoria lineage of influenza B virus. Int J Infect Dis 11: 40–47
21. Matsuzaki Y, Abiko C, Mizuta K, Sugawara K, Takashita E, Muraki Y, Suzuki H, Mikawa M, Shimada S, Sato K et al (2007) A nationwide epidemic of influenza C virus infection in Japan in 2004. J Clin Microbiol 45:783–788
22. Yuanji G, Desselberger U (1984) Genome analysis of influenza C viruses isolated in 1981/82 from pigs in China. J Gen Virol 65(Pt 11):1857–1872
23. Wright PF, Neumann G, Kawaoka Y (2007) Orthomyxoviruses. In: Knipe DM, Howley PM (eds) Fields virology, 5th edn. Lippincott Williams & Wilkins, Philadephia, pp 1714–1715
24. Eisen MB, Sabesan S, Skehel JJ, Wiley DC (1997) Binding of the influenza A virus to cell-surface receptors: structures of five hemagglutinin-sialyloligosaccharide complexes determined by X-ray crystallography. Virology 232:19–31
25. Skehel JJ, Wiley DC (2000) Receptor binding and membrane fusion in virus entry: the influenza hemagglutinin. Annu Rev Biochem 69:531–569
26. Glaser L, Stevens J, Zamarin D, Wilson IA, Garcia-Sastre A, Tumpey TM, Basler CF, Taubenberger JK, Palese P (2005) A single amino acid substitution in 1918 influenza virus hemagglutinin changes receptor binding specificity. J Virol 79:11533–11536
27. Stevens J, Blixt O, Tumpey TM, Taubenberger JK, Paulson JC, Wilson IA (2006) Structure and receptor specificity of the hemagglutinin from an H5N1 influenza virus. Science 312:404–410
28. Childs RA, Palma AS, Wharton S, Matrosovich T, Liu Y, Chai W, Campanero-Rhodes MA, Zhang Y, Eickmann M, Kiso M et al (2009) Receptor-binding specificity of pandemic influenza A (H1N1) 2009 virus determined by carbohydrate microarray. Nat Biotechnol 27:797–799
29. Maines TR, Jayaraman A, Belser JA, Wadford DA, Pappas C, Zeng H, Gustin KM, Pearce MB, Viswanathan K, Shriver ZH et al (2009) Transmission and pathogenesis of swine-origin 2009 A(H1N1) influenza viruses in ferrets and mice. Science 325:484–487
30. Van Hoeven N, Pappas C, Belser JA, Maines TR, Zeng H, Garcia-Sastre A, Sasisekharan R, Katz JM, Tumpey TM (2009) Human HA and polymerase subunit PB2 proteins confer transmission of an avian influenza virus through the air. Proc Natl Acad Sci USA 106: 3366–3371
31. Sorrell EM, Wan H, Araya Y, Song H, Perez DR (2009) Minimal molecular constraints for respiratory droplet transmission of an avian-human H9N2 influenza A virus. Proc Natl Acad Sci USA 106:7565–7570
32. Sieczkarski SB, Whittaker GR (2005) Viral entry. Curr Top Microbiol Immunol 285:1–23
33. Matlin KS, Reggio H, Helenius A, Simons K (1981) Infectious entry pathway of influenza virus in a canine kidney cell line. J Cell Biol 91:601–613
34. Nunes-Correia I, Eulalio A, Nir S, Pedroso de Lima MC (2004) Caveolae as an additional route for influenza virus endocytosis in MDCK cells. Cell Mol Biol Lett 9:47–60

35. Takeda M, Pekosz A, Shuck K, Pinto LH, Lamb RA (2002) Influenza a virus M2 ion channel activity is essential for efficient replication in tissue culture. J Virol 76:1391–1399
36. Pinto LH, Holsinger LJ, Lamb RA (1992) Influenza virus M2 protein has ion channel activity. Cell 69:517–528
37. Wang P, Palese P, O'Neill RE (1997) The NPI-1/NPI-3 (karyopherin alpha) binding site on the influenza a virus nucleoprotein NP is a nonconventional nuclear localization signal. J Virol 71:1850–1856
38. Cros JF, Garcia-Sastre A, Palese P (2005) An unconventional NLS is critical for the nuclear import of the influenza A virus nucleoprotein and ribonucleoprotein. Traffic 6:205–213
39. Area E, Martin-Benito J, Gastaminza P, Torreira E, Valpuesta JM, Carrascosa JL, Ortin J (2004) 3D structure of the influenza virus polymerase complex: localization of subunit domains. Proc Natl Acad Sci USA 101:308–313
40. Jung TE, Brownlee GG (2006) A new promoter-binding site in the PB1 subunit of the influenza A virus polymerase. J Gen Virol 87:679–688
41. Perez DR, Donis RO (2001) Functional analysis of PA binding by influenza a virus PB1: effects on polymerase activity and viral infectivity. J Virol 75:8127–8136
42. Poole E, Elton D, Medcalf L, Digard P (2004) Functional domains of the influenza A virus PB2 protein: identification of NP- and PB1-binding sites. Virology 321:120–133
43. Fechter P, Mingay L, Sharps J, Chambers A, Fodor E, Brownlee GG (2003) Two aromatic residues in the PB2 subunit of influenza A RNA polymerase are crucial for cap binding. J Biol Chem 278:20381–20388
44. Krug RM, Bouloy M, Plotch SJ (1980) RNA primers and the role of host nuclear RNA polymerase II in influenza viral RNA transcription. Philos Trans R Soc Lond B Biol Sci 288:359–370
45. Labadie K, Dos Santos AE, Rameix-Welti MA, van der Werf S, Naffakh N (2007) Host-range determinants on the PB2 protein of influenza A viruses control the interaction between the viral polymerase and nucleoprotein in human cells. Virology 362:271–282
46. Kawaguchi A, Naito T, Nagata K (2005) Involvement of influenza virus PA subunit in assembly of functional RNA polymerase complexes. J Virol 79:732–744
47. Fodor E, Smith M (2004) The PA subunit is required for efficient nuclear accumulation of the PB1 subunit of the influenza A virus RNA polymerase complex. J Virol 78:9144–9153
48. Li X, Palese P (1994) Characterization of the polyadenylation signal of influenza virus RNA. J Virol 68:1245–1249
49. Luo GX, Luytjes W, Enami M, Palese P (1991) The polyadenylation signal of influenza virus RNA involves a stretch of uridines followed by the RNA duplex of the panhandle structure. J Virol 65:2861–2867
50. Zheng H, Lee HA, Palese P, Garcia-Sastre A (1999) Influenza A virus RNA polymerase has the ability to stutter at the polyadenylation site of a viral RNA template during RNA replication. J Virol 73:5240–5243
51. Ye Q, Krug RM, Tao YJ (2006) The mechanism by which influenza A virus nucleoprotein forms oligomers and binds RNA. Nature 444:1078–1082
52. Vreede FT, Brownlee GG (2007) Influenza virion-derived viral ribonucleoproteins synthesize both mRNA and cRNA in vitro. J Virol 81:2196–2204
53. Baudin F, Petit I, Weissenhorn W, Ruigrok RW (2001) In vitro dissection of the membrane and RNP binding activities of influenza virus M1 protein. Virology 281:102–108
54. Akarsu H, Burmeister WP, Petosa C, Petit I, Muller CW, Ruigrok RW, Baudin F (2003) Crystal structure of the M1 protein-binding domain of the influenza A virus nuclear export protein (NEP/NS2). EMBO J 22:4646–4655
55. Gallagher PJ, Henneberry JM, Sambrook JF, Gething MJ (1992) Glycosylation requirements for intracellular transport and function of the hemagglutinin of influenza virus. J Virol 66:7136–7145
56. Enami M, Sharma G, Benham C, Palese P (1991) An influenza virus containing nine different RNA segments. Virology 185:291–298

57. Bancroft CT, Parslow TG (2002) Evidence for segment-nonspecific packaging of the influenza a virus genome. J Virol 76:7133–7139
58. Watanabe T, Watanabe S, Noda T, Fujii Y, Kawaoka Y (2003) Exploitation of nucleic acid packaging signals to generate a novel influenza virus-based vector stably expressing two foreign genes. J Virol 77:10575–10583
59. de Wit E, Spronken MI, Rimmelzwaan GF, Osterhaus AD, Fouchier RA (2006) Evidence for specific packaging of the influenza A virus genome from conditionally defective virus particles lacking a polymerase gene. Vaccine 24:6647–6650
60. Fujii K, Fujii Y, Noda T, Muramoto Y, Watanabe T, Takada A, Goto H, Horimoto T, Kawaoka Y (2005) Importance of both the coding and the segment-specific noncoding regions of the influenza A virus NS segment for its efficient incorporation into virions. J Virol 79: 3766–3774
61. Liang Y, Hong Y, Parslow TG (2005) Cis-acting packaging signals in the influenza virus PB1, PB2, and PA genomic RNA segments. J Virol 79:10348–10355
62. Fujii Y, Goto H, Watanabe T, Yoshida T, Kawaoka Y (2003) Selective incorporation of influenza virus RNA segments into virions. Proc Natl Acad Sci USA 100:2002–2007
63. Gog JR, Afonso ED, Dalton RM, Leclercq I, Tiley L, Elton D, von Kirchbach JC, Naffakh N, Escriou N, Digard P (2007) Codon conservation in the influenza A virus genome defines RNA packaging signals. Nucleic Acids Res 35:1897–1907
64. Schmitt AP, Lamb RA (2005) Influenza virus assembly and budding at the viral budozone. Adv Virus Res 64:383–416
65. Chen BJ, Takeda M, Lamb RA (2005) Influenza virus hemagglutinin (H3 subtype) requires palmitoylation of its cytoplasmic tail for assembly: M1 proteins of two subtypes differ in their ability to support assembly. J Virol 79:13673–13684
66. Zhang J, Pekosz A, Lamb RA (2000) Influenza virus assembly and lipid raft microdomains: a role for the cytoplasmic tails of the spike glycoproteins. J Virol 74:4634–4644
67. Barman S, Adhikary L, Chakrabarti AK, Bernas C, Kawaoka Y, Nayak DP (2004) Role of transmembrane domain and cytoplasmic tail amino acid sequences of influenza a virus neuraminidase in raft association and virus budding. J Virol 78:5258–5269
68. Carrasco M, Amorim MJ, Digard P (2004) Lipid raft-dependent targeting of the influenza A virus nucleoprotein to the apical plasma membrane. Traffic 5:979–992
69. Bourmakina SV, Garcia-Sastre A (2003) Reverse genetics studies on the filamentous morphology of influenza A virus. J Gen Virol 84:517–527
70. Elleman CJ, Barclay WS (2004) The M1 matrix protein controls the filamentous phenotype of influenza A virus. Virology 321:144–153
71. Palese P, Tobita K, Ueda M, Compans RW (1974) Characterization of temperature sensitive influenza virus mutants defective in neuraminidase. Virology 61:397–410
72. Mitnaul LJ, Matrosovich MN, Castrucci MR, Tuzikov AB, Bovin NV, Kobasa D, Kawaoka Y (2000) Balanced hemagglutinin and neuraminidase activities are critical for efficient replication of influenza A virus. J Virol 74:6015–6020
73. Colman PM (1994) Influenza virus neuraminidase: structure, antibodies, and inhibitors. Protein Sci 3:1687–1696
74. Fitch WM, Leiter JM, Li XQ, Palese P (1991) Positive Darwinian evolution in human influenza A viruses. Proc Natl Acad Sci USA 88:4270–4274
75. Smith DJ, Lapedes AS, de Jong JC, Bestebroer TM, Rimmelzwaan GF, Osterhaus AD, Fouchier RA (2004) Mapping the antigenic and genetic evolution of influenza virus. Science 305:371–376
76. National Institute of Allergy and Infectious Diseases NIH (2007) http://www3.niaid.nih.gov/news/focuson/flu/illustrations/timeline/
77. Horimoto T, Kawaoka Y (2005) Influenza: lessons from past pandemics, warnings from current incidents. Nat Rev Microbiol 3:591–600
78. Wang H, Feng Z, Shu Y, Yu H, Zhou L, Zu R, Huai Y, Dong J, Bao C, Wen L et al (2008) Probable limited person-to-person transmission of highly pathogenic avian influenza A (H5N1) virus in China. Lancet 371:1427–1434

79. Webster RG, Govorkova EA (2006) H5N1 influenza-continuing evolution and spread. N Engl J Med 355:2174–2177
80. Uyeki TM (2008) Global epidemiology of human infections with highly pathogenic avian influenza A (H5N1) viruses. Respirology 13(Suppl 1):S2–S9
81. Dawood FS, Jain S, Finelli L, Shaw MW, Lindstrom S, Garten RJ, Gubareva LV, Xu X, Bridges CB, Uyeki TM (2009) Emergence of a novel swine-origin influenza A (H1N1) virus in humans. N Engl J Med 360:2605–2615
82. Bean WJ, Schell M, Katz J, Kawaoka Y, Naeve C, Gorman O, Webster RG (1992) Evolution of the H3 influenza virus hemagglutinin from human and nonhuman hosts. J Virol 66:1129–1138
83. Ferguson NM, Galvani AP, Bush RM (2003) Ecological and immunological determinants of influenza evolution. Nature 422:428–433
84. Plotkin JB, Dushoff J, Levin SA (2002) Hemagglutinin sequence clusters and the antigenic evolution of influenza A virus. Proc Natl Acad Sci USA 99:6263–6268
85. Suzuki Y, Ito T, Suzuki T, Holland RE Jr, Chambers TM, Kiso M, Ishida H, Kawaoka Y (2000) Sialic acid species as a determinant of the host range of influenza A viruses. J Virol 74:11825–11831
86. Vines A, Wells K, Matrosovich M, Castrucci MR, Ito T, Kawaoka Y (1998) The role of influenza A virus hemagglutinin residues 226 and 228 in receptor specificity and host range restriction. J Virol 72:7626–7631
87. Rogers GN, D'Souza BL (1989) Receptor binding properties of human and animal H1 influenza virus isolates. Virology 173:317–322
88. Rogers GN, Paulson JC (1983) Receptor determinants of human and animal influenza virus isolates: differences in receptor specificity of the H3 hemagglutinin based on species of origin. Virology 127:361–373
89. Rogers GN, Pritchett TJ, Lane JL, Paulson JC (1983) Differential sensitivity of human, avian, and equine influenza A viruses to a glycoprotein inhibitor of infection: selection of receptor specific variants. Virology 131:394–408
90. Thompson CI, Barclay WS, Zambon MC, Pickles RJ (2006) Infection of human airway epithelium by human and avian strains of influenza a virus. J Virol 80:8060–8068
91. Matrosovich MN, Matrosovich TY, Gray T, Roberts NA, Klenk HD (2004) Human and avian influenza viruses target different cell types in cultures of human airway epithelium. Proc Natl Acad Sci USA 101:4620–4624
92. Shinya K, Ebina M, Yamada S, Ono M, Kasai N, Kawaoka Y (2006) Avian flu: influenza virus receptors in the human airway. Nature 440:435–436
93. Ibricevic A, Pekosz A, Walter MJ, Newby C, Battaile JT, Brown EG, Holtzman MJ, Brody SL (2006) Influenza virus receptor specificity and cell tropism in mouse and human airway epithelial cells. J Virol 80:7469–7480
94. van Riel D, Munster VJ, de Wit E, Rimmelzwaan GF, Fouchier RA, Osterhaus AD, Kuiken T (2006) H5N1 virus attachment to lower respiratory tract. Science 312:399
95. de Jong MD, Simmons CP, Thanh TT, Hien VM, Smith GJ, Chau TN, Hoang DM, Chau NV, Khanh TH, Dong VC et al (2006) Fatal outcome of human influenza A (H5N1) is associated with high viral load and hypercytokinemia. Nat Med 12:1203–1207
96. Kandun IN, Wibisono H, Sedyaningsih ER, Yusharmen HW, Purba W, Santoso H, Septiawati C, Tresnaningsih E, Heriyanto B et al (2006) Three Indonesian clusters of H5N1 virus infection in 2005. N Engl J Med 355:2186–2194
97. Oner AF, Bay A, Arslan S, Akdeniz H, Sahin HA, Cesur Y, Epcacan S, Yilmaz N, Deger I, Kizilyildiz B et al (2006) Avian influenza A (H5N1) infection in eastern Turkey in 2006. N Engl J Med 355:2179–2185
98. Butt KM, Smith GJ, Chen H, Zhang LJ, Leung YH, Xu KM, Lim W, Webster RG, Yuen KY, Peiris JS et al (2005) Human infection with an avian H9N2 influenza A virus in Hong Kong in 2003. J Clin Microbiol 43:5760–5767
99. Koopmans M, Wilbrink B, Conyn M, Natrop G, van der Nat H, Vennema H, Meijer A, van Steenbergen J, Fouchier R, Osterhaus A et al (2004) Transmission of H7N7 avian influenza

A virus to human beings during a large outbreak in commercial poultry farms in the Netherlands. Lancet 363:587–593

100. Fouchier RA, Schneeberger PM, Rozendaal FW, Broekman JM, Kemink SA, Munster V, Kuiken T, Rimmelzwaan GF, Schutten M, Van Doornum GJ et al (2004) Avian influenza A virus (H7N7) associated with human conjunctivitis and a fatal case of acute respiratory distress syndrome. Proc Natl Acad Sci USA 101:1356–1361

101. Uiprasertkul M, Puthavathana P, Sangsiriwut K, Pooruk P, Srisook K, Peiris M, Nicholls JM, Chokephaibulkit K, Vanprapar N, Auewarakul P (2005) Influenza A H5N1 replication sites in humans. Emerg Infect Dis 11:1036–1041

102. Nicholls JM, Chan MC, Chan WY, Wong HK, Cheung CY, Kwong DL, Wong MP, Chui WH, Poon LL, Tsao SW et al (2007) Tropism of avian influenza A (H5N1) in the upper and lower respiratory tract. Nat Med 13:147–149

103. Yamada S, Suzuki Y, Suzuki T, Le MQ, Nidom CA, Sakai-Tagawa Y, Muramoto Y, Ito M, Kiso M, Horimoto T et al (2006) Haemagglutinin mutations responsible for the binding of H5N1 influenza A viruses to human-type receptors. Nature 444:378–382

104. Gambaryan AS, Tuzikov AB, Bovin NV, Yamnikova SS, Lvov DK, Webster RG, Matrosovich MN (2003) Differences between influenza virus receptors on target cells of duck and chicken and receptor specificity of the 1997 H5N1 chicken and human influenza viruses from Hong Kong. Avian Dis 47:1154–1160

105. Wan H, Perez DR (2006) Quail carry sialic acid receptors compatible with binding of avian and human influenza viruses. Virology 346:278–286

106. Shu LL, Lin YP, Wright SM, Shortridge KF, Webster RG (1994) Evidence for interspecies transmission and reassortment of influenza A viruses in pigs in southern China. Virology 202:825–833

107. Castrucci MR, Donatelli I, Sidoli L, Barigazzi G, Kawaoka Y, Webster RG (1993) Genetic reassortment between avian and human influenza A viruses in Italian pigs. Virology 193: 503–506

108. Call SA, Vollenweider MA, Hornung CA, Simel DL, McKinney WP (2005) Does this patient have influenza? JAMA 293:987–997

109. Bhat N, Wright JG, Broder KR, Murray EL, Greenberg ME, Glover MJ, Likos AM, Posey DL, Klimov A, Lindstrom SE et al (2005) Influenza-associated deaths among children in the United States, 2003–2004. N Engl J Med 353:2559–2567

110. Nolte KB, Alakija P, Oty G, Shaw MW, Subbarao K, Guarner J, Shieh WJ, Dawson JE, Morken T, Cox NJ et al (2000) Influenza A virus infection complicated by fatal myocarditis. Am J Forensic Med Pathol 21:375–379

111. Davis MM, Taubert K, Benin AL, Brown DW, Mensah GA, Baddour LM, Dunbar S, Krumholz HM (2006) Influenza vaccination as secondary prevention for cardiovascular disease: a science advisory from the American Heart Association/American College of Cardiology. J Am Coll Cardiol 48:1498–1502

112. Guarner J, Paddock CD, Shieh WJ, Packard MM, Patel M, Montague JL, Uyeki TM, Bhat N, Balish A, Lindstrom S et al (2006) Histopathologic and immunohistochemical features of fatal influenza virus infection in children during the 2003–2004 season. Clin Infect Dis 43:132–140

113. Chan PK (2002) Outbreak of avian influenza A(H5N1) virus infection in Hong Kong in 1997. Clin Infect Dis 34:S58–S64

114. Beigel JH, Farrar J, Han AM, Hayden FG, Hyer R, de Jong MD, Lochindarat S, Nguyen TK, Nguyen TH, Tran TH et al (2005) Avian influenza A (H5N1) infection in humans. N Engl J Med 353:1374–1385

115. Chotpitayasunondh T, Ungchusak K, Hanshaoworakul W, Chunsuthiwat S, Sawanpanyalert P, Kijphati R, Lochindarat S, Srisan P, Suwan P, Osotthanakorn Y et al (2005) Human disease from influenza A (H5N1), Thailand, 2004. Emerg Infect Dis 11:201–209

116. Morens DM, Taubenberger JK, Fauci AS (2008) Predominant role of bacterial pneumonia as a cause of death in pandemic influenza: implications for pandemic influenza preparedness. J Infect Dis 198:962–970

117. Finelli L, Fiore A, Dhara R, Brammer L, Shay DK, Kamimoto L, Fry A, Hageman J, Gorwitz R, Bresee J et al (2008) Influenza-associated pediatric mortality in the United States: increase of *Staphylococcus aureus* coinfection. Pediatrics 122:805–811

118. Reed C, Kallen AJ, Patton M, Arnold KE, Farley MM, Hageman J, Finelli L (2009) Infection with community-onset *Staphylococcus aureus* and influenza virus in hospitalized children. Pediatr Infect Dis J 28:572–576

119. CDC (2009) Surveillance for pediatric deaths associated with 2009 pandemic influenza A (H1N1) virus infection – United States, April–August 2009. MMWR Morb Mortal Wkly Rep 58:941–947

120. Jamieson DJ, Honein MA, Rasmussen SA, Williams JL, Swerdlow DL, Biggerstaff MS, Lindstrom S, Louie JK, Christ CM, Bohm SR et al (2009) H1N1 2009 influenza virus infection during pregnancy in the USA. Lancet 374:451–458

121. Massey PD, Pearce G, Taylor KA, Orcher L, Saggers S, Durrheim DN (2009) Reducing the risk of pandemic influenza in Aboriginal communities. Rural Remote Health 9:1290

122. Baker MG, Wilson N, Huang QS, Paine S, Lopez L, Bandaranayake D, Tobias M, Mason K, Mackereth GF, Jacobs M et al (2009) Pandemic influenza A(H1N1)v in New Zealand: the experience from August 2009. Euro Surveill 14(34) pii:19319

123. Public Health Agency of Canada (2009) http://www.phac-aspc.gc.ca/fluwatch/08-09/w33_09/index-eng.php. Accessed 3 Sept 2009

124. Ng WF, To KF, Lam WW, Ng TK, Lee KC (2006) The comparative pathology of severe acute respiratory syndrome and avian influenza A subtype H5N1 – a review. Hum Pathol 37:381–390

125. To KF, Chan PK, Chan KF, Lee WK, Lam WY, Wong KF, Tang NL, Tsang DN, Sung RY, Buckley TA et al (2001) Pathology of fatal human infection associated with avian influenza A H5N1 virus. J Med Virol 63:242–246

126. Chan MC, Cheung CY, Chui WH, Tsao SW, Nicholls JM, Chan YO, Chan RW, Long HT, Poon LL, Guan Y et al (2005) Proinflammatory cytokine responses induced by influenza A (H5N1) viruses in primary human alveolar and bronchial epithelial cells. Respir Res 6:135

127. Le Goffic R, Balloy V, Lagranderie M, Alexopoulou L, Escriou N, Flavell R, Chignard M, Si-Tahar M (2006) Detrimental contribution of the Toll-like receptor (TLR)3 to influenza A virus-induced acute pneumonia. PLoS Pathog 2:e53

128. Shahangian A, Chow EK, Tian X, Kang JR, Ghaffari A, Liu SY, Belperio JA, Cheng G, Deng JC (2009) Type I IFNs mediate development of postinfluenza bacterial pneumonia in mice. J Clin Invest 119:1910–1920

129. van der Sluijs KF, van Elden LJ, Nijhuis M, Schuurman R, Pater JM, Florquin S, Goldman M, Jansen HM, Lutter R, van der Poll T (2004) IL-10 is an important mediator of the enhanced susceptibility to pneumococcal pneumonia after influenza infection. J Immunol 172:7603–7609

130. Sun K, Metzger DW (2008) Inhibition of pulmonary antibacterial defense by interferon-gamma during recovery from influenza infection. Nat Med 14:558–564

131. McAuley JL, Hornung F, Boyd KL, Smith AM, McKeon R, Bennink J, Yewdell JW, McCullers JA (2007) Expression of the 1918 influenza A virus PB1-F2 enhances the pathogenesis of viral and secondary bacterial pneumonia. Cell Host Microbe 2:240–249

132. Peltola VT, Murti KG, McCullers JA (2005) Influenza virus neuraminidase contributes to secondary bacterial pneumonia. J Infect Dis 192:249–257

133. Hatta M, Gao P, Halfmann P, Kawaoka Y (2001) Molecular basis for high virulence of Hong Kong H5N1 influenza A viruses. Science 293:1840–1842

134. Salomon R, Franks J, Govorkova EA, Ilyushina NA, Yen HL, Hulse-Post DJ, Humberd J, Trichet M, Rehg JE, Webby RJ et al (2006) The polymerase complex genes contribute to the high virulence of the human H5N1 influenza virus isolate A/Vietnam/1203/04. J Exp Med 203:689–697

135. Maines TR, Lu XH, Erb SM, Edwards L, Guarner J, Greer PW, Nguyen DC, Szretter KJ, Chen LM, Thawatsupha P et al (2005) Avian influenza (H5N1) viruses isolated from humans in Asia in 2004 exhibit increased virulence in mammals. J Virol 79:11788–11800

136. Massin P, van der Werf S, Naffakh N (2001) Residue 627 of PB2 is a determinant of cold sensitivity in RNA replication of avian influenza viruses. J Virol 75:5398–5404

137. Steel J, Lowen AC, Mubareka S, Palese P (2009) Transmission of influenza virus in a mammalian host is increased by PB2 amino acids 627K or 627E/701N. PLoS Pathog 5: e1000252

138. Zamarin D, Garcia-Sastre A, Xiao X, Wang R, Palese P (2005) Influenza virus PB1-F2 protein induces cell death through mitochondrial ANT3 and VDAC1. PLoS Pathog 1:e4

139. Zamarin D, Ortigoza MB, Palese P (2006) Influenza A virus PB1-F2 protein contributes to viral pathogenesis in mice. J Virol 80:7976–7983

140. Garcia-Sastre A, Biron CA (2006) Type 1 interferons and the virus-host relationship: a lesson in detente. Science 312:879–882

141. Hale BG, Randall RE, Ortin J, Jackson D (2008) The multifunctional NS1 protein of influenza A viruses. J Gen Virol 89:2359–2376

142. Li Z, Jiang Y, Jiao P, Wang A, Zhao F, Tian G, Wang X, Yu K, Bu Z, Chen H (2006) The NS1 gene contributes to the virulence of H5N1 avian influenza viruses. J Virol 80: 11115–11123

143. Obenauer JC, Denson J, Mehta PK, Su X, Mukatira S, Finkelstein DB, Xu X, Wang J, Ma J, Fan Y et al (2006) Large-scale sequence analysis of avian influenza isolates. Science 311:1576–1580

144. Li S, Schulman J, Itamura S, Palese P (1993) Glycosylation of neuraminidase determines the neurovirulence of influenza A/WSN/33 virus. J Virol 67:6667–6673

145. Kawaoka Y, Webster RG (1988) Sequence requirements for cleavage activation of influenza virus hemagglutinin expressed in mammalian cells. Proc Natl Acad Sci USA 85:324–328

146. Horimoto T, Kawaoka Y (1994) Reverse genetics provides direct evidence for a correlation of hemagglutinin cleavability and virulence of an avian influenza A virus. J Virol 68: 3120–3128

147. Senne DA, Panigrahy B, Kawaoka Y, Pearson JE, Suss J, Lipkind M, Kida H, Webster RG (1996) Survey of the hemagglutinin (HA) cleavage site sequence of H5 and H7 avian influenza viruses: amino acid sequence at the HA cleavage site as a marker of pathogenicity potential. Avian Dis 40:425–437

148. Itoh Y, Shinya K, Kiso M, Watanabe T, Sakoda Y, Hatta M, Muramoto Y, Tamura D, Sakai-Tagawa Y, Noda T et al (2009) In vitro and in vivo characterization of new swine-origin H1N1 influenza viruses. Nature 460:1021–1025

149. Lofgren E, Fefferman N, Naumov YN, Gorski J, Naumova EN (2006) Influenza seasonality: underlying causes and modeling theories. J Virol 81:5429–5436

150. Stone L, Olinky R, Huppert A (2007) Seasonal dynamics of recurrent epidemics. Nature 446:533–536

151. Tellier R (2006) Review of aerosol transmission of influenza A virus. Emerg Infect Dis 12:1657–1662

152. Brankston G, Gitterman L, Hirji Z, Lemieux C, Gardam M (2007) Transmission of influenza A in human beings. Lancet Infect Dis 7:257–265

153. Lowen AC, Mubareka S, Tumpey TM, Garcia-Sastre A, Palese P (2006) The guinea pig as a transmission model for human influenza viruses. Proc Natl Acad Sci USA 103:9988–9992

154. Mubareka S, Lowen AC, Steel J, Coates AL, Garcia-Sastre A, Palese P (2009) Transmission of influenza virus via aerosols and fomites in the guinea pig model. J Infect Dis 199:858–865

155. Lowen AC, Mubareka S, Steel J, Palese P (2007) Influenza virus transmission is dependent on relative humidity and temperature. PLoS Pathog 3:1470–1476

156. Lowen AC, Steel J, Mubareka S, Palese P (2008) High temperature (30°C) blocks aerosol but not contact transmission of influenza virus. J Virol 82:5650–5652

157. Lowen AC, Steel J, Mubareka S, Carnero E, Garcia-Sastre A, Palese P (2009) Blocking interhost transmission of influenza virus by vaccination in the guinea pig model. J Virol 83:2803–2818

The Epidemiology of Influenza and Its Control

Lone Simonsen, Cécile Viboud, Robert J. Taylor, and Mark A. Miller

Abstract In this chapter, we highlight how recent advances in influenza epidemiology can inform strategies for disease control. Given the challenge of direct measurement, influenza epidemiology has benefited greatly from statistical inference from the analysis of large datasets regarding hospitalization, mortality, and outpatient visits associated with seasonal circulation of influenza viruses. These data have allowed comparison of the impact of influenza in various climates and the evaluation of the direct and indirect benefits of vaccination, the latter through the vaccination of "transmitter populations" such as school children, to achieve herd immunity. Moreover, the resolution of influenza epidemiology has undergone a leap to the molecular level due to the integration of new antigenic and viral genomic data with classical epidemiological indicators. Finally, the new data have led to an infusion of quantitative studies from the fields of evolutionary biology, population genetics, and mathematics. Molecular influenza epidemiology is providing deeper insight into temporal/spatial patterns of viruses, the important role of reassortment in generating genetic novelty, and global diffusion of virus variants – including the role of the tropics, as a source of new variants. Higher resolution, contemporary, and historic epidemiological data provide a more detailed picture of the effect of age and other host characteristics on outcomes, as well as better estimates of the transmissibility of pandemic and seasonal influenza viruses. New epidemiologic and virologic data from the current A/H1N1pdm 2009 pandemic improve our understanding of the emergence and establishment of new viral subtypes in

L. Simonsen
George Washington University School of Public Health and Health Services, Washington, DC, USA
Fogarty International Center, National Institutes of Health, Bethesda, MD, USA

C. Viboud and M.A. Miller (✉)
Fogarty International Center, National Institutes of Health, Bethesda, MD, USA
e-mail: millermar@mail.nih.gov

R.J. Taylor
SAGE Analytica, LLC, Bethesda, MD, USA

G. Del Giudice and R. Rappuoli (eds.), *Influenza Vaccines for the Future*, 2nd edition,
Birkhäuser Advances in Infectious Diseases,
DOI 10.1007/978-3-0346-0279-2_2, © Springer Basel AG 2011

human populations and their mortality and morbidity burden in the first years of circulation. Re-examination of observational studies of vaccine effectiveness in seniors is leading to reconsideration of seasonal and pandemic vaccine priorities, while mathematical modelers have developed tools to explore optimal strategies for mitigating on-going and future pandemics. The field of influenza epidemiology has rapidly progressed in the past decade and become truly multidisciplinary. Progress could be sustained in the next decade by further interdisciplinary studies between virology, evolutionary biology, immunology, and clinical outcomes.

1   Introduction

Influenza viruses evolve continuously, challenging mammalian and avian hosts with new variants and causing complex epidemic patterns with regard to age, place, and time. Human influenza viruses cause disease through a variety of direct and indirect pathological effects. The direct effects include destruction of infected cells, damage to respiratory epithelium, and immunological responses that cause general malaise and pneumonia. Indirect consequences of infection include secondary bacterial infections as a result of tissue damage and exacerbation of underlying comorbid conditions such as cardiovascular disease, renal disease, diabetes, or chronic pulmonary disease [1, 2]. Given the lack of the conduct of laboratory tests, the morbidity and mortality associated with influenza is frequently classified into broad disease categories, such as pneumonia and influenza (P&I), respiratory illness, or all-cause (AC) mortality determined through statistical inference, based on seasonal coincidence of virus circulation and disease outcomes [3–5].

Given the difficulty of directly measuring influenza morbidity and mortality, time series models have been developed to elucidate patterns of disease within various age groups and populations [5–13]. Such models allow for quantification of disease burden by season and severity of circulating strains [9]. Historical data have also elucidated the links between influenza transmission across geographic regions and population movements [14] and allowed comparison of the impact and transmissibility of past pandemics and epidemics in multiple countries [15–24]. Similar models applied to prospective syndromic surveillance data have allowed the study of the epidemiological signature of recurring and reemerging strains of influenza on populations [25]. Mathematical modeling and statistical analyses of influenza activity in tropical countries have rekindled interest into the seasonal drivers of influenza and offered new insights into the circulation patterns of this virus at the global and regional scales [26–28] (Fig. 1).

The field of influenza epidemiology has recently undergone a quantum leap in resolution due to the increased availability of antigenic and viral genomic data and the integration of these data with classical epidemiological indicators [29–32]. The emerging field of molecular influenza epidemiology and evolution, or "phylodynamics" [29], has provided a much clearer picture of the complex dynamics of global influenza virus circulation and reassortment patterns. The growing number of available influenza genome sequences from specimens collected around the world has started to

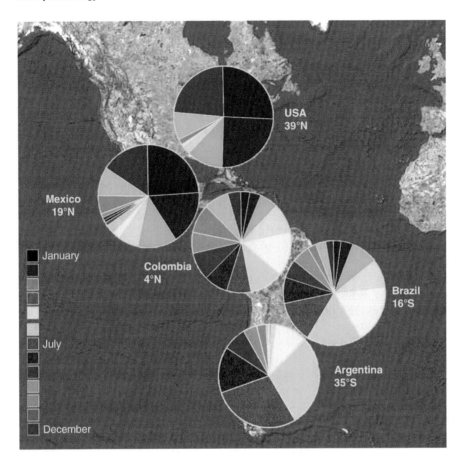

Fig. 1 Comparison of influenza virus seasonal patterns in temperate and tropical countries in the Americas. *Pie charts* represent the percent distribution of influenza virus isolation by month as compiled from WHO data between 1997 and 2005 (*color bar*). Note the transition in seasonal patterns from north to south. The latitude of the capital city is indicated for each country in the *legend*. Adapted from Viboud et al. [27]

create a more coherent picture of the global epidemiology of influenza, in particular the interplay between virus evolution, population immunity, and impact.

We highlight how influenza epidemiology through statistical inference tools has helped refine existing strategies for influenza control. We begin by examining the spatial and temporal spread of seasonal influenza, and how old and new analytical tools are reshaping quantitative thinking in influenza epidemiology and control. We examine historical patterns of disease observed during the three pandemics of the twentieth century and discuss the epidemiology of the recent avian A/H5N1 influenza threat and the current A/H1N1pdm 2009 pandemic. We review what is known about the impact of vaccine in older age groups – the group with the greatest influenza-related mortality burden – and a discussion of the implications of influenza epidemiology for pandemic planning. We conclude with a short discussion of the epidemiology of "H1N1pdm," the virus behind the current pandemic. Readers looking for a more comprehensive treatment of the vast field of influenza epidemiology may consider supplementing this chapter with some of the classical reviews published over the last decades [2, 33–37].

2 Seasonal Influenza: New Insights

The disease burden of annual influenza epidemics varies greatly in terms of hospitalizations and deaths. In the USA, clinical illness affects 5–20% of the population and asymptomatically infects a larger number [36]. Infants, who are exposed to influenza epidemics as a novel antigenic challenge after maternal antibodies decline, may have attack rates as high as 30–50% in their first year of life, depending on the frequency of contacts with older siblings [38]. For reasons not fully understood, influenza viruses cause seasonal epidemics in the northern and southern hemisphere during their respective winters. In the tropics, the timing of activity is less defined, with sometimes year-round circulation or bi-seasonal peaks during the year (Fig. 1) [27, 28, 39–42].

2.1 Methods Used to Estimate the Mortality Burden of Influenza

Estimates of the number of influenza-related deaths are typically inferred through statistical analysis. The syndromic diagnosis "influenza-like illness" is rarely laboratory confirmed and is often caused by non-influenza respiratory viruses. Moreover, influenza may be an inciting factor that brings about death from secondary bacterial pneumonia or an underlying chronic disorder. In these cases, the secondary infection or underlying disorders are typically identified as the cause of death which may occur weeks after the initial viral infection. Because of these ascertainment problems, determining the magnitude of influenza-related deaths requires indirect approaches in which mathematical or statistical models are applied to broad death categories. This approach was first used in 1847 by William Farr to characterize an influenza epidemic in London and was further developed and extensively used throughout the twentieth century. The refinements include Serfling-like cyclical regression models [6, 12, 18, 21, 43–46] and Arima models [7, 8, 47, 48], which are applied to monthly or weekly time series of P&I or AC mortality. Overall, investigators from at least 17 countries have used variants of these Serfling-type models to estimate the mortality burden of influenza. Similar issues and statistical approaches apply to the estimation of the influenza burden on hospitalization [10, 11, 49]. The various statistical approaches all attribute "excess" health outcomes (deaths or hospitalizations) in winter months to influenza. Such seasonal approaches are not suited to studying disease burden of influenza in countries with tropical climates because they require an annual seasonal pattern of viral activity interrupted by influenza-free periods.

More recently, the US Centers for Disease Control and Prevention (CDC) has used an approach to measure hospitalization and mortality burden based on a new generation of seasonal regression models integrating laboratory surveillance data on influenza and respiratory syncytial virus (RSV) [5, 11]. In such models, winter seasonal increases in deaths or hospitalizations are directly proportional to the magnitude of respiratory virus activity. In the USA between 1980 and 2001, Thompson et al. [5, 11] estimated that seasonal influenza epidemics were associated with 17 deaths per 100,000 on average (range 6–28 per 100,000) depending on the severity of the circulating strains. Reassuringly, different model approaches, with and without the quantification of the number of viral isolates, yield similar average estimates of the influenza mortality burden in the USA [13, 50, 51]. Estimates from Europe and Canada are similar to those from the USA [44, 52, 53]. Viral surveillance data with the integration of hospitalization or death indicators are particularly useful for the study of influenza in the tropics where there is less seasonality.

2.2   Age and Time Variability in Influenza-Related Mortality in Temperate Climates

Influenza-related deaths contribute ~5% (range 0–10%) of all winter mortality in persons over 65 years of age in the USA, with similar proportion in Italy and Canada [12, 53, 54]. Seasons dominated by the influenza A/H3N2 subtype are typically associated with 2–3-fold higher mortality than seasons dominated by A/H1N1 and influenza B viruses from the 1980s to 2009. The pattern is not always uniform; there have been influenza A/H3N2-dominated seasons with little excess mortality (e.g., 2005–2006 northern hemisphere season). The age-specific risk of influenza-related (excess) mortality rates rises sharply past age 65 years (Fig. 2). People aged ≥80 years are at approximately 11-fold higher risk than people aged 65–69 years. Moreover, in recent decades about 90% of all influenza-related deaths occurred among seniors ≥65 years, 75% occurred among seniors aged ≥70 years, and 55% occurred among seniors over 80 years [12]. As the population in the USA and other developed countries has aged substantially over the last decades, the crude number of influenza-related deaths has been rising. Because the risk of influenza-related death increases exponentially with age in the later decades of life, it is essential to standardize for age when comparing mortality impact in different countries and over time [12, 54].

2.3   Burden and Circulation Patterns of Influenza in the Tropics

Because most seasonal influenza models ("Serfling approaches") depend on winter seasonality in the data, they are not generally useful for tropical countries. However,

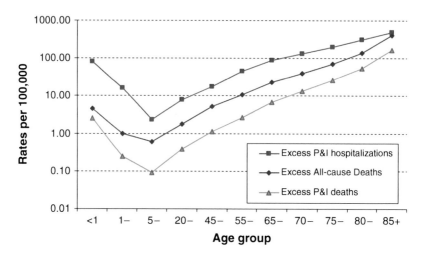

Fig. 2 Average age-specific rates of influenza-related excess deaths and hospitalizations for ten seasons during 1990–2001 in the USA (estimated from Serfling regression models). Note the characteristic U-shape of severe disease burden by age that characterizes seasonal influenza. Data source: Vital Statistics from the National Center for Health Statistics (NCHS) and hospital discharge data from Agency for Health Care Research and Quality (AHRQ)

such models can be used for unusually severe epidemics and pandemics, where the excess disease burden is many fold greater than in average years [43]. Integration of viral surveillance data with death or hospitalization indicators is the most useful approach in tropical settings, although long-term historical surveillance data are usually lacking [27]. A series of studies in Hong Kong and Singapore recently found that annual influenza-related hospitalization and mortality rates in wealthy (sub) tropical locations are similar to those in temperate countries [39–42]. In Hong Kong, as in many other countries, the influenza is associated not only with pneumonia outcomes but also with a wide range of chronic health conditions such as diabetes and cardiovascular diseases [42]. In addition, influenza-related hospitalization rates in Hong Kong vary with age as a U-shaped curve [41], in which young infants and elderly people are at highest risk of severe outcomes, reminiscent of the age pattern of epidemic influenza in the USA and other temperate countries.

The spread of influenza in the tropics has also proven to be an enigma. Influenza seasonality in the southernmost temperate regions is 6 months out of phase with the northern hemisphere. A study from Brazil found seasonal influenza activity starting early in remote, less densely populated equatorial regions of the north (March–April) and traveling in ~3 months to the more temperate areas of the south during their winter season (June–July) [28]. This finding was contrary to what was expected, given that the larger, well-connected, densely populated cities are located in the south. If population movements were a driving factor like in the USA [14], then the opposite traveling wave would have been expected. This study has inspired further studies to investigate the circulation of specific influenza virus subtypes during a season based on analysis of viral genomics data. Finding firm evidence of this unusual circulation pattern suggested from analysis of regional mortality data also bears on considerations of use of southern or northern hemisphere vaccine formulation and timing. Because of this study, Brazil is considering changing the timing of vaccination in the north of Brazil to accommodate the early occurrence of influenza in that area.

2.4   The Burden of Influenza in Infants and Young Children

For age groups other than those over 65 years of age, it can be difficult to measure the relatively low seasonal impact of influenza mortality above the expected baseline. However, for occasional severe seasons, a surge in P&I deaths can often be seen in children and young adults. For example, the 2003–2004 season was dominated by a new antigenic variant of A/H3N2 viruses (A/Fujian/2003) and was unusually severe; in the USA, 153 children with documented influenza infections died of primary or secondary pneumonia and sepsis [55]. Surprisingly, 47% of the children who died had no known underlying risk conditions. The reason for this unusual epidemic of pediatric deaths has not been resolved. As a result of this experience, the CDC enhanced their influenza surveillance system with a reporting system for children hospitalized with laboratory-confirmed influenza.

2.5   The Impact of Influenza on Morbidity

Very few quantitative data on mild influenza morbidity with known population denominators are available. The most careful studies using the longest existing time series come from the Royal Network of General Practitioners in the UK, which has reported influenza-like illnesses on a weekly basis since 1966 [52, 56]. Such long-term morbidity records are unique and have allowed the study of the 1968–1969 influenza pandemic transmission patterns based on case data [57]. In addition to the UK, several countries have national sentinel surveillance systems in place (USA, France, Netherlands, Australia, and New Zealand are examples). These are used to detect the onset and peak timing of influenza epidemics, as well as the magnitude of morbidity impact relative to surrounding seasons. In the USA, emergency room visit time series are now being analyzed in the context of biodefense research and have shed light on interannual and age-specific variability in influenza impact [25, 58].

In contrast, quantitative burden studies using samples of national hospital discharge data and estimation approaches similar to those used for excess mortality are more widely available, in particular since the 1970s [11, 49, 59, 60]. The patterns of excess hospitalizations are quite similar to those of excess mortality, with a U-shaped incidence reflecting the highest values in young children and seniors (Fig. 2).

2.6   The Relative Contribution of Influenza and RSV

One controversy in the literature concerns the relative contributions of influenza and RSV to the winter increase in respiratory hospitalizations and deaths, especially among seniors. The current CDC modeling approach simultaneously estimates the influenza and RSV burden by correlating periods of excess mortality with their respective period and magnitude of viral activity [5]. In the overall US population, the CDC investigators estimate that the average seasonal RSV burden is approximately one-third of that of influenza for all seasons during the 1990s. However, the relative contribution of RSV and influenza varies greatly with age.

For US infants of <12 months of age, the RSV contribution to mortality is more than twofold greater than that of influenza (5.5 vs. 2.2 deaths per 100,000) based on the CDC model [5]. Above 5 years of age, mortality due to influenza predominates in the US data, similar to the age pattern of respiratory deaths in the UK [61]. For seniors over age 65, the CDC model estimates the average seasonal RSV burden at ~10,000 deaths, which is approximately one-third the estimated deaths attributed to influenza over the same period. But others disagree; several observational studies set in the UK by Fleming et al. [62, 63] have argued that RSV has replaced influenza as the major cause of respiratory mortality and hospitalization, in particular in the elderly. Further, in a recent laboratory-based study set in a large cohort of seniors hospitalized with pneumonia, twice as many hospitalizations were attributed to RSV as influenza [64]. But because influenza-related pneumonia is most often due to secondary bacterial infections (quite distinct from primary RSV pneumonia) that occur long after the triggering influenza infection has been cleared, it is possible that this study substantially underestimated the influenza burden [65].

Two recent studies carefully delineated the relative burden of influenza and RSV in children, using seasonality in pediatric respiratory hospitalizations and focusing the analysis on seasons when the influenza and RSV epidemics occurred at different times [39, 66]. The authors subtracted hospitalization rates during periods of high influenza circulation from baseline "peri-influenza" winter periods when neither influenza nor RSV was circulating (Fig. 3). Using this approach, the authors attributed a similar number of hospitalizations to RSV and influenza in children under 5 years in the USA [66]. In a parallel study from Hong Kong, investigators attempted to delineate the burden of RSV, influenza, and other respiratory pathogens in various age groups in this subtropical setting with less clear seasonality [39]. Although influenza burden estimates in Hong Kong were similar to those of the USA in most age groups [27], children under 5 years appeared to have approximately tenfold higher rates of hospitalization in Hong Kong than in the USA [39]. Such large discrepancies may reflect true geographical differences in influenza transmission and impact, although they are perhaps more likely to result from differences in access to hospital care. Indeed, young children in Hong Kong tend to be rushed to the hospital when they have respiratory symptoms (Malik Peiris, personal communication).

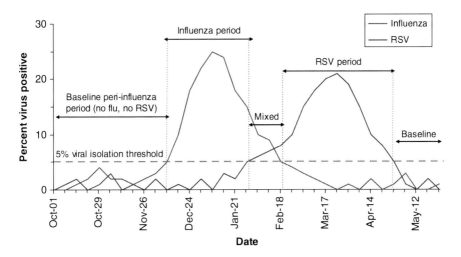

Fig. 3 An analytic approach to estimate influenza-related hospitalization rates in USA and Hong Kong children [39, 66]. This method relies on identifying the precise periods of influenza and RSV viral circulation for each season studied. Influenza-related excess rates were calculated as the difference in observed rates between periods of influenza and RSV circulation and those with low circulation of both influenza and RSV

Finally, there are numerous studies on respiratory virus isolates from children hospitalized with respiratory symptoms in tropical and subtropical settings. A review of these studies attributes a substantial proportion of pediatric respiratory hospitalizations to influenza A and B viruses [67]. Unfortunately, it is difficult to compare findings across studies because they are often carried out using different laboratory techniques and are set in different study years, seasons, and clinical settings. These studies frequently present a systematic age pattern that suggests that RSV is more important in infancy, with a gradual shift to influenza by about 5 years of age as the pathogen more likely to cause severe respiratory illness.

2.7 Observational Transmission Studies

The transmission patterns of influenza were carefully documented in classic virus surveillance studies that meticulously followed all respiratory illness episodes in a large number of families in Cleveland, Ohio, Tecumseh, Michigan, Seattle, and Washington in the 1950s through the 1970s [35, 68, 69]. Unfortunately, such careful studies have not been repeated in contemporary populations, so little is known about the consequences of increasing population movements and changes in intrafamilial interactions. The result is that the existing mathematical models employed to "forecast" the likely patterns and spread of a pandemic influenza virus rely largely on parameter values of transmission and age group dynamics that are decades old and may not reflect current realities.

In parallel to careful family studies tracking the infection status of each individual, time series mortality data aggregated at the scale of cities, regions, or countries can also be used as a proxy to estimate the transmissibility of influenza [16, 19, 20, 23, 24, 70–73]. Two crucial factors, the basic reproductive number, $R0$, and the effective reproductive number, R, have been estimated for pandemic and epidemic influenza. $R0$ measures the average number of secondary infections per primary case for a new pathogen invading a fully susceptible population (e.g., a pandemic influenza virus), whereas R measures a similar quantity for a recurrent pathogen re-invading a partially susceptible population (e.g., seasonal influenza virus). Current estimates of $R0$ and R are in the range of 1.7–5.4 for pandemics and 1.0–2.1 for seasonal influenza epidemics. While these estimates of transmissibility are not as high as for other respiratory viruses (e.g., for measles R is ~15), the generation time for influenza is relatively short, on the order of 2–4 days. Consequently, in a 60-day period, there could be $R^{(60/4)}$ to $R^{(60/2)}$ infections.

Overall, the use of time series of population-level data (hospitalizations, mortality) in large populations has provided a more complete picture of the transmissibility of influenza through space and time. One study correlated mortality peaks in US influenza seasons for the last 30 years with daily transportation data and found that epidemics spread across the country in an average of about 6 weeks and that transmission was correlated with adult work travel patterns [14].

2.8 Syndromic Surveillance and Its Contributions to Influenza Epidemiology

Use of real-time syndromic surveillance data is another area with substantial promise in influenza epidemiology. Information technology now allows for the rapid compilation and analysis of electronic health records from emergency rooms, inpatient hospitals, and outpatient clinics. Syndromic surveillance efforts have already provided a new level of insight into age and geographic patterns of impact of influenza epidemics. In particular, a recent study that combined time series analysis of age-specific emergency room visits with laboratory-confirmed timing of influenza and RSV periods in New York City demonstrated that the burden of a contemporary influenza epidemic varies greatly at the level of age cohorts in children and adults, perhaps as a consequence of different historical exposures to influenza [25].

2.9 Influenza Genomics and Molecular Epidemiology

Phylogenetic and antigenic studies of influenza viruses have increased our understanding of the emergence and spread of new influenza drift variants both locally and globally. Begun in 2004, the Influenza Genome Sequencing Project, as well as an increased number of sequences published by other contributors, has resulted in the publication of over 80,000 influenza genes from isolates around the world isolated from numerous species. These data have led directly to advances in molecular influenza epidemiology [31]. Studies emerging from this project have demonstrated a high frequency of gene segment reassortment in A/H3N2 viruses, perhaps more frequent around the time of transition to new antigenic variants [30]. Specifically, one possible mechanism leading to the emergence of antigenic novelty is reassortment between dominant and subdominant lineages of past seasons. Further, each A/H3N2-dominated season features multiple genetically distinct cocirculating lineages that may or may not have similar antigenic properties [32]. Studies of recent epidemics of A/H3N2 in New York City and New Zealand have shown that next season's viruses are seeded by importation either from the opposite global hemisphere or from the tropics and that there is no preferred hemisphere leading the circulation of viruses [74]. This rapidly emerging area of molecular influenza epidemiology has increased our understanding of viral circulation patterns around the globe, and the genesis and spread of drift variants.

3 Pandemic Influenza: Lessons from Historical Data and Modeling

Historic experience with influenza pandemics in the twentieth century has been a prelude to the current pandemic with the global spread of novel A/H1N1pdm virus [75]. The three pandemics of the twentieth century – the 1918 A/H1N1 "Spanish influenza," the 1957 A/H2N2 "Asian influenza," and 1968 A/H3N2 "Hong Kong influenza" – were highly variable in terms of mortality impact (Table 1). The catastrophic 1918 pandemic resulted in 0.2% to as much as 8% mortality in various countries around the world and an estimated global mortality of ~50 million people [76]. The relatively mild 1968 pandemic, however, was not appreciably worse than

Table 1 Mortality impact and patterns of three most recent pandemics, compared with the contemporary impact of seasonal influenza

Pandemic and virus subtype	Evolutionary history (segments involved)	Approximate global mortality impact	Proportion of deaths in persons <65 years of age
1918–1919 A(H1N1)	All avian (all eight segments)	~50 M	~95%
1957–1958 A(H2N2)	Reassortant HA + NA + PB1	~1–2 M	~40%
1968–1969 A(H3N2)	Reassortant HA + PB1	~0.5–1 M	~50%
Contemporary H3N2 seasons	No shift – only gradual genetic drift	~0.5–1 M	~10%

HA hemagglutinin, NA neuraminidase, PB1 polymerase, M million

other severe seasonal epidemics in terms of total influenza-related deaths, whereas the 1957 A/H2N2 pandemic was moderately severe [15, 18]. As of September 2009, in the northern hemisphere autumn season, the impact of the A/H1N1 pandemic virus appears relatively mild, though it has an uncertain future of mutating to a more virulent strain.

3.1 History Lessons from the Field of Archaeo-Epidemiology

Recent efforts to re-examine the 1918 Spanish influenza pandemic [77], as well as that of later pandemics, have allowed for a more comprehensive view of pandemics and highlighted their diversity in time and space. Historical vital statistics data have been analyzed to provide a quantitative analysis of the last century's three pandemics. For each of these pandemics, there was a quantitative and qualitative change in the mortality patterns, as compared to seasonal influenza epidemics. The shift of the mortality burden to younger ages has been a "signature" of each pandemic and stands in marked contrast to the low mortality burden among young people during typical influenza epidemics [15, 78]. This age shift was most pronounced in the 1918 pandemic but occurred in all three pandemics for which age group mortality data have been studied (Table 1; Fig. 4). During the initial outbreak of the novel H1N1pdm virus (April 2009), a shift of morbidity and mortality toward younger age groups was observed in Mexico [79] and remains a characteristic of this virus.

Sero-archaeology studies of collections of serum from blood donors have been informative about preexisting influenza antibodies and therefore indicate the past circulation of historical pandemic viruses, even in tropical populations. These

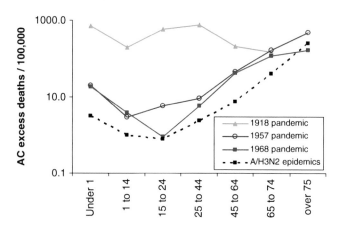

Fig. 4 Age-specific mortality impact of three historical pandemics contrasted with the average impact of recent A/H3N2 epidemics in the 1990s. Based on a Serfling model applied to US all-cause excess mortality data (and presented on a logarithmic scale)

studies provide interesting pieces of the puzzle but have unfortunately fallen out of fashion lately. For example, one collection of serum gathered before the 1968 pandemic showed that people born before 1892 had antibodies to the hemagglutinin A/H3 antigen; this may partially explain the fact that seniors older than 77 years were only at a decreased risk of death during that pandemic [15, 80]. In another example, a sero-epidemiology study looking at influenza antibodies in a population of women in Ghana following the 1968 pandemic showed that in the tropics, most had become infected 5 years after the emergence of the A/H3N2 subtype [81].

A similar antigen recycling phenomenon may explain the low rates of morbidity and mortality observed in people over the age of 50 years in the early months of A/H1N1pdm virus circulation [79]. For almost all persons born from 1918 to 1957 (~52- to 91-year olds in 2009), the first exposure to an influenza A virus was to the strains containing A(H1); those born from 1957 to 1968 (~41 to ~52) to A(H2); those born since the 1968 pandemic (<41 years of age), most likely first saw A(H3). Indeed, the A(H1) subtype was reintroduced in 1977 but rarely dominates [5], suggesting that most people born after 1977 were first exposed to A(H3) viruses. This is important because the concept of "original antigenic sin" postulates that the first encounter with an influenza virus, likely in childhood, provides the strongest immunity in later years [82]. Therefore, people born before 1957 may have the greatest natural immunity to the currently circulating A/H1N1pdm pandemic virus in 2009 [79].

Looking back to the 1918 A/H1N1 pandemic suggests that antigen recycling may have also played a role and could partly explain the extreme case of mortality age shift associated with this pandemic. In this pandemic, seniors were completely spared, in stark contrast to the extreme mortality impact in the young adults, as shown by age-specific mortality surveillance from New York City (Fig. 5) [17, 78]. This was further confirmed in an additional study of age-detailed mortality time series from Copenhagen [22]. This phenomenon could be explained by immune protection conferred by prior exposure (recycling) of an H1Nx virus in the late nineteenth century. Alternatively, the atypical mortality spike in young adults in the 1918 pandemic may be explained by an unusual immune dysfunction causing a "cytokine storm" [83, 84], which primarily affected young adults. These two possibilities – recycling and immune pathology – cannot be resolved without further experimental and epidemiological studies. This unfortunately leaves us with a great unknown as we attempt to deal with the current pandemic: if the pandemic virus contains a hemagglutinin antigen that has not previously circulated in human populations – such as the current avian A/H5N1 virus in Asia – then the recycling hypothesis would suggest seniors could be at great risk, as suggested by one author [85]. In contrast, the immune pathology hypothesis suggests that immune senescence might mitigate the full impact among seniors, leaving young adults at highest risk of dying.

Comparative studies of pandemic influenza in multiple countries have revealed many interesting insights. For example, a recent study used annual mortality data from multiple countries to estimate the mortality burden of the 1918–1920 influenza pandemic and uncovered substantial geographical differences in influenza-related

Fig. 5 Comparison of age mortality patterns during the 1918 pandemic and a severe interpandemic season, New York City. (**a**) Influenza season-attributable excess deaths are plotted for the 1915–1916 interpandemic seasons (○), the pandemic herald wave (epidemic months March and April 1918; ▲), and the main fall pandemic wave (September 1918 to April 1919, ■). (**b**) Relative risk of death is plotted by age group on a log10 scale for the herald (▲) and fall pandemic waves (■), relative to the severe interpandemic season. Adapted from Olson et al. [17]

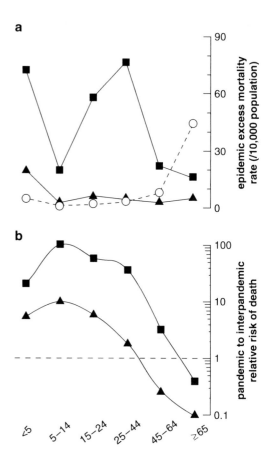

mortality rates. The percentage of the population that died varied from 0.2% in Scandinavia to 8% in some areas of India, representing a 40-fold difference in mortality risk in these settings [76]. The underlying reasons for this substantial variability are not well understood but might be revealed by additional historical pandemic studies.

In a second example, analysis of excess mortality data from several countries put a surprising spin on the 1968 pandemic [18]. An unexpected pattern of a "smoldering" mortality impact in European and some Asian countries was revealed – a relatively mild first wave of the emerging virus in the 1968–1969 season, followed by a very severe 1969–1970 season. This is different from the classical impression based on the North American experience that most of the impact occurs with the first exposure to pandemic strains. It may be more common than previously thought that the first wave of a pandemic virus results in low mortality, only to be followed by a more dramatic impact a few months later. Indeed, this intriguing pattern was not only observed in some countries during the mild 1968 pandemic but also consistent with the herald wave experience in New York City and Scandinavia during the catastrophic 1918 pandemic [17, 22]. Further, historical mortality data from the less-studied 1889–1892 pandemic in England also suggest a pattern of successive pandemic waves where the first encounter was not the most lethal [86]. The reasons for this "smoldering" (or herald wave) pattern are still unknown and may be partly related to on-going adaptation in newly emerged pandemic viruses and preexisting population immunity.

3.2   Transmission Models Used to Predict Future Pandemic Scenarios

Mathematical transmission models have been employed to simulate in detail the possible spread of a new pandemic virus in a susceptible human population (e.g., [70, 71, 87, 88]). These models seek to predict the spatiotemporal dynamic of a hypothetical pandemic virus and the effectiveness of intervention strategies such as vaccination before an outbreak with a partially matched, low-efficacy vaccine, distribution of antivirals for prophylaxis or treatment, school closure, case isolation, and household quarantine. These models generally agree that a combination of measures, if implemented early and with sufficient compliance, might bring about a meaningful level of mitigation and substantially slow geographic spread. Subsequent studies found that early, targeted, and layered use of nonpharmaceutical interventions could greatly reduce the overall pandemic attack rate, provided the intrinsic transmissibility (basic reproductive number, $R0$) of the emerging virus is not greater than two [89–91]. Mathematical models can be useful to estimate the potential impact of interventions assuming a wide range of parameters. Furthermore, they can prioritize research by highlighting the most sensitive and uncertain parameters for a desired outcome. Simulation models currently used for pandemic planning still need to be tested against real disease data, and for this we must continue to gather data on influenza morbidity, mortality, and viral genetic sequences in both pandemic and seasonal influenza scenarios [92].

3.3   Predicting the Impact of Pandemics

Until spring 2009, concern has focused on the highly pathogenic variant of A/H5N1 influenza that emerged in Hong Kong in 1997 and remerged in 2003. A/H5N1 has now spread to avian populations in more than 30 countries. It is present endemically in Southeast Asia, causing regular die-offs in poultry and wild birds, and occasionally affects humans. As of August 31, 2007, the World Health Organization (WHO) had counted 327 laboratory-confirmed H5N1 cases and noted a very high case fatality of ~61% (http://www.who.int/topics/avian_influenza/en/ 2007). While H5N1 continues to be an economic problem in Asia, Africa, Europe, and the Middle

East, the critical question for public health is whether it will gain the ability to effectively transmit among humans. This could occur in one of the two ways: by gradual mutations of avian H5N1 viruses, or by reassortment with circulating human influenza A viruses (H3N2 or H1N1), in humans or another animal. Several comprehensive discussions of the threat of an avian influenza pandemic have been published (e.g., [85, 93–96]).

There are still many uncertainties about the pandemic potential of the circulating avian H5N1 virus, including its potential to effectively transmit between humans and the evolutionary mechanisms that may concurrently affect its virulence. The classical belief is that extremely pathogenic viruses are not well adapted to their hosts – moribund patients do not transmit viruses as easily as those who remain mobile. Further, the pathogenesis of novel pandemic viruses remains unclear, in particular the proportions of severe disease caused by immune-mediated patholog-ical responses, secondary bacterial infections (for which treatments exist), and exacerbation of chronic illnesses. Modern medicine can mitigate some of the pathological mechanisms and control secondary bacterial infections to a certain extent; however, there is undoubtedly a different proportion of persons living with chronic comorbid conditions now than was the case during previous pandemics. Finally, we do not know the degree of cross-protection afforded by early exposure to other influenza virus antigens [97]. If one simply applies the 1918 mortality experience to today's population, anywhere from 0.2% to 8% of a country's population could die, and the highest burden would be suffered by developing countries [76].

The emergence of the H1N1pdm virus poses the threat of a potentially severe pandemic in the months to come. Research efforts have intensified and a vaccine has been developed, but many questions remain unanswered. We do not know whether the H1N1pdm virus will reassort with seasonal influenza viruses. We do not know what proportions of severe disease caused by immune-mediated patho-logical responses, secondary bacterial infections, and exacerbation of chronic ill-nesses it will cause nor do we know how well medical interventions will mitigate the impact. In terms of mortality, although age groups with severe disease tend to be under 60 years of age, there are more people living with comorbid conditions than during previous pandemics. Thus, for the moment transmission dynamics, morbid-ity and mortality impact, and the degree of immunity remain obscure.

4   Epidemiology and the Control of Influenza

Influenza vaccines were originally developed for use by the military and have been shown to be highly effective in preventing infection in healthy adults [98]. Most countries that use seasonal influenza vaccine have adopted a policy of targeting influenza vaccination efforts to those at "high risk" of severe outcomes, including those age 65 years and older, persons with certain chronic diseases and their close contacts. Although current policy continues to emphasize vaccination of seniors, the "gold standard" evidence that this strategy effectively reduces influenza-related mortality in that age group is not strong [99]. It has recently become evident that influenza-related mortality has not decreased in at least some countries despite major gains in vaccination coverage among people at highest risk [5, 12, 54, 100]. Because "gold standard" evidence from randomized clinical trials is scarce, epidemiological tools and studies constitute the vast majority of the evidence base for whether vaccination programs are beneficial. Paradoxically, observational studies have consistently argued that about 50% of all winter deaths in seniors are preventable with influenza vaccination despite the relatively low immune response to vaccine in this population [101].

4.1 The Scarce Evidence from Clinical Trials

Langmuir, who originally formulated the policy of targeting seniors and high-risk population for vaccination, questioned whether the vaccine would really be effective in seniors who respond less vigorously to the vaccine than younger adults [102]. Only a single randomized placebo-controlled clinical trial set in young healthy seniors is available. It showed that vaccination effectively prevents influenza illness in seniors aged 60–69 years but could not document significant benefits in seniors ≥70 years [103]. The authors expressed concern that their nonsignificant finding of 23% efficacy in seniors ≥70 years old indicated immune senescence (a decline in immune response with age), although they also noted limitations on the statistical power of their study to address this question. As both T-cell and B-cell responses are impaired in older individuals, it is plausible that the vaccine antibody response to the drifting influenza viruses and vaccine components is less vigorous in seniors [104]. Consequently, immunologists have long perceived a need for more effective vaccine formulations for this population, including the need for adjuvants and a move back to whole-cell vaccine products. The recent emergence of novel avian strains and development of vaccines against them has reopened many of the discussions of immunogenicity and correlates of protection.

4.2 Evidence from Observational Studies

In the near-absence of randomized clinical trials, these cohort studies have long provided the evidence base that supports influenza vaccine policy. Paradoxically, the concerns about immune senescence and vaccine failure have existed in parallel with cohort studies reporting extraordinarily large mortality benefits in vaccinated seniors [105–107]. In these studies, comparison of vaccinated and unvaccinated seniors indicates that vaccination could prevent fully 50% of all deaths among during winter months, implying that influenza causes half of all winter deaths among seniors. Instead, meta-analyses consolidated the findings and produced estimates with tight confidence intervals. But only about 5% of all winter deaths can be attributed to influenza in an average season according to excess mortality studies [5, 12, 54]. Even in the 1968 A/H3N2 pandemic and in more recent seasons such as 1997–1998, when the vaccine was completely mismatched to the new circulating variant of A/H3N2, the proportion of all deaths attributed to influenza never exceeded 10% of all winter deaths among seniors [12].

A few researchers subsequently addressed this paradox directly and investigated the possibility that unrecognized bias has led the majority of cohort studies to systematically overestimate influenza vaccine benefits. In 2006, two published reports clearly demonstrated that the senior cohort study findings are largely a result of systematic mismeasurements [108, 109]. First, they showed that the greatest mortality reductions occurred in early winter before influenza ever circulated and were not specifically associated with the peak influenza period. Second, they showed that the analytical adjustment techniques typically used in cohort studies actually magnified the mismeasurement rather than reducing it. The authors concluded that the magnitude of the unadjusted bias detected was sufficient to account entirely for the observed benefit of 50% mortality reduction during the entire winter period. This problem in the evidence base was also highlighted in a recent Cochrane review and an editorial [106, 110]. The source of bias may be a subset of frail seniors who are undervaccinated in the fall months for that season and subsequently contribute substantially to mortality in the early winter months [99]. Studies have substantiated that frail elderly are indeed vaccinated less often than their healthy peers [111, 112]. Controlling for these biases yields far more modest estimates of mortality reductions [113].

In summary, the emerging picture is a mixture of that residual selection bias, counter-productive adjustment efforts, and low-specificity endpoints has led to systematic overestimation in virtually all cohort studies published over the last decades. Adjustments for selection bias may be possible, but only if high specificity endpoints are studied. Beyond that, a commonly agreed set of standards for carrying out and reporting observational studies that includes a framework for detection of bias would be helpful. Also, previously published observational studies could undergo reanalysis, guided by such expectations as that vaccine benefits should be highest in peak influenza periods and for well-matched influenza vaccines. We have recently proposed such a framework [99].

4.3   Revisiting the Evidence Base Supporting Strategies for Protecting Populations with Vaccine

If we discount the biased cohort studies, the remaining studies suggest that the benefits of the vaccine are in fact much less than previously thought to be – probably lower than 30% in seniors >70 years of age. This assessment is based on the "gestalt" of results from the randomized placebo-controlled clinical trial

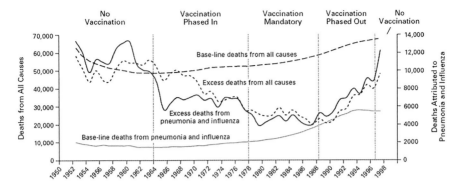

Fig. 6 Herd immunity and influenza vaccination. Encouraging evidence from the Japanese experience of vaccinating school children between 1964 and 1996. The graph compares the different phases of the vaccination program with baseline total death rates, rates of excess deaths from all causes and pneumonia and influenza, in Japan, 1950–1998. Adapted from Reichert et al. [117, 122]

described earlier [103], a nested case–control study using laboratory-confirmed endpoints from an RSV study [64, 65], and the excess mortality studies showing little decline in mortality as vaccine coverage rose [12, 54]. None of these studies are conclusive, but if these findings hold up in future studies, then there is ample room for improvement of influenza vaccines, including better vaccine formulations, adjuvants, or higher doses or combinations of live and killed vaccine doses [114–116].

Japan is the only country that has implemented a policy of vaccinating school children, with a strategy of reducing transmission in the community and thereby indirectly protecting high-risk populations. Although Japan abandoned this policy in 1994, an excess mortality study found evidence that it was associated with substantially reduced excess mortality in elderly people for the decades it was in place [117] (Fig. 6). Other studies have examined the value of inducing greater herd immunity based on local community trials or mathematical models [118–121], but unfortunately none have thus far proved conclusive enough to extend the policy of school children vaccination nationally. To fully investigate the indirect benefits of a school children vaccination program, it would be necessary to conduct a large cluster-randomized study across the country; such a study has been proposed but has not yet been undertaken [122].

4.4   Vaccines for the Control of Pandemic Influenza

Prior to spring 2009, a great deal of effort had been expended to develop and clinically test several types of vaccines against H5N1 influenza, including inactivated, live-attenuated, and DNA vaccine preparations. Several countries had stockpiled million doses of "prepandemic" inactivated vaccines based on H5N1 strains.

During May to August, 2009, a first vaccine against the H1N1pdm virus has been developed with plans to vaccinate populations in the northern hemisphere autumn months. National planning documents have set forth priorities for how to deploy an effective vaccine as it becomes available, and detailed logistical plans have been laid for vaccine distribution. WHO and national pandemic plans are reviewed in Uscher-Pines et al. [123].

But uncertainties abound. We still do not know which age groups will be most at risk, although a shift in mortality toward younger people is very likely. Whether this shift will put young adults at greatest absolute risk (as was the case in 1918–1919), or just higher relative risk (as in 1957–1958 and 1968–1969), cannot be predicted. Although effective vaccines against H1N1pdm have already been developed and are being manufactured in large quantities, just how quickly the billions of doses required to vaccinate a substantial portion of the world's population will be available is unknown. Resource-poor countries fear that they will be able to obtain vaccine for their populations only after wealthy countries have covered their own – a fear that had already exacerbated tensions over sharing of H5N1 data and samples [124]. For all these reasons, it is not clear that policy makers' hopes that vaccines will play a major role in limiting the global impact of the next pandemic will be realized.

5   Remaining Questions in Influenza Epidemiology and Considerations About the 2009 Pandemic

Many unsolved questions about influenza epidemiology remain [86, 125–127]. Solving these riddles will depend on the successful integration of many separate fields, including immunology, phylogenetics, virology, and clinical ascertainment. Exciting progress has recently been made in areas where mathematical modelers and phylogenetic researchers have entered the influenza field [29, 31, 128, 129]. This cross-fertilization has, for example, produced useful new findings in molecular influenza epidemiology, which may in turn lead to improved tools for the selection of vaccine strains [130].

Regarding pandemic influenza, for more than a decade the world had been bracing for a pandemic emerging from an avian H5N1 virus. Preparedness efforts anticipated that a pandemic would likely originate in Asia and focused strongly on surveillance of wild and domestic birds. Instead, the pandemic H1N1pdm virus emerged in Mexico, displaying a complex evolutionary lineage drawn from gene segments found in human, avian, and swine populations.

Fortunately, the H1N1pdm pandemic has thus far proved to be relatively mild, and the mortality impact of the summer 2009 northern hemisphere wave was not severe. Unlike seasonal outbreaks, however, the mortality and morbidity patterns of H1N1pdm show the "signature age shift" typical of influenza pandemics. Adults aged 20–50 years are at highest risk of severe morbidity and mortality [79], whereas children experience high rates of illness but relatively few severe outcomes. Seniors are largely spared from both illness and death, perhaps because of childhood exposure to H1N1 viruses circulating during 1918–1956.

Taken together, these features – a mild summer wave with elevated mortality in young adults and sparing of seniors – resemble the first wave pattern of the 1918 pandemic in the USA [17] and Europe [22]. Of note, morbidity impact in the first wave varied a great deal among US cities and regions [131]. For example, about 7% of New Yorkers have experienced influenza-like illness during the early weeks of the epidemic May 1–20, 2009, based on a phone survey [132], whereas other cities experienced little or delayed elevation of influenza-like illness [131]. Such spatial-temporal heterogeneity in timing of local epidemics remains unexplained.

Southern hemisphere countries, however, had the first encounter with the H1N1pdm virus under typical winter conditions. Reports from Argentina, New Zealand, and Australia suggest that pandemic impact is heterogeneous. Argentina experienced an emergency situation with severe overcrowding in hospitals and intensive care units, whereas New Zealand or Australia experienced no more than the equivalent of a severe A/H3N2 seasonal influenza epidemic [133]. Such variability between countries occurred during the 1918–1920 pandemic and was attributed to differences in access to care and overall mortality risk among developing countries [76]. In this on-going outbreak, however, it is still too early to quantify differences in disease burden with precision.

The case fatality rate is a key indicator of the severity of the H1N1pdm pandemic and an important decision parameter for determining pandemic response. But it is difficult to make an accurate estimate early in a pandemic. Because most H1N1pdm cases are not confirmed by laboratory testing and therefore not included in the "confirmed" tally, the case fatality rate tends to be greatly overestimated. In New Zealand, a combined strategy integrating epidemiological surveillance and modeling led to a case fatality rate estimate of 0.005% [134], far lower than earlier estimates based on early data from Mexico [135] and lower than the typical seasonal case fatality rate of ~0.2%. It is important to consider that while the case fatality rate – and perhaps even the total number of H1N1pdm-related deaths – may be lower than in a typical seasonal influenza epidemic, the higher proportion of deaths occurring in young adults results in a much higher burden of life years lost than in a typical influenza season, where 90% of deaths occur in those over 65 years of age [12].

Even though the similarities in the epidemiology of H1N1pdm and the 1918 pandemic are worrisome, as of September 2009, the pandemic is still relatively mild. We simply cannot know whether the virus will cause more severe waves in the coming months and years. A likely challenge will be the constant, dynamic real-time reassessment of benefit/risk of vaccinating atypical target groups during a pandemic. While policy makers plan to target vaccines to various groups, the perceived benefits from individuals will be based on severity of illness and real or temporally associated adverse reactions identified through surveillance and the media. Rapid reassessments of risks and benefits will be crucial for the viability of a vaccination program.

One impending question regarding vaccines, however, is whether the H1N1pdm virus will replace either or both of the influenza A viruses that had been circulating previously, H3N2 and H1N1. If all three cocirculate in the next season, the new H1N1pdm could be added as a component in the seasonal vaccine. But even if the new H1N1pdm thoroughly dominates the 2009–2010 season in one country, the other subtypes should probably still be included until the long-term pattern becomes clear. For example, it is not unusual for influenza A/H3N2 viruses to account for >99% of influenza specimens isolated in a country on a given year, only to become uncommon the next year, when seasonal A/H1N1 or influenza B virus might dominate. To avoid dropping any component too soon, it will be necessary to track subtype distribution globally over at least a few years. If history is a guide, as immunity builds up in younger population, the H1N1pdm virus will cause seasonal epidemics, with a proportionate shift in mortality to the older age groups.

Whatever the scenario, the epidemiological characteristics of a pandemic directly affect the ethical principles that should be invoked when allocating limited vaccine doses [136, 137]. For that reason, it is absolutely essential that real-time surveillance data from the early phase of a pandemic continue to be freely shared and rapidly interpreted to determine who is at risk and where scarce resources such as pandemic vaccine and antivirals could best be used. Moreover, pandemic planners should build sufficient flexibility into their plans to allow rapid shifts in planned control strategies, as key epidemiological insights hopefully become available in the early pandemic phase. Continued influenza surveillance efforts in temperate and tropical regions, combined with international sharing of epidemiological and viral sequence data, are our best hope for limiting the impact of current and future influenza pandemics.

Acknowledgments We are enormously grateful to our many colleagues – nationally and internationally through the Multinational Influenza Seasonal Mortality Study network – for the many inspiring conversations we have had over the years about the "mysteries" of influenza epidemiology. We thank the Department of Health and Human Services OGHA and the Department of Homeland Security RAPIDD program for funding support. We also thank Marta Balinska who provided editorial assistance to this manuscript.

References

1. Nicholson K, Hay A (1998) Textbook of influenza. Blackwell, Oxford
2. Schoenbaum S (1996) Impact of influenza in persons and populations. In: Brown L, Hampson A, Webster A (eds) Options for the control of influenza III. Elsevier, Amsterdam, pp 17–25
3. Reichert TA, Simonsen L, Sharma A, Pardo SA, Fedson DS, Miller MA (2004) Influenza and the winter increase in mortality in the United States, 1959–1999. Am J Epidemiol 160:492–502
4. Simonsen L (1999) The global impact of influenza on morbidity and mortality. Vaccine 17 (Suppl 1):S3–S10
5. Thompson WW, Shay DK, Weintraub E, Brammer L, Cox N, Anderson LJ, Fukuda K (2003) Mortality associated with influenza and respiratory syncytial virus in the United States. JAMA 289:179–186

6. Serfling R (1963) Methods for current statistical analysis of excess pneumonia-influenza deaths. Public Health Rep 78:494–506
7. Choi K, Thacker SB (1981) An evaluation of influenza mortality surveillance, 1962–1979. I. Time series forecasts of expected pneumonia and influenza deaths. Am J Epidemiol 113:215–226
8. Carrat F, Valleron AJ (1995) Influenza mortality among the elderly in France, 1980–90: how many deaths may have been avoided through vaccination? J Epidemiol Community Health 49:419–425
9. Simonsen L, Clarke MJ, Williamson GD, Stroup DF, Arden NH, Schonberger LB (1997) The impact of influenza epidemics on mortality: introducing a severity index. Am J Public Health 87:1944–1950
10. Simonsen L, Fukuda K, Schonberger LB, Cox NJ (2000) The impact of influenza epidemics on hospitalizations. J Infect Dis 181:831–837
11. Thompson WW, Shay DK, Weintraub E, Brammer L, Bridges CB, Cox NJ, Fukuda K (2004) Influenza-associated hospitalizations in the United States. JAMA 292:1333–1340
12. Simonsen L, Reichert TA, Viboud C, Blackwelder WC, Taylor RJ, Miller MA (2005) Impact of influenza vaccination on seasonal mortality in the US elderly population. Arch Intern Med 165:265–272
13. Dushoff J, Plotkin JB, Viboud C, Earn DJ, Simonsen L (2006) Mortality due to influenza in the United States – an annualized regression approach using multiple-cause mortality data. Am J Epidemiol 163:181–187
14. Viboud C, Bjornstad ON, Smith DL, Simonsen L, Miller MA, Grenfell BT (2006) Synchrony, waves, and spatial hierarchies in the spread of influenza. Science 312: 447–451
15. Simonsen L, Olson D, Viboud C, Miller M (2004) Pandemic influenza and mortality: past evidence and projections for the future. In: Knobler S, Oberholtzer K (eds) Forum on microbial threats. Pandemic influenza: assessing capabilities for prevention and response. Institute of Medicine, The National Academy of Sciences, Washington
16. Mills CE, Robins JM, Lipsitch M (2004) Transmissibility of 1918 pandemic influenza. Nature 432:904–906
17. Olson DR, Simonsen L, Edelson PJ, Morse SS (2005) Epidemiological evidence of an early wave of the 1918 influenza pandemic in New York City. Proc Natl Acad Sci USA 102: 11059–11063
18. Viboud C, Grais RF, Lafont BA, Miller MA, Simonsen L (2005) Multinational impact of the 1968 Hong Kong influenza pandemic: evidence for a smoldering pandemic. J Infect Dis 192:233–248
19. Chowell G, Ammon CE, Hengartner NW, Hyman JM (2006) Transmission dynamics of the great influenza pandemic of 1918 in Geneva, Switzerland: assessing the effects of hypothetical interventions. J Theor Biol 241:193–204
20. Viboud C, Tam T, Fleming D, Handel A, Miller MA, Simonsen L (2006) Transmissibility and mortality impact of epidemic and pandemic influenza, with emphasis on the unusually deadly 1951 epidemic. Vaccine 24:6701–6707
21. Viboud C, Tam T, Fleming D, Miller MA, Simonsen L (2006) 1951 influenza epidemic, England and Wales, Canada, and the United States. Emerg Infect Dis 12:661–668
22. Andreasen V, Viboud C, Simonsen L (2007) Epidemiologic characterization of the summer wave of the 1918 influenza pandemic in Copenhagen: implications for pandemic control strategies. J Infect Dis 197:270–278
23. Chowell G, Miller MA, Viboud C (2007) Seasonal influenza in the United States, France, and Australia: transmission and prospects for control. Epidemiol Infect 2:1–13
24. Chowell G, Nishiura H, Bettencourt LM (2007) Comparative estimation of the reproduction number for pandemic influenza from daily case notification data. J R Soc Interface 4: 155–166

25. Olson DR, Heffernan RT, Paladini M, Konty K, Weiss D, Mostashari F (2007) Monitoring the impact of influenza by age: emergency Department fever and respiratory complaint surveillance in New York City. PLoS Med 4:e247

26. Dushoff J, Plotkin JB, Levin SA, Earn DJ (2004) Dynamical resonance can account for seasonality of influenza epidemics. Proc Natl Acad Sci USA 101:16915–16916

27. Viboud C, Alonso WJ, Simonsen L (2006) Influenza in tropical regions. PLoS Med 3:e89

28. Alonso WJ, Viboud C, Simonsen L, Hirano EW, Daufenbach LZ, Miller MA (2007) Seasonality of influenza in Brazil: a traveling wave from the Amazon to the subtropics. Am J Epidemiol 165:1434–1442

29. Grenfell BT, Pybus OG, Gog JR, Wood JL, Daly JM, Mumford JA, Holmes EC (2004) Unifying the epidemiological and evolutionary dynamics of pathogens. Science 303: 327–332

30. Holmes EC, Ghedin E, Miller N, Taylor J, Bao Y, St George K, Grenfell BT, Salzberg SL, Fraser CM, Lipman DJ, Taubenberger JK (2005) Whole-genome analysis of human influenza A virus reveals multiple persistent lineages and reassortment among recent H3N2 viruses. PLoS Biol 3:e300

31. Nelson MI, Holmes EC (2007) The evolution of epidemic influenza. Nat Rev Genet 8:196–205

32. Nelson MI, Simonsen L, Viboud C, Miller MA, Taylor J, George KS, Griesemer SB, Ghedin E, Sengamalay NA, Spiro DJ et al (2006) Stochastic processes are key determinants of short-term evolution in influenza a virus. PLoS Pathog 2:e125

33. Cox NJ, Subbarao K (2000) Global epidemiology of influenza: past and present. Annu Rev Med 51:407–421

34. Glezen WP (1982) Serious morbidity and mortality associated with influenza epidemics. Epidemiol Rev 4:25–44

35. Monto AS (2002) Epidemiology of viral respiratory infections. Am J Med 112(Suppl 6A): 4S–12S

36. Noble G (1982) Epidemiological and clinical aspects of influenza. CRC Press, Boca Raton

37. Stuart-Harris C (1979) Epidemiology of influenza in man. Br Med Bull 35:3–8

38. Glezen WP, Taber LH, Frank AL, Gruber WC, Piedra P (1997) Influenza virus infections in infants. Pediatr Infect Dis J 16:1065–1068

39. Chiu SS, Lau YL, Chan KH, Wong WH, Peiris JS (2002) Influenza-related hospitalizations among children in Hong Kong. N Engl J Med 347:2097–2103

40. Chow A, Ma S, Ling AE, Chew SK (2006) Influenza-associated deaths in tropical Singapore. Emerg Infect Dis 12:114–121

41. Wong CM, Yang L, Chan KP, Leung GM, Chan KH, Guan Y, Lam TH, Hedley AJ, Peiris JS (2006) Influenza associated weekly hospitalization in a subtropical city. PLoS Med 3:e89

42. Wong CM, Chan KP, Hedley AJ, Peiris JS (2004) Influenza-associated mortality in Hong Kong. Clin Infect Dis 39:1611–1617

43. Assaad F, Cockburn WC, Sundaresan TK (1973) Use of excess mortality from respiratory diseases in the study of influenza. Bull World Health Organ 49:219–233

44. Rizzo C (2007) Trends for influenza-related deaths during pandemic and epidemic seasons, Italy, 1969–2001. Emerg Infect Dis 13:694–699

45. Rocchi G, Ragona G, De Felici A, Muzzi A (1974) Epidemiological evaluation of influenza in Italy. Bull World Health Organ 50:401–406

46. Viboud C, Boelle PY, Pakdaman K, Carrat F, Valleron AJ, Flahault A (2004) Influenza epidemics in the United States, France, and Australia, 1972–1997. Emerg Infect Dis 10: 32–39

47. Imaz MS, Eimann M, Poyard E, Savy V (2006) Influenza associated excess mortality in Argentina: 1992–2002. Rev Chilena Infectol 23:297–306

48. Stroup DF, Thacker SB, Herndon JL (1988) Application of multiple time series analysis to the estimation of pneumonia and influenza mortality by age 1962–1983. Stat Med 7: 1045–1059

49. Barker WH (1986) Excess pneumonia and influenza associated hospitalization during influenza epidemics in the United States, 1970–78. Am J Public Health 76:761–765
50. Simonsen L, Taylor R, Viboud C, Dushoff J, Miller M (2006) US flu mortality estimates are based on solid science. Br Med J 332:177–178
51. Thompson W, Weintraub E, Cheng P et al (2007) Comparing methods for estimating influenza-associated deaths in the United States: 1976/1977 through 2002/2003 respiratory seasons. In: Katz JM (ed) Options for the control of influenza VI, International Medical Press, London
52. Fleming DM (2000) The contribution of influenza to combined acute respiratory infections, hospital admissions, and deaths in winter. Commun Dis Public Health 3:32–38
53. Schanzer DL, Tam TW, Langley JM, Winchester BT (2007) Influenza-attributable deaths, Canada 1990–1999. Epidemiol Infect 135:1109–1116
54. Rizzo C, Viboud C, Montomoli E, Simonsen L, Miller MA (2006) Influenza-related mortality in the Italian elderly: no decline associated with increasing vaccination coverage. Vaccine 24:6468–6475
55. Bhat N, Wright JG, Broder KR, Murray EL, Greenberg ME, Glover MJ, Likos AM, Posey DL, Klimov A, Lindstrom SE et al (2005) Influenza-associated deaths among children in the United States, 2003–2004. N Engl J Med 353:2559–2567
56. Elliot AJ, Fleming DM (2006) Surveillance of influenza-like illness in England and Wales during 1966–2006. Euro Surveill 11:249–250
57. Hall IM, Gani R, Hughes HE, Leach S (2007) Real-time epidemic forecasting for pandemic influenza. Epidemiol Infect 135:372–385
58. Brownstein JS, Kleinman KP, Mandl KD (2005) Identifying pediatric age groups for influenza vaccination using a real-time regional surveillance system. Am J Epidemiol 162:686–693
59. Crighton EJ, Elliott SJ, Moineddin R, Kanaroglou P, Upshur RE (2007) An exploratory spatial analysis of pneumonia and influenza hospitalizations in Ontario by age and gender. Epidemiol Infect 135:253–261
60. Fleming DM, Zambon M, Bartelds AI, de Jong JC (1999) The duration and magnitude of influenza epidemics: a study of surveillance data from sentinel general practices in England, Wales and the Netherlands. Eur J Epidemiol 15:467–473
61. Fleming DM, Pannell RS, Cross KW (2005) Mortality in children from influenza and respiratory syncytial virus. J Epidemiol Community Health 59:586–590
62. Fleming DM, Cross KW (1993) Respiratory syncytial virus or influenza? Lancet 342:1507–1510
63. Fleming DM, Elliott AJ, Cross KW (2007) Is routine seasonal influenza vaccination of elderly people an effective community policy? In: Katz JM (ed) Options for the control of influenza VI, International Medical Press, London
64. Falsey AR, Hennessey PA, Formica MA, Cox C, Walsh EE (2005) Respiratory syncytial virus infection in elderly and high-risk adults. N Engl J Med 352:1749–1759
65. Simonsen L, Viboud C (2005) Respiratory syncytial virus infection in elderly adults. N Engl J Med 353:422–423
66. Izurieta HS, Thompson WW, Kramarz P, Shay DK, Davis RL, DeStefano F, Black S, Shinefield H, Fukuda K (2000) Influenza and the rates of hospitalization for respiratory disease among infants and young children. N Engl J Med 342:232–239
67. Weber MW, Mulholland EK, Greenwood BM (1998) Respiratory syncytial virus infection in tropical and developing countries. Trop Med Int Health 3:268–280
68. Monto AS (1994) Studies of the community and family: acute respiratory illness and infection. Epidemiol Rev 16:351–373
69. Monto AS, Cavallaro JJ (1971) The Tecumseh study of respiratory illness. II. Patterns of occurrence of infection with respiratory pathogens, 1965–1969. Am J Epidemiol 94:280–289
70. Ferguson NM, Cummings DA, Fraser C, Cajka JC, Cooley PC, Burke DS (2006) Strategies for mitigating an influenza pandemic. Nature 442:448–452

71. Ferguson NM, Cummings DAT, Cauchemez S, Fraser C, Riley S, Meeyai A, Iamsirithaworn S, Burke DS (2005) Strategies for containing an emerging influenza pandemic in Southeast Asia. Nature 437:209–214

72. Spicer CC (1979) The mathematical modelling of influenza epidemics. Br Med Bull 35: 23–28

73. Spicer CC, Lawrence CJ (1984) Epidemic influenza in Greater London. J Hyg (Lond) 93:105–112

74. Nelson MI, Simonsen L, Viboud C, Miller MA, Holmes EC (2007) Phylogenetic analysis reveals the global migration of seasonal influenza a viruses. PLoS Pathog 3:1220–1228

75. Kilbourne ED (1997) Perspectives on pandemics: a research agenda. J Infect Dis 176(Suppl 1): S29–S31

76. Murray CJ, Lopez AD, Chin B, Feehan D, Hill KH (2006) Estimation of potential global pandemic influenza mortality on the basis of vital registry data from the 1918–20 pandemic: a quantitative analysis. Lancet 368:2211–2218

77. Barry J (2004) The great influenza: the epic story of the deadliest plague in history. Viking Penguin, New York

78. Simonsen L, Clarke MJ, Schonberger LB, Arden NH, Cox NJ, Fukuda K (1998) Pandemic versus epidemic influenza mortality: a pattern of changing age distribution. J Infect Dis 178:53–60

79. Chowell G, Bertozzi SM, Colchero MA, Lopez-Gatell H, Alpuche-Aranda C, Hernandez M, Miller MA (2009) Severe respiratory disease concurrent with the circulation of H1N1 influenza. N Engl J Med 361:674–679

80. Simonsen L, Reichert TA, Miller M (2003) The virtues of antigenic sin: consequences of pandemic recycling on influenza-associated mortality. In: Kawaoka Y (ed) Options for the control of influenza V. International Congress Series, no. 1263. Elsevier, Okinawa, pp 791–794

81. McGregor IA, Schild GC, Billewicz WZ, Williams K (1979) The epidemiology of influenza in a tropical (Gambian) environment. Br Med Bull 35:15–22

82. Francis T Jr (1960) On the doctrine of original antigenic sin. Proc Am Philos Soc 104(6): 572–578

83. Kash JC, Tumpey TM, Proll SC, Carter V, Perwitasari O, Thomas MJ, Basler CF, Palese P, Taubenberger JK, García-Sastre A et al (2006) Genomic analysis of increased host immune and cell death responses induced by 1918 influenza virus. Nature 443:578–581

84. Kobasa D, Jones SM, Shinya K, Kash JC, Copps J, Ebihara H, Hatta Y, Kim JH, Halfmann P, Hatta M et al (2007) Aberrant innate immune response in lethal infection of macaques with the 1918 influenza virus. Nature 445:319–323

85. Palese P (2004) Influenza: old and new threats. Nat Med 10:S82–S87

86. Stuart-Harris CH (1970) Pandemic influenza: an unresolved problem in prevention. J Infect Dis 122:108–115

87. Germann TC, Kadau K, Longini IM Jr, Macken CA (2006) Mitigation strategies for pandemic influenza in the United States. Proc Natl Acad Sci USA 103:5935–5940

88. Longini IM Jr, Nizam A, Xu S, Ungchusak K, Hanshaoworakul W, Cummings DA, Halloran ME (2005) Containing pandemic influenza at the source. Science 309:1083–1087

89. Bootsma MC, Ferguson NM (2007) From the cover: the effect of public health measures on the 1918 influenza pandemic in U.S. cities. Proc Natl Acad Sci USA 104:7588–7593

90. Glass K, Barnes B (2007) How much would closing schools reduce transmission during an influenza pandemic? Epidemiology 18:623–628

91. Glass RJ, Glass LM, Beyeler WE, Min HJ (2006) Targeted social distancing design for pandemic influenza. Emerg Infect Dis 12:1671–1681

92. Smith DJ (2006) Predictability and preparedness in influenza control. Science 312:392–394

93. Peiris JS, de Jong MD, Guan Y (2007) Avian influenza virus (H5N1): a threat to human health. Clin Microbiol Rev 20:243–267

94. Subbarao K, Luke C (2007) H5N1 viruses and vaccines. PLoS Pathog 3:e40

95. Taubenberger JK, Morens DM, Fauci AS (2007) The next influenza pandemic: can it be predicted? JAMA 297:2025–2027
96. Webster RG, Hulse-Post DJ, Sturm-Ramirez KM, Guan Y, Peiris M, Smith G, Chen H (2007) Changing epidemiology and ecology of highly pathogenic avian H5N1 influenza viruses. Avian Dis 51:269–272
97. Bermejo-Martin JF, Kelvin DJ, Guan Y, Chen H, Perez-Breña P, Casas I, Arranz E, de Lejarazu RO (2007) Neuraminidase antibodies and H5N1: geographic-dependent influenza epidemiology could determine cross-protection against emerging strains. PLoS Med 4:e212
98. Demicheli V, Rivetti D, Deeks JJ, Jefferson TO (2004) Vaccines for preventing influenza in healthy adults. Cochrane Database Syst Rev 3:CD001269
99. Simonsen L, Taylor RJ, Viboud C, Miller MA, Jackson LA (2007) Mortality benefits of influenza vaccination in elderly people: an ongoing controversy. Lancet Infect Dis 7: 658–666
100. Reichert TA, Pardo SA, Valleron AJ et al (2007) National vaccination programs and trends in influenza-attributable mortality in four countries. In: Katz JM (ed) Options for the control of influenza VI, International Medical Press, London
101. Goodwin K, Viboud C, Simonsen L (2005) Antibody response to influenza vaccination in the elderly: a quantitative review. Vaccine 24:1159–1169
102. Langmuir AD, Henderson DA, Serfling RE (1964) The epidemiological basis for the control of influenza. Am J Public Health Nations Health 54:563–571
103. Govaert TM, Thijs CT, Masurel N, Sprenger MJ, Dinant GJ, Knottnerus JA (1994) The efficacy of influenza vaccination in elderly individuals. A randomized double-blind placebo-controlled trial. JAMA 272:1661–1665
104. Vallejo AN (2007) Immune remodeling: lessons from repertoire alterations during chronological aging and in immune-mediated disease. Trends Mol Med 13:94–102
105. Gross PA, Hermogenes AW, Sacks HS, Lau J, Levandowski RA (1995) The efficacy of influenza vaccine in elderly persons. A meta-analysis and review of the literature. Ann Intern Med 123:518–527
106. Jefferson T, Rivetti D, Rivetti A, Rudin M, Di Pietrantonj C, Demicheli V (2005) Efficacy and effectiveness of influenza vaccines in elderly people: a systematic review. Lancet 366:1165–1174
107. Vu T, Farish S, Jenkins M, Kelly H (2002) A meta-analysis of effectiveness of influenza vaccine in persons aged 65 years and over living in the community. Vaccine 20:1831–1836
108. Jackson LA, Jackson ML, Nelson JC, Neuzil KM, Weiss NS (2006) Evidence of bias in estimates of influenza vaccine effectiveness in seniors. Int J Epidemiol 35:337–344
109. Jackson LA, Nelson JC, Benson P, Neuzil KM, Reid RJ, Psaty BM, Heckbert SR, Larson EB, Weiss NS (2006) Functional status is a confounder of the association of influenza vaccine and risk of all cause mortality in seniors. Int J Epidemiol 35:345–352
110. Jefferson T (2006) Influenza vaccination: policy *versus* evidence. Br Med J 333:912–915
111. Bratzler DW, Houck PM, Jiang H, Nsa W, Shook C, Moore L, Red L (2002) Failure to vaccinate Medicare inpatients: a missed opportunity. Arch Intern Med 162:2349–2356
112. Fedson DS, Wajda A, Nicol JP, Roos LL (1992) Disparity between influenza vaccination rates and risks for influenza-associated hospital discharge and death in Manitoba in 1982–1983. Ann Intern Med 116:550–555
113. Örtqvist Å, Granath F, Askling J, Hedlund J (2007) Influenza vaccination and mortality: prospective cohort study of the elderly in a large geographical area. Eur Respir J 30:414–422
114. Keitel WA, Atmar RL, Cate TR, Petersen NJ, Greenberg SB, Ruben F, Couch RB (2006) Safety of high doses of influenza vaccine and effect on antibody responses in elderly persons. Arch Intern Med 166:1121–1127
115. Minutello M, Senatore F, Cecchinelli G, Bianchi M, Andreani T, Podda A, Crovari P (1999) Safety and immunogenicity of an inactivated subunit influenza virus vaccine combined with MF59 adjuvant emulsion in elderly subjects, immunized for three consecutive influenza seasons. Vaccine 17:99–104

116. Treanor JJ, Mattison HR, Dumyati G, Yinnon A, Erb S, O'Brien D, Dolin R, Betts RF (1992) Protective efficacy of combined live intranasal and inactivated influenza A virus vaccines in the elderly. Ann Intern Med 117:625–633

117. Reichert TA, Sugaya N, Fedson DS, Glezen WP, Simonsen L, Tashiro M (2001) The Japanese experience with vaccinating school children against influenza. N Engl J Med 344:889–896

118. Glezen WP (2006) Herd protection against influenza. J Clin Virol 37:237–243

119. Longini IM Jr, Halloran ME (2005) Strategy for distribution of influenza vaccine to high-risk groups and children. Am J Epidemiol 161:303–306

120. Monto AS, Davenport FM, Napier JA, Francis T Jr (1970) Modification of an outbreak of influenza in Tecumseh, Michigan by vaccination of school children. J Infect Dis 122:16–25

121. Halloran ME, Longini IM Jr (2006) Public health. Community studies for vaccinating school children against influenza. Science 311:615–616

122. Reichert TA, Sugaya N, Fedson DS, Glezen WP, Simonsen L, Tashiro M (2001) Vaccinating Japanese school children against influenza: author reply. N Engl J Med 344:1948

123. Uscher-Pines L, Omer SB, Barnett DJ, Burke TA, Balicer RD (2006) Priority setting for pandemic influenza: an analysis of national preparedness plans. PLoS Med 3:e436

124. Enserink M (2007) Data sharing. New Swiss influenza database to test promises of access. Science 315:923

125. Earn D, Dushoff J, Levin S (2002) Ecology and evolution of the flu. Trends Ecol Evol 37:334–340

126. Hope-Simpson RE (1992) The transmission of epidemic influenza. Plenum Press, New York

127. Thacker SB (1986) The persistence of influenza A in human populations. Epidemiol Rev 8:129–142

128. Ferguson NM, Galvani AP, Bush RM (2003) Ecological and immunological determinants of influenza evolution. Nature 422:428–433

129. Smith DJ, Lapedes AS, de Jong JC, Bestebroer TM, Rimmelzwaan GF, Osterhaus AD, Fouchier RA (2004) Mapping the antigenic and genetic evolution of influenza virus. Science 305:371–376

130. Plotkin JB, Dushoff J, Levin SA (2002) Hemagglutinin sequence clusters and the antigenic evolution of influenza A virus. Proc Natl Acad Sci U S A 99:6263–6268

131. Distribute Network. http://isds.cirg.washington.edu/distribute/index.php

132. New-York City Department of Health. http://www.nyc.gov/html/doh/downloads/pdf/cd/h1n1_citywide_survey.pdf

133. New-Zealand influenza surveillance. http://www.moh.govt.nz/moh.nsf/indexmh/influenza-a-h1n1-situation

134. Baker MG, Wilson N, Huang QS, Paine S, Lopez L, Bandaranayake D, Tobias M, Mason K, Mackereth GF, Jacobs M, Thornley C, Roberts S, McArthur C (2009) Pandemic influenza A (H1N1)v in New Zealand: the experience from April to August 2009. Euro Surveill 14 (34):1–6

135. Fraser C, Donnelly CA, Cauchemez S, Hanage WP, Van Kerkhove MD, Hollingsworth TD, Griffin J, Baggaley RF, Jenkins HE, Lyons EJ, Jombart T, Hinsley WR, Grassly NC, Balloux F, Ghani AC, Ferguson NM, Rambaut A, Pybus OG, Lopez-Gatell H, Alpuche-Aranda CM, Chapela IB, Zavala EP, Guevara DM, Checchi F, Garcia E, Hugonnet S, Roth C, WHO Rapid Pandemic Assessment Collaboration (2009) Pandemic potential of a strain of influenza A (H1N1): early findings. Science 324(5934):1557–1561

136. Emanuel EJ, Wertheimer A (2006) Public health. Who should get influenza vaccine when not all can? Science 312:854–855

137. Gostin LO (2006) Medical countermeasures for pandemic influenza: ethics and the law. JAMA 295:554–556

Epidemiology of Influenza in Tropical and Subtropical Low-Income Regions

W. Abdullah Brooks and Mark C. Steinhoff

Abstract Influenza appears to be a major contributor to morbidity, hospitalization, and likely mortality in the tropical and subtropical low-income countries; however, its contribution has been largely underestimated due to a lack of data from these regions. Limited available data indicate that influenza circulation in the tropics differs in two respects from that in the temperate northern and southern hemispheres. First, while seasonal influenza tends to occur primarily in the late fall and winter in temperate zones, it appears to circulate year-round in tropics, with seasonal influenza A peaks between March and September in many of tropical settings, complementing temperate zone seasonality. This prolonged circulation may partly account for its apparent higher incidence in those countries where data are available. Virus circulation in East and Southeast Asia may determine seasonal reintroduction and circulation elsewhere. Second, the fraction of infections resulting in clinically important illness, particularly childhood pneumonia, appears to be higher in the tropics. Influenza may be responsible for a substantial fraction of the childhood pneumonia and pneumonia-related mortality, both from primary infection and from interaction with respiratory bacterial agents in the tropical belt. Introduction of influenza vaccine as a means to control influenza-related pneumonia in young children may be warranted. Indeed, control of childhood pneumonia may provide a mechanism for influenza vaccine uptake in these countries with wider benefits to both disease burden and mortality reduction, as well as surge capacity for vaccine production during pandemics. Concern about pandemic influenza has

W.A. Brooks (✉)
Head Infectious Diseases Unit, International Centre for Diarrhoeal Disease Research, Bangladesh (ICDDR,B), Mohakhali, GPO Box 128, Dhaka 1000, Bangladesh
Johns Hopkins Bloomberg School of Public Health, 615 North Wolfe Street, Suite E8132, Baltimore, MD 21205, USA
e-mail: abrooks@jhsph.edu
M.C. Steinhoff
Global Health Center, Cincinnati Children's Hospital Medical Center, 3333 Burnet Avenue, ML 2048, Cincinnati, OH 45229, USA
e-mail: mark.steinhoff@gmail.com

G. Del Giudice and R. Rappuoli (eds.), *Influenza Vaccines for the Future*, 2nd edition,   55
Birkhäuser Advances in Infectious Diseases,
DOI 10.1007/978-3-0346-0279-2_3, © Springer Basel AG 2011

increased interest in vaccine use, including increased seasonal vaccine use, and initiation of vaccine production in some countries. Continued and enhanced surveillance in the tropics, particularly in East and Southeast Asia, is warranted both to monitor burden and the impact of interventions, such as vaccination, and to identify emergence and spread of novel viruses.

1 Purpose and Background

The purpose of this chapter is to summarize available influenza data from largely underrepresented, high respiratory disease-endemic regions of the tropical and subtropical belt. This will allow a comparison of the epidemiology between regions, and in so doing, reveal important knowledge gaps in global influenza epidemiology. At the same time, this comparison will permit, where data are available, identification of clinical management and disease control opportunities. Finally, it will facilitate identifying outstanding research and intervention needs to the research community, policy makers, funding agencies, and other stakeholders.

A key focus of this review is the burden of childhood pneumonia, which is the leading cause of child mortality worldwide, and which in 2000 caused 1.9 million deaths in this age group [1]. This may be an underestimate due to misclassification of neonatal deaths and inadequate surveillance in high pneumonia-endemic regions. Importantly, over 90% of these global deaths occur in 40 developing countries, and two thirds occur in just ten tropical and subtropical countries [2] (Fig. 1). The role of invasive bacterial disease has been well described by both disease burden and vaccine trials [2–8], leading to recommendations for vaccination against *Haemophilus influenzae* type b and *Streptococcus pneumoniae* infections [9, 10]. The contribution of influenza, as well as other respiratory viruses, to childhood pneumonia is not well described [11], partly because influenza historically has been perceived as a mild disease that is uncommon in the tropical belt [12]. The threat of pandemic influenza, often arising in tropical regions, has increased interest in influenza virus surveillance, defining the influenza disease burden and approaches to vaccine intervention in this part of the world. This chapter will focus on nonpandemic, seasonal influenza disease as the chapter by Simonsen et al. reviews pandemic influenza disease.

2 Geographical Distribution

Global influenza disease burden data are spotty, as many developing countries do not have an influenza surveillance system or adequate laboratory capacity for virus detection. However, sites in sub-Saharan Africa, Latin America, as well as southern and northern Asia have recently added influenza surveillance programs [13].

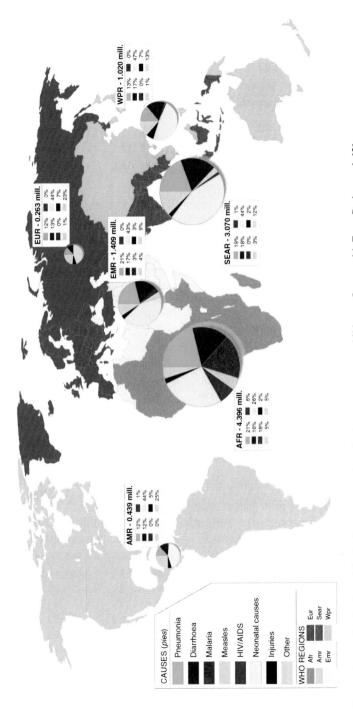

Fig. 1 Global distribution of deaths from pneumonia and other causes among children <5 years old. From Rudan et al. [2]

A major factor in increasing surveillance capability has been the introduction of nucleic acid technologies, including RT-PCR for virus detection, and reduced reliance on cell culture. Despite these changes, most of the recent data are from local outbreaks or passive sentinel hospital- or clinic-based studies, with very little from population-based surveillance, which can provide denominators of at-risk populations, and therefore incidence.

2.1   North/South East Asia and Pacific (1.99 Billion; 29% of Global Population[1])

A recent review of published data on human influenza from January 1980 through December 2006 [14] identified 35 publications with sufficient detail to assess data but found laboratory-confirmed data from only 9 of the 18 eligible countries from the region, including Thailand, Taiwan, Hong Kong, Japan, Korea (South), Indonesia, Myanmar, Malaysia, and Singapore (Table 1). Notably, no English-language reports were found from the Republic of China or Vietnam though subsequent to this survey there have been reports from these countries [15–17]. Only 5 out of 35 eligible studies from this selection included true incidence data for influenza. The remainder of these studies concentrated on pneumonia etiological assessment and/ or hospitalizable illness, outpatient visits, febrile seizures, mortality, or other outcomes. Two studies from Hong Kong and Singapore used indirect statistical modeling methods to estimate influenza disease burden from local databases of hospital discharge diagnosis, cause of death, and virological surveillance. Of the 15 studies that reported influenza-related pneumonia hospitalization, 14 (93%) used cell culture for virus detection and reported a range of 0–12% of laboratory-confirmed influenza among cases. All 13 outpatient studies used cell culture and reported laboratory-confirmed influenza among 11–26% of tested patients.

Using passive surveillance, a report from Thailand estimated the annual influenza incidence at 64–91 episodes/100,000 persons [14]. Among the other studies with incidence rates, not all reported population-based estimates. A study in Hong Kong reported an average of 10.5% of patients per week being influenza positive (29.3 cases/100,000 hospital admissions), while another study from Hong Kong estimated 4,051 excess hospitalizations for pneumonia and influenza and 15,873 for respiratory and circulatory diseases, with year-round influenza circulation with seasonal peaks occurring from January to March [14]; data from Hong Kong (Yap et al.) reported influenza admissions among persons ≥65 years in the range of 58.5 episodes/10,000 persons ≥65 years. Another Hong Kong-based study from Chiu et al. estimated the attributable hospitalization risk to be over 280 episodes/ 10,000 child-years among children <1 year, over 200 episodes/10,000 for children 1–2 years, 77 episodes/10,000 for children 2–5 years, and nearly 21 episodes/

---

[1]Country population estimates from UNICEF, 2008 with global total = 6.734 billion.

Table 1 Influenza-associated hospitalization or clinic pneumonia incidence rates/1,000 person-years[a]

Country	Laboratory method	Study type	0–5 years	5–21 years	≥65 years	All age
East Asia						
Thailand[b]						0.6–0.9
Hong Kong[c]		Statistical model			5.8	0.3
Hong Kong[b]		Statistical model, excess hospitalization by season	<1 year 28.8; 1–2 years 20.9; 2–5 years 7.7	5–10 years 20.9		
Hong Kong[d]	DFA, culture	Sentinel two hospitals surveillance 2003–2006	<1 year 3.9–7.8; 1 year 4.1–9.6; 2–4 years 3.8–6.0	5–17 years 0.2–2.2		
Vietnam[e]	PCR	Sentinel hospital surveillance	0–1 year 16.9; 1–2 years 18.4; 2–3 years 6.9; 3–4 years 3.4; 4–5 years 2.0; <5 years 8.7			
South Asia						
India[f]	DFA	2001–2004 active rural population-based surveillance	0–3 years 141			
Bangladesh[g]	Culture, RT-PCR	Active urban population-based surveillance	0–5 years 102			≥5 years 55.2[h]
Near and Middle East						
Middle East[i]		Hospital-based	0–4 years 13.9	5–14 years 51.0	5.7	13.3
Sub-Saharan Africa						
Sub-Saharan Africa[j]						
Americas						
Central and South America[k]		Longitudinal cohort data	2–12 years 16.2	5–7 years 5.4		
North America (USA)[l]		Clinic and hospital-based sentinel surveillance	6–23 months 22.0			
USA[m]	PCR	Prospective population-based hospital surveillance 2000–2004	0–5 months 2.4–7.2; 6–23 months 0.6–1.5; 24–59 months 0.04–0.6			

[a] Revised and updated from Simmerman and Uyeki [14]
[b] Simmerman and Uyeki [14]
[c] Adapted from Simmerman and Uyeki [14]
[d] Chiu et al. [64]. Rates are for influenza A only
[e] Yoshida [16]
[f] Broor et al. [29]
[g] Brooks et al. [36]
[h] Brooks, for persons ≥5 years old (unpublished data)
[i] Peled et al. [39]
[j] No published population-based estimates
[k] Gordon et al. [57]
[l] Fiore et al. [60]
[m] Poehling et al. [63]

10,000 for children 5–12 years in 1999 [14]. Regarding mortality rates, a study from Hong Kong estimated 3–16% of all deaths among persons ≥65 years to be influenza related, while another study from Singapore estimated influenza was associated with 14.8 episodes/100,000 person-years for all-cause mortality or 3.8% of all deaths. These hospitalization rates are substantially higher than those reported in the USA, although mortality rates (Singapore) are similar [18, 19].

Regarding laboratory-confirmed illness, 11–26% of outpatients with influenza-like illness were confirmed to have influenza in these studies. In terms of seasonality, those in more northern latitudes (Taiwan, Japan) reported winter seasonal peaks, while those in the tropics reported year-round circulation with peaks during the rainy seasons (May to September) [20–22] (Fig. 2).

Despite the relative wealth of the SE Asian region, until recently only half of the countries had influenza-related illness or mortality data, and of these, a minority reported disease burden rates with laboratory-confirmed influenza. Methodological issues, including case definitions and spectrum bias in patient selection for passive surveillance, make intercountry comparisons difficult. Importantly, the laboratory diagnostic techniques used in these studies (typically tissue culture) substantially underestimate true burden compared to newer assays, like multiplex polymerase chain reaction (PCR) [23].

2.2   South Asia (1.5 Billion; 24% of World Population)

Published data from South Asia are more limited than for East Asia. Studies published before 2004, the year when CDC began supporting influenza research in the region [24], have been sporadic passive sentinel hospital-based studies, typically studying the etiology of hospitalized febrile and respiratory illnesses.

There have been several reports from India reporting the prevalence of influenza among inpatients and outpatients, and primarily relying on tissue culture for virus identification [25–28]. There has been a recent report of a 3-year prospective surveillance of respiratory disease in a cohort of children in rural North India, utilizing a fluorescent antibody detection assay. This project reported an incidence of influenza A respiratory infection of 141 (95% CI 108–179)/1,000 child-years in children 0–3 years [29].

Recent prospective passive surveillance project carried out from 2004 to 2007 in young children less than 3 years of age with respiratory illness in a peri-urban region of Nepal showed that influenza virus circulation was perennial and present for 9 months of each of the years [30]. Of 2,219 cases of World Health Organization (WHO)-defined clinical pneumonia presenting to the study clinic, 11% of the cases were associated with influenza virus as determined by multiplex PCR performed on nasal aspirates compared to 15% with RSV. These two viruses accounted for two thirds of all viruses detected [30, 31].

In Bangladesh, there have been several early and small-scale reports. Two of the earliest were hospital-based reports on the prevalence of influenza among patients

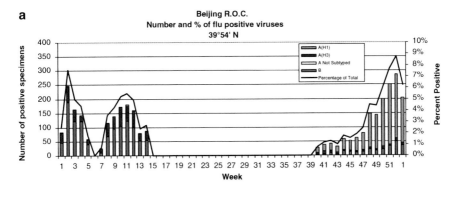

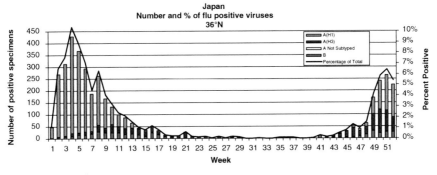

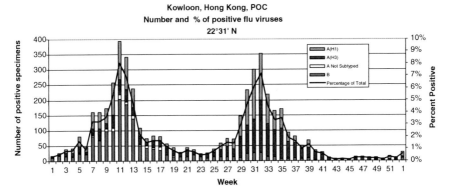

Fig. 2 (continued)

at a diarrhea hospital with influenza [32, 33]. One reported that there was no discernable seasonal pattern to the viral infections [32] but neither provided burden estimates. Among two more recent prevalence studies in children, one reported neglible prevalence of influenza among a cohort of 252 newborns by RT-PCR [34], while a pilot population-based prevalence study testing banked acute and convalescent serum by hemagluttinin inhibition (HI) among children <13 years under surveillance for febrile diseases reported an acute influenza infection prevalence

b

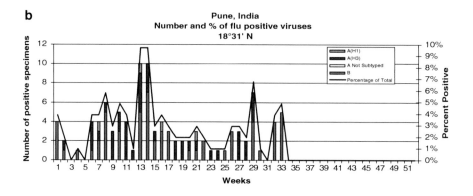

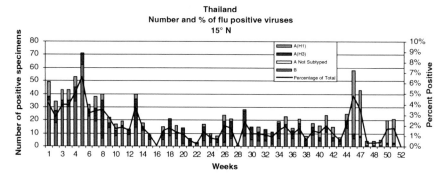

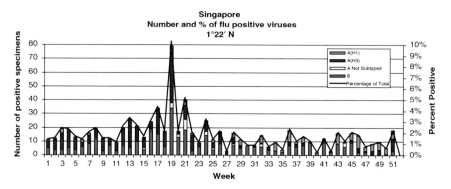

Fig. 2 Seasonal distribution of influenza in Asia. (**a**) Northeastern Asia. Source: WHO Global Atlas of Infectious Diseases (http://gamapserver.who.int/GlobalAtlas/PDFFactory/FluNet/index.asp?act= rmvCountries&rptGrp= 1). (**b**) South and Southeast Asia.
Source: WHO Global Atlas of Infectious Diseases (http://gamapserver.who.int/GlobalAtlas/ PDFFactory/FluNet/index.asp?act=rmvCountries &rptGrp=1)

of 16% (21 of 128) among children with fever and cough [24]. Since then, both hospital- and population-based surveillances have been initiated and have generated new reports. Among 3,699 inpatients and outpatients at 12 hospitals across Bangladesh, 385 (10%) were influenza positive by RT-PCR [35]. The highest influenza prevalence among patients visiting hospital occurred among those between 6 and 20 years old. Population-based active surveillance in Dhaka using tissue culture isolation and RT-PCR, on the other hand, demonstrated that for seasonal influenza, 50% of acute infections in 2008 occurred among children <5 years and 80% occurred among children <12 years, indicating that the primary burden of influenza occurs among the young. The incidence for influenza among children <5 years is reported at 102 episodes/1,000 child-years [36] and is 55.2 episodes/1,000 person-years among all ages.

Importantly, data from Bangladesh indicate that 28% of <5 years with influenza infection develop clinical pneumonia and that nearly two thirds of these cases occur among children <2 years [36]. Of the influenza viruses, there appears to be a nearly threefold greater association between pneumonia and influenza A (H3N2) than between seasonal influenza A (H1N1) and influenza B viruses, although all three are substantially associated with pneumonia.

Both studies reported perennial virus circulation but with peak influenza A seasonality occurring during the months between April and September.

Substantiating the surveillance data on childhood influenza disease burden, a recent randomized vaccine trial in which 170 pregnant mothers were given trivalent-inactivated influenza vaccine during the third trimester demonstrated a 63% reduction in rapid test-proven influenza respiratory illness among infants in their first 6 months of life [37], strongly supporting a role for influenza in childhood pneumonia and other respiratory illnesses.

All of these lines of evidence indicate substantial circulation of influenza in South Asia, and a contribution to respiratory disease burden, particularly in young children.

2.3 Middle East/North Africa (405 Million; 6% of World Population)

Most reports on influenza burden from the Middle East have come from Israel, often via telephone surveys, and fewer studies have tried to estimate population-based incidence combined with laboratory data. A study in Tel Aviv reported that among 9,300 children during the 1997–1998 season, 38% had ILI symptoms, with the highest incidence in the 3–15 year age group, specifically kindergarten and school-aged children [38], who comprised 83% of all cases. On the basis of laboratory-confirmed influenza from viral cultures of nasal swabs, incidence among all children was estimated at 22 episodes/1,000 children/year.

A separate Israeli report from the same period covering 23 November 1997–27 March 1998 stated that among 18,684 individuals enrolled in two clinics, 5,947 (18.1%) were enrolled for ARI-like symptoms [39]. Influenza was associated with 21.6% of all patients tested during the period, with the highest incidence among children 5–14 years and adults ≥ 65 years. The seasonal peak occurred during January. The incidence of influenza was highest among children 5–14 years (51.0 episodes/1,000 person-years), followed by the 0–4 years age group (13.9 episodes/ 1,000 person-years). Cocirculation of both influenza A viruses (H3N2 and H1N1) [39] as well as both lineages of influenza B viruses [40] has been reported.

A report among Lebanese children for the 2007–2008 season also confirmed cocirculation of influenza A and B viruses [41], and a January peak, but could not be used to estimate burden.

These studies suggest seasonal influenza circulation, with cocirculation of all virus types, a higher disease burden among children, primarily school-aged, and a winter peak similar to the remainder of the northern hemisphere.

2.4   Sub-Saharan Africa (821 Million; 12% of World Population)

Historically, data from sub-Saharan Africa consisted of outbreak reports, such as those in Madagascar and the Democratic Republic of Congo [42].

Until recently, routine influenza surveillance including characterization of influenza isolates has been conducted in only two African countries, Senegal and South Africa [43], although the Gambia has reported influenza and other respiratory viral infections in hospitalized children [44, 45]. There has been a report on circulating influenza viruses in Kenya [46], and a recent study from Kenya in 2006–2007 shows the circulation of nine antigenically different influenza A H3N2 viruses in a single season [47].

Within South Africa, surveillance has been conducted in three locations, Cape Town, Durban, and Johannesburg, which were instrumental in documenting an influenza outbreak in 1998 and have continued to document seasonal influenza strains. In addition, there have been other reports of outbreaks of influenza A (H3N2) [48] and of the complex pattern of influenza B virus circulation [49] in southern Africa, but these reports do not provide population burden estimates. Hospital data from HIV-positive and -negative children indicate that influenza contributes substantially to severe lower respiratory tract infections in South Africa [50], while pneumococcal vaccine trials suggest substantial burden from both primary influenza infection and coinfection from bacterial pathogens [51]. Influenza disease burden in the WHO African region remains underappreciated [43], and expanded surveillance that provides disease burden estimates is needed.

2.5   Central and South America
(570 Million; 8% of World Population)

Thirteen national influenza centers are reported to exist in nine Latin American countries, and all are reported to have the capacity for viral isolation and subtyping by HAI using WHO reference antisera [52]. Brazil, Chile, and Argentina employ surveillance networks within their countries to capture representative data. These sites in turn share samples with CDC for vaccine surveillance. Although a seeding hierarchy of influenza A (H3N2) introduction into South America, by way of Europe and North America from East and Southeast Asian strains, has been hypothesized [53], based on antigenic and genetic analysis of hemagglutinin, published data on influenza virus circulation from the region are limited. Early CDC reports from the influenza reporting centers in Central and South America indicated circulation of both influenza A and B viruses, with a peak season of May to July [54, 55]. A later report from Brazil confirmed a seasonal southern traveling wave of infection, beginning in March to April at the equator and traveling southward toward the temperate areas during May to July [56]. Although this study did not report incidence and morbidity per se, it attributed 0.03% of all 19 million deaths that occurred in Brazil between 1979 and 2001 to influenza (i.e., 570,000 deaths or approximately 17,812 deaths/year). Apart from a recent study from Nicaragua that reported an influenza incidence in 2007 among a cohort of children 2–12 years of 16.2/100 person-years [57], based on RT-PCR, there are no recent incidence estimates from Central America.

From Brazil, a study of 184 children hospitalized with pneumonia reported that 9% of the cases were associated with influenza virus detection by PCR [58]. A recent report from Peru of multihospital sentinel surveillance for influenza-like illness during 2006–2008 reported an overall isolation of influenza in 35% and influenza A in 25% of 6,835 ill patients [59]. This surveillance project also showed the continuous presence of influenza virus during the 3 years of surveillance.

These limited data suggest that influenza circulates through the tropical and subtropical belt of Central and South America and contributes substantively to disease burden and mortality. Representative data from more countries in the region, allowing comparisons between tropical and temperate areas, would be helpful in better determining burden and seasonality, and approaches to vaccine utilization.

2.6 Comparison of Influenza Burden Between Tropical/ Subtropical Countries and North America/Europe [North America (452 Million) and Europe (502 Million) Together Equal About 14% of World Population]

Although the current viruses influenza A (H3N2), and pre-2009 (H1N1) and influenza B have been in global circulation since 1977 [60], influenza epidemiology in temperate and tropical/subtropical regions appears to differ in at least two important respects.

First, there are differences in seasonality (Table 2). In North America and Europe, influenza has an annual seasonal epidemic pattern, typically circulating during winter, from November through March, peaking in January and February [60].

Table 2 Global chronology of influenza disease and vaccine strategies

Issue	Region		
	North	South	Tropical/subtropical
Influenza virus circulates	November to April	May to October	~12 months (perennial)
Vaccine strain composition announced	February, same year	September, previous year	North or south[a]
Timing of influenza immunization	October to December	April to June	No recommendation

[a]From WHO recommendation for 2008 influenza vaccine: "Epidemiological considerations will influence which recommendation (September 2008 or February 2008) is more appropriate for countries in equatorial regions."
Source: http://www.who.int/csr/disease/influenza/recommendations2008south/en/index.html (accessed October 2009)

The Middle East may be similar to the temperate north, with a winter seasonal peak [41]. The southern hemisphere shows a March to September seasonality, including Australia and New Zealand [61] which follow the temperate zone winter pattern. In the tropics/subtropics, currently available data indicate that influenza infections have a perennial pattern, often circulating year-round [12, 14, 21, 30, 35, 36] with peaks between March and September [14, 16, 36, 56]. Thus, some of the northern hemisphere tropical and subtropical areas show a complementary peak season to the temperate north.

Second, there appear to be differences in disease burden (Table 1) both in incidence and in the fraction of persons with lower airway complications, like pneumonia, at least between the lower income tropical countries with higher overall respiratory disease burden when compared to wealthier temperate countries. Although based on limited data, there appears to be a consistently higher incidence in the tropical belt than is typically reported in temperate areas. The discrepancy between regions is most noteworthy among children. Average seasonal influenza infection incidence among children in the northern hemisphere is estimated at 4.6% per year for children 0–19 years and 9.5% per year for children <5 years [62]. Among children 6–23 months in the USA, influenza incidence is estimated at 22.0/1,000 child-years and for children 5–7 years it is 5.4/1,000 child-years [60].

While the higher incidence in the tropics may be partially related to the perennial circulation of influenza virus, resulting in greater exposure to influenza viruses, it does not explain the higher fraction of lower respiratory complications. Indeed, most interest lies with the more severe infections, measured by hospitalization and mortality. Estimates around hospitalization are sometimes divided between "pneumonia and influenza" and "respiratory and circulatory" [18, 19], the former being a subset of the latter. US data indicate a mean (SD) all-age pneumonia and influenza hospitalization rate of 52.0 (25.2)/100,000 person-years and a rate of 114.8 (43.6)/100,000 for respiratory and circulatory hospitalization associated with influenza [18]. The highest rates exist for children <5 years, which is 113.9/100,000 person-years, and for the elderly ≥65 years, for whom rates increase with increasing age from 229.7/100,000 for persons 65–69 years to 1,669.2/100,000 for persons ≥85 years [18].

Hospitalizations, particularly in children, represent, however, only a fraction of the total influenza burden. One study estimated the burden of outpatient pneumonia among US children to be 50 clinic visits and 6 emergency department (ED) visits/ 1,000 children in 2002–2003 and 95 and 27 clinic and ED visits in 2003–2004, making outpatient influenza disease burden among young children at least tenfold greater than hospitalizations [63].

Published surveillance data on hospitalization or laboratory-confirmed influenza pneumonia incidence are not available for Central and Latin America or Africa. Hospitalization rates derived through statistical modeling reported by middle and higher income centers in East and Southeast Asia are similar to those in the USA [14]. These findings also reflect data from Korea and Japan, two of the world's wealthiest countries. Hospitalization rates among children <2 years in a recent Hong Kong study based on viral culture data were overall four to six times greater than those among comparable age groups in US studies, although they were only marginally higher than US rates for children <5 years [64]. Importantly, hospitalization rates were highest (103.8 cases/10,000 child-years) for children <1 year during circulation of a novel variant of H3N2 [64]. Notably, this rate is substantially lower than a previous estimate [65]. Differences in methodology as well as vaccination rates between the Hong Kong and US studies may partially explain the variation.

Population-based data from Bangladesh in children <5 years show an incidence of 102 infections/1,000 child-years [36] or an annual 10.2% incidence. Prospective serologic surveillance during a single year in a cohort of 140 Bangladeshi infants (0–6 months) has shown attack rates of 32/100 infants (Henkle et al. submitted). These data suggest that Bangladeshi children less than 5 years old have nearly five times the infection rate of those less than 2 years old in the USA [60]. The rate for influenza-associated pneumonia among children <5 years in Bangladesh is 28.6 episodes/1,000 child-years [36]. Hospitalization for childhood pneumonia is uncommon in Bangladesh [5, 36, 66]; however, approximately 13% of all pneumonias are severe and should be hospitalized [67], representing a conservative estimate for hospitalization rates. Using this figure would result in a hospitalization rate for influenza pneumonia of 371.8/100,000 for South Asian children <5 years compared with 113.9 for all influenza respiratory hospitalizations for children in the USA, which is a 3.3-fold higher rate. On the basis of these data, it can be deduced that the severe complication rate for influenza-associated lower airway obstruction is substantially higher among Southeast Asian children than among US children and is likely higher among children throughout the tropical belt than in temperate zones. Importantly, routine influenza vaccine is either not in regular use or not even available in most of these settings [14, 36].

The rates of flu-related pneumonia for tropical regions lacking reported data are not likely to be lower than those for the USA and Europe and are likely higher. Given the population sizes, these would represent a substantial contribution to childhood pneumonia burden, as well as hospitalization and mortality for all ages, as suggested by the Brazilian data [56]. Although there is little information from prospective studies on the effect of HIV and other immunocompromising comorbidities [68], at least one study suggests that children with HIV may have a higher burden than non-HIV infected children and benefit more from preventive measures [51].

Given the high incidence of other pneumonia-causing pathogens in these regions, notably *S. pneumoniae* and *H. influenzae* type b [69, 70], and the possibility that their interaction with influenza may exacerbate pneumonia severity [71], influenza may have an even greater impact on childhood pneumonia burden in the tropical belt.

Recent data from Thailand indicate that not only the burden but also the costs of influenza are substantial, resulting in up to 20% of household monthly income per illness episode [72], which translates in up to over $62 million annually in economic losses, of which decreased productivity accounts for 56% of the total cost [73]. Additional economic analyses are needed from lesser developed countries for a better understanding of the economic impact of influenza.

Data regarding groups at high risk for influenza disease are limited from tropical regions, but it is likely that high-risk groups described in temperate regions also experience high risk in the tropics. Studies should be undertaken among the very young, the elderly, and those with chronic illnesses, and among healthy pregnant women to assess the increased risk associated with nonpandemic influenza infection.

3 Global Influenza Circulation

The epidemiology of influenza in the tropics appears to play an important role in the generation and dissemination of new variant influenza viruses. Until recently how influenza virus subtypes spread around the world, factors underlying seasonality, and even virus subtype distribution have been poorly understood [12, 40, 74, 75]. There has been debate as to whether seasonal epidemics result from persistence of viruses from the previous season or from introduction from other regions. Examining the phylogenetic relationships between influenza A (H3N2) viruses isolated between 1999 and 2005 in New Zealand and Australia and those from New York, one study concluded that global viral migration was a major factor in seasonal emergence at least for influenza A (H3N2) [61], although regional temporal relationships were not established. A study involving global antigenic and genetic analysis of hemagglutinin (HA) from influenza A (H3N2) viruses isolated between 2002 and 2007 demonstrated temporally overlapping viral epidemics in the East and Southeast Asian tropical region that create a regional viral network of continuously circulating influenza viruses, one of which subsequently seeds Oceania, North America and Europe, and South America along major air travel routes [53] (Fig. 3).

Together, these data argue against local reemergence of persistent influenza virus from prior seasons and in favor of introduction from other regions. They also underscore the importance of Asia in the overall ecology of influenza, at least for the A (H3N2) virus, and that seasonality is a global and interactive

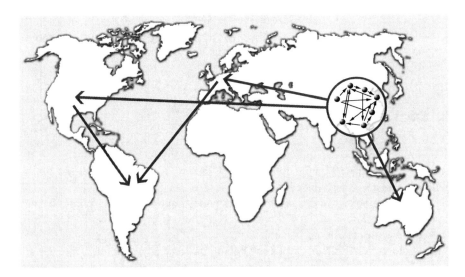

Fig. 3 Schematic of the dominant seeding hierarchy of seasonal influenza A (H3N2) viruses. Modified from Russell et al. [53]

phenomenon. These data also provide an explanatory mechanism for the generation, selection, and dissemination of novel antigenically drifted and pandemic influenza viruses [53, 76].

4 Vaccine Strategies for Tropical/Subtropical Regions

Public health authorities in most tropical countries have generally not considered influenza vaccine a high priority, partly because of the cost of the vaccine in relation to other public health vaccines, and the need for annual distribution of influenza vaccine [13, 77]. There is a growing interest in consideration of strategies for seasonal and pandemic influenza immunization, and several countries have begun production of influenza vaccines. Currently available data indicate that standard trivalent-inactivated [15, 37] and -attenuated live vaccines [78–81] are safe, immunogenic, and effective in tropical regions.

The details of immunization strategies are complex in the tropical setting, including the selection of vaccine strains to include in a vaccine and the timing of annual immunization [82–84] (Table 2). One study from Bangladesh showed that immunization of ten sequential monthly cohorts of young adults, who were then followed for at least 6 months, resulted in a 36% overall reduction of clinical febrile influenza-like respiratory illnesses during the 15-month project [37]. These preliminary data from year-round immunization in a setting of perennial influenza virus circulation demonstrate overall clinical effectiveness equal to that reported during seasonal influenza immunization in temperate regions [60]. Another strategy, given the growing evidence of influenza burden in childhood illness in developing countries, might be to use influenza vaccine as a means of controlling pneumonia in young children [85] rather than seasonal influenza.

5   Summary

Influenza appears to be a major contributor to morbidity, hospitalization, and likely mortality in the tropical and subtropical low-income regions; however, its contribution has been largely underestimated due to a lack of data from these regions [11]. It is also likely responsible for a substantial fraction of the childhood pneumonia and pneumonia-related mortality, both from primary infection and from interaction with respiratory bacterial agents. Introduction of influenza vaccine as a means to control influenza-related pneumonia in young children may be warranted [85] and may provide a mechanism for influenza vaccine uptake in these countries with wider benefits to both disease burden and mortality reduction, as well as surge capacity for vaccine production during pandemics.

Concern about pandemic influenza has increased interest in vaccine use, including increased seasonal vaccine use, and initiation of vaccine production in some countries. Continued and enhanced surveillance in the tropics, particularly in East and Southeast Asia, is warranted both to monitor burden and the impact of interventions, such as vaccination, and to identify emergence and spread of novel viruses.

References

1. Bryce J, Boschi-Pinto C, Shibuya K, Black RE (2005) WHO estimates of the causes of death in children. Lancet 365:1147–1152
2. Rudan I, Boschi-Pinto C, Biloglav Z, Mulholland K, Campbell H (2008) Epidemiology and etiology of childhood pneumonia. Bull World Health Organ 86:408–416
3. Adegbola RA, Secka O, Lahai G, Lloyd-Evans N, Njie A, Usen S, Oluwalana C, Obaro S, Weber M, Corrah T et al (2005) Elimination of *Haemophilus influenzae* type b (Hib) disease from the Gambia after the introduction of routine immunisation with a Hib conjugate vaccine: a prospective study. Lancet 366:144–150
4. Berman S (1991) Epidemiology of acute respiratory infections in children of developing countries. Rev Infect Dis 13(Suppl 6):S454–S462
5. Brooks WA, Breiman RF, Goswami D, Hossain A, Alam K, Saha SK, Nahar K, Nasrin D, Ahmed N, El Arifeen S et al (2007) Invasive pneumococcal disease burden and implications for vaccine policy in urban Bangladesh. Am J Trop Med Hyg 77:795–801
6. Cutts FT, Zaman SM, Enwere G, Jaffar S, Levine OS, Okoko JB, Oluwalana C, Vaughan A, Obaro SK, Leach A et al (2005) Efficacy of nine-valent pneumococcal conjugate vaccine against pneumonia and invasive pneumococcal disease in the Gambia: randomised, double-blind, placebo-controlled trial. Lancet 365:1139–1146

7. Klugman KP, Madhi SA, Huebner RE, Kohberger R, Mbelle N, Pierce N (2003) A trial of a 9-valent pneumococcal conjugate vaccine in children with and those without HIV infection. N Engl J Med 349:1341–1348

8. Levine OS, Lagos R, Munoz A, Villaroel J, Alvarez AM, Abrego P, Levine MM (1999) Defining the burden of pneumonia in children preventable by vaccination against *Haemophilus influenzae* type b. Pediatr Infect Dis J 18:1060–1064

9. World Health Organisation (2007) Pneumococcal conjugate vaccine for childhood immunization – WHO position paper. Wkly Epidemiol Rec 82:93–104

10. World Health Organisation (2006) WHO position paper on *Haemophilus influenzae* type b conjugate vaccines. (Replaces WHO position paper on Hib vaccines previously published in the *Weekly Epidemiological Record*). Wkly Epidemiol Rec 81:445–452

11. Scott JA, Brooks WA, Peiris JS, Holtzman D, Mulholland EK (2008) Pneumonia research to reduce childhood mortality in the developing world. J Clin Invest 118:1291–1300

12. Viboud C, Alonso WJ, Simonsen L (2006) Influenza in tropical regions. PLoS Med 3:e89

13. Higgs ES, Hayden FG, Chotpitayasunondh T, Whitworth J, Farrar J (2008) The Southeast Asian Influenza Clinical Research Network: development and challenges for a new multilateral research endeavor. Antiviral Res 78:64–68

14. Simmerman JM, Uyeki TM (2008) The burden of influenza in East and South-east Asia: a review of the English language literature. Influenza Other Respir Viruses 2:81–92

15. Liang XF, Wang HQ, Wang JZ, Fang HH, Wu J, Zhu FC, Li RC, Xia SL, Zhao YL, Li FJ et al (2010) Safety and immunogenicity of 2009 pandemic influenza A H1N1 vaccines in China: a multicentre, double-blind, randomised, placebo-controlled trial. Lancet 375:56–66

16. Yoshida LM, Suzuki M, Yamamoto T, Nguyen HA, Nguyen CD, Nguyen AT, Oishi K, Vu TD, Le TH, Le MQ et al (2010) Viral pathogens associated with acute respiratory infections in central vietnamese children. Pediatr Infect Dis J 29:75–77

17. Zhu FC, Wang H, Fang HH, Yang JG, Lin XJ, Liang XF, Zhang XF, Pan HX, Meng FY, Hu YM et al (2009) A novel influenza A (H1N1) vaccine in various age groups. N Engl J Med 361:2414–2423

18. Thompson WW, Shay DK, Weintraub E, Brammer L, Bridges CB, Cox NJ, Fukuda K (2004) Influenza-associated hospitalizations in the United States. JAMA 292:1333–1340

19. Thompson WW, Shay DK, Weintraub E, Brammer L, Cox N, Anderson LJ, Fukuda K (2003) Mortality associated with influenza and respiratory syncytial virus in the United States. JAMA 289:179–186

20. Shek LP, Lee BW (2003) Epidemiology and seasonality of respiratory tract virus infections in the tropics. Paediatr Respir Rev 4:105–111

21. Simmerman JM, Chittaganpitch M, Levy J, Chantra S, Maloney S, Uyeki T, Areerat P, Thamthitiwat S, Olsen SJ, Fry A et al (2009) Incidence, seasonality and mortality associated with influenza pneumonia in Thailand: 2005–2008. PLoS ONE 4:e7776

22. Waicharoen S, Thawatsupha P, Chittaganpitch M, Maneewong P, Thanadachakul T, Sawanpanyalert P (2008) Influenza viruses circulating in Thailand in 2004 and 2005. Jpn J Infect Dis 61:321–323

23. Lee WM, Grindle K, Pappas T, Marshall DJ, Moser MJ, Beaty EL, Shult PA, Prudent JR, Gern JE (2007) High-throughput, sensitive, and accurate multiplex PCR-microsphere flow cytometry system for large-scale comprehensive detection of respiratory viruses. J Clin Microbiol 45:2626–2634

24. Abdullah Brooks W, Terebuh P, Bridges C, Klimov A, Goswami D, Sharmeen AT, Azim T, Erdman D, Hall H, Luby S et al (2007) Influenza A and B infection in children in urban slum, Bangladesh. Emerg Infect Dis 13:1507–1508

25. Chatterjee S, Mukherjee KK, Mondal MC, Chakraborty MS (1996) A study of influenza A virus in the city of Calcutta, India, highlighting the strain prevalence. Acta Microbiol Pol 45:279–283

26. John TJ, Cherian T, Steinhoff MC, Simoes EA, John M (1991) Etiology of acute respiratory infections in children in tropical southern India. Rev Infect Dis 13(Suppl 6):S463–S469

27. Rao BL, Yeolekar LR, Kadam SS, Pawar MS, Kulkarni PB, More BA, Khude MR (2005) Influenza surveillance in Pune, India, 2003. Southeast Asian J Trop Med Public Health 36: 906–909
28. Yeolekar LR, Kulkarni PB, Chadha MS, Rao BL (2001) Seroepidemiology of influenza in Pune, India. Indian J Med Res 114:121–126
29. Broor S, Parveen S, Bharaj P, Prasad VS, Srinivasulu KN, Sumanth KM, Kapoor SK, Fowler K, Sullender WM (2007) A prospective three-year cohort study of the epidemiology and virology of acute respiratory infections of children in rural India. PLoS ONE 2:e491
30. Mathisen M, Strand TA, Sharma BN, Chandyo RK, Valentiner-Branth P, Basnet S, Adhikari RK, Hvidsten D, Shrestha PS, Sommerfelt H (2009) RNA viruses in community-acquired childhood pneumonia in semi-urban Nepal; a cross-sectional study. BMC Med 7:35
31. Mathisen M, Strand TA, Sharma BN, Chandyo RK, Valentiner-Branth P, Basnet S, Adhikari RK, Hvidsten D, Shrestha PS, Sommerfelt H (2010) Clinical presentation and severity of viral community-acquired pneumonia in young Nepalese children. Pediatr Infect Dis J 29:e1–e6
32. Huq F, Rahman M, Nahar N, Alam A, Haque M, Sack DA, Butler T, Haider R (1990) Acute lower respiratory tract infection due to virus among hospitalized children in Dhaka, Bangladesh. Rev Infect Dis 12(Suppl 8):S982–S987
33. Rahman M, Huq F, Sack DA, Butler T, Azad AK, Alam A, Nahar N, Islam M (1990) Acute lower respiratory tract infections in hospitalized patients with diarrhea in Dhaka, Bangladesh. Rev Infect Dis 12(Suppl 8):S899–S906
34. Hasan K, Jolly P, Marquis G, Roy E, Podder G, Alam K, Huq F, Sack R (2006) Viral etiology of pneumonia in a cohort of newborns till 24 months of age in rural Mirzapur, Bangladesh. Scand J Infect Dis 38:690–695
35. Zaman RU, Alamgir AS, Rahman M, Azziz-Baumgartner E, Gurley ES, Sharker MA, Brooks WA, Azim T, Fry AM, Lindstrom S et al (2009) Influenza in outpatient ILI case-patients in national hospital-based surveillance, Bangladesh, 2007–2008. PLoS ONE 4:e8452
36. Brooks WA, Goswami D, Rahman M, Nahar K, Fry AM, Balish A, Iftekharuddin N, Azim T, Xu X, Klimov A et al (2010) Influenza is a major contributor to childhood pneumonia in a tropical developing country. Pediatr Infect Dis J 29:216–221
37. Zaman K, Roy E, Arifeen SE, Rahman M, Raqib R, Wilson E, Omer SB, Shahid NS, Breiman RE, Steinhoff MC (2008) Effectiveness of maternal influenza immunization in mothers and infants. N Engl J Med 359:1555–1564
38. Kiro A, Robinson G, Laks J, Mor Z, Varsano N, Mendelson E, Amitai ZS (2008) [Morbidity and the economic burden of influenza in children in Israel – a clinical, virologic and economic review]. Harefuah 147:960–965, 1031
39. Peled T, Weingarten M, Varsano N, Matalon A, Fuchs A, Hoffman RD, Zeltcer C, Kahan E, Mendelson E, Swartz TA (2001) Influenza surveillance during winter 1997–1998 in Israel. Isr Med Assoc J 3:911–914
40. Chi XS, Bolar TV, Zhao P, Rappaport R, Cheng SM (2003) Cocirculation and evolution of two lineages of influenza B viruses in Europe and Israel in the 2001–2002 season. J Clin Microbiol 41:5770–5773
41. Zaraket H, Dbaibo G, Salam O, Saito R, Suzuki H (2009) Influenza virus infections in Lebanese children in the 2007–2008 season. Jpn J Infect Dis 62:137–138
42. Nicholson KG, Wood JM, Zambon M (2003) Influenza. Lancet 362(9397):1733–1745
43. Schoub BD, McAnerney JM, Besselaar TG (2002) Regional perspectives on influenza surveillance in Africa. Vaccine 20(Suppl 2):S45–S46
44. Forgie IM, O'Neill KP, Lloyd-Evans N, Leinonen M, Campbell H, Whittle HC, Greenwood BM (1991) Etiology of acute lower respiratory tract infections in Gambian children: II. Acute lower respiratory tract infection in children ages one to nine years presenting at the hospital. Pediatr Infect Dis J 10:42–47
45. Mulholland EK, Ogunlesi OO, Adegbola RA, Weber M, Sam BE, Palmer A, Manary MJ, Secka O, Aidoo M, Hazlett D et al (1999) Etiology of serious infections in young Gambian infants. Pediatr Infect Dis J 18:S35–S41

46. Gachara G, Ngeranwa J, Magana JM, Simwa JM, Wango PW, Lifumo SM, Ochieng WO (2006) Influenza virus strains in Nairobi, Kenya. J Clin Virol 35:117–118
47. Bulimo WD, Garner JL, Schnabel DC, Bedno SA, Njenga MK, Ochieng WO, Amukoye E, Magana JM, Simwa JM, Ofula VO et al (2008) Genetic analysis of H3N2 influenza A viruses isolated in 2006–2007 in Nairobi, Kenya. Influenza Other Respi Viruses 2:107–113
48. Besselaar TG, Botha L, McAnerney JM, Schoub BD (2004) Antigenic and molecular analysis of influenza A (H3N2) virus strains isolated from a localised influenza outbreak in South Africa in 2003. J Med Virol 73:71–78
49. Besselaar TG, Botha L, McAnerney JM, Schoub BD (2004) Phylogenetic studies of influenza B viruses isolated in southern Africa: 1998–2001. Virus Res 103:61–66
50. Madhi SA, Ramasamy N, Bessellar TG, Saloojee H, Klugman KP (2002) Lower respiratory tract infections associated with influenza A and B viruses in an area with a high prevalence of pediatric human immunodeficiency type 1 infection. Pediatr Infect Dis J 21:291–297
51. Madhi SA, Klugman KP (2004) A role for *Streptococcus pneumoniae* in virus-associated pneumonia. Nat Med 10:811–813
52. Savy V (2002) Regional perspectives on influenza surveillance in South America. Vaccine 20 (Suppl 2):S47–S49
53. Russell CA, Jones TC, Barr IG, Cox NJ, Garten RJ, Gregory V, Gust ID, Hampson AW, Hay AJ, Hurt AC et al (2008) The global circulation of seasonal influenza A (H3N2) viruses. Science 320:340–346
54. Centers for Disease Control and Prevention (CDC) (1995) Update: influenza activity – worldwide, 1995. MMWR Morb Mortal Wkly Rep 44:644–645, 651–652
55. Centers for Disease Control and Prevention (CDC) (1994) Update: influenza activity – worldwide, 1994. MMWR Morb Mortal Wkly Rep 43:691–693
56. Alonso WJ, Viboud C, Simonsen L, Hirano EW, Daufenbach LZ, Miller MA (2007) Seasonality of influenza in Brazil: a traveling wave from the Amazon to the subtropics. Am J Epidemiol 165:1434–1442
57. Gordon A, Saborio S, Kuan G, Videa E, Ortega O, Reingold A, Balmaseda A, Harris E (2009) A prospective cohort study of the seasonality and burden of pediatric influenza in Nicaragua XI International Symposium on Respiratory Viral Infections. The Macrae Group, Bangkok
58. Nascimento-Carvalho CM, Ribeiro CT, Cardoso MR, Barral A, Araujo-Neto CA, Oliveira JR, Sobral LS, Viriato D, Souza AL, Saukkoriipi A et al (2008) The role of respiratory viral infections among children hospitalized for community-acquired pneumonia in a developing country. Pediatr Infect Dis J 27:939–941
59. Laguna-Torres VA, Gomez J, Ocana V, Aguilar P, Saldarriaga T, Chavez E, Perez J, Zamalloa H, Forshey B, Paz I et al (2009) Influenza-like illness sentinel surveillance in Peru. PLoS ONE 4:e6118
60. Fiore AE, Shay DK, Broder K, Iskander JK, Uyeki TM, Mootrey G, Bresee JS, Cox NJ (2009) Prevention and control of seasonal influenza with vaccines: recommendations of the Advisory Committee on Immunization Practices (ACIP), 2009. MMWR Recomm Rep 58:1–52
61. Nelson MI, Simonsen L, Viboud C, Miller MA, Holmes EC (2007) Phylogenetic analysis reveals the global migration of seasonal influenza A viruses. PLoS Pathog 3:1220–1228
62. Bueving HJ, van der Wouden JC, Berger MY, Thomas S (2005) Incidence of influenza and associated illness in children aged 0–19 years: a systematic review. Rev Med Virol 15:383–391
63. Poehling KA, Edwards KM, Weinberg GA, Szilagyi P, Staat MA, Iwane MK, Bridges CB, Grijalva CG, Zhu Y, Bernstein DI et al (2006) The underrecognized burden of influenza in young children. N Engl J Med 355:31–40
64. Chiu SS, Chan KH, Chen H, Young BW, Lim W, Wong WH, Lau YL, Peiris JS (2009) Virologically confirmed population-based burden of hospitalization caused by influenza A and B among children in Hong Kong. Clin Infect Dis 49:1016–1021
65. Chiu SS, Lau YL, Chan KH, Wong WH, Peiris JS (2002) Influenza-related hospitalizations among children in Hong Kong. N Engl J Med 347:2097–2103

66. Naheed A, Saha SK, Breiman RF, Khatun F, Brooks WA, El Arifeen S, Sack D, Luby SP (2009) Multihospital surveillance of pneumonia burden among children aged <5 years hospitalized for pneumonia in Bangladesh. Clin Infect Dis 48(Suppl 2):S82–S89

67. World Health Organization/UNICEF (2000) Management of the child with a serious infection or severe malnutrition: guidelines for care at the first-referral level in developing countries. Department of Child and Adolescent Health and Development, World Health Organization, Geneva

68. Kunisaki KM, Janoff EN (2009) Influenza in immunosuppressed populations: a review of infection frequency, morbidity, mortality, and vaccine responses. Lancet Infect Dis 9:493–504

69. O'Brien KL, Wolfson LJ, Watt JP, Henkle E, Deloria-Knoll M, McCall N, Lee E, Mulholland K, Levine OS, Cherian T (2009) Burden of disease caused by *Streptococcus pneumoniae* in children younger than 5 years: global estimates. Lancet 374:893–902

70. Watt JP, Wolfson LJ, O'Brien KL, Henkle E, Deloria-Knoll M, McCall N, Lee E, Levine OS, Hajjeh R, Mulholland K et al (2009) Burden of disease caused by *Haemophilus influenzae* type b in children younger than 5 years: global estimates. Lancet 374:903–911

71. McCullers JA (2006) Insights into the interaction between influenza virus and pneumococcus. Clin Microbiol Rev 19:571–582

72. Clague B, Chamany S, Burapat C, Wannachaiwong Y, Simmerman JM, Dowell SF, Olsen SJ (2006) A household survey to assess the burden of influenza in rural Thailand. Southeast Asian J Trop Med Public Health 37:488–493

73. Simmerman JM, Lertiendumrong J, Dowell SF, Uyeki T, Olsen SJ, Chittaganpitch M, Chunsutthiwat S, Tangcharoensathien V (2006) The cost of influenza in Thailand. Vaccine 24:4417–4426

74. Schweiger B, Zadow I, Heckler R (2002) Antigenic drift and variability of influenza viruses. Med Microbiol Immunol 191:133–138

75. Barr IG, Komadina N, Hurt A, Shaw R, Durrant C, Iannello P, Tomasov C, Sjogren H, Hampson AW (2003) Reassortants in recent human influenza A and B isolates from Southeast Asia and Oceania. Virus Res 98:35–44

76. Rambaut A, Pybus OG, Nelson MI, Viboud C, Taubenberger JK, Holmes EC (2008) The genomic and epidemiological dynamics of human influenza A virus. Nature 453:615–619

77. Macroepidemiology of Influenza Vaccination (MIV) Study Group (2005) The macroepidemiology of influenza vaccination in 56 countries, 1997–2003. Vaccine 23:5133–5143

78. Breiman RF, Brooks WA, Goswami D, Lagos R, Borja-Tabora C, Lanata CF, Londono JA, Lum LC, Rappaport R, Razmpour A et al (2009) A multinational, randomized, placebo-controlled trial to assess the immunogenicity, safety, and tolerability of live attenuated influenza vaccine coadministered with oral poliovirus vaccine in healthy young children. Vaccine 27:5472–5479

79. Forrest BD, Pride MW, Dunning AJ, Capeding MR, Chotpitayasunondh T, Tam JS, Rappaport R, Eldridge JH, Gruber WC (2008) Correlation of cellular immune responses with protection against culture-confirmed influenza virus in young children. Clin Vaccine Immunol 15:1042–1053

80. Lum LC, Borja-Tabora CF, Breiman RF, Vesikari T, Sablan BP, Chay OM, Tantracheewathorn T, Schmitt HJ, Lau YL, Bowonkiratikachorn P et al (2010) Influenza vaccine concurrently administered with a combination measles, mumps, and rubella vaccine to young children. Vaccine 28:1566–1574

81. Tam JS, Capeding MR, Lum LC, Chotpitayasunondh T, Jiang Z, Huang LM, Lee BW, Qian Y, Samakoses R, Lolekha S et al (2007) Efficacy and safety of a live attenuated, cold-adapted influenza vaccine, trivalent against culture-confirmed influenza in young children in Asia. Pediatr Infect Dis J 26:619–628

82. Centers for Disease Control and Prevention (CDC) (2009) Use of northern hemisphere influenza vaccines by travelers to the southern hemisphere. MMWR Morb Mortal Wkly Rep 58:312

83. de Mello WA, de Paiva TM, Ishida MA, Benega MA, Dos Santos MC, Viboud C, Miller MA, Alonso WJ (2009) The dilemma of influenza vaccine recommendations when applied to the tropics: the Brazilian case examined under alternative scenarios. PLoS ONE 4:e5095

84. Pontoriero AV, Baumeister EG, Campos AM, Savy VL, Lin YP, Hay A (2003) Antigenic and genomic relation between human influenza viruses that circulated in Argentina in the period 1995–1999 and the corresponding vaccine components. J Clin Virol 28:130–140

85. Brooks WA (2009) A four-stage strategy to reduce childhood pneumonia-related mortality by 2015 and beyond. Vaccine 27:619–623

The Origin and Evolution of H1N1 Pandemic Influenza Viruses

Robert G. Webster, Richard J. Webby, and Michael Perdue

Abstract Despite extensive planning for the next influenza pandemic in humans, nature has once again confounded the influenza experts. The emergence and development of an H1N1 pandemic strain while an H1N1 virus was still circulating in humans is an unprecedented event. Here, we examine the emergence of H1N1 influenza viruses in the USA, Europe, and Asia from the natural aquatic bird reservoir through intermediate hosts including pigs and turkeys to humans. There were some remarkable parallel evolutionary developments in the swine influenza viruses in the Americas and in Eurasia. Classical swine influenza virus in the USA emerged either before or immediately after the Spanish influenza virus emerged in humans in 1918. Over the next 50 plus years this swine influenza virus became increasingly attenuated in pigs but occasionally transmitted to humans causing mild clinical infection but did not consistently spread human to human. The remarkable parallel evolution was the introduction of avian influenza virus genes independently in swine influenza viruses in Europe and the USA, with almost simultaneous acquisition of genes from seasonal human influenza. Influenza in pigs in both Eurasia and America became more aggressive necessitating the production of vaccines, and the incidence of transmission of clinical influenza to humans increased. Eventually the different triple reassortants with gene segments from

R.G. Webster (✉)
Department of Human and Health Services (HHS), Biomedical Advanced Research and Development Authority (BARDA), 330 Independence Avenue, SW Rm G640, Washington, DC 20201, USA
Department of Infectious Diseases, St. Jude Children's Research Hospital, 262 Danny Thomas Place, Memphis, TN 38105, USA
e-mail: Robert.Webster@STJUDE.org
R.J. Webby,
Department of Infectious Diseases, St. Jude Children's Research Hospital, 262 Danny Thomas Place, Memphis, TN 38105, USA
M. Perdue
Department of Human and Health Services (HHS), Biomedical Advanced Research and Development Authority (BARDA), 330 Independence Avenue, SW Rm G640, Washington, DC 20201, USA

G. Del Giudice and R. Rappuoli (eds.), *Influenza Vaccines for the Future*, 2nd edition,
Birkhäuser Advances in Infectious Diseases,
DOI 10.1007/978-3-0346-0279-2_4, © Springer Basel AG 2011

avian, swine, and human influenza viruses in pigs in Europe and America met and mated and developed into the 2009 pandemic H1N1 influenza that is highly transmissible in people, pigs, and turkeys. Whether this occurred in Mexico or in Asia is currently unknown. The failure of the experts was to not recognize the importance of pigs in the evolution and host range transmission of influenza viruses with pandemic potential.

1 Introduction

Despite intensive pandemic planning and analysis of the available scientific knowledge of influenza viruses, none of the experts forecast the emergence of an H1N1 influenza virus as the causative agent of the first pandemic of the twenty-first century. Influenza once again confounded the experts. Since an H1N1 subtype of influenza was circulating and causing seasonal influenza in humans, this subtype was not on the "probable list." Focus was on the H5N1 influenza virus that emerged and spread to humans in 1997 [1, 2], also on the "possible list" were H2, H6, H7, and H9 for these subtypes were either transmitted to intermediate hosts, occasionally infected humans, or had caused a pandemic in humans previously.

The emergence of the H1N1 2009 pandemic influenza virus means that a paradigm shift in our thinking must occur regarding the antigenic distance that will permit a circulating subtype to reemerge, successfully transmit, and cause a pandemic. Although we failed to predict the H1N1 2009, our preparedness for an influenza pandemic has permitted a rapid response to the novel virus.

Although we did not "get it right" from the perspective of subtype, we do have a much better understanding of the ultimate reservoirs of influenza in the aquatic waterfowl of the world, of the probable roles of swine as the intermediate host, novel strategies to produce vaccines and antivirals, and molecular markers of pathogenicity. Here, we will consider the ultimate reservoirs of influenza virus in the wild aquatic migratory waterfowl of the world and the interplay between influenza in that reservoir and in pigs and people in the emergence of pandemic H1N1 influenza viruses.

2 The Ultimate Reservoirs

There is general consensus that the wild migratory aquatic birds of the world are the ultimate reservoirs of all influenza A viruses [3–5]. The generally benign infection of their natural host without apparent disease signs together with intestinal replication, transmission through water, and thermal stability in water are all indicators of viruses that are in equilibrium with their natural hosts [6]. One feature that is less well understood is the phylogenetic separation of the 16 different HA subtypes of influenza viruses in the world into two superfamilies – one in the Americas and the other in Eurasia [4]. This geographical separation is surprising because more than six million aquatic birds are known to migrate between Eurasia and the Americas through the Alaskan region (http://alaska.usgs.gov/science/biology/avian_influenza/migrants_tables.html).

Each of the pandemics of the past century including the H1N1 Spanish 1918, H2N2 Asian 1957, and H3N2 Hong Kong 1968 acquired a novel HA gene, as well as a novel PB1 gene from the aquatic bird reservoir [7, 8]. Novel neuraminidase (NA) genes were acquired less frequently in 1918 Spanish H1N1 and 1957 Asian H2N2. While we understood the role of the novel HA and NA in circumventing the immune response of the host, the importance of the PB1 gene is still largely unresolved.

The H1N1 Russian 1977 was a genuine reintroduction of a virus that had been completely genetically conserved for 27 years [9] indicating that it had to have been preserved in a frozen state. The novel H1N1 influenza virus that emerged in 2009 (see below) is a complex reassortant that obtained gene segments from avian influenza viruses (presumably from the ultimate reservoir species), swine influenza viruses, and human influenza viruses (Fig. 1); six of the gene segments were from the American lineage viruses and two were from the Eurasian lineage.

3 Intermediate Hosts

Influenza viruses in their natural avian hosts replicate at a higher temperature (40–42°C) than in mammalian species (37°C) and have an avian-type receptor specificity preferentially binding to α2-3 terminal sialic acid that is different from the receptor specificity of mammalian viruses (α2-6 terminal sialic acid). While influenza A viruses have been demonstrated to transmit directly from some avian species to humans (e.g., H5N1 transmitted to humans in Azerbaijan, killing three members of the family harvesting "down" from dead wild swans), a majority of these transmissions have been transitory and have not led to the emergence of transmissible viruses.

On the basis of epidemiological evidence, it was proposed that pigs may serve as intermediate hosts in the transmission of influenza viruses from the wild bird reservoir to humans [10]. Studies on the respiratory tract of pigs found both α2-3 and α2-6 sialic acid receptors [11] and pigs have a body temperature of 39°C. Subsequent studies showed that all of the subtypes of avian influenza tested could replicate in the respiratory tract of the pig [12].

Studies on the types of receptors for influenza viruses in avian species showed that ducks possess mainly α2-3 sialic acid receptors [11], while some other species such as the quail, pheasant, and turkey have dual receptor specificity [13, 14]. In the live poultry market system that is common in Southeast Asia where ducks, chickens, quail, pigeons, chukar, and pheasants are housed together, conditions are optimal for interspecies spread and reassortment of influenza viruses. Thus, live poultry markets plus backyard pig and poultry farming provide optimal conditions for interspecies transmission.

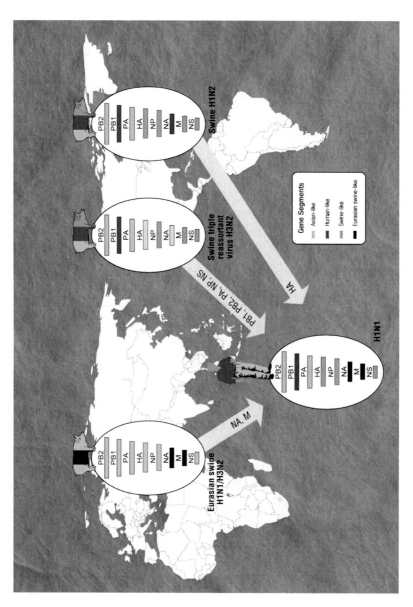

Fig. 1 Genesis of the H1N1 pandemic influenza virus 2009. The H1N1 pandemic influenza virus that emerged in 2009 is a complex reassortant influenza virus containing five gene segments from the triple reassortant (Fig. 3) (PB1, PB2, PA, NP, NS) swine influenza virus from North America and the H1 hemagglutinin (HA) also of swine origin from North America plus the neuraminidase (NA) and matrix (M) genes from European swine influenza viruses

4 Transmission

Live poultry markets (wet markets) are the ultimate man-made "mixing vessel" where domestic waterfowl can introduce the influenza viruses after exposure to wild migratory waterfowl and provide avian hosts with different receptor specificity and permit rampant reassortment to occur [15]. Epidemiological studies in Hong Kong establish that wet markets are a risk factor in the genesis and transport of novel viruses back to the farms [16]. The banning of waterfowl (ducks and geese) and later quail from Hong Kong live poultry markets in 1998 vastly reduced the subtypes of influenza viruses that were detectable and by 2005 the only subtype found in prospective surveillance was H9N2 that was associated with subclinical infection in chickens. The recognition of the impact of the live poultry market in the genesis of influenza virus has led to the decision that they should be closed and that poultry would be provided chilled or frozen. The acceptance of the biological vulnerabilities of the wet market system (in the USA as well as in Asia) and its importance in the evolution of influenza viruses has been slow. The ceremonial use of live poultry at festival occasions is part of Asian culture and is hard to change. There is slow acceptance of the high risk of genetic reassortment of influenza viruses in live poultry markets, but Taiwan decided to close all wet markets in 2009 and the number of markets in Hong Kong and Shanghai, China is being reduced. It is somewhat ironical that the number of live markets in the USA has increased in 2009 and that the keeping of backyard chickens is being approved in southern cities of the USA in 2009.

5 H1N1 Influenza in Pigs, People, and Poultry in USA 1918–1998

The H1N1 influenza virus that caused the 1918 Spanish influenza pandemic emerged in swine in the USA either before 1918 or in 1918 [17]. It is probable that multiple reassortant events were involved in the emergence of the 1918 Spanish influenza virus, and it is unknown if the virus emerged from pigs to people or vice versa [18]. The early descriptions of swine influenza on midwestern farms in the USA were of a serious respiratory disease that occurred in the winter months. Dr Richard Shope who initially isolated the classical swine influenza virus [19] was convinced that the virus disappeared from the pig population of USA during the summer months. Subsequent studies showed that the classical H1N1 swine influenza virus circulates year-round in pigs [20] but caused clinical disease signs only in the cooler months. By the 1960s and later swine influenza had become very mild and was considered almost a nonevent and did not merit the use of vaccine in the swine industry [20, 21].

Despite the mild clinical nature of swine influenza in pigs, there were intermittent transmissions of the classical swine influenza virus to humans. From 1974 to 2005, there were 43 confirmed cases of classical swine H1N1 in humans in the USA with six fatalities [22]. The transmission that received the most attention occurred in 1976 and is referred to as the Fort Dix incident [23]. In that incident, a young soldier at Fort Dix military camp was infected with swine influenza and died. The virus transmitted to 13 soldiers with mild disease signs and subsequent serological studies indicated that at least 200 soldiers were infected. The scientific and public health officials in the USA were greatly concerned that an outbreak of H1N1 influenza with catastrophic health impact similar to Spanish influenza was imminent. A vaccine was prepared and a national vaccination program was initiated in the USA.

In retrospect the response to the Fort Dix episode was an overreaction by the health authorities. The 1976 H1N1 influenza virus failed to spread beyond the initial focus, and the 1976 swine influenza vaccine was associated with a rare occurrence Guillain–Barre syndrome (GBS) – an ascending paralysis. Consequently, the vaccine program was stopped. The association between the 1976 H1N1 vaccine and the GBS has not been satisfactorily resolved. There were some 40 million persons vaccinated with the swine influenza vaccine; there were 500 cases of GBS recorded with 25 deaths. The incidence of GBS was later determined to be from 4.9 to 5.9 per million [24, 25] and became a major issue when the threat of influenza disappeared and the risk from GBS outweighed any benefit resulting in the cessation of the vaccine program. Extensive investigation failed to establish an association between any vaccine lots and the GBS cases. About 20% of the vaccine used was whole-inactivated virus and the remainder was subunit vaccine. At that time, vaccine was much less pure than current 2009 vaccines and was not standardized for antigen content. GBS continues to occur after a number of different virus infections but at a very low level and has subsequently not been associated with influenza vaccination.

In the early 1980s, classical swine influenza was reported to cause infection in domestic turkeys with mild infection and decrease in egg production [26] (Fig. 2). Surveillance in turkeys from 1980 to 1989 in the USA recorded the presence of avian-like H1N1 influenza viruses, classical swine influenza viruses, and the first reported double reassortants of swine and avian origin influenza viruses in the USA [27]. Preliminary characterization of the genotypes of these reassortants suggested that they possessed the replication complex (PB2, PB1, PA, NP) from avian sources and the remaining gene segments from classical swine influenza viruses. It could be postulated that the turkey serves as the intermediate host for the introduction of avian H1N1 genes into pigs for it is likely that transmission occurs in both directions between turkeys and pigs.

The first reported transmission of human H3N2 to pigs in the USA with production of a double reassortant was in 1998 when A/Swine/North Carolina/35922/98 (H3N2) was isolated from pigs with respiratory disease in North Carolina [28]. This virus possessed the PB1, HA, and NA from the then current human strain [A/Nanchang/933/95 (H3N2)] and the other gene segments from classical swine influenza virus (PB2, PA, NP, NS, M1, M2). While this virus caused respiratory diseases in pigs, it did not establish a stable lineage and disappeared. However, in the same year, a "triple reassortant" virus with gene segments from the circulating H3N2

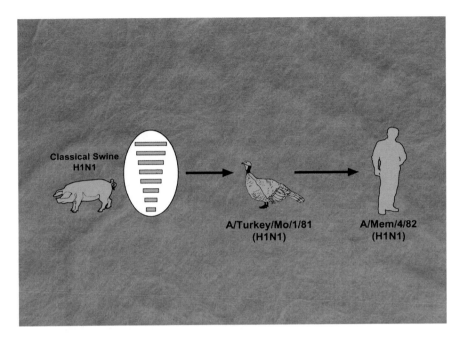

Fig. 2 Transmission of swine influenza to turkeys. Classical swine influenza virus transmitted to turkeys in the USA in the early 1980s. The H1N1 virus from turkeys retained the ability to infect humans [27]

human virus (PB1, HA, NA), classical swine influenza virus (NP, M, NS), and avian influenza virus (PB2, PA) [28] emerged in pigs in USA (Fig. 3). It is tempting to speculate that the double reassortant from North Carolina was a precursor of the triple reassortant for the same gene package from A/Nanchang/933/95 (H3N2) (PB1, HA, NA) appeared in the reassortant. The triple reassortant was highly transmissible and spread rapidly to pigs throughout the USA [29] and caused disease of sufficient severity to merit production and use of vaccines in the swine industry.

Thus, from 1918 to 1998, classical swine influenza virus remained antigenically and molecularly stable without introduction of novel gene segments. In 1998 or slightly before, the "monogamous" nature of the classical swine influenza changed dramatically with the tendency to mate with both avian and human influenza viruses. Consequently a number of different H1 and H3 influenza viruses with either N1 or N2 NAs on the "triple reassortant" backbone emerged in pigs and spread locally in the USA from 1998 to 2009.

6 European Swine Influenza

Classical H1N1 swine influenza of US origin had been introduced in Italy sometime before 1976 [30]. Swine influenza in Europe evolved along similar lines but was

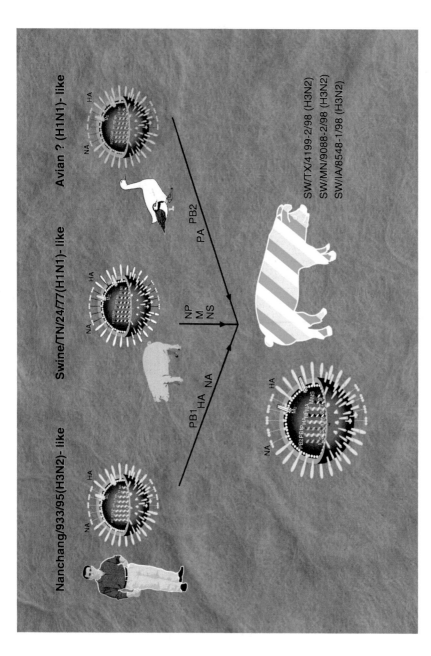

Fig. 3 Genesis of the triple reassortant swine influenza virus in the USA. From 1918 to 1998, the classical swine influenza virus had transmitted to turkeys and people, but gene segments from influenza viruses in people and poultry had not been detected in pigs. In 1998, a triple reassortant emerged with gene segments from humans (PB1, HA, NA), swine (NP, M, NS), and avian sources (PB2, PA) [28]. The triple reassortant was highly transmissible and rapidly spread to pigs throughout North America [29]

different from that in the USA and was characterized by the introduction of a novel wholly avian H1N1 virus in 1979 [31]. This avian H1N1 influenza virus was very successful in pigs in Europe, established a permanent lineage, and replaced classical swine H1N1 influenza virus. Shortly after the introduction of avian H1N1 influenza virus into pigs, reassortants with human H3N2 influenza viruses were detected in pigs in Italy [A/swine/Italy/526/83 (H3N2)]. These double reassortants possessed the HA and NA from human influenza virus and the remaining gene segments from the avian 1979 influenza virus [32]. Both the avian-like swine influenza virus and the double reassortant possessing the human HA and NA continued to circulate in pigs in Italy through 1979 and had a tendency to reassort with human H1N1 and H3N2. Although no clinically apparent human infections were reported in humans working with pigs in Italy, serological studies showed that 20% of them had serological evidence of infection with A/Port Chalmers/1/73 (H3N2) – a virus that had not circulated in humans for 20 years. A control group of persons not working with swine did not show these antibodies [33]. Infection of two children with mild respiratory diseases in the Netherlands in 1993 was caused by an H3N2 virus antigenically like A/Port Chalmers/1/73 with the avian-swine-like genome [34].

Reassortant influenza viruses possessing the HA from the circulating human H1N1 virus, the N2 from swine, and the internal genes from the circulating avian influenza virus in pigs were isolated from swine in Britain in 1994 [35]. In Germany, an H1N2 that was a reassortant between swine H1N2 and swine H3N2 was isolated from pigs in 2005 [36].

Thus, a different lineage of avian H1N1 was present in pigs in Europe but like the American swine virus with avian genes had a propensity to reassort with the currently circulating human influenza viruses.

7 Asian Swine Influenza Viruses

Each of the swine influenza viruses that established stable transmissible lineages in the USA and Europe has been detected in pigs in Asia. In addition, several swine influenza lineages unique to Asia have been detected.

The first transmission of human H3N2 to swine was detected in Taiwan soon after it appeared in humans in 1968 [37]. This transmission of H3N2 virus to swine was very successful and its descendants have been maintained in pigs in China through the present time [38]. It is noteworthy that these H3N2 viruses remained antigenically conserved with little change in over 30 years presumably due to the short life span of the majority of the pigs and the absence of immune selection.

Classical swine influenza virus and the triple reassortant from USA were introduced into Asia – presumably by importation of American swine breeding stock [39]. A novel H1N1 influenza virus containing all genes of avian origin was detected in pigs in Southern China in 1996, but this lineage has apparently died out [40].

In Thailand, reassortants of classical swine influenza virus and the Eurasian avian-like swine lineage have been reported. The HA, NA, and NS genes were from classical swine influenza virus and the remaining gene segments were from the European avian-like swine influenza virus [41, 42]. It is noteworthy that these reassortants were isolated from humans in Thailand and the Philippines [43]. Thus, the recent influenza viruses from swine in Asia have properties similar to influenza viruses from America in that they reassorted freely with human H3N2 viruses and had the capacity to infect humans.

In addition to the European and American swine lineage influenza viruses in pigs in Asia, two avian influenza viruses have been isolated; these include H5N1 influenza viruses [44, 45] and H9N2 viruses [46]. Both of these viruses (H5N1 and H9N2) have transiently transmitted to humans but probably directly from avian sources [2]. Neither the H5N1 nor the H9N2 influenza viruses have established stable lineages in pigs or have consistently transmitted in pigs or people.

8   Pandemics in Humans

From the above considerations, it appears that avian H1N1 influenza viruses have a propensity to transmit to pigs. This occurred during the emergence of the H1N1 1918 Spanish influenza virus [8], in 1979 during the emergence of the Eurasian avian influenza virus that became established in pigs [31], and prior to 1993 in Asia that did not persist in pigs [40]. The precursors of the avian influenza virus genes in the American triple reassortant that established itself in pigs in the USA in 1998 are unresolved; whether the H1N1 viruses from turkeys in the USA were involved remains to be established.

Whether transmission to pigs is a reoccurring intermediate step in the transmission of H1N1 influenza virus to humans or from humans to pigs is unresolved. Regardless it is apparent that the influenza viruses of pigs and people frequently exchange (Fig. 4) and participation of a PB1 gene of avian origin is involved. The human pandemics and epidemics caused by H1N1 in the past century include the 1918 Spanish influenza pandemic, the H1N1 Russian epidemic, and the 2009 pandemic that is ongoing. The 1918 Spanish influenza has been described as the "mother of all pandemics" [17] having killed directly or indirectly between 20 and 50 million persons worldwide. An initial mild wave in the spring of 1918 was replaced by a lethal wave in the fall. Determination of the complete nucleotide sequence of the 1918 Spanish influenza virus by Jeffrey Taubenberger and associates has permitted reconstruction of the 1918 influenza virus and establishment of its biological properties in mice, ferrets, and macaques [47–49]. However, to date no human archeological material of the mild 1918 wave has been sequenced; consequently we have no knowledge of the molecular changes that occurred or which gene segments were involved.

The reemergence of the H1N1 virus that disappeared in 1957 after the emergence of the H2N2 Asian influenza virus was in all likelihood a laboratory accident.

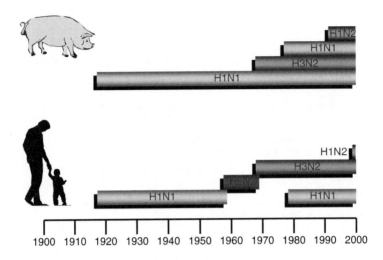

Fig. 4 Influenza in people and pigs. Each of the pandemic influenza viruses of humans since 1900 has spread to pigs. The exception may be the H2N2 Asian 1957 virus. The index human case of H2N2 in South Central China reported that pigs in the village were sick

The reintroduced Russian 1977 H1N1 virus affected mainly children and young adults born during the 27 years that the virus was frozen. The surprising feature of the reintroduced H1N1 virus was that it competed with the then circulating H3N2 virus and established a successful parallel lineage.

The detection of two cases of swine-like H1N1 in humans in Southern California in April 2009 although unusual was not unprecedented. However, when the virus was characterized as a novel H1N1 influenza virus and associated with widespread respiratory diseases in humans in Mexico, a pandemic threat was declared by the World Health Organization (WHO) [50, 51]. The novel H1N1 possessed the HA and the internal gene segments from the descendants of the triple reassortant influenza virus circulating in pigs in the USA and the neuraminidase (NA) and matrix (M) gene of Eurasian avian-like swine influenza virus (Fig. 1). This virus spread rapidly in humans and by June 11, 2009, the WHO declared a pandemic situation. The novel H1N1 virus of swine origin rapidly spread globally causing a summer wave of illness in the USA and Europe and rapidly became the dominant influenza virus strain in the southern hemisphere. Humans of middle and younger age groups are most susceptible to infection, whereas those over 60 are less affected and those over 80 are immune [52]. In healthy middle-aged people, the disease signs are generally similar to seasonal influenza, but persons with health complications as well as pregnant women or obese people are at increased risk. Information from the Australian experience with the first winter wave of pandemic H1N1 influenza indicated that the virus killed twice as many children less than 10 years than seasonal influenza – 61% of the children had no underlying medical condition.

The hospitalization rate was 45.7 per 100,000 for boys under 5 years and 35.4 for girls. There was a marked increase in hospitalization in women aged 15–34 due to the vulnerability of pregnant women. One of the two characteristics of the pandemic H1N1 virus that is different from seasonal H1N1 is the capacity to replicate deep in the lungs which can result in pneumonia as well as long-term virus shedding and increased patient loads in hospital intensive care facilities.

Antigenic and molecular characterization of the pandemic H1N1 influenza virus indicates that the virus has remained antigenically stable and essentially identical with the prototype A/California/4/09 (H1N1) virus [50, 51]. None of the molecular markers in the HA, NA, PB2, or NS genes associated with high pathogenicity of the 1918 Spanish influenza or H5N1 influenza has been detected in the pandemic H1N1 [50, 51]. Why then is the pandemic H1N1 killing more young people and pregnant women? It is clear that there are other characteristics of high pathogenicity that remain to be elucidated and are likely multigenic.

9   Pandemic H1N1 2009 in Pigs and Poultry

Phylogenetic analysis of the pandemic H1N1 using Bayesian molecular clock methods indicates that the closest ancestors of the virus existed 9.2–17.2 years ago [18]. This indicates that the ancestors of the current pandemic H1N1 virus have been circulating in pigs for over a decade. Despite intensive planning for the current pandemic, there is an enormous gap in our knowledge of influenza in swine globally. Consideration should be given to the establishment of a global prospective surveillance system in pigs similar to the Global Influenza Surveillance Network (GISN) in humans.

Perhaps the best ongoing influenza surveillance in pigs was initiated in Hong Kong in 1998 after the emergence of the novel H5N1 avian influenza virus that transmitted to 18 humans. In that program, 526 nasal and tracheal swab samples and 100 sera from apparently healthy pigs collected at the central slaughterhouse are analyzed virologically and serologically monthly. Although the virus isolation rate is low (~1%), the serological rate approaches 50%. Virological analysis has provided a gold mine of information on the genesis of swine influenza virus in Asia [18]. These studies confirmed the presence of European swine influenza viruses, American swine and human-like swine lineage viruses in pigs in Southeast Asia. A novel reassortant A/Swine/Hong Kong/415/04 (H1N2) possessing the triple reassortant swine influenza backbone from the USA and the matrix gene from the European swine lineage was isolated in 2004 indicating reassortment between European and American swine lineage influenza viruses.

The lack of swine influenza virus surveillance in South and Central America leaves the place of origin of the 2009 novel H1N1 virus open to speculation. Has the European swine influenza lineage been circulating in pigs in South and Central America together with the triple reassortant from North America swine? Alternatively was the novel H1N1 2009 virus generated in Asia and carried by inapparent infection in humans to Mexico? Future studies of the swine influenza viruses from Hong Kong and virological surveillance in South America should answer these questions.

Transmission of the novel H1N1 from humans to pigs has already occurred in multiple countries including Canada, Australia, Argentina, and Ireland. Experimental and field studies indicate that in pigs the virus causes moderate respiratory disease similar to classical and triple reassortant swine influenza. As in humans, the virus tends to replicate deep in the lungs and cause pneumonia but is not isolated outside the respiratory tract or from the intestines. Virus shedding tends to be longer than for classical or triple reassortant swine influenza viruses and has been detected for up to 16 days (Ian Brown, personal communication).

The novel 2009 H1N1 virus has also been isolated from turkeys in Chile and Canada where it causes mild infection and a drop in egg production (http://www. oie.int/wahis/public.php?page=weekly_report_index&admin=0). Thus, the novel H1N1 2009 is behaving much like earlier swine influenza viruses, and it is inevitable that this virus will spread to pigs and turkeys globally.

There is reluctance by the pork industry in the USA to initiate prospective surveillance of apparently healthy pigs for novel 2009 H1N1 influenza virus. The difficulty is related to a possible drop in pork consumption as occurred in Asia in 2009. The reports of the novel H1N1 in pigs referred to in Asia as swine influenza caused a two-thirds reduction in the consumption of pork. The novel H1N1 2009 virus is still referred to as swine influenza in Asia and the purchase of pork has normalized. Pork is perfectly safe to consumers and the problem is one of education and public relations for swine influenza has been part of the pork industry in the USA for nearly 100 years. It is necessary from a public health perspective to initiate prospective surveillance in apparently healthy pigs globally, along the line of the GISN program of human surveillance. It would be a catastrophe if variation in virulence or antigenicity of the H1N1 pandemic influenza virus occurred in pigs and was not detected until humans again served as their own sentinels.

10 Perspective

Influenza in people and pigs is closely intermingled with exchange of viruses in both directions. The optimal strategy for the control of influenza in people is the use of vaccines which are covered in detail in other chapters. Since transmission of the novel 2009 H1N1 virus from people to pigs has occurred in multiple countries, persons working with pigs should be included in a high priority group to receive the novel H1N1 vaccine. If the disease signs in pigs remain mild with mortality less than 1%, it is unlikely that a vaccine will be widely used. However, if the morbidity in pigs approaches 100% and if the severity of diseases increases, then a vaccine will be sought.

The major unresolved issues continue to be whether the novel 2009 H1N1 will:

• Become more virulent as happened with the 1918 Spanish H1N1 influenza strain

- Acquire resistance to oseltamivir and zanamivir
- Show rapid antigenic drift
- Evolve "silently" in swine or poultry and go undetected

Although the pandemic 2009 H1N1 influenza virus has remained antigenically stable, the detection of multiple oseltamivir-resistant variants that to date are sensitive to zanamivir and are not establishing transmissible lineages indicates that variants are occurring but as yet have no survival advantage. History has taught us that each of the above scenarios is possible either by mutation or by reassortant. The presently circulating pandemic H1N1 is more severe in both healthy children and medically compromised individuals. It is essential that virus surveillance and characterization is done both in humans and in pigs and poultry at the human animal interface so that we do not again fail to detect what is ongoing in the lower animal reservoir.

Acknowledgments This study was supported by contract HHSN266200700005C from the National Institute of Allergy and Infectious Diseases, National Institutes of Health, Department of Health and Human Services, and the American Lebanese Syrian Associated Charities (ALSAC). The authors thank James Knowles for help with manuscript preparation and Elizabeth Stevens for the figures. Biomedical Advanced Research and Development Authority (BARDA) provided support for Robert G. Webster and Michael Perdue.

References

1. de Jong JC, Claas EC, Osterhaus AD, Webster RG, Lim WL (1997) A pandemic warning? Nature 389:554
2. Peiris JS, de Jong MD, Guan Y (2007) Avian influenza virus (H5N1): a threat to human health. Clin Microbiol Rev 20:243–267
3. Slemons RD, Johnson DC, Osborn JS, Hayes F (1974) Type A influenza viruses from wild free-flying ducks in California. Avian Dis 18:119–124
4. Webster RG, Bean WJ, Gorman OT, Chambers TM, Kawaoka Y (1992) Evolution and ecology of influenza A viruses. Microbiol Rev 56:152–179
5. Olsen B, Munster VJ, Wallensten A, Waldenström J, Osterhaus AD, Fouchier RA (2006) Global patterns of influenza A virus in wild birds. Science 312:384–388
6. Stallknecht DE, Kearny SSM, MT ZPJ (1990) Persistence of avian influenza viruses in water. Avian Dis 34:406–411
7. Kawaoka Y, Krauss S, Webster RG (1989) Avian-to-human transmission of the PB1 gene of influenza A viruses in the 1957 and 1968 pandemics. J Virol 63:4603–4608
8. Taubenberger JK, Reid AH, Fanning TG (2000) The 1918 influenza virus: a killer comes into view. Virology 274:241–245
9. Nakajima K, Desselberger U, Palese P (1978) Recent human influenza A (H1N1) viruses are closely related genetically to strains isolated in 1950. Nature 274:334–339
10. Scholtissek C (1990) Pigs as the "mixing vessel" for the creation of new pandemic influenza A viruses. Med Princ Pract 2:65–71
11. Ito T, Couceiro JN, Kelm S, Baum LG, Krauss S, Castrucci MR, Donatelli I, Kida H, Paulson JC, Webster RG, Kawaoka Y (1998) Molecular basis for the generation in pigs of influenza A viruses with pandemic potential. J Virol 72:7367–7373

12. Kida H, Ito T, Yasuda J, Shimizu Y, Itakura C, Shortridge KF, Kawaoka Y, Webster RG (1994) Potential for transmission of avian influenza viruses to pigs. J Gen Virol 75:2183–2188

13. Matrosovich MN, Krauss S, Webster RG (2001) H9N2 influenza A viruses from poultry in Asia have human virus-like receptor specificity. Virology 281:156–162

14. Humberd J, Guan Y, Webster RG (2006) Comparison of the replication of influenza A viruses in Chinese ring-necked pheasants and chukar partridges. J Virol 80:2151–2161

15. Webster RG (2004) Wet markets – a continuing source of severe acute respiratory syndrome and influenza? Lancet 363:234–236

16. Kung NY, Guan Y, Perkins NR, Bissett L, Ellis T, Sims L, Morris RS, Shortridge KF, Peiris JS (2003) The impact of a monthly rest day on avian influenza virus isolation rates in retail live poultry markets in Hong Kong. Avian Dis 47:1037–1041

17. Taubenberger JK, Morens DM (2006) 1918 Influenza: the mother of all pandemics. Emerg Infect Dis 12:15–22

18. Smith GJ, Bahl J, Vijaykrishna D, Zhang J, Poon LL, Chen H, Webster RG, Peiris JS, Guan Y (2009) Dating the emergence of pandemic influenza viruses. Proc Natl Acad Sci USA 106:11709–11712

19. Shope RE (1931) Swine influenza. III. Filtration experiments and aetiology. J Exp Med 54:373–385

20. Easterday BC, Hinshaw VS (1992) Swine influenza. In: Leman AD, Straw BE, Mengeling WL, D'Allaire SD, Taylor DJ (eds) Diseases of swine, 7th edn. Iowa State University Press, Ames, pp 349–357

21. Easterday BC (1981) Swine influenza. In: Leman AD, Glock RD, Mengeling WL, Penny RHC, Scholl E, Straw B (eds) Diseases of swine, 5th edn. Iowa State University Press, Ames, pp 184–194

22. Myers KP, Olsen CW, Gray GC (2007) Cases of swine influenza in humans: a review of the literature. Clin Infect Dis 44:1084–1088

23. Top FH Jr, Russell PK (1977) Swine influenza A at Fort Dix, New Jersey (January–February 1976). IV. Summary and speculation. J Infect Dis 136:S376–S380

24. Langmuir AD (1979) Guillain–Barré syndrome: the swine influenza virus vaccine incident in the United States of America, 1976–77: preliminary communication. J R Soc Med 72:660–669

25. Langmuir AD, Bregman DJ, Kurland LT, Nathanson N, Victor M (1984) An epidemiologic and clinical evaluation of Guillain–Barré syndrome reported in association with the administration of swine influenza vaccines. Am J Epidemiol 119:841–879

26. Hinshaw VS, Webster RG, Bean WJ, Downie J, Senne DA (1983) Swine influenza-like viruses in turkeys: potential source of virus for humans? Science 220:206–208

27. Wright SM, Kawaoka Y, Sharp GB, Senne DA, Webster RG (1992) Interspecies transmission and reassortment of influenza A viruses in pigs and turkeys in the United States. Am J Epidemiol 136:488–497

28. Zhou NN, Senne DA, Landgraf JS, Swenson SL, Erickson G, Rossow K et al (1999) Genetic reassortment of avian, swine, and human influenza A viruses in American pigs. J Virol 73:8851–8856

29. Webby RJ, Swenson SL, Krauss SL, Gerrish PJ, Goyal SM, Webster RG (2000) Evolution of swine H3N2 influenza viruses in the United States. J Virol 74:8243–8251

30. Nardelli L, Pascucci S, Gualandi GL, Loda P (1978) Outbreaks of classical swine influenza in Italy in 1976. Zentralbl Veterinärmed B 25:853–857

31. Pensaert M, Ottis K, Vandeputte J, Kaplan MM, Bachmann PA (1981) Evidence for the natural transmission of influenza A virus from wild ducts to swine and its potential importance for man. Bull World Health Organ 59:75–78

32. Castrucci MR, Donatelli I, Sidoli L, Barigazzi G, Kawaoka Y, Webster RG (1993) Genetic reassortment between avian and human influenza A viruses in Italian pigs. Virology 193:503–506

33. Campitelli L, Donatelli I, Foni E, Castrucci MR, Fabiani C, Kawaoka Y, Krauss S, Webster RG (1997) Continued evolution of H1N1 and H3N2 influenza viruses in pigs in Italy. Virology 232:310–318

34. Claas EC, Kawaoka Y, de Jong JC, Masurel N, Webster RG (1994) Infection of children with avian–human reassortant influenza virus from pigs in Europe. Virology 204:453–457

35. Brown IH, Harris PA, McCauley JW, Alexander DJ (1998) Multiple genetic reassortment of avian and human influenza A viruses in European pigs, resulting in the emergence of an H1N2 virus of novel genotype. J Gen Virol 79:2947–2955

36. Zell R, Motzke S, Krumbholz A, Wutzler P, Herwig V, Durrwald R (2008) Novel reassortant of swine influenza H1N2 virus in Germany. J Gen Virol 89:271–276

37. Kundin WD (1970) Hong Kong A-2 influenza virus infection among swine during a human epidemic in Taiwan. Nature 228:857

38. Yu H, Zhang GH, Hua RH, Zhang Q, Liu TQ, Liao M, Tong GZ (2007) Isolation and genetic analysis of human origin H1N1 and H3N2 influenza viruses from pigs in China. Biochem Biophys Res Commun 356:91–96

39. Lee CS, Kang BK, Kim HK, Park SJ, Park BK, Jung K, Song DS (2008) Phylogenetic analysis of swine influenza viruses recently isolated in Korea. Virus Genes 37:168–176

40. Guan Y, Shortridge KF, Krauss S, Li PH, Kawaoka Y, Webster RG (1996) Emergence of avian H1N1 influenza viruses in pigs in China. J Virol 70:8041–8046

41. Chutinimitkul S, Thippamom N, Damrongwatanapokin S, Payungporn S, Thanawongnuwech R, Amonsin A et al (2008) Genetic characterization of H1N1, H1N2 and H3N2 swine influenza virus in Thailand. Arch Virol 153:1049–1056

42. Takemae N, Parchariyanon S, Damrongwatanapokin S, Uchida Y, Ruttanapumma R, Watanabe C et al (2008) Genetic diversity of swine influenza viruses isolated from pigs during 2000 to 2005 in Thailand. Influenza Other Respi Viruses 2:181–189

43. Komadina N, Roque V, Thawatsupha P, Rimando-Magalong J, Waicharoen S, Bomasang E et al (2007) Genetic analysis of two influenza A (H1) swine viruses isolated from humans in Thailand and the Philippines. Virus Genes 35:161–165

44. Choi YK, Nguyen TD, Ozaki H, Webby RJ, Puthavathana P, Buranathal C et al (2005) Studies of H5N1 influenza virus infection of pigs by using viruses isolated in Vietnam and Thailand in 2004. J Virol 79:10821–10825

45. Takano R, Nidom CA, Kiso M, Muramoto Y, Yamada S, Shinya K et al (2009) A comparison of the pathogenicity of avian and swine H5N1 influenza viruses in Indonesia. Arch Virol 154:677–681

46. Cong YL, Pu J, Liu QF, Wang S, Zhang GZ, Zhang XL et al (2007) Antigenic and genetic characterization of H9N2 swine influenza viruses in China. J Gen Virol 88:2035–2041

47. Tumpey TM, Basler CF, Aguilar PV, Zeng H, Solórzano A, Swayne DE, Cox NJ, Katz JM, Taubenberger JK, Palese P, García-Sastre A (2005) Characterization of the reconstructed 1918 Spanish influenza pandemic virus. Science 310:77–80

48. Tumpey TM, García-Sastre A, Taubenberger JK, Palese P, Swayne DE, Pantin-Jackwood MJ, Schultz-Cherry S, Solórzano A, Van Rooijen N, Katz JM, Basler CF (2005) Pathogenicity of influenza viruses with genes from the 1918 pandemic virus: functional roles of alveolar macrophages and neutrophils in limiting virus replication and mortality in mice. J Virol 79:14933–14944

49. Kobasa D, Jones SM, Shinya K, Kash JC, Copps J, Ebihara H, Hatta Y, Kim JH, Halfmann P, Hatta M, Feldmann F, Alimonti JB, Fernando L, Li Y, Katze MG, Feldmann H, Kawaoka Y (2007) Aberrant innate immune response in lethal infection of macaques with the 1918 influenza virus. Nature 445:319–323

50. Garten RJ, Davis CT, Russell CA, Shu B, Lindstrom S, Balish A, Sessions WM, Xu X, Skepner E, Deyde V, Okomo-Adhiambo M, Gubareva L, Barnes J, Smith CB, Emery SL, Hillman MJ, Rivailler P, Smagala J, de Graaf M, Burke DF, Fouchier RA, Pappas C, Alpuche-Aranda CM, López-Gatell H, Olivera H, López I, Myers CA, Faix D, Blair PJ, Yu C, Keene KM, Dotson PD Jr, Boxrud D, Sambol AR, Abid SH, St George K, Bannerman T,

Moore AL, Stringer DJ, Blevins P, Demmler-Harrison GJ, Ginsberg M, Kriner P, Waterman S, Smole S, Guevara HF, Belongia EA, Clark PA, Beatrice ST, Donis R, Katz J, Finelli L, Bridges CB, Shaw M, Jernigan DB, Uyeki TM, Smith DJ, Klimov AI, Cox NJ (2009) Antigenic and genetic characteristics of swine-origin 2009 A(H1N1) influenza viruses circulating in humans. Science 325:197–201
51. Neumann G, Noda T, Kawaoka Y (2009) Emergence and pandemic potential of swine-origin H1N1 influenza virus. Nature 459:931–939
52. Centers for Disease Control and Prevention (CDC) (2009) Serum cross-reactive antibody response to a novel influenza A (H1N1) virus after vaccination with seasonal influenza vaccine. MMWR Morb Mortal Wkly Rep 58:521–524

The Emergence of 2009 H1N1 Pandemic Influenza

Benjamin Greenbaum, Vladimir Trifonov, Hossein Khiabanian, Arnold Levine, and Raul Rabadan

Abstract The emergence of a novel H1N1 virus in Mexico and the USA in spring 2009 and its rapid spread around the globe has led the World Health Organization to declare the first pandemic of the twenty-first century. Employing almost real-time sequencing technologies and disseminating this information freely and widely has permitted the most intensive investigation of the origins and evolution of an influenza pandemic in the history of this disease. The small levels of sequence diversity of the first isolates permitted a realistic estimate of when the 2009 H1N1 virus first entered the human population. The rate of change in influenza RNA sequences permitted several groups to trace the origins of this virus to swine and a reassortment of North American and Eurasian swine influenza. These virus strains in turn have been traced back to swine, avian, and human virus reassortments occurring years ago in swine, all the way back to the 1918–1930 H1N1 viruses. The influenza virus sequence information spans the dimensions of time (90 years), space (locations all over the world), and hosts (birds, humans, swine, etc.). The high evolutionary rate of this virus and the growing amount of information is allowing researchers to follow its changes in the search for possible factors that could contribute to an increase in its virulence.

1 Introduction

In March 2009 a number of cases of acute respiratory illness were identified in Mexico as a novel H1N1 influenza strain, currently referred to as S-OIV H1N1, H1N1 pdm, 2009 H1N1 or swine flu in the media. Along with this news came

B. Greenbaum and A. Levine
The Simons Center for Systems Biology, Institute for Advanced Study, Princeton, NJ, USA
V. Trifonov, H. Khiabanian and R. Rabadan (✉)
Department of Biomedical Informatics, Center for Computational Biology and Bioinformatics, Columbia University College of Physicians and Surgeons, New York, NY, USA
e-mail: rabadan@dbmi.columbia.edu

G. Del Giudice and R. Rappuoli (eds.), *Influenza Vaccines for the Future*, 2nd edition, 95
Birkhäuser Advances in Infectious Diseases,
DOI 10.1007/978-3-0346-0279-2_5, © Springer Basel AG 2011

various, sometimes conflicting, reports about the new strain's origins and virulence. These were soon followed by reports of a number of cases in the USA during April, initially in California and Texas [1, 2], but quickly throughout the rest of the country. By mid-June 76 countries and all American states had reported cases of the novel strain with particularly high concentrations in a set of densely populated areas [3]. Simultaneous community level outbreaks in multiple regions around the world raised the possibility that a new pandemic had emerged. On June 11, 2009, the World Health Organization (WHO) declared the first influenza pandemic of the twenty-first century [4]. In the USA, a number of cases of the novel strain continued to occur over the summer, while the overall number of flu cases decreased [5]. Almost all of these anomalous summer cases were identified as the new strain. Simultaneously, during the southern hemisphere's winter flu season there has been substantial circulation of the virus with some reports indicating that it may be the dominant circulating strain in regions closely monitored by the WHO.

Unlike the three pandemics of the twentieth century, this strain has emerged during an age when two important tools are available. Firstly, viral genomes can be rapidly sequenced, allowing the RNA genomes of new isolates to be available for researchers within days, even hours. Hence, one can track the genetic evolution of the virus in an almost real-time fashion, especially when compared to previous emerging diseases. Secondly, researchers have access to sequence data through the internet and large databases, such as the Influenza Virus Resource at the National Center for Biotechnology Information (NCBI) [6]. Within this resource are recent sequences from the new strain as well as multiple sequences from current and historical strains that have been circulating over the past 90 years in several different hosts. The historical strains include multiple avian and swine isolates, along with early isolates from the emergence of the human 1918 H1N1, 1957 H2N2, and 1968 H3N2 pandemics, with many samples of those strains' descendants as they continued to evolve in the human population. This project also coordinates dedicated efforts in a group of major cities in the northern and southern hemispheres for the 2009 H1N1 strain, providing numerous, frequently updated high-quality samples of the new strain.

As a consequence, this pandemic is the first to take place in the genomic information age, where viruses can be rapidly sequenced and, just as importantly, compared to both currently circulating and historical influenza strains in multiple host species. This has allowed researchers over the critical first few months of the pandemic to address the twin questions of where the virus came from and where it may be headed with tools that were completely nonexistent in previous pandemics. This is essentially an empirical test of previously held theories of how the virus evolves, in addition to addressing related practical questions about vaccination and surveillance. Many ideas about emergence and pandemics that were generated as a consequence of recent sequence data and sequencing of historical strains will now be utilized to understand this emerging pandemic. In this work we examine and review information about the new influenza strain during the first few months of the pandemic so as to understand its emergence and see how it compares with what was expected from studies of previous pandemic strains. Throughout this work, we highlight how these new tools have provided a background for studying an emerging influenza virus, while at the same time noting some holes in our current tools and information, whose repair would improve the response to future pandemics.

2 The Origins

From the first release of the viral sequences from California, Texas, and Mexico in April 2009 many groups applied different techniques to understand the origins of the new virus [7–10]. These techniques include clustering, different phylogenetic methods, and sequence alignment and similarity. The premises of all these methods are similar: compare the new sequences with the ones deposited in databases, e.g., the aforementioned NCBI database.

2.1 Ancestral Strains

Influenza is a segmented, single-stranded, negative-sense RNA virus with eight segments. When two influenza viruses coinfect the same cell, new viruses could be generated containing segments from both parental strains, known as reassortment. As a consequence of the reassortment process, the eight segments can have different evolutionary histories. Figure 1 shows a phylogenetic tree of the HA segment in the context of other H1 viruses. The HA of the recent H1N1 viruses is related to viruses that have been circulating in pigs since 1930, and probably dating back to 1918.

For the 2009 H1N1 virus a comparison of all segments with strains in the database shows a dual geographic origin, though both arms of this tree come from a swine origin, rather than from human or avian sources as had been speculated [7]. Six of the eight segments of the H1N1 2009 virus are closest to swine viruses that had circulated in North America, while segments encoding the neuraminidase and matrix proteins were most similar to swine strains circulating in Eurasia. This indicated that the virus was most likely to have originated as a reassortment between two circulating swine viruses from these regions.

Looking further back in history to the most closely related ancestors of these two sources for 2009 H1N1 among deposited strains, a more complicated picture begins to emerge (Fig. 2) [11]. Within the six segments related to strains found in North America the closest similarity is with H1N2 and H3N2 swine viruses isolated in several parts of the USA and Canada around the turn of the twenty-first century [12]. Swine H1N2 viruses were isolated since 1999 and were the result of a reassortment between swine H3N2 viruses and classic swine H1N1 viruses. Swine H3N2 viruses were the result of a triple reassortment between avian, human, and swine viruses. The segments of swine origin from classical H1N1 viruses either descended directly from or had a common ancestor to the 1918 pandemic [13–15]. Classic swine H1N1 strains dominate recorded strains from

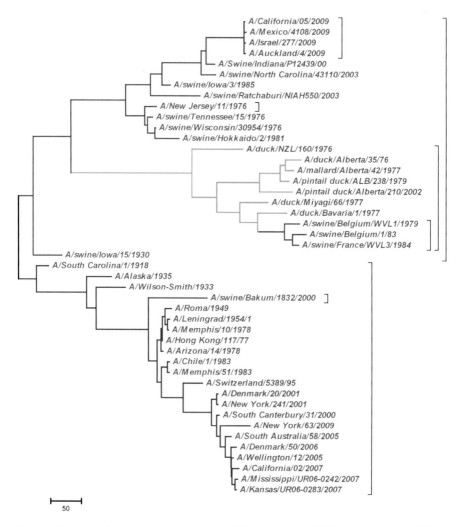

Fig. 1 Unrooted phylogenetic distance tree of HA segment in H1N1 viruses since 1918 HA isolated from humans is colored in blue, swine in pink, and birds in green

the earliest sequenced swine flu genomes, dating back from the first influenza isolates in the 1930s until the mid-1990s. The second component of the triple reassortant swine H3N2 is closest to human H3N2, originating in the 1968 pandemic, which was a reassortment between avian H3 influenza and the human H2N2 strain of the 1957 pandemic (itself a reassortment of descendents from the 1918 H1N1 and avian H2N2) [16]. The final components of this strain are avian in origin and are found in the polymerase complex segments PB2 and PA.

Since 1998, in addition to the classical H1N1 viruses, reassortant H3N2 and H1N2 viruses have been circulating in North American swine [17, 18]. The recent discovery of these swine influenza strains may be due to the fact that the number of

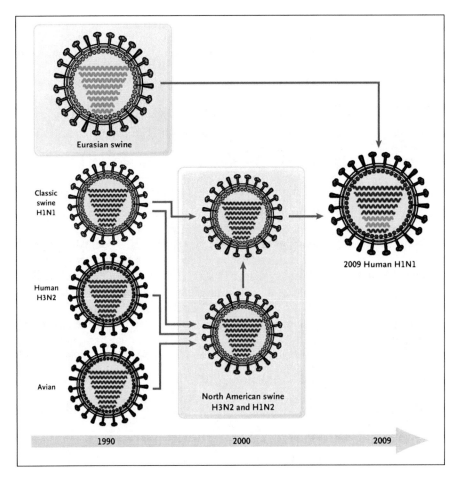

Fig. 2 The history of the recent ancestors of the 2009 H1N1 pandemic strain H1N1. The recent ancestors were isolated in swine in last two decades. Figure from Trifonov et al. [11]

sequences collected starting in the mid-1990s has increased exponentially as a result of greater surveillance in swine populations. It is clear that in recent years, all three strains have been cocirculating in swine. There have been sporadic cases of human infection with swine viruses without a major outbreak, typically among people in contact with pigs. These cases have been mostly asymptomatic compared to seasonal influenza, but with higher recorded incidences of diarrhea (three out of ten patients) than is usually expected. Diarrhea has also been reported in about 24% of American S-OIV cases to date [19].

The Eurasian ancestors of pandemic 2009 H1N1 are H1N1 swine viruses that have been circulating in swine since the end of the 1970s [20]. The origin of several segments of these viruses was probably avian. It is interesting to note that the relationship between pandemic H1N1 and Eurasian H1N1 swine viruses is distant, with the closest relatives dating to the 1990s. It is still unclear how these segments have passed unnoticed for more than a decade, probably reflecting the lack of systematic surveillance of swine viruses on a global scale, as shown in Fig. 3 [11].

As the recent history of the ancestors of pandemic H1N1 show, viruses reassort very frequently, especially swine viruses [21]. Pigs are documented to allow productive replication of human, avian, and swine influenza viruses. This picture of multiple cocirculating swine strains entering the human population after reassorting has increased the interest in the "mixing vessel" theory of swine influenza [22, 23]. This hypothesis asserts that, because swine can become infected with swine, human, and avian strains, it offers the greatest opportunity to generate diversity through reassortment. One possible explanation for the role of swine in reassortment events is due to the fact that epithelial cells in the upper respiratory tract of swine expresses both human and avian receptors [24].

The idea that emergent influenza viruses in humans typically come directly from pigs, as a mediator of human, avian, and swine strains, has been suggested for the origins for the 1957 and 1968 pandemic viruses, but is controversial for the 1918 pandemic strain, which appears to have a possible avian origin [15]. The current H1N1 pandemic also supports the swine origin of a mixed influenza virus strain with ancestors in swine, avian, and human strains. Hopefully, one outcome of the attention drawn by this pandemic will be better surveillance of swine viruses so that the appearance of new strains in swine and the frequency of reassortant strains in

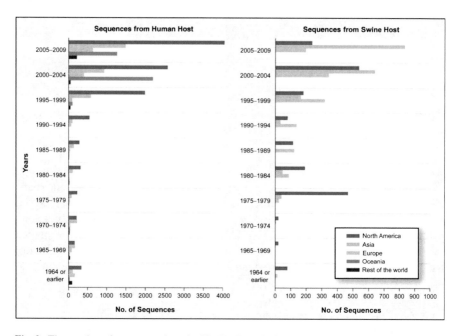

Fig. 3 The number of sequences deposited in GenBank indicate the geographic and time diversity of influenza isolates since the 1960s. Figure from Trifonov et al. [11]

swine can be better observed in real time. The 2009 H1N1 strain was probably circulating in swine populations prior to the outbreak but went undetected. The final answer to where and when this virus emerged in humans will only be solved when more data becomes available.

2.2   Recent Emergence

Influenza viruses, like many other single stranded RNA viruses, have very high evolutionary rates. An examination of the genome sequences from the first isolates of 2009 H1N1 from California, Mexico, and Texas showed a high degree of similarity, suggesting a very recent common ancestor. One can estimate the time to the most recent common ancestor by evaluating evolutionary rates that have been determined in the past and comparing the genomes of the different 2009 H1N1 isolates. To get an idea of the order of magnitude of this numbers we can compare two of the most distant early sequences, A/California/04/2009 and A/New York/18/2009 which were isolated on the 1st and the 25th of April 2009, respectively. The end of January 2009 can be considered to be a rough estimate of their most recent common ancestor, simply by observing that there are 23 differences between their 13 kb genomes and that previous estimates of evolutionary rates of change are 4–5×10^{-3} nucleotides per year [9, 15, 25, 26]. Employing Bayesian phylogenetic methods, which provide a good way of estimating the time to the most recent ancestor, these results suggest a common ancestor for April pandemic isolates dating to January or February 2009 [25–27]. This is compatible with the initial reports from Mexico of the start of the outbreak.

Phylogenetic and clustering techniques clearly show the initial formation of clades, as expected from the natural propagation of the virus but with the caveat that the sequences that are available are coming from only a few places in the world due to limited sampling. A second interesting aspect of the early stages of the epidemic was the branching pattern in sequences that occurred at different geographic locations. Namely a set of California strains segregated away from the others in sequence distributions indicating that, even at that early stage in the pandemic, a geographic segregation among early isolated strains was already beginning. As a result, the hallmark strain A/California/07/2009 always appears as part of a distinct group. As the number of viruses has continued to grow since the emergence of the strain, this segregation, which was apparent early on, has continued to be observed and expanded [28]. The A/California/07/2009 strain is more closely related to several circulating Mexican strains then it is to the New York strains, causing speculation that the California strain may be the start of the pandemic or at least part of the original "clade" and the New York strains may be from a somewhat later cluster. Nonetheless, this early appearance of differentiation teaches a valuable lesson about influenza's ability to mutate rapidly, which we will address again later.

2.3   The Increasing Diversity of the Pandemic Virus

As the virus spreads and mutates, the viral population diversifies. Figure 4 shows the increase in the number of viral isolates that were deposited in GenBank since late March 2009 (left) and how it corresponds to the diversity, measured as the number of polymorphic sites in hemagglutinin. The number of polymorphic positions generated in a segment per unit length, S, should be proportional to size of the viral population N:

$$\frac{\partial S}{\partial t} = \kappa N(t)(1 - S)$$

This diversification of the whole genome is illustrated by the phylogenetic tree shown in Fig. 5.

The amount of variation is somewhat different for different segments. In particular, HA seems to accumulate substitutions faster than other segments, suggesting that selection is playing a role in this protein. Table 1 shows a list of site-by-site differences for the eight chromosomes among the two aforementioned strains isolated in April 2009: New York/18 and California/04. As previously noted, there are 23 changes between the two isolates, with 14 of those changes being neutral (no amino acid changes). There are two or three differences per segment, with the majority causing nonsynonymous changes in fairly established patterns. The majority of the seven nonsynonymous changes occurred in HA. All four of the coding changes recorded here lie in the HA1 domain, which encodes the exposed and epitope containing portion of HA, corroborating previous studies across influenza that this region is subject to a greater degree of positive selection than the rest of the virus [29]. This includes one of the few observed transitions at position 658.

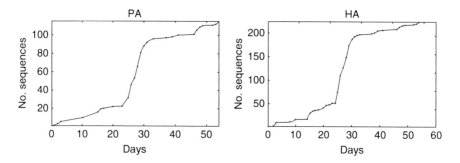

Fig. 4 The increase in the number of sequences corresponds to the increase in polymorphic sites

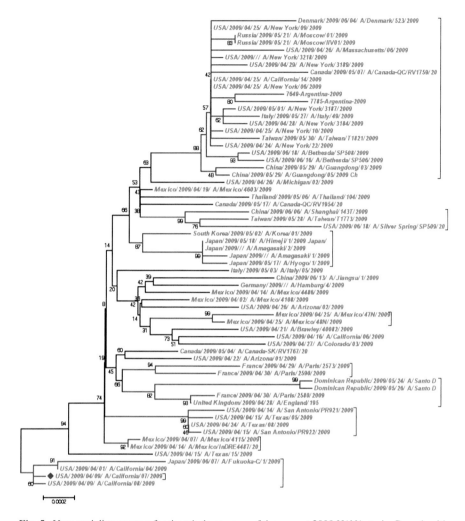

Fig. 5 Unrooted distance tree for the whole genome of the current 2009 H1N1 strain. Branch with more than 85% confidence from bootstrapping are highlighted in red. The aforementioned California/07 strain is marked by a blue diamond

3 Pathogenicity

The pathogenicity of the new virus was not clear when it was first identified. Although much remains to be revealed about the causes of pathogenicity for influenza A viruses in general, it has become clear that it depends on multiple genes and differs between hosts. Since the HA protein mediates the binding of viral particles to the host cell, the interplay between receptors exposed on the cell surface and the receptor-binding specificity of the HA protein play an important role in the

Table 1 List of nucleotide and amino acid differences between 2009 H1N1 strains New York/18 and California/04.

Segment	Position	Mutation New York/ 18-California/04	Amino acid
PB2	1218	A-G	–
	1872	T-C	–
	2163	A-G	–
PB1	1758	G-A	–
	2033	A-G	Asn-Ser
PA	670	T-C	Ser-Pro
	1986	G-T	–
HA	298	T-C	Ser-Pro
	640	G-A	Ala-Thr
	658	A-T	Thr-Ser
	891	G-A	–
	1012	G-A	Val-Il
	1408	T-C	–
NP	298	A-G	Il-Val
	1143	A-G	–
	1248	A-G	–
NA	317	A-G	Il-Val
	742	G-A	Asp-Asn
	1044	A-G	–
MP	492	A-G	–
	600	A-G	–
NS	366	G-A	–
	443	A-G	–

viral infectivity [30–36]. Another known source of pathogenicity related to the HA protein is its ability to be cleaved, as this event is required for viral infection and plays an important role in the release of the virus from the cell [37]. One might also suspect that the efficiency of the viral genome replication mechanism could be an important source of increased viral titer and virulence. Indeed, multiple sites in the PB2 gene have been confirmed as contributing to the infectivity of the influenza A virus [38–42].

A mechanism by which influenza counteracts the host innate immune response is via its NS1 protein, an interferon antagonist [43–47]. This is counteracted by specific, stimulatory nucleotide sequences in RNA segments that trigger the innate immune response to produce interferon and other cytokines, whose potentially toxic overstimulation could lead to increased virulence. As an example, recent studies suggest that the human innate immune system may induce selection against CpG in a sequence specific context in human influenza segments by stimulating innate receptors, while the innate immune system of birds does this less well, if at all [48, 49]. The result is that avian-like influenza viruses infecting humans likely produce more interferon and cytokines, inducing selection for viruses that avoid this trigger. While 1918 H1N1 and avian H5N1 present a high number of immunostimulatory motifs, the 2009 H1N1 pandemic virus shows a similar number to the previously seasonal H1N1 viruses, suggesting that the lack of these motifs could be part of the low virulence observed in the pandemic virus.

Another possible factor in pathogenicity is a protein in the PB1 segment called PB1-F2 encoded in the +1 reading frame [50–52]. The absence of a full-length PB1-F2 protein has been suggested to account for the low pathogenicity of 2009 H1N1 [53]. An analysis of the context this protein sequence within PB1 using Kozak's optimization rules for initiation of translation [54] shows that its poor expression is probably due to inefficient translation initiation. Changes in this sequence that regulate initiation of translation could enhance production of this protein. PB1-F2 induces apoptosis in human CD8[+] T cells and alveolar macrophages by binding to mitochondria [51, 55] and increases the severity of primary viral and secondary bacterial infections in mice [56, 57]. In isolates obtained since 1947 it is truncated and inactivated by the presence of stop codons in classical swine H1N1 virus and human H1N1 virus, as well as in 2009 H1N1. Its varying length leads one to question its significance to the evolutionary fitness of influenza. The evolution of PB1-F2 can be compared to PB1 and control reading frames within the same segment (Fig. 6) that do not appear to encode proteins [58]. The length of the controls is as conserved as PB1-F2. Furthermore, the probability of a long subsequence without stop codons in the +1 reading frame of a PB1 segment generated at random is more than 0.9. PB1, PB1-F2 and the control segments show similar

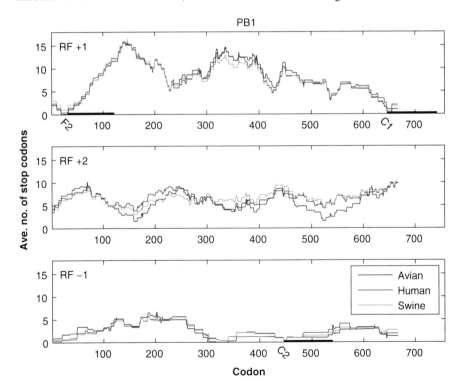

Fig. 6 Average number of stop codons in a window of length 90 for reading frames +1, +2, and -1 of the PB1 segment. Reading frame +1 contains PB1-F2 (codons 31-121) and contrl region PB1-C1 (codons 646-743). Reading frame -1 contains the control PB1-C2 (codons 446-540)

nucleotide evolutionary rates but very different rates at the amino acid level. This can be explained entirely by negative selection in PB1, as previously observed [59–61], implying that PB1-F2 has a similar contribution to the fitness of the virus as other, nontranslated, sequences and so is of little or no evolutionary significance.

4 Previous Immunity

When the seasonal H1N1 human influenza virus reemerged in 1977, the main concern was a lack of resistance in the population born after 1957, the year H1N1 was replaced by H2N2, as they had never been exposed to this viral subtype. As expected, the spreading epidemic was almost entirely restricted to this population. Even after more than two decades, when the publicly available extensive information on the age distribution of patients who show symptomatic disease from the two subtypes of seasonal human influenza across multiple geographical locations and seasons is pooled, striking differences emerge, indicating that symptomatic flu due to seasonal H1N1 virus is distributed mainly in a younger population relative to the seasonal H3N2 virus [62]. These observations can potentially explain why two different influenza strains can both circulate in the human population at the same time. The partitioning of the population into young and old hosts may have permitted both influenza stains to cocirculate. The more strains that can cocirculate in a population and move from human to pig and avian to pig, the greater the diversity of influenza strains generated and ultimately tested in the human population. The analysis here of the origins of the 2009 H1N1 virus reflects this diversity with contributions from H1N1 strains and H3N2 human and swine strains, as well as H1N2 swine. These distinct characteristic age groups are possibly carried over from previous pandemics and provide a recurring pattern of similar viruses that depends upon the generation of the host and the appearance of a young population that never was exposed to that virus strain. Perhaps there will come a time when a vaccine can be constructed that will anticipate this recurrent pattern of virus strains and break this pattern. If so it is likely a new pattern will nonetheless emerge as these viruses rapidly evolve in response to their host's immunity.

The preliminary studies regarding the age distribution of patents showing symptomatic flu from S-OIV H1N1 virus indicate a similar distribution to the seasonal H1N1, with the greater disease burden on the population younger than 25 years of age [63, 64]. It also has been suggested that older populations may have preexisting immunity to the novel virus. Likewise, the results from serological studies indicate that 33% of those aged more than 60 have cross-reactive antibody responses to 2009 H1N1, even before vaccination against the seasonal flu. A similar cross-reactive antibody response is observed only in 6% of those aged between 18 and 40, and no response exists among children. In addition, the vaccination against the seasonal flu from the past four seasons does not change the amount of this response in any of these age groups [65].

5   Conclusion

Current data implies that the closest ancestors of 2009 H1N1 human infuenza virus came from pigs a few years ago. Having circulated in pigs for several years the 2009 H1N1 strain probably emerged in humans very recently. The great similarity between the circulating strains of this virus in humans suggests a recent common ancestor which first emerged in January or February of 2009, though it remains unclear how and where the pandemic started. This enigma can only be solved if we fill in the gaps in swine influenza strains and more distant 2009 H1N1-like sequences are isolated from humans and swine. The analysis of the sequences collected during this pandemic has clearly demonstrated the imperfections of the surveillance system, especially in swine. Although pandemics could start anywhere on the planet, a more comprehensive surveillance system could help to quickly identify repeated isolations of potential strains that are candidates for breaking into a human population that has no immunity to that strain.

The more distant origins of the swine H1N1 virus that led to the human 2009 H1N1 virus have also been traced. The H1N1 swine influenza strains derived from a reassortment where six out of eight viral segments came from a North American swine influenza virus and two segments (neuraminidase and matrix) came from a Eurasian H1N1 strain of swine influenza virus. The former was itself a reassortment of avian, human, and swine influenza. This evolutionary history supports the role of swine in the reassortment of influenza virus strains from human, avian, and other swine.

Continued observation of the evolution and diversification of the new virus can alert us to possible changes that can affect its pathogenicity. Likewise, monitoring different viral clusters, particularly in vaccine target areas, can help improve the vaccine for the upcoming season. Mutations and reassortments in influenza make it unpredictable. Any infectious disease is the result of a complicated interplay of different factors, including the pathogen, the host, other possible pathogens that coinfect the same host, and the environment. The unprecedented amount of genomic information is the first step in understanding these complex host and pathogen dynamics. The availability of electronic health records will allow the integration of patient history into this picture [66]. Many new techniques are becoming available, including high-throughput RNA sequencing directly from a host without intervention of replication in culture or eggs imposing new selective forces upon viruses. In addition, sequencing procedures of total host and viral RNA species along with expression arrays can tell us a great deal about the response of the host, the innate immune system, and how differences between viruses impact upon the host. This type of procedure will detect single nucleotide polymorphisms or copy number variations that can alter the host immune response and result in the evolution of different viruses within that host. All of these new technologies and methods will allow researches to generate an integrated genetic and molecular picture of the disease beyond that provided by the traditional disciplines of virology, immunology, and epidemiology.

Acknowledgments B. Greenbaum would like to acknowledge the support of Eric and Wendy Schmidt. R. Rabadan and H. Khiabanian would like to acknowledge support from Eureka (Exceptional, Unconventional Research Enabling Knowledge Acceleration) grant number 1R01LM010140-01.

References

1. CDC (2009) Swine influenza A (H1N1) infection in two children – Southern California, March–April 2009. MMWR 58:400–402
2. CDC (2009) Update: swine influenza A (H1N1) infections – California and Texas, April 2009. MMWR 58(Dispatch):1–3
3. http://www.who.int/csr/don/2009_06_15/en/index.html
4. http://www.who.int/mediacentre/news/statements/2009/h1n1_pandemic_phase6_20090611/en/index.html
5. http://www.cdc.gov/flu/weekly/pdf/overview.pdf
6. Bao Y, Bolotov P, Dernovoy D, Kiryutin B, Zaslavsky L, Tatusova T, Ostell J, Lipman D (2008) The influenza virus resource at the National Center for Biotechnology Information. J Virol 82:596–601
7. Trifonov V, Khiabanian H, Greenbaum B, Rabadan R (2009) The origin of the recent swine influenza A(H1N1) virus infecting humans. Euro Surveill 14(17):pii=19193
8. Solovyov A, Palacios G, Briese T, Lipkin WI, Rabadan R (2009) Cluster analysis of the origins of the new influenza A(H1N1) virus. Euro Surveill 14(21):pii=19224
9. Smith GJ, Vijaykrishna D, Bahl J, Lycett SJ, Worobey M, Pybus OG, Ma SK, Cheung CL, Raghwani J, Bhatt S, Peiris JS, Guan Y, Rambaut A (2009) Origins and evolutionary genomics of the 2009 swine-origin H1N1 influenza A epidemic. Nature 459:1122–1125
10. Novel Swine-Origin Influenza A (H1N1) Virus Investigation Team, Dawood FS, Jain S, Finelli L, Shaw MW, Lindstrom S, Garten RJ, Gubareva LV, Xu X, Bridges CB, Uyeki TM (2009) Emergence of a novel swine-origin influenza A (H1N1) virus in humans. N Engl J Med 360:2605–2615, Erratum in: N Engl J Med 2009 361:102
11. Trifonov V, Khiabanian H, Rabadan R (2009) Geographic dependence, surveillance, and origins of the 2009 influenza A (H1N1) virus. N Engl J Med 361:115–119
12. Shinde V, Bridges CB, Uyeki TM et al (2009) Triple-reassortant swine influenza A (H1) in humans in the United States, 2005–2009. N Engl J Med 360:2616–2625
13. Chun J (1919) Influenza including its infection among pigs. Natl Med J 5:34–44
14. Dorset M, McBryde CN, Niles WB (1922) Remarks on hog flu. J Am Vet Med Assoc 62:162–171
15. Smith GJ, Bahl J, Vijaykrishna D, Zhang J, Poon LL, Chen H, Webster RG, Peiris JS, Guan Y (2009) Dating the emergence of pandemic influenza viruses. Proc Natl Acad Sci USA 106:11709–11712
16. Lindstrom SE, Cox N, Klimov A (2004) Evolutionary analysis of human H2N2 and early H3N2 viruses: evidence for genetic divergence and multiple reassortment among H2N2 and H3N2 viruses. Int Congr Ser 1263:184–190
17. Olsen CW (2002) The emergence of novel swine influenza viruses in North America. Virus Res 85:199–210
18. Vincent AL, Ma W, Lager KM, Janke BH, Richt JA (2008) Swine influenza viruses: a North American perspective. Adv Virus Res 72:127–154
19. http://www.cdc.gov/h1n1flu/surveillanceqa.htm
20. Pensaert M, Ottis K, Vandeputte J, Kaplan MM, Bachmann PA (1981) Evidence for the natural transmission of influenza A virus from wild ducks to swine and its potential importance for man. Bull World Health Organ 59:75–78

21. Khiabanian H, Trifonov V, Rabadan R (2009) Reassortment patterns in swine influenza viruses. PLoS ONE 4(10):e7366

22. Ma W, Kahn RE, Richt JA (2009) The pig as a mixing vessel for influenza viruses: human and veterinary implications. J Mol Genet Med 3:158–166

23. Scholtissek C (1990) Pigs as "mixing vessels" for the creation of new pandemic influenza A viruses. Med Princ Pract 2:65–71

24. Gambaryan AS, Karasin AI, Tuzikov AB, Chinarev AA, Pazynina GV, Bovin NV, Matrosovich MN, Olsen CW, Klimov AI (2005) Receptor-binding properties of swine influenza viruses isolated and propagated in MDCK cells. Virus Res 114:15–22

25. Rambaut A, Holmes E (2009) The early molecular epidemiology of the swine-origin A/H1N1 human influenza pandemic. PLoS Curr Influenza:RRN1003

26. Rambaut A, Pybus OG, Nelson MI, Viboud C, Taubenberger JK, Holmes EC (2008) The genomic and epidemiological dynamics of human influenza A virus. Nature 453:615–619

27. Lemey P, Suchard M, Rambaut A (2009) Reconstructing the initial global spread of a human influenza pandemic: a Bayesian spatial-temporal model for the global spread of H1N1pdm. PLoS Curr Influenza:RRN1031

28. Parks DH, MacDonald NJ, Beiko RG (2009) Tracking the evolution and geographic spread of Influenza A. PLoS Curr Influenza:RRN1014

29. Nelson MI, Holmes EC (2008) The evolution of epidemic influenza. Nat Genet 8:196–205

30. Rogers GN, Paulson JC (1983) Receptor determinants of human and animal influenza virus isolates: differences in receptor specificity of the H3 hemagglutinin based on species of origin. Virology 127:361–373

31. Ito T, Couceiro JN, Kelm S, Baum LG, Krauss S, Castrucci MR, Donatelli I, Kida H, Paulson JC, Webster RG, Kawaoka Y (1998) Molecular basis for the generation in pigs of influenza A viruses with pandemic potential. J Virol 72:7367–7373

32. Matrosovich M, Zhou N, Kawaoka Y, Webster R (1999) The surface glycoproteins of H5 influenza viruses isolated from humans, chickens, and wild aquatic birds have distinguishable properties. J Virol 73:1146–1155

33. Shinya K, Ebina M, Yamada S, Ono M, Kasai N, Kawaoka Y (2006) Avian flu: influenza virus receptors in the human airway. Nature 440:435–436

34. van Riel D, Munster VJ, de Wit E, Rimmelzwaan GF, Fouchier RA, Osterhaus AD, Kuiken T (2006) H5N1 virus attachment to lower respiratory tract. Science 312:399

35. Stevens J, Corper AL, Basler CF, Taubenberger JK, Palese P, Wilson IA (2004) Structure of the uncleaved human H1 hemagglutinin from the extinct 1918 influenza virus. Science 303:1866–1870

36. Tumpey TM, Maines TR, Van Hoeven N, Glaser L, Solórzano A, Pappas C, Cox NJ, Swayne DE, Palese P, Katz JM, García-Sastre A (2007) A two-amino acid change in the hemagglutinin of the 1918 influenza virus abolishes transmission. Science 315:655–659

37. Kawaoka Y, Webster RG (1988) Sequence requirements for cleavage activation of influenza virus hemagglutinin expressed in mammalian cells. Proc Natl Acad Sci USA 85:324–328

38. Subbarao EK, London W, Murphy BR (1993) A single amino acid in the PB2 gene of influenza A virus is a determinant of host range. J Virol 67:1761–1764

39. Gabriel G, Abram M, Keiner B, Wagner R, Klenk HD, Stech J (2007) Differential polymerase activity in avian and mammalian cells determines host range of influenza virus. J Virol 81:9601–9604

40. Van Hoeven N, Pappas C, Belser JA, Maines TR, Zeng H, García-Sastre A, Sasisekharan R, Katz JM, Tumpey TM (2009) Human HA and polymerase subunit PB2 proteins confer transmission of an avian influenza virus through the air. Proc Natl Acad Sci USA 106:3366–3371

41. Geiss GK, Salvatore M, Tumpey TM, Carter VS, Wang X, Basler CF, Taubenberger JK, Bumgarner RE, Palese P, Katze MG, García-Sastre A (2002) Cellular transcriptional profiling in influenza A virus-infected lung epithelial cells: the role of the nonstructural NS1 protein in

the evasion of the host innate defense and its potential contribution to pandemic influenza. Proc Natl Acad Sci USA 99:10736–10741

42. Steel J, Lowen AC, Mubareka S, Palese P (2009) Transmission of influenza virus in a mammalian host is increased by PB2 amino acids 627K or 627E/701N. PLoS Pathog 5: e1000252

43. Garcia-Sastre A (2001) Inhibition of interferon-mediated antiviral responses by influenza A viruses and other negative-strand RNA viruses. Virology 279:375–384

44. Pichlmair A, Schulz O, Tan CP, Näslund TI, Liljeström P, Weber F, Reis e Sousa C (2006) RIG-I-mediated antiviral responses to single-stranded RNA bearing 5′-phosphates. Science 314:997–1001

45. Diebold SS, Kaisho T, Hemmi H, Akira S, Reis e Sousa C (2004) Innate antiviral responses by means of TLR7-mediated recognition of single-stranded RNA. Science 303:1529–1531

46. Imai Y, Kuba K, Neely GG, Yaghubian-Malhami R, Perkmann T, van Loo G, Ermolaeva M, Veldhuizen R, Leung YH, Wang H, Liu H, Sun Y, Pasparakis M, Kopf M, Mech C, Bavari S, Peiris JS, Slutsky AS, Akira S, Hultqvist M, Holmdahl R, Nicholls J, Jiang C, Binder CJ, Penninger JM (2008) Identification of oxidative stress and Toll-like receptor 4 signaling as a key pathway of acute lung injury. Cell 133:235–249

47. Jackson D, Hossain MJ, Hickman D, Perez DR, Lamb RA (2008) A new influenza virus virulence determinant: the NS1 protein four C-terminal residues modulate pathogenicity. Proc Natl Acad Sci USA 105:4381–4386

48. Greenbaum BD, Levine AJ, Bhanot G, Rabadan R (2008) Patterns of evolution and host gene mimicry in influenza and other RNA viruses. PLoS Pathog 4:e1000079

49. Greenbaum BD, Rabadan R, Levine AJ (2009) Patterns of oligonucleotide sequences in viral and host cell RNA identify mediators of the host innate immune system. PLoS ONE 4:e5969

50. Chen W, Calvo PA, Malide D, Gibbs J, Schubert U, Bacik I, Basta S, O'Neill R, Schickli J, Palese P, Henklein P, Bennink JR, Yewdell JW (2001) A novel influenza A virus mitochondrial protein that induces cell death. Nat Med 7:1306–1312

51. Conenello GM, Palese P (2007) Influenza A virus PB1-F2: a small protein with a big punch. Cell Host Microbe 2:207–209

52. Conenello GM, Zamarin D, Perrone LA, Tumpey T, Palese P (2007) A single mutation in the PB1-F2 of H5N1 (HK/97) and 1918 influenza A viruses contributes to increased virulence. PLoS Pathog 3:e141

53. Taia T, Wang R, Palese P (2009) Unraveling the mystery of swine influenza virus. Cell 137:983–985

54. Kozak M (1991) Structural features in eukaryotic mRNAs that modulate the initiation of translation. J Biol Chem 266:19867–19870

55. Zamarin D, Garcia-Sastre A, Xiao X, Wang R, Palese P (2005) Influenza virus PB1-F2 protein induces cell death through mitochondrial ANT3 and VDAC1. PLoS Pathog 1:e4

56. Zamarin D, Ortigoza MB, Palese P (2006) Influenza A virus PB1-F2 protein contributes to viral pathogenesis in mice. J Virol 80:7976–7983

57. McAuley JL, Hornung F, Boyd KL, Smith AM, McKeon R, Bennink J, Yewdell JW, McCullers JA (2007) Expression of the 1918 influenza A virus PB1-F2 enhances the pathogenesis of viral and secondary bacterial pneumonia. Cell Host Microbe 2:240–249

58. Trifonov V, Racaniello V, Rabadan R (2009) The contribution of the PB1-F2 protein to the fitness of influenza A viruses and its recent evolution in the 2009 influenza A (H1N1) pandemic virus. PLoS Curr Influenza:RRN1006

59. Obenauer JC, Denson J, Mehta PK, Su X, Mukatira S, Finkelstein DB, Xu X, Wang J, Ma J, Fan Y, Rakestraw KM, Webster RG, Hoffmann E, Krauss S, Zheng J, Zhang Z, Naeve CW (2006) Large-scale sequence analysis of avian influenza isolates. Science 311:1576–1580

60. Obenauer JC, Fan Y, Naeve CW (2006) Response to comment on "Large-scale sequence analysis of avian influenza isolates". Science 313:1573

61. Holmes EC, Lipman DJ, Zamarin D, Yewdell JW (2006) Comment on "Large-scale sequence analysis of avian influenza isolates". Science 313:1573

62. Khiabanian H, Farrell G, St. George K, Rabadan R (2009) Differences in patient age distribution between influenza A subtypes. PLoS ONE 4(8):e6832
63. CDC (2009) Novel H1N1 flu: facts and figures. CDC, Atlanta. Available at http://www.cdc.gov/H1N1FLU/surveillanceqa.htm
64. Kelly H, Grant K, Williams S, Smith D (2009) H1N1 swine origin influenza infection in the United States and Europe in 2009 may be similar to H1N1 seasonal influenza infection in two Australian states in 2007 and 2008. Influenza Other Respir Viruses 3:183–188
65. CDC (2009) Serum cross-reactive antibody response to a novel influenza A (H1N1) virus after vaccination with seasonal influenza vaccine. MMWR Morb Mortal Wkly Rep 58(19): 521–524
66. Rabadan R, Mostashari F, Calman N, Hripcsak G (2009) Next generation syndromic surveillance: molecular epidemiology, electronic health records and the pandemic influenza A (H1N1) virus. PLoS Curr Influenza:RRN1012

Part II
Immunity and Vaccine Strategies

Influenza Vaccines Have a Short but Illustrious History of Dedicated Science Enabling the Rapid Global Production of A/Swine (H1N1) Vaccine in the Current Pandemic

John Oxford, Anthony Gilbert, and Robert Lambkin-Williams

Abstract Vaccines for the swine flu pandemic of 2009 have been produced in an exquisitely short time frame. This speed of production comes because of 50 years of hard work by virologists worldwide in pharma groups, research laboratories, and government licensing units. The present chapter presents the background framework of influenza vaccine production and its evolution over 50 years. Isolation of the causative virus of influenza in 1933, followed by the discovery of embryonated hen eggs as a substrate, quickly led to the formulation of vaccines. Virus-containing allantoic fluid was inactivated with formalin. The phenomenon of antigenic drift of the virus HA was soon recognized and as WHO began to coordinate the world influenza surveillance, it became easier for manufacturers to select an up-to-date virus. Influenza vaccines remain unique in that the virus strain composition is reviewed yearly, but modern attempts are being made to free manufacturers from this yolk by investigating internal virus proteins including M2e and NP as "universal" vaccines covering all virus subtypes. Recent technical innovations have been the use of Vero and MDCK cells as the virus cell substrate, the testing of two new adjuvants, and the exploration of new presentations to the nose or epidermal layers as DNA or antigen mixtures. The international investment into public health measures for a global human outbreak of avian H5N1 influenza together with a focus of swine influenza H1N1 is leading to enhanced production of conventional vaccine and to a new research searchlight on T-cell epitope vaccines, viral live-attenuated carriers of influenza proteins, and even more innovative substrates to cultivate virus, including plant cells.

J. Oxford (✉), A. Gilbert, and R. Lambkin-Williams
London Bioscience Innovation Centre, Retroscreen Virology Ltd, 2 Royal College Street, London NW1 ONH, UK
e-mail: j.oxford@retroscreen.com

G. Del Giudice and R. Rappuoli (eds.), *Influenza Vaccines for the Future*, 2nd edition, 115
Birkhäuser Advances in Infectious Diseases,
DOI 10.1007/978-3-0346-0279-2_6, © Springer Basel AG 2011

1   Introduction

When the influenza A virus first emerged from a presumed avian reservoir at the end of the ice age 10,000 or so years ago, there was a distinct difficulty in finding new human victims. For example, at that time, only a few hundred settlers were in the London region near the Royal London Hospital, now a community of four million people. At that time a traveler would have to walk a 100 miles to find another small settlement, perhaps at Stonehenge near Salisbury.

Nowadays we have a truly global community of six billion people, linked so that two million people are moving each day by plane, while perhaps ten million are journeying in their homelands. Influenza, like all viruses, is opportunistic. In 1918, it had the unprecedented opportunity to spread at the end of the first global war. Ten million soldiers began the move homewards, and every steamship was packed as they fanned out from France to England, Europe, the USA, Canada, Australia, India, and SE Asia [1–3]. How perfect for a virus spread by aerosol droplets, close contact, and contamination of towels, cups, and every day utensils. A virgin population, which had never before encountered the avian virus (H1N1), was on the stage of this theater of infection. Perhaps, a billion people were infected in the next 18 months, and 50–60 million died, making this by far the biggest outbreak of infectious diseases ever recorded, with an impact many times greater than the so-called bubonic plague outbreaks in Medieval Europe. However, more than two billion people survived. The overall mortality was less than 1%, although in a few semi-closed societies of hunter-gatherers in the Arctic, the mortality from the disease and subsequent starvation as young hunters died and husky dogs attacked and ate the survivors exceeded 90% [4–7]. It is well to remember that when H1N1 emerged in 1916/1917 and became pandemic in 1918 everyone except for the over 70s were fully susceptible. This is different from today where most people on planet earth have immune memory to the H1N1 family of viruses and by definition to A/Swine flu. This explains why the current H1N1 vaccine is immunogenic. While most people in the world were infected, we are forced to view the innate protective power of our immune system with awe [8, 9]. We are equipped with 100,000 genes, seven million years of evolution, and 80,000 years of specialization since our emergence from Africa. In contrast, influenza is a miniscule eight-gene vehicle. A recent study [10] of the reproductive number ($R0$) of the 1918 virus suggests that, unexpectedly, it may have been quite low, not exceeding three persons infected with a single case. The current pandemic A/Swine H1N1 virus is not so different. This would place pandemic influenza not far above the lowly group of viruses such as small pox and SARS and not reaching the heights that measles has attained. However, this unexpected theoretical analysis, if it is not flawed, gives us more practical opportunities to break a chain of infection of a pandemic with antivirals, hygiene, and vaccines [11–13]. We are experimenting with these approaches at the present moment.

The new world of the twenty-first century, although harboring in some countries a few old-fashioned attitudes, akin to "influenza and pneumonia is the old person's friend" nevertheless has the capability for the first time to defend itself against

Mother Nature and her threat of influenza. For the first time in history, intense surveillance by the World Health Organization (WHO), early identification of a new pandemic influenza virus by molecular diagnostics, application of vaccination and antiviral chemoprophylaxis, and possible quarantine and masks could actually prevent a pandemic arising. For the expressed intention of WHO and the world community of infectious disease researchers is to deflect the first wave of the first pandemic of the twenty-first century. In this endeavor, our huge resources of natural innate immunity, assisted by new vaccines, are already helping us. The formulation of the vaccines and their stockpiling alongside antineuraminidase (NI) antivirals has needed significant investment of time and money, and this started with a three billion Euro investment from the USA and EU. We are presently gathering the fruits of this investment with the outbreak of A/Swine (H1N1) virus.

Baroness Findlay of Glandaff put the epidemiology of influenza H5N1 situation succinctly in the House of Lords Report of Pandemic Influenza [14] "We believe the risk of a pandemic of human-to-human transmissible virus is to be taken very seriously. We believe that it may not happen in the very short time. To explain why we came to this stance; we believe that the problem, if it does emerge is more likely to emerge in Asia. Asia is where fire fighting must be done today." The Baroness had just heard the background science that China alone holds 700 million domestic ducks, a possible Trojan Horse of virus persistence, which approximates to 70% of the world's domestic duck population. Expert evidence from FAO had summarized that China, Indonesia, and Vietnam represented the core of the problem, but only 160 million dollars were available at that point in 2005/2006 to help, and biosecurity is not imposed strictly, while veterinary services are haphazard. The current pandemic virus emerged from pigs but a continuing threat is another reassortant event with H5N1 most likely in a coinfected child in Egypt or SE Asia where H5N1 viruses are endemic and where swine H1N1 viruses are spreading.

We are not the first generation of virologists to recognize the influenza pandemic threat, but we are the first to have the knowledge of the avian and pig reservoir and the tools to deal with the problem in a scientific manner. The world capacity for influenza vaccine today of one billion doses did not arrive by accident: it came to us from the hard work and dedication of four generations of dedicated scientists and doctors. The intention here is to give just tribute to these pioneers and their new discoveries. Using the vaccine methods developed over six decades, we can for the first time confront influenza as it emerges, surround it, and actually prevent a pandemic. We no longer need to be passive observers at a theater of infection. Churchill coined the phrase "Give us the tools and we will finish the job." Well, we now have them and we will. Such is the essence and spirit of this chapter.

2   A Snapshot of the First Six Decades of Influenza Virology

The serendipitous discovery of infection of ferrets, which produce clinical signs, and the cross-infection of a student from a ferret was the first technology foundation stone [9]. Ferrets are used today as a key model to investigate new vaccines.

The two most important technologies, which form the granite-like foundation of influenza vaccine research, are the hemagglutination inhibition test (Fig. 1) and the cultivation of virus in embryonated hen's eggs (Fig. 2), first reported in 1941 and 1946, respectively [15, 16].

If one adds two other vital scientific observations that of Hobson et al. [17] who correlated a HI titer of 40 with protective efficacy in volunteers in 1972 and then the discovery of a single radial diffusion for standardization of the hemagglutinin (HA) content of vaccines by Schild in 1973, it is quite apparent that the technologies are all now well tried and tested [18]. The elucidation of the structure of the fragmented influenza genome [19] has quickly led to techniques, genetic reassortment, and correlation of functions with certain genes (Fig. 3). From a practical viewpoint, some old much passaged viruses such as A/PR/8/34 (H1N1) grew to extraordinary infectious titers in the egg allantoic cavity, exceeding a new wild-type virus by 100-fold or more. Why not create a reassortant in the laboratory with six replicative genes of A/PR/8/34 to give high replication while having the two new HA and neuraminidase (NA) genes of the new epidemic virus? This technique proved to be a masterstroke and in the last quarter of a century three laboratories, CSL in

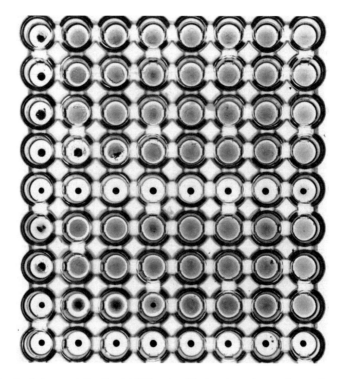

Fig. 1 The classic hemagglutination inhibition test. The test depends upon interaction of eight HA units of virus that would normally agglutinate 0.5% turkey red blood cells. Preincubation of this standard virus with dilutions of serum antibody abrogates the agglutinating property of the virus (*vertical rows* 5 and 9). No antibody is detectable in *rows* 1–4, 6–8

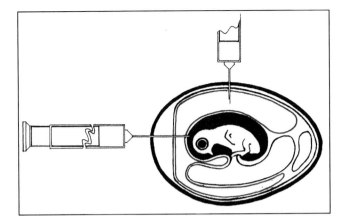

Fig. 2 Inoculation of embryonated hen's eggs to grow influenza virus for vaccine. Virus is inoculated through the shell of a 10-day-old embryonated hen's egg and more rarely in the research laboratory into the amniotic cavity (*top*). After 2 days of incubation at 37°C, the clear fluids are removed and titrated for HA by hemagglutination

Fig. 3 The influenza genome is in eight fragments. The genome could be labeled with 32P extracted and separated on polyacrylamide gels

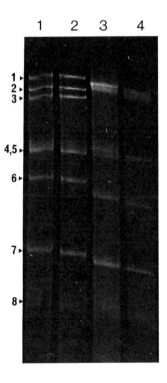

Melbourne, NIBSC in London, and Ed Kilbourne's laboratory in New York, have rushed each year to produce the new candidate vaccine viruses prefixed IVR-, NIB-, and X-, respectively. The almost made-to-order technique of gene reassortment

with influenza was also central to producing host range mutants with attenuation genes for live vaccines. Some of the starter and seed viruses for the current production of A/California/4/09 H1N1 (Swine) vaccine used this biological technology, while others used reverse genetics to make a GM starter virus.

Undoubtedly the simultaneous discovery of the reverse genetics [20, 21] by the three laboratories in New York, Wisconsin, and Oxford was a masterstroke in technical advance, which has enabled mutations to be placed, at will, into the genomes of the negative-strand viruses. The conjunction of older and newer techniques with the licensing of the mammalian cell lines from monkey kidney (Vero) [22], dog kidney (MDCK) [23], or human tissue (PER-6) has led directly to the newly emerging influenza vaccines of the twenty-first century. We are using all these techniques of the last 50 years to produce the A/Swine H1N1 vaccines of 2009 for the current pandemic.

3 The Historical Steps in Killed Vaccine Development

The first experiments on the attempted immunization of animals were made in the USA by Francis Magill [24] and in England by Andrewes and Smith in 1937 [25]. The model is still vital today and the first experimental assessment of A/California/ 04/09 H1N1 vaccine was made in this model. Mouse lung suspensions or filtrates were used after inactivation with formaldehyde, and it was found relatively easy to protect mice against intranasal infection with influenza. Immunization experiments in man were accelerated when allantonic fluid preparations of virus formed the starting material soon after the technique of allantoic inoculation of fertile hen's eggs was discovered [16]. The first field trial demonstrating short-term protection by inactivated vaccine took place in the USA during a sharp epidemic of influenza in 1943 (Commission Influenza 1944) [26].

Progress with the development of purer, more potent vaccines has proceeded steadily since those early days, and technical advances with ultracentrifugation and chromatography, by methods producing richer cultures and chemical inactivation avoiding too great a modification of the surface HA and NA antigens have all helped. To avoid the relatively high rate of local and general systemic reactions caused by the older egg-grown inactivated whole-virus vaccines, chemical treatment to disrupt the particle and to separate the wanted antigens (HA and NA) from other constituents of the virus has led to a variety of different split or subunit vaccines (Figs. 4–6). Ether extraction [27, 28], deoxycholate treatment [29], and treatment with other detergents have been introduced. Some methods have provided subunit vaccines causing fewer clinical side reactions than the older whole-virus particle vaccines, but drawbacks have appeared, including that of reduced antigenicity. Adjuvants of oily emulsions promised potent vaccines with excellent antibody responses, and a few reactions were first encountered. However, a rare abscess at the site of inoculation caused much distress and this early approach had to be abandoned. In spite of attempts to develop safer materials, none have yet

Fig. 4 Whole-virus vaccine.
Influenza viruses are
pleomorphic with a fringe of
HA and NA spikes

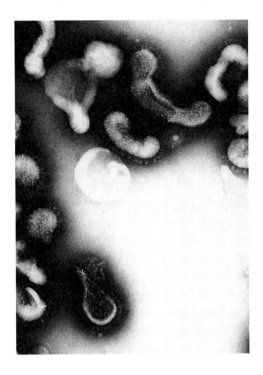

Fig. 5 Split influenza virus
vaccine. The whole virus is
disrupted with detergent,
which dissolves the lipid
membrane releasing HA, NA,
and internal NP, seen as
"lamb tails"

Fig. 6 Subunit influenza
virus vaccine. The split virus
is fractioned in a sucrose
gradient, and the HA and NA
subunits are separated from
NP and M, and standardized
by SRD and used for vaccine

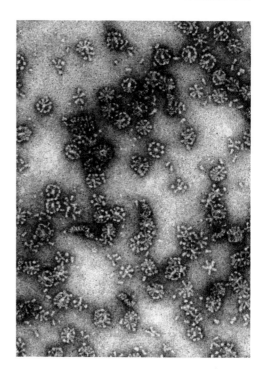

been developed commercially until very recently when MF59 and A50 have been formulated. Thus, after 60 years of work, the hope of an ideal inactivated vaccine free from the induction of clinical reactions and yet potent immunogenically has just been fulfilled with pandemic H5N1 vaccines and swine H1N1 vaccines.

In 1946, a major antigenic deviation of influenza A virus occurred with the appearance of A/CAM/46 (H1N1) virus in Australia. In the USA and Europe, outbreaks of influenza occurred early in 1947, which were due to the same virus; some communities previously receiving vaccine containing PR8 and Weiss viruses (H0N1 in the old classification and now reclassified H1N1) were attacked. This time the vaccine did not protect against the new virus typified by the prototype A/FM/1/47 (H1N1) [30, 31], and this led to realization of the enormous importance of the updated antigenic make-up of inactivated vaccine.

Yet other difficulties have become appreciated, one of which is the inappropriate antibody response occurring sometimes after inoculation, when the vaccine induces cross-reacting antibody to heterologous viruses or the first virus in the subtype which the vaccine first experienced, rather than that appropriate to the specific antigen, HA, of the vaccine virus. This response is probably allied to the phenomenon of "original antigenic sin." Sometimes this aberrant response can be useful as with A/Swine H1N1 vaccine. It is likely that the over 65s will produce recall antibody to H1N1 viruses which infected them in the 1940s and that this virus is somewhat related to the current A/Swine virus.

4 Vaccine Purification Historical and Present

The starting materials for almost all types of inactivated vaccine are allantoic fluids from fertile hen's eggs previously inoculated with a seed culture, the yield of which is enhanced using a recombinant virus, one parent of which is a high-yielding laboratory strain (A/PR8/34) and the other acts as the donor of the requisite surface HA and NA antigens from a wild-type virus [32]. The A/PR/8/34 virus donates six genes and the wild-type virus two genes: the ensuing reassortant high growth viruses are called 6/2 reassortants. Purification from unwanted egg material is accomplished by ultracentrifugation on a zonal ultracentrifuge [33]. Whole-virus particles thus separated are inactivated by formalin or β-propiolactone, the HA content being as high as possible commensurate with the necessity to avoid febrile reactions after inoculation. Children were sensitive to the older egg-grown whole-virus vaccines; as many as 30% under 2 years developed fever after 0.25 ml of vaccine and up to 8% of 6-year-old children were similarly affected after 0.5 ml [34]. The precise constituent producing the fever was not clearly identified, but the viral proteins were believed to be concerned [35, 36]. More modern whole-cell virus vaccines produced in cell culture are more purified and produce fewer side reactions.

Separation of the HA and NA by means of detergents such as Tween 80 or Triton N101 produced split-virus or subunit vaccine, and general experience suggested that these materials are less pyrogenic, but less immunogenic, than whole-virus vaccine [37]. This was particularly well demonstrated by studies during the swine influenza campaign in the USA in 1976, when many observers reported results, which ultimately led to the recommended use in children of two doses of split-type rather than whole-virus vaccines. Such recommendations continue at the present time. In adults, too, the older egg-grown whole-virus vaccines gave a higher proportion of febrile reactions than split virus [38]. However, this situation is changing as whole-virus vaccines produced in Vero cells for example come to the fore.

5 Early Progress: The Standardization of Potency, Composition, and Dosage of Inactivated Vaccines

Former methods for assays of the potency of inactivated vaccine depended on measuring the HA activities of the vaccines with erythrocyte suspensions using the Salk pattern technique of Miller and Stanley [15]. In retrospect, this technique was not hugely accurate especially for subunit and split viruses. In a major technological breakthrough, Schild et al. [18] proposed a method of assay based on single radial immunodiffusion (SRD) (Fig. 7). The HA antigen content of vaccines was estimated using SRD tests in agarose gels containing specific HI antibodies. The SRD method was modified and refined by Wood et al. [39]. It may

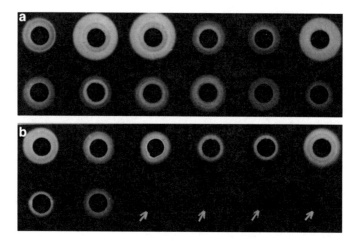

Fig. 7 Single radial diffusion (SRD) test to standardize HA. Vaccine antigen is pipetted into 3-mm wells in an agar plate containing specific anti-HA, -NA, and -NP antibodies. After a few hours incubation, a zone of precipitation is quantified and the area is proportional to the quantity of HA in the vaccine

be gradually replaced now by HPMC technologies. The SRD technique was valid for both whole-virus and split-virus vaccines and was quickly adopted for international use and is still the gold standard. In this test, vaccine virus preparations and reference antigen calibrated in terms of micrograms of HA are disrupted with detergent, and dilutions of the treated antigens are introduced into wells in SRD immunoplates. The size of the precipitation ring obtained for the vaccine is compared with that obtained with a reference antigen of calibrated HA content titrated on the same plate. The vaccine potency is measured in terms of micrograms of HA per vaccine dose. Inactivated influenza vaccines frequently contain two or more virus strains and the HA content of each component (15 μg) is assayed independently.

6   HA Dosage of Vaccines and Relationship to HI Antibody Response

It has been known for many years that the serological response to inactivated vaccine depends on the previous experience of the recipient to infection by viruses of the same subtype of influenza A virus as that present in the vaccine. Although a single subcutaneous injection of (H1N1) vaccine gave as good a response as two doses prior to 1957, the advent of the new pandemic A/Asian (H2N2) virus produced a different effect. Thus, Holland et al. [40] demonstrated that two doses at an interval of two or more weeks produced a better response to one dose and in this regard the vaccine-induced immune response was much inferior to that noted before the change in virus subtype. Such an experience was again noted during the first year of circulation of A/Hong Kong (H3N2) virus and also when the A/New Jersey/76 (Hsw1N1) vaccine was used in children and young adults. Also, in the circumstances of 1977–1978, when most persons under 25 years of age had no previous antibody to the recirculating H1N1 virus, a two-dose regimen for children and young adults produced a more satisfactory response than a single injection [41]. To reiterate, in 1977 an "old" H1N1 virus from the 1950s was accidentally released from a laboratory and established itself as an epidemic virus. It was called a "pseudo pandemic." Everyone over 24 years had previous immunity. The contrast between the effects of a single dose of vaccine in persons infected with H1N1 viruses at least 20 years earlier was very striking. These data have immediate relevance today in terms of H5N1 vaccine and of course with the A/Swine vaccine. The world is full of immune virgins as regards H5N1 but not in the case of A/Swine H1N1. Most persons have immune memory to the H1N1 family, and therefore it comes as no surprise that vaccines induce high levels of HI antibody.

Several factors are of importance in the determination of the quantity and the precise composition of the antibody response to the surface antigens of the virus present in inactivated vaccine. First and foremost, the quantities of HI and NI antibodies induced by vaccine are broadly related to the quantity of antigen present in a single dose. Second, the precise composition of the antibodies formed in response to influenza A virus is important. Thus, reinforcement of previously acquired antibodies by the orientation of the B-lymphocyte response to the first infection by the particular subtype of virus experienced in childhood or later may take precedence over the strain-specific antibody response to the vaccine virus. Third, the precise response is influenced by the route by which the vaccine is presented to the body's immune system.

First then, several earlier studies reported a graded relationship between the quantity of antigen inoculated and the antibody response that results. This was so in the study of Mostow et al. [42], who gave increasing doses of vaccine in a single injection containing 300–4,600 chick cell agglutination (CCA) units containing A/Japan/57 (H2N2) virus groups of volunteers. The serum HI response was tested with four different H2N2 viruses isolated 1962–1967 and also the homologous virus. With more than a tenfold increase in HA from the least to the highest dose, the geometric mean titer (GMT) of antibody increased only fivefold. Similar results were obtained by Potter et al. [43], who inoculated student volunteers with vaccines ranging in dosage from 5 to 400 IU and containing A/Port Chalmers/73 (H3N2) virus. The vaccine was a surface-antigen detergent-treated material [44] adsorbed to aluminum hydroxide gel. GMT HI serum titers increased against homologous virus from 8- to 174-fold with the increase in dose of vaccine HA. Three other H3N2 strains and A/Singapore/57 (H2N2) virus were also tested, and all three H3N2 viruses showed graded HI antibody responses proportional in magnitude to increase in antigen dose, as did the homologous virus.

The Pandemic Working Group of the MRC Committee on Influenza Vaccine [45] gave graded doses of whole-virus vaccine containing the A/New Jersey/76 (Hsw1N1) strain to groups of volunteers in 1976. Those less than 44 years of age,

who did not possess significant serum HI antibody to the virus before immunization, showed a postvaccination antibody titer ranging from 64 to 148 GMT with a
nearly eightfold increase in dose from 8 to 61 µg of HA. Above this age, in those
45–64 with preexisting Hsw1 antibody, there was an increase in antibody titer from
7 to 36 times (GMT) with a change in HA concentration from 4 to 61 µg. Thus, the
effect of increasing the potency of this vaccine on the antibody response was much
greater in those sera, which indicated that they had been exposed to the antigen,
presumably by infection with a related virus, than in those with no such exposure.
Both whole and detergent-split-virus vaccines showed a relatively poor HI response
in volunteers less than 25 years of age whose initial serum had no significant
amount of prevaccination or postinfection HI antibody. In this group of subjects,
two doses of vaccine gave a better antibody response than did one, but the resultant
postvaccination GMT was half that obtained with a single dose of the vaccine over
25 years of age. This historical data is very relevant to us today as we analyze the HI
data from the current batches of A/Swine H1N1 vaccine where in the over 5 years a
single dose of vaccine is sufficient because of wide preexposure to members of the
H1N1 family of viruses. The younger groups had no prior immune memory to the
H1N1 family of virus, their experience being more orientated to H2N2 and H3N2
families.

These examples underline the practical importance of a considerable degree of
antigenic drift within a subtype comprising HI antibody response. Also, the recall of
antibodies induced by previous infection illustrates the general rule that an up-to-
date monovalent vaccine reinforces antibodies against former members of the
subtype, while also inducing specific antibodies to the vaccine virus. This was
clearly shown by direct comparison of monovalent and polyvalent vaccines such as
the MRC Committee on Influenza Vaccine's trials [46–49].

The quantitative dose response already described for HI is also found with NI
antibody but is less consistent. Thus, Potter et al. [50] noted that there was a two- to
sixfold increase in NI antibody as vaccine potency was increased from 5 to 400 IU
of HA. Yet the trial of A/New Jersey/76 (Hsw1N1) vaccine conducted by the
Pandemic Working Group of the MRC Influenza Vaccine Committee [45] found
only a slight increase in NI antibody after an increased dose from 100 to 200 IU
using 100 IU of HA in the vaccine. Nicholson et al. [41] gave a whole-virus vaccine
of the A/USSR/77 (H1N1) virus, which ranged in potency up to sixfold, and found,
in those under 25, a threefold increase in NI antibody. However, in those over
25 years of age, an increase in dose of vaccine had a less constant effect on NI
antibody formation. One possible reason for the variation in the effect of different
vaccines on the NI antibody is the lack of consistency in the NA content [51];
however, another possibility may be that immunological priming to the HA in the
vaccine can in some way suppress the immunogenicity of the NA antigen, which
may be physically associated with the HA.

The second important variable in the immune response to inactivated vaccine
arises from the relative amounts of cross-reactive and strain-specific antibodies that
are generated. The differentiation of these require special techniques such as SRD
and the adsorption studies. Webster et al. [52] compared, in adults, the response to

an A/Port Chalmers/73 (H3N2) subunit vaccine to homologous and heterologous H3N2 viruses. Most of the antibody was cross-reactive with A/Hong Kong/68 virus but when higher doses of the vaccines were used, strain-specific A/Port Chalmers/ 73 antibody was produced in addition to that against heterologous virus. Oxford et al. [53, 54] compared whole- and split-virus vaccines containing A/Victoria/75 or A/Scotland/74 viruses and using single radial hemolysis and adsorption techniques showed that in an immunized adult, cross-reactive antibody was induced much more frequently than specific antibody against homologous virus. They showed the same phenomenon in adults during infection with A/Port Chalmers/73 virus, who frequently also developed antibody rises to A/Hong Kong antigens from 1968. Oxford et al. [54] used similar techniques to analyze sera from children aged 3–6 years immunized with a surface-antigen vaccine containing A/Victoria/75 (H3N2) antigens. Most children produced a strain-specific serum antibody to the vaccine antigens, whereas adults similarly vaccinated tended to produce antibody cross-reacting with all variants of the H3N2 subtype tested. Postepidemic sera from those of various ages recently infected by A/Texas/77-like strain showed cross-reactive antibody in adults but in contrast mostly strain-specific responses in children. Strain-specific antibody is considered to be more protective.

7   The Route of Vaccination

The influence of the route of immunization with inactivated vaccine has been studied in the past by many observers. The chief alternative to the subcutaneous–intramuscular route is intradermal injection using a reduced amount of vaccine. The advantages of this route are economy and the avoidance of febrile reaction. The principal disadvantage is the fact that the antibody response is less consistent. It was shown by Appleby et al. [55] that the GMT after intradermal vaccine was less than half that obtained with subcutaneous vaccine, and this seemed logical in that only one-tenth of the vaccine dose was given intradermally. McCarroll and Kilbourne [56] found little difference in the antibody responses to intradermal and subcutaneous vaccines in equivalent doses. Tauraso et al. [57] reinvestigated the question using a two-dose regime before the arrival of the A/Hong Kong/68 (H3N2) epidemic. In the equivalent amount of 0.1 ml of vaccine, antibodies formed in higher titer after intradermal than subcutaneous vaccine. However, the titers after 0.5 ml of vaccine subcutaneously were little different from intradermal injection of 0.1 ml. It is considered advisable, however, in practice to limit intradermal vaccination when the vaccine is in short supply or when, in children or the aged, reactions after subcutaneous vaccine might pose problems.

The nasal route of inoculation either by instillation of drops or by spray was first studied in detail by Waldman et al. [58]. Compared with the subcutaneous vaccine in a dose of 0.5 ml, antibodies capable of neutralizing the virus A/Taiwan/64 (H3N2) increased to a greater extent in sputum and nasal secretions after repeated nasal inoculation with a total volume of 3.6 ml vaccine. In contrast, the intranasal vaccine produced a much lower rise in serum antibody, the GMT being only one-sixth that after subcutaneous vaccine. Waldman et al. [59], using an aerosol spray, found that a better serum antibody response occurred with a small-sized particle spray than a larger one, but the nasal antibody response was better after the latter or with nasal drops. Absorption studies showed that a majority of the secretory antibody (IgA) response in nasal secretion was cross-reactive with heterologous viruses (A/Hong Kong/68 H3N2). Phillips et al. [60] compared subcutaneous or intradermal vaccine in nurses with vaccine dropped intranasally. The subcutaneous route produced the best serum antibody rises, and intradermal vaccine was superior to the intranasal route in terms of antibody response. The nasal antibody titers after immunization by either subcutaneous or respiratory routes paralleled those in serum.

The fact that nasal antibodies increase after subcutaneous vaccine [61, 62] is important because the lack of a good response in serum antibody in those given the same vaccine intranasally is a limitation hardly offset by local nasal secretory changes. Challenge of immunized groups of persons by live-attenuated virus also supports the view that nasal antibodies play a supplementary role to serum HI antibody [63].

8 Early Quantification of Side Reactions to Vaccines: Whole-Virus Versus Split and Subunit

The field trials of inactivated influenza A H1N1 vaccines in 1976 and 1977 added to knowledge concerning the reactogenicity of different preparations. The split-virus type of vaccine then used unquestionably caused fewer systemic febrile responses in both children and adults. The fact that reactions with whole-virus vaccines used at the time were unpleasantly severe for those without serum antibodies to the vaccine virus before inoculation had not been fully appreciated. In the case of children aged 6–18 in the American trials of A/New Jersey/76 (Hsw1N1) virus, the most potent vaccines caused fever in up to 63% of vaccines. In the UK, the Pandemic Working Group of the MRC Committee on Influenza Vaccine found that a dose of 61 µg of HA (1,000 IU) of whole-virus vaccine with the same Hsw1N1 strain produced, in adults, local reactions in 50% and systemic effects in over 60% of volunteers. Even the lower doses of 18–27 µg of HA caused local reactions in 50% and systemic effects in 40%. The A/USSR/77 (H1N1) virus vaccine trial in 1978 in Britain showed that adsorbed or aqueous split-virus vaccine produced fewer reactions than did whole virus [51]. After a second dose of the same vaccine, fewer volunteers experienced reactions than seen after the first dose. Later studies of the endotoxin content of various pools of inactivated type A or B vaccines using the limulus lysate test gave no hint of a parallel between the occurrence of general reactions and the endotoxin content [64].

Neurological illness is a recognized sequel to immunization with a variety of vaccines but had not previously been observed with any frequency after influenza virus vaccines. Wells [65] noted the rare instance of Guillain–Barré syndrome (GBS), which appeared in excess among the persons vaccinated with A/Swine vaccine compared with the numbers in unvaccinated individuals. Of 1,098 persons with GBS reported from October 1, 1976, to the January 31, 1977, 532 had received vaccine before the onset of neurological symptoms. The overall risk of GBS was calculated as ten cases per million vaccinated. The rate of occurrence during the 10-week swine vaccine period was five to six times greater than in unvaccinated persons. However, the excess in number was greater in the second and third weeks after inoculation than either the first or subsequent weeks. As reported by Langmuir [66], GBS was not associated with a particular variety of vaccine or age group. However, that numbers were slightly greater in those aged 25–44 than in middle aged or elderly persons, which appears to rule out the possibility that the syndrome was, in some way, related to the absence of antibodies to the swine virus before immunization, for most of those aged over 45 would have been exposed to antigens of this virus many years before. After the swine influenza campaign was terminated, surveillance was continued, and during the period 1978–1979, when 12.5 million doses of ordinary inactivated vaccine were estimated to have been used, the related risk of GBS was 1.4 times the incidence in unvaccinated persons. This risk was regarded as not significant [67]. No clue to the cause of the marginally increased risk of GBS in immunized persons in 1976 has yet been obtained but could be virus strain related. No untoward effects have been noted in the billions of vaccines used since 1979 to the present day.

9   Advent of the 1968 (H3N2) Pandemic Virus and Use of Inactivated Vaccines

At the time when A/Hong Kong/68 (H3N2) virus was spreading in Asia, plans were made by the MRC Committee on Influenza Vaccine to protect children in residential schools and other groups in a controlled manner. Inactivated polyvalent vaccine containing two H2N2 viruses (A/England/64 and A/England/66) and a B strain were compared with an H3N2 A/Hong Kong whole or deoxycholate-treated virus vaccine in initial serological trials. Antibody formation even in those without detectable serum HI antibody gave GMTs over 100 in those receiving A/Hong Kong vaccine intramuscularly. However, controlled trials in two boarding schools showed no convincing evidence of protection. In uncontrolled trials in other schools either the polyvalent or the A/Hong Kong vaccine were given or no vaccine at all. There were 12 schools where epidemics of influenza occurred in January and February 1969 but no evidence of protection was found in those receiving A/HK vaccines. The only clue obtained concerning the vaccine failure was first that only one dose of vaccine had been given, and this is known to be inadequate to give a satisfactory antibody response in previously seronegative persons, and second, there was an interval between vaccine administration and infection of 2–4 months. These two factors may have combined to explain the absence of protection because of the inadequacy of the antibody response at the time of challenge. It would be fair to add that others [68, 69] did obtain protection from A/Hong Kong/68 whole-virus vaccine during the first outbreak of influenza due to this virus in the USA. The use of modern adjuvanted H5N1 vaccine in two doses is anticipated to give protective effects. The current A/Swine vaccines produce protective HI antibody (>40) in most persons over 5 years of age following a single dose. As emphasized above, this reassuring situation is because most persons have prior immunity to the H1N1 family of viruses which circulated between 1918 and 1957 and then again from 1977 to the present day.

10   First Studies with Live Influenza Vaccines

The use of living but attenuated virus as an immunizing agent developed slowly from the initial studies of Mawson and Swan [70] in Australia and the USSR. The major difficulty of the lack of a laboratory test to indicate that cultured virus had lost its pathogenicity, while retaining infectivity for man, meant that deliberate intranasal inoculation of volunteers furnished the only way to select a suitable strain for infection without causing clinical reaction. In spite of the widespread adoption of live vaccines selected by this method and given as an intranasal spray in the USSR, little interest was exhibited in most other countries. From 1956 onwards, trials took place in volunteers in England and Wales to provide evidence of safety and immunogenicity of cultured viruses and the drawback of a reduced infectivity of well-attenuated viruses handicapped progress. The necessity to observe a match between the antigens of epidemic viruses and those present in the vaccine was a further drawback until the technique of reassortment of characters between two strains, one of which was of proven attenuation, was utilized to yield seed viruses with the desirable clinical and antigenic properties. Other disadvantages of live viruses appeared during the intensive researches of the 1980s particularly in the USA and in England [71, 72]. It cannot yet be claimed that the ideal live-attenuated virus vaccine has been formulated, but reverse genetics and increased knowledge of virulence genes have now lead to a resurgence of interest.

In the 1980s, genetic studies were intensively pursued in attempts first to define the particular gene or combination of genes, donated by the attenuated virus that confers the property of attenuation upon the reassortant strain. It was found that the biological properties of excreted virus may be altered compared with those of the original virus in the vaccine and the manner of this alteration was also studied genetically. Such work is essential in achieving the goal of an effective and safe vaccine virus for human use. Experimental inoculations were carried out initially in small-scale tests in volunteers under semi-isolation to permit close observation (see below).

11   Host Range Virus Mutants as Live Vaccines

Multiple cultivation and passage of viruses either in animal hosts, such as ferrets and mice, or in developing chick embryos or tissue cultures had been practiced even before the use of temperature-sensitive (*ts*) or cold-adapted (*ca*) mutants was suggested. Early workers in Britain used the PR8/34 virus as a host range mutant, which, although noninfective for man, has retained animal pathogenicity even after many passages in eggs. As a donor parent with good powers of multiplication in the laboratory, PR8 was mated with various strains of wild-type influenza A viruses to obtain recombinants with up-to-date surface HA and NA antigens. This method was preferable to simple laboratory cultivation because some viruses failed to alter in pathogenicity after as many as 30 serial passes in cultures [73], although other virus strains appeared to become attenuated with only a few passages in eggs.

PR8 virus was chosen also by workers in Belgium who prepared reassortants from a number of viruses, some of which were licensed for human use [74]. To select recombinants with as high proportion of RNA components as possible derived from the host range mutant PR8, Florent et al. [75] used RNA–RNA hybridization to identify gene origins. Later the gene constellation of four of the candidate vaccine viruses was determined, and Florent [76] found that some clones of Beare and Hall's [77] recombinants of PR8 and A/Englannd/69 (H3N2) containing five genes from PR8 were satisfactorily attenuated. However, one clone though containing six PR8 genes was nevertheless clinically virulent to volunteers. A further genetic study of PR8 host range recombinants using viruses tested clinically by Beare and Reed [78] was made by Oxford et al. [79]. It was again found that recombinants from PR8 and A/England/69 viruses could contain only the surface HA and NA genes from wild-type virus and yet retain virulence for man.

Additional attempts to stabilize the attenuation of candidate viruses were made both by Beare at the Medical Research Council's laboratories at Salisbury and the RIT workers by rendering the virus resistant to an inhibitor present in normal horse serum. This property was present in the RIT series of recombinants. It seems strange that stabilization has not been pursued since nor has cultivation of host range mutant viruses, such as PR8, at abnormally low temperatures, such as 25°C. This method was found by Sabin [80] to be preferable to normal temperatures when attenuating polio viruses, and it was exploited by both workers in the USA and USSR.

Marker tests, which can be equated with attenuation of virulence for man, were sought with relatively variable results. One such test used weanling rats that were inoculated intranasally first with virus and later with cultures of *Haemophilus influenzae*. Virulent virus induces bacteremia and meningitis, and using this method Jennings et al. [81] successfully separated a number of reassortant viruses and obtained some correlation with clinical virulence. Yet the host range mutant parent PR8/34 and RIT 4050, which are both attenuated in man, were classed as virulent by the rat.

A new approach at that time used an avian (duck) virus, which was found to have only low pathogenicity for squirrel monkeys inoculated intranasally and was proposed as a donor of attenuation. A reassortant with a virulent human A/Udorn/72 (H3N2) virus behaved as did the avian parent in the squirrel monkey, although immunizing the latter against the virulent parent. Clinical trials have suggested that this virus is attenuated for man and is immunogenic but has not been investigated since [82].

12   Temperature-Sensitive Virus Mutants as Live Vaccines

Most work on the development of viruses with restricted multiplication at temperatures above the normal range for cultivation has been affected by Chanock, Murphy, and associates at the National Institutes of Health, Bethesda [83]. The technique used chemically produced mutation in virus RNA by cultivation in the presence of the mutagenic agent 5-fluorouracil. After cultivation and plaquing at 33°C, 37°C, and 38°C, mutant viruses with the requisite temperature sensitivity were obtained. Intranasal inoculation of hamsters confirmed temperature restriction, in that much lower titers of virus were found in the hotter lungs than in the cooler upper respiratory tract.

Spread from inoculated volunteers to adults in contact was not observed, and no evidence of a change in virulence was found in viruses recovered from adult recipients of vaccine [84]. However, in seronegative children, the A/Hong Kong/68-ts-1 [E] virus produced mild febrile reactions and a virus that had lost its properties was recovered from some who were infected.

A second series of ts-1a2 was then developed by combining two defective ts viruses, each of which belonged to a different complementation group in respect of the genetic defect. The progeny exhibited greater temperature restriction than the ts-1[E] line of viruses. It was termed A/Udorn/72 ts-1A2, and it was recombined with three further viruses; wild-type A/Victoria/3/75, A/Alaska/77 (H3N2), and A/Hong Kong/77 (H1N1). These ts-1A2 viruses were highly immunogenic and exhibited temperature restriction of multiplication in cell cultures and reduced replication in the hamster lung. The A/Victoria/3/75-ts-1A2 recombinant retained its ts properties after inoculation into doubly seronegative children. Unfortunately, when the A/Alaska/77-ts-1A2 virus was similarly tested in a single child after tests in adults had shown genetic stability, the nasal secretions of the vaccine yielded a ts-positive virus that produced plaques at 39°C even though the child had shown no symptoms or fever. The recombinant 1A2 virus with A/HongKong/77 (H1N1) parent exhibited a capacity to infect 70% of doubly seronegative adults and was attenuated compared with the wild-type parent. Nevertheless, it appeared possible that a virus such as the A/Alaska-ts-1A2 might, if transferred to contacts from an inoculated child, result in clinical illness, and clinical studies with this particular virus were not pursued.

13 Cold-Adapted Virus Mutants as Live Vaccines

Beginning with a strain of H2N2 virus recovered in Ann Arbor, Michigan, in 1960 by cultivation of throat washings in tissue cultures at 36°C, Maassab [85, 86] evolved a virus, A/Ann Arbor/6/60 (H2N2), which has acted as a donor of attenuation to other viruses by genetic reassortment. Earlier passages were made in chick kidney tissue cultures followed by intranasal passages in mice and then a gradual adaptation to lower temperatures, in tissue cultures and in developing hens' eggs inoculated allantoically, led to a virus with good powers of multiplication at 25°C. The ca variant was found to retain the infectivity of the original strain for both the mouse and the ferret, although it produced no deaths in mice and no fever or turbinate lesions in ferrets, whereas the original virus was pathogenic for both species. The virus proved to be temperature sensitive with a shut-off temperature of 37°C [87]. Recombinants with wild-type viruses of both H2N2 and H3N2 subtypes were prepared, studied in the laboratory and in volunteers, and analyzed genetically. The original A/Ann Arbor/6/60 (H2N2) virus was not, however, tested in fully susceptible persons presumably because of the difficulty in that period of finding seronegative adults. A few persons with low titers of serum neutralizing antibodies (1:4 to 1:6) were inoculated and as judged by antibody responses, became infected without undergoing clinical illnesses. More rigorous clinical studies have been pursued with recombinants, in particular, those with H3N2 antigens, and details of the results have been brought together and earlier data summarized by Kendal [72]. The donor ca parent has been more recently reassorted with H5N1 genes.

It is clear that infectivity and immunogenicity were fully retained for seronegative adults of whom 111 received H3N2 recombinants. Among those receiving three of four recombinants, clinical reactions were minimal or negligible but with the fourth, derived from the A/Scotland/74 parent, in 4 of 12 volunteers receiving $10^{8.5}$ and in 1 receiving $10^{7.5}$ TCID$_{50}$, there were clinical illnesses. Viruses re-isolated from the vaccines retained *ts* properties and so did those given recombinants of A/Victoria/75 (H3N2) and A/Alaska/77 (H3N2). However, some loss of *ca* restriction was found in virus re-isolated from volunteers given the A/Scotland/74 recombinant.

Cold-adapted recombinants with A/USSR/77 (H1N1)-like virus have also been studied in adult volunteers and found to be less immunogenic as judged by HI antibody responses. A better response was obtained by Wright et al. [88] in children in Nashville given $10^{6.5}$ TCID$_{50}$ of strain CR 35 (H1N1) and none of 11 children developed adverse clinical reactions even though eight became infected. All re-isolated viruses retained the ts phenotype. The failure to elicit serum antibody response in adults given this same virus recombinant is puzzling. Using the ELISA enzyme-linked assay, Murphy et al. [89] found that by this more sensitive method antibody rises could be demonstrated and the results tallied better with the ability to re-isolate viruses from the inoculated volunteers than did the serum HI responses.

The Leningrad group of workers led by Smorodinstev [90] was the first to obtain a virus indirectly attenuated by cultivation at 25°C. The group used strains selected by inoculating volunteers with several viruses derived from cultures repeatedly incubated at 25–26°C to speed up attenuation. Approximately 5–7 months were required for the preparation and production of new strains even using genetic recombination to incorporate new surface HA and NA antigens. Although Alexieva et al. [91] found that cold cultivation was not successful in producing reliably attenuated viruses for use in children, the technique was adopted for general use. Genetic studies of the Leningrad viruses are described briefly by Kendal et al. [72], and these parent ca viruses are currently the center of new interest for attenuated H5N1 vaccines.

Usually, preliminary studies were made in the USSR in 18–21-year-old sero-negative adults who receive virus twice at intervals of 10–14 days administrated by nasal spray. Viruses were attenuated by passage for varying periods at 25°C and both donor viruses and recombinants proved temperature sensitive. In 1961–1964, when H2N2 viruses were circulating, 5,165 children aged from 1 to 6 received the ca A/Leningrad/57 (H2N2) virus. Some febrile reactions occurred but only in less than 1% of the children. Further studies of recombinants with H3N2 or H1N1 antigens and the same Leningrad H2N2 parent after 47 serial passages under cold conditions of cultivation (25°C) were conducted in children, half of whom had no detectable serum antibody to the vaccine strain. No reactions occurred and over 90% of the children responded with antibody production. It is clear from the earlier papers by Alexieva et al. [91, 92] that intranasal administration of children aged 7–15 were too reactogenic and that this is the reason why the peroral route has been chosen for routine administration in the USSR.

A Japanese virus recovered in 1957, A/Okuda/57(H2N2), was found to be attenuated for children and served as a donor of attenuation both in Japan and in England. Zhilova et al., Japanese workers, [92] developed a recombinant virus (KO-1) from ultraviolet-irradiated A/Okuda/57 and wild-type A/Kumamoto/22/76 (H3N2). Serial passaging in eggs in the presence of normal horse serum was followed by plaque purification and later clinical tests in a few children. The M (membrane) gene was found to have been donated by the Okuda parent. From reassortants with other human viruses, a candidate WRL 105 virus was selected and underwent clinical trials without harmful clinical effects [93] but has been little investigated since that time.

14   Mammalian Cell Culture Vaccines

Cultivation of influenza viruses in mammalian cells rather than eggs initially encouraged two manufactures to invest in cell culture fermenters for vaccine production [22, 23]. Many more groups are now using these technologies to produce the current A/Swine H1N1 vaccine. Capacity can be increased to cope with a surge in demand for a pandemic virus vaccine. Moreover, the final vaccine has the theoretical advantage of the absence of egg proteins. The cell culture vaccine virus is also easier to purify. Where clinical isolates of influenza viruses are cultivated in mammalian cells and eggs in parallel, different antigenic variants may be selected [94]. The biological variants have amino acid substitutions in the receptor binding site in proximity to an antigenic site on the HA, and an amino acid change in this region can alter antigenicity. Of the two virus subpopulations that can be selected, the virus which is grown on MDCK (or Vero) cells rather than in eggs appears more closely related to the wild-type clinical virus. There is some indication that cell-grown virus vaccines offer greater protection in animal models than the corresponding egg-grown vaccine. These are all powerful arguments in favor of the new generation of influenza vaccines being cultivated currently in Vero [22] or MDCK [23] or Per 6 cells.

15   The Current Pandemic of A/Swine H1N1 and Vaccine Production and Efficacy

Alongside 50 years of experience producing an immunogenic and safe vaccine, the world capacity for influenza monovalent vaccine manufacture has expanded to the present two billion doses. The preparation work and investment for H5N1 are showing rewards with the current pandemic of A/Swine H1N1. Most manufacture is still located within the EU, but production is increasing in the USA, Korea, Japan, China, and most recently, India.

The international collaboration in face of the outbreak of A/California/4/09 (H1N1) in Mexico around Christmas 2008 to the present, the exchange of clinical data and viruses enabled vaccine manufacture to start production by May/June 2009. By October 2009, the production of hundreds of millions of doses of a monovalent vaccine containing 15 µg of HA and immunogenic after a single ion dose in the over 5-year olds is a quite remarkable achievement. Many countries have started to immunize at-risk groups, namely younger people <65 years of age with diabetes, obesity, chronic heart, or lung problems, and the immunosuppressed including pregnant women. It is forecast that up to 40–50% of some countries could volunteer for the vaccine. It is especially important that medical and nursing staff take the vaccine to protect both themselves and their patients. However, such large vaccination campaigns open schisms in modern societies,which on the one hand become very concerned about young persons dying but on the other hand have prejudices about vaccines in general. In the first winter wave the over 65s, unusually for a pandemic virus, are protected by prior experience of the H1N1 family and hence overall mortality is likely to be less than a seasonal year but the mortality is likely to be in younger persons, thus exposing our Achilles heel. Additionally nearly half the deaths to date have been in young persons without comorbidities.

Finally, within a year the virus is likely to mutate to allow it to infect the over 65 group, so, paradoxically, mortality in the second pandemic year could easily exceed the first.

16 Unlike Historical Vaccines Could Newly Developed Twenty-First Century Vaccines Induce Protection Across the Different Virus Subtypes?

There are 16 known subtypes of the HA of influenza A virus. Only three subtypes have caused pandemics in humans, H1, H2, and H3, while H5, H7, and H9 predominantly circulating in birds have crossed the species barrier into humans and caused human outbreaks. We do not know whether these latter three subtypes could mutate into human-to-human transmitters and thereby acquire pandemic potential. At present, H5N1 is causing considerable concern in SE Asia. An important question therefore is whether a vaccine could be engineered to give so-called heterotypic or cross-subtype immunity to protect against all these potentially pandemic viruses. It is well known that the internal proteins of influenza A virus such as M1, M2 and NP are shared by all influenza A viruses. These internally situated proteins are certainly immunogenic (particular NP) but could the immunity induced, either T cell or antibody, be broadly reacting?

To back up the central core of this approach, it has been known for 40 years that mice infected with an influenza A (H1N1) virus would later resist a lethal challenge from an influenza A (H3N2) virus. Given the lack of genetic and antigenic related-ness between the H1 and H3 proteins, or indeed the corresponding N1 and N2 proteins, this strong cross-immunity was attributed to an internal protein such as NP or M. However, it has been difficult to construct a solid database and there has been a lingering doubt about this so-called cross-protective immunity. Most virologists deduced, virtually by elimination, that a cross-reactive portion of the HA (HA2) could have provided the cross protection. Furthermore, this cross protection is particularly seen in the mouse model, leading some to conclude that the mouse recognized cross protection epitopes that perhaps humans did not.

Fundamental studies to correlate the genetics and immunology of NP and M established the cytotoxic T-cell response to portions of these proteins. However, the work clearly showed that M2 could be a cross-reactive immunogen, although a relatively weak one [95]. The M2 protein is an integral membrane protein of influenza A viruses that is expressed at the plasma membrane of virus-infected cells and is also present in small amounts on virions. The important extracellular domain, potentially targeted by antibodies and T cells, is conserved by virtually all influenza A viruses. Even the 1918 pandemic virus differs only in one amino acid. The first indication that the M2 was immunologically active was the observation that an anti-M2 monoclonal antibody reduced the spread of virus cell culture. Not unexpectedly, the antibody reacted with the extracellular domain of M2. Even more excitingly, the antibody reduced the replication of virus in mouse lungs. Immunization studies with M2 constructs, however, have given more mixed results. Immunization of mice with DNA plasmid of M1 and M2 gene gave protection mainly via T-helper cell activity. An alternative approach utilized a hepatitis B core and M2 fusion protein. The cross protection resided in antibodies, although M2-specific antibodies did not neutralize the virus in vitro. Presumably, protection was mediated by an indirect mechanism such as complement-mediated cytotoxicity or antibody-dependant cytotoxicity. However, the protection induced in the mouse model was considerably less than that induced by a conventional sub unit HA/NA vaccine.

It could be argued that weak heterotypic immunity may be present already in the community and that this is helping to prevent the emergence of chicken influenza A (H5N1) in SE Asia [96]. Certainly with evidence of tens of millions of domestic birds infected since late 2003 in 13 countries in SE Asia, with only a handful of human infections and only human-to-human transmission in family groups, there is a possibility that the unique cocirculation since 1977 of two influenza A viruses (H1N1 and H3N2) may have enhanced heterotypic immunity in most communities, which in turn abrogates the emergence of chicken influenza A (H5N1) into humans. It would be foolhardy, though, to take this argument to a fuller conclusion and relax preparations for a new pandemic influenza A virus.

17 The Historical Use of Volunteers to Study Influenza and Vaccines

At present, with the unprecedented research investment into influenza vaccines, there are new discoveries of adjuvants and vaccine formulations to be tested as well as fundamentals of virus transmission, infectiousness, and pathogenicity. The ultimate test is in influenza-infected volunteers. This specialized work was initiated over 60 years ago.

During the great pandemic of 1918, when the precise nature of the causative microbe of the Spanish influenza had not been established, a group of American scientists asked for young volunteers from the army and navy. The quest was to probe the nature of the microbe that was already causing devastation in their own country and where, by 1919, 500,000 young people were to die. However, this was not the first study into the precise nature of the microbe. The infection had first been documented a year earlier as a herald wave in the great city-sized military base and encampment of Etaples [6, 12]. Here the British army constructed the largest establishment [97] in its history, where 100,000 newly recruited soldiers each day intermingled with thousands of wounded soldiers, pigs and, in the nearby villages and markets, with ducks, domestic chickens, and geese. These are now recognized as the necessary biological features of an epicenter for the creation of a pandemic virus. We surmise, in retrospect, that an avian virus from a silently infected goose or duck could have crossed species either to a pig or to a soldier already infected with a human strain of influenza. This is the mixing bowl hypothesis. Indeed, common epidemic influenza was known to be circulating in the winter of 1916–1917 in Etaples. Another factor in Etaples could have been the hundreds of tons of gases of 25 varieties contaminating the landscape of the nearby Somme battlefield, as well as many of the wounded soldiers brought by the night trains into the 12 hospitals on site and causing respiratory distress. A group of pathologists there and at Abbeville, led by G. Gibson, raised the question of the nature of the microbe. Could it be a Gram-negative bacterium such as *H. influenzae*, already described by Pfeiffer as the cause of the previous influenza pandemic of 1889? Or could it be a virus? Viruses were rather unknown entities at that time but had been identified by their filter-passing nature. Hence, Gibson's experiment was quite simply to take sputum from a soldier victim and filter it through a Berkefield candle filter, which would hold back any known bacterium but allow the passage of the much smaller ultra-filterable virus. But what then? Gibson had not even considered that a human volunteer would receive the filtrate. In fact, he gave it to a series of macaques and, inadvertently, to himself. He died and the macaques became ill. His premature discovery of new virus influenza has lain undiscovered and hitherto unquoted in the archives of the First World War [98].

Meanwhile, in the USA, a more vigorous decision had been taken, and army and navy volunteers were infected intranasally with filtered material from Spanish influenza victims. Some volunteers were placed 0.5 m from dying servicemen, who coughed in their faces. The incredible result of this heroic endeavor is that not a single volunteer became ill, whereas all around the USA their companions were dying. It is more than possible that the volunteers had already been subclinically infected in the early summer outbreak of 1918, which was less virulent than the autumn virus and would be expected to give cross-immunity.

18   The MRC Common Cold and Influenza Quarantine Unit in Salisbury (UK)

As soon as the Second World War was over, the Medical Research Council in the UK established the Common Cold Unit in Salisbury at the Harvard Hospital. The hospital was a donation from the USA to cope with expected bomb casualties from London. In the event, this fully equipped multibuilding facility was used as an acute surgical hospital for servicemen. With Christopher Andrewes as its first chief scientist, the unit recruited volunteers to unravel the virological mysteries of respiratory disease. For the next 40 years, a small team of virologists and clinicians infected volunteers and discovered the first human coronavirus, the common cold virus, and were the first to describe the clinical effects of interferons. Essentially similar units were set up in the USA and USSR.

19   Estimates of Vaccine Protection Obtained in the Past by Deliberate Challenge in Quarantine Units

The considerable difficulties encountered in mounting field trials led to experiments in which immunized volunteers were subjected to deliberate inoculation with live virus in the form of either attenuated strain or modified wild-type strain. This protocol was suggested by Henle et al. [99], who immunized a group of children with inactivated influenza A (H1N1) virus vaccine and then inoculated them with egg-cultured virus of the same subtype but recently isolated, by inhalation of an aerosol. High rates of infection (75%) were produced in 28 unimmunized children of whom 10 became ill. Those receiving vaccine either escaped subsequent infection or developed serological changes; only 1 child of the 42 vaccinated children thus challenged became ill. Although this study illustrated the outstanding success of the immunized protocol, there are probably few observers today who would be prepared to submit their children to a similar risk of deliberately induced illness. Ideally young adults 18–45 are used for quarantine experiments. Such a risk is, of course, experienced during epidemics and Bell et al. [100] undertook a similar experiment in adult volunteers some of whom were immunized with a single dose of inactivated A/Japan/305/57 (H2N2) virus vaccine soon after the A/Asian epidemic began. The volunteers were isolated before being given intranasally pooled nasopharyngeal washings from patients with influenza and this caused clinical illness in 87% of volunteers previously given a placebo. As 50% of the vaccinated volunteers developed fever after challenge in this experiment, the single injection of inactivated vaccine proved relatively ineffective, presumably because of its inadequate immunogenicity.

The information obtained by deliberate challenge of immunized volunteers has been explored in the past using modified attenuated virus strains. Beare et al. [73] did this in their comparison of inactivated or live influenza B vaccines in which a challenge from the live virus B strain was used to assess the comparative efficacy of the two vaccines. Reinoculation with live virus was resisted better by those receiving the same material a month previously than by those injected with inactivated vaccine.

Couch [101] has reported a number of trials in volunteers after inactivated vaccine using a low dose of an essentially unmodified H3N2 virus that had received one or two passages in human embryonic kidney culture. It was first established by Greenberg et al. [102] that previous infection by homotypic H3N2 virus gave protection against deliberate exposure for up to 4 years after the original infection. Comparison of inactivated vaccine A/HongKong/68 (H3N2) given intranasally or subcutaneously showed that following challenge with live virus only those who had developed a serum antibody response after vaccine by either route resisted infection.

In a further trial of an anti-NA inactivated vaccine made from an Heq1N2 virus, it was shown that a reduced frequency of illness and a reduced titer of virus in nasal wash specimens resulted following live H3N2 virus challenge compared with the findings in control subjects. The number of those who contracted infection was also reduced somewhat by the inactivated NA vaccine, thus supporting the suggestion of Schulman et al. [103] that NA antibody, although incapable of neutralizing viral infectivity, could limit the extent of viral replication. Beutner et al. [104] also immunized children with an NA-specific vaccine and noted that antibody to NA had a role protecting against illness rather than against infection. Slepushkin et al. [105] and Monto and Kendal [106] came to similar conclusions with regard to NA vaccine and the clinical evidence of protection from illness.

A series of experiments on volunteers, designed to obtain evidence of protection from vaccines containing viruses that were homotypic or heterologous to the challenge virus, is important in relation to the determination of the best composition of inactivated vaccine. Potter et al. [43] gave one of four inactivated monovalent H3N2 virus vaccines to groups of students, measured their pre- and postimmunization antibodies by HI and NI tests, and later challenged all the groups with a live intranasal H3N2 virus (WRL 105). This virus was antigenically nearest to the A/Port Chalmers/73 virus and vaccine from this latter strain and also that containing A/Scotland/74 virus gave better protection against infection than earlier H3N2 virus vaccines; the result thus correlated with the induced HI antibody titers.

Larson et al. [107] also challenged the immunity produced by inactivated vaccine made from A/Port Chalmers/73 (H3N2) virus with that from a strain developed by the Pasteur Institute [108]. This virus (30c) with an antigen closely similar to A/England/72 (H3N2) was selected in the laboratory by a method analogous to natural selection by antigenic drift, and thus represents the first human attempt to anticipate antigen variation in nature. Challenge of those immunized with one or the other vaccines showed that protection by the heterologous 30c virus was about one-quarter as effective as that produced by the homologous A/Port Chalmers/73 virus.

Experiences related by Couch also confirm [101] that antibody effective against the homologous HA of the challenging virus is more protective than that formed by heterologous antigen. Protection was also compared after inactivated vaccine by intranasal or subcutaneous routes, which showed that the important mediator of immunity was the serum IgG content of anti-HA rather than the respiratory secretion content of specific IgA.

20   A New Retroscreen Quarantine Unit in London

We have established a new quarantine unit, based in London (http://www.retroscreen. com), but very much centered upon the experience and ethos of the Common Cold Unit of the past [109]. In a series of experiments over the past 2 years, we have infected over 250 young volunteers with influenza A (H3N2), influenza B, and influenza (H1N1) virus and more recently respiratory syncytial virus, and we now have fully characterized virus pools [110]. In the USA, a quarantine unit had already been established in Virginia and also at Baylor and pioneered work into the new

Fig. 8 A volunteer room at the Common Cold and Influenza Unit, Harvard Hospital, Salisbury, in the 1980s. Volunteers would stay for 2 weeks in this country-placed unit to be infected and carefully studied for clinical symptoms

NA inhibitors of influenza using an influenza A virus isolated in 1991 [111]. So far our own unit has focused on evaluating new influenza vaccines [112]. We use groups of 20 young volunteers and quarantine them in a student hostel or hotel or phase I clinical unit along with clinicians and scientists (Fig. 8). The MRC Common Cold Unit was rooted strongly in the postwar era with deck chairs, free run rabbits, country walks, afternoon cream teas, and two-course English meals. Our new unit reflects a more diverse community, so chicken tikka is as common on the menu as roast lamb and baked potatoes, but the wish of many of the volunteers is the same: to contribute to knowledge.

21   Conclusion

Influenza A virus has a proven record as a "bioterrorist" virus but driven not in Churchill's words by the "evil forces of perverted science" but by the vast unfathomable laws of nature and emergence, reemergence, and resurgence of natural disease. We are experiencing the attacks on pregnant women and younger persons at the present moment with A/Swine H1N1 [113–115]. Information from the human genome project, whereby a significant proportion of the 30,000 active genes are already known to be involved in innate and acquired immunity, provides reassurance that the immune system will continue to provide some protection against new viruses. This is excellently illustrated with A/Swine where most of the population has immune memory to this H1N1 family of viruses.

Gauguin in his last great painting "Who are we, where have we come from, where are we going?" asks crucial questions about the future of humankind. But it was the medieval painter Breugel who asked the major question, yet to be answered in the twenty-first century. His medieval painting "The Triumph of Death" shows a horseman on a white charger scything at random and gathering souls during an outbreak of *Pasteurella pestis* in medieval times. The question haunting the painting is "why do some persons survive while others die." Even in 1918 in most communities 99% of persons infected with the virus survived. But why did some die and exactly how were they killed by such a minute and fragile form of life that we know as the orthomyxovirus influenza? Was the immune reaction and ensuing cytokine storm overwhelming or was virus replication in the endothelial cells of the air sacs more important?

An extraordinary clear message is emerging, which tells us to build our public health infrastructure and continue and expand our epidemiological vigilance and surveillance against all these infectious viruses and bacteria. The virus cannot be permanently dislodged from its avian and swine reservoir. For pandemic influenza, every country needs a detailed and practical plan and a supply of antiviral drugs and new vaccines at hand for an emergence of H5N1. This virus will be a lot more difficult to deal with than A/Swine H1N1. We would then be "at the end of the beginning" as regards protection of all citizens. Influenza was the twentieth century's weapon of mass destruction. Nature is the greatest bioterrorist of our world and emerging viruses could do for us all, as easily and as quickly, or even more so, than the Great Influenza of 1918, except for the fact that we now have the ammunition to fight back: knowledge of virus transmission and how to break it with disinfectants and social distancing, and effective antivirals and vaccines. The current A/Swine H1N1 pandemic has exposed flaws in pandemic plans and also has exposed many countries that have no preparation whatsoever.

Acknowledgments We are pleased to receive grant income from the EU to develop new influenza vaccines.

References

1. Phillips H, Killingray D (2002) The Spanish influenza pandemic of 1918–1919: new perspectives. Routledge Social History of Medicine Series. Routledge, UK
2. Churchill WS (1993) The Great War, vol 1 and 2. George Newnes Ltd, London
3. Crosby AW (1918) America's forgotten pandemic. Cambridge University Press, New York
4. Medical Research Committee (1919) Studies of influenza in hospitals of the British armies in France, 1918. Special report series no 36. HM Stationery Office, London, p 112
5. Ministry of Health (1920) Reports on the pandemic of influenza 1918–1919. Reports on public health and medical subjects, no. 4. Stationery Office, London
6. Oxford JS (2000) Influenza A pandemics of the 20th century with special reference to 1918: virology, pathology and epidemiology. Rev Med Virol 10:119–133

7. Macpherson WG, Herringham WP, Elliott TR, Balfour A (1927) Medical services diseases of the war. Medical aspects of aviation and gas warfare and gas poisoning, vol 2. HMSO, London
8. Collier L, Oxford JS (2007) Human virology: a text for students of medicine. Oxford University Press, Oxford
9. Stuart-Harris CH, Schild GC, Oxford JS (1983) Influenza: the viruses and the disease. Edward Arnold, London
10. Ferguson NM, Cummings DA, Cauchemez S, Fraser C, Riley S, Meeyai A, Iamsirithaworn S, Burke DS (2005) Strategies for containing an emerging influenza pandemic in Southeast Asia. Nature 437:209–214
11. Barry JM (2004) The great influenza, the epic story of the deadliest plague in history. Viking, New York
12. Oxford JS (2005) Preparing for the first influenza pandemic of the 21st century. Lancet Infect Dis 5:129–132
13. Oxford JS, Lambkin-Williams R, Sefton A, Daniels R, Elliot A, Brown R, Gill D (2005) A hypothesis: the conjunction of soldiers, gas, pigs, ducks, geese and horses in Northern France during the Great War provided the conditions for the emergence of the "Spanish" influenza pandemic of 1918–1919. Vaccine 23:940–945
14. House of Lords Report on Pandemic Influenza (2005) HMSO, London
15. Miller GL, Stanley WM (1944) Quantative aspects of the red blood cell agglutination test for influenza virus. J Exp Med 79:185
16. Burnet FM (1941) Growth of influenza virus in the allantoic cavity of the chick embryo. Aust J Exp Biol Med Sci 19:291
17. Hobson D, Curry RL, Beare AS, Word-Gardner A (1972) The role of serum HI antibody in protection against challenge infection with influenza A and B viruses. J Hyg 70:767–777
18. Schild GC, Wood TM, Newman RW (1975) A single radial immunodiffusion technique for the assay of haemagglutinin antigen. Bull World Health Organ 52:223–231
19. Palese P, Schulman JL (1976) Mapping of the influenza virus genome: identification of the haemagglutinin and neuraminidase genes. Proc Natl Acad Sci USA 73:2142–2146
20. Hoffman E, Neumann G, Kawaoka Y, Hoborn G, Webster RG (2000) A DNA transfection system for generation of influenza A virus from eight plasmids. Proc Natl Acad Sci USA 97:6108–6113
21. Schickli JH, Flandorfer A, Nakaya T, Martinez-Sobrido L, Garcia-Sastre A, Palese P (2001) Plasmid-only rescue of influenza A virus vaccine candidates. Philos Trans R Soc Lond B Biol Sci 356:1965–1973
22. Kistner O, Barrett PN, Mundt W, Reiter M, Schober-Bendixen S, Dorner F (1998) Development of a mammalian cell (Vero) derived candidate influenza virus vaccine. Vaccine 16:960–968
23. Palache AM, Brands R, van Scharrenburg G (1997) Immunogenicity and reactogenicity of influenza subunit vaccines produced in MDCK cells or fertilised chicken eggs. J Infect Dis 176(Suppl 1):S20–S23
24. Francis T Jr, Nagill TP (1935) Immunological studies with the virus of influenza. J Exp Med 62:505
25. Andrewes CH, Smith W (1937) Influenza: further experiments on the active immunisation of mice. Br J Exp Pathol 18:43
26. Commission on Influenza, Board of Influenza and Other Epidemic Diseases in the Arm (1944) A clinical evaluation of vaccination against influenza. J Am Med Assoc 124:982
27. Davenport FM, Hennessy AV, Brandon FM, Webster RG, Barrett CD Jr, Lease GO (1964) Comparisons of serological and febrile responses in humans to vaccination with influenza viruses or their haemagglutinins. J Lab Clin Med 63:5–13
28. Brandon FB, Cox F, Lease GO, Timm EA, Quinn E, McLean IW Jr (1967) Respiratory virus vaccines. III. Some biological properties of sephadex-purified ether-extracted influenza virus antigens. J Immunol 98:800–805

29. Duxbury AE, Hampson AW, Sievers JGM (1968) Antibody response in humans to deoxycholate-treated influenza virus vaccines. J Immunol 101:62–67
30. Francis T Jr, Salk JE, Quilligan JJ Jr (1947) Experience with vaccination against influenza in the spring of 1947. Am J Public Health 37:1013–1016
31. Loosli CG, Schoenberger J, Barnett G (1948) Results of vaccination against influenza during the epidemic of 1947. J Lab Clin Med 33:789
32. Kilbourne ED (1969) Future influenza vaccines and use of genetic recombinants. Bull World Health Organ 41:643–645
33. Reimer CB, Baker RS, van Frank RM, Newlin TE, Cline GB, Anderson NG (1967) Purification of large quantities of influenza virus by density-gradient centrifugation. J Virol 1:1207–1216
34. Glezen WP, Loda FA, Denny FW (1969) A field evaluation of inactivated, zonal-centrifuged influenza vaccines in children in Chapel Hill, North Carolina, 1968–1969. Bull World Health Organ 41:566–569
35. Zakstelskaja LJa, Yakhno MA, Isacenko VA, Molibg EV, Hlustov SA, Antonova IV et al (1978) Influenza in the USSR in 1977: recurrence of influenza virus A subtype H1N1. WHO Bulletin 56:919
36. Salk JE (1948) Reactions to concentrated influenza vaccines. J Immunol 58:369
37. Potter CW, Jennings R, Clark A (1977) The antibody response and immunity to challenge infection induced by whole inactivated and Tween-ether split influenza vaccines. Dev Biol Stand 39:323–328
38. Ennis FA, Mayner RE, Barry DW, Manischewitz JE, Dunlap RC, Verbonitz MW, Bozeman RM, Schild GC (1977) Correlation of laboratory studies with clinical responses to A/New Jersey influenza vaccines. J Infect Dis 136 (Suppl):S397–S406
39. Wood JM, Schild GC, Newman RW, Seagroatt V (1977) Application of an improved single-radial-immunodiffusion technique for the assay of influenza haemagglutinin antigen content of whole virus and subunit vaccines. Dev Biol Stand 39:193–200
40. Holland WW, Isaacs A, Clarke SKR, Heath RB (1958) A serological trial of Asian influenza vaccine after the autumn epidemic. Lancet 271:820–822
41. Nicholson KG, Tyrrell DAJ, Harrison P, Potter CW, Jennings R, Clark A (1979) Clinical studies of monovalent inactivated whole virus and subunit A/USSR/77 (H1N1) vaccine; serological responses and clinical reactions. J Biol Stand 7:123–136
42. Mostow SR, Schoenbaum SC, Dowdle WR, Coleman MT, Kaye HS, Hierholzer JC (1970) Studies on inactivated influenza vaccines. II. Effect of increasing dosage on antibody with resistance to influenza in man. Am J Med 92:248–256
43. Potter CW, Jennings R, Nicholson K, Tyrrell DAJ, Dickinson KG (1977) Immunity to attenuated influenza virus WRL 105 infection induced by heterologous, inactivated influenza A virus vaccines. J Hyg (Lond) 79:321–332
44. Brady MI, Furminger IGS (1976) A surface antigen influenza vaccine. 1. Purification of haemagglutinin and neuraminidase proteins. 2. Pyrogenicity and antigenicity. J Hyg (Camb) 77:161–172
45. Pandemic Working Group of Medical Research Council's Committee on Influenza and Other Respiratory Virus Vaccines (1977) Antibody responses and reactogenicity of graded doses of inactivated influenza A/New Jersey/76 whole-virus vaccine in humans. J Infect Dis 136: S475
46. Medical Research Council Committee on Influenza Vaccine (1953) Clinical trials of influenza vaccine. Br Med J 2:1–7
47. Medical Research Council Committee on Influenza Vaccine (1957) Clinical trials of influenza vaccine. Br Med J 2:1–7
48. Medical Research Council Committee on Influenza Vaccine (1958) Trials of an Asian influenza vaccine. Br Med J 1:415–418
49. Medical Research Council Committee on Influenza Vaccine (1964) Clinical trials of oil-adjuvant influenza vaccine, 1960–3. Br Med J 2:267–271

50. Potter CW, Jennings R, Phair JP, Clarke A, Stuart-Harris CH (1977) Dose-response relationship after immunisation of volunteers with a new surface-antigen-adsorbed influenza virus vaccine. J Infect Dis 135:423–431

51. Kendal AP, Bozeman FM, Ennis FA (1980) Further studies of the neuraminidase content of inactivated influenza vaccines and the neuraminidase antibody responses after vaccination of immunologically primed and unprimed populations. Infect Immun 29:966–971

52. Webster RG, Kasel JA, Couch RB, Laver WG (1976) Influenza virus subunit vaccines. II. Immunogenicity and original antigenic sin in humans. J Infect Dis 134:48–58

53. Oxford JS, Schild GC, Potter C, Jennings R (1979) The specificity of the antihaemagluttinin antibody response induced in man by inactivated vaccines and by natural infection. J Hyg (Camb) 82:51–56

54. Oxford JS, Haaheim LR, Slepushkin A, Werner J, Kuwert E, Schild GC (1981) Strain specificity of serum antibody to the haemagglutinin of influenza A (H3N2) viruses in children following immunisation or natural infection. J Hyg (Camb) 86:17–26

55. Appleby JC, Himmelweit F, Stuart-Harris CH (1951) Immunisation with influenza virus a vaccines: comparison of intradermal and subcutaneous routes. Lancet 257:1384–1387

56. McCarroll JR, Kilbourne ED (1958) Immunisation with Asian strain influenza vaccine – equivalence of the subcutaneous and intradermal routes. N Engl J Med 259:618–621

57. Tauraso NM, Gleckman R, Pedreira FA, Sabbaj J, Yahwak R, Madoff MA (1969) Effect of dosage and route of inoculation upon antigenicity of inactivated influenza virus vaccine (Hong Kong strain) in man. Bull World Health Organ 41:507–516

58. Waldman RH, Case JA, Fulk RV, Togo Y, Hornick RB, Heiner GG, Dawkin Jun AT, Mann JJ (1968) Influenza antibody in human respiratory secretions after subcutaneous or respiratory immunisation with inactivated virus. Nature 218:594–595

59. Waldman RH, Wigley FM, Small PA Jr (1970) Specificity of respiratory secretion antibody against influenza virus. J Immunol 105:1477–1483

60. Phillips CA, Forsythe BR, Christmas WA, Gump DW, Whorton EB, Rogers I, Rudin A (1970) Purified influenza vaccine; clinical and serological response to varying doses and different routes of immunisation. J Infect Dis 122:26–32

61. Potter CW, Stuart-Harris CH, McClaren C (1972) Antibody in respiratory secretions following immunisation with influenza virus vaccines. In: Perkins FT, Regamey RHS (eds) International symposium series immunological standardisation, vol 20. Karger, Basel, p 198

62. Ruben FL, Potter CW, Stuart-Harris CH (1975) Humoral and secretory antibody responses to immunisation with low and high dosage split influenza virus vaccines. Arch Virol 47:157–166

63. Downie JC, Stuart-Harris CH (1970) The production of neutralising activity in serum and nasal secretions following immunisation with influenza B virus. J Hyg (Camb) 68:233–244

64. Ennis FA, Dowdle WR, Barry DW, Hochstein HD, Wright PF, Karzon DT, Marine WM, Meyer HM Jr (1977) Endotoxin content and clinical reactivity to influenza vaccines. J Biol Stand 5:165–167

65. Wells CEC (1971) A neurological note on vaccinations against influenza. Br Med J 3:755–756

66. Langmuir AD (1979) Guillain-Barré syndrome: the swine influenza virus vaccine incident in the United States of America, 1976–77. J R Soc Med 72:660–669

67. Hurwitz ES, Schonberger LB, Nelson DB, Holman RC (1981) Guillain-Barré syndrome and the 1978–1979 influenza vaccine. N Engl J Med 304:1557–1561

68. Mogabgab WJ, Liederman E (1970) Immunogenicity of 1967 polyvent and 1968 Hong Kong influenza vaccines. J Am Med Assoc 211:1672–1676

69. Knight V, Couch RB, Douglas RG, Tauraso NM (1971) Serological responses and results of natural infectious challenge of recipients of zonal ultracentrifuged influenza.A2/AICHI/2/68 vaccine. Bull World Health Organ 45:767–771

70. Mawson J, Swan C (1943) Intranasal vaccination of humans with living attenuated influenza virus strains. Med J Aust 1:394

71. Stuart-Harris CH (1980) Present status of live influenza virus vaccine. J Infect Dis 142:784

72. Kendal AP, Maasab HF, Alexandrova GI, Ghendon YZ (1981) Development of cold-adapted recombinant live attenuated influenza A vaccines in the USA and USSR. Antiviral Res 1:339

73. Beare AS, Bynoe ML, Tyrrell DAJ (1968) Investigation into attenuation of influenza viruses by serial passage. Br Med J 4:482–484

74. Huygelen C, Petermans J, Vascoboinic E, Berge E, Colinet G (1973) Live attenuated influenza virus vaccine in vitro and in vivo properties. In: Perkins FT, Regamey RHS (eds) International symposium on influenza vaccines for man and horses. Series immunobiological standards, vol 20. Karger, Basel, p 152

75. Florent G, Lobmann M, Beare AS, Zygraich N (1977) RNA's of influenza virus recombinants derived from parents of known virulence for man. Arch Virol 54:19–28

76. Florent G (1980) Gene constellation of live influenza A vaccines. Arch Virol 64:171–173

77. Beare AS, Hall TS (1971) Recombinant influenza A viruses as live vaccine for man. Lancet 298:1271–1273

78. Beare AS, Reed S (1977) The study of antiviral compounds in volunteers. In: Oxford JS (ed) Chemoprophylaxis and viral infections of the respiratory tract, vol 2. CRC, Cleveland, p 27

79. Oxford JS, McGeoch DJ, Schild GC, Beare AS (1978) Analysis of virion RNA segments and polypeptides of influenza A virus recombinants of defined virulence. Nature 273:778–779

80. Poliomyelitis Congresses (1948–1961) Papers and discussions at 1st, 2nd, 3rd, 4th and 5th international poliomyelitis congresses 1951, 1954, 1957 and 1961. Lippincott, Philadelphia

81. Jennings R, Potter CW, Teh CZ, Mahmud MI (1980) The replication of type A influenza viruses in the infant rat: a marker for virus attenuation. J Gen Virol 49:343–354

82. Murphy BR, Clements ML, Maasab HF, Buckler-White AJ, Tian S-F, London WT, Chanock RM (1984) The basis of attenuation of virulence of influenza virus for man. In: Stuart-Harris CH, Potter CW (eds) Molecular virology and epidemiology of influenza. Academic, London, p 211

83. Chanock RM, Murphy BR (1979) Genetic approaches to control of influenza. Perspect Biol Med 22:S37

84. Richman DD, Murphy BR, Chanock RM, Gwaltney JM Jr, Douglas RG, Betts RF, Blacklow NR, Rose FB, Parrino TA, Levine MM, Caplan ES (1976) Temperature-sensitive mutants of influenza A virus XII. Safety, antigenicity, transmissibility and efficacy of influenza A/Udorn/72-*ts*-1[E] recombinant viruses in human adults. J Infect Dis 134:585–594

85. Maassab HF (1967) Adaptation and growth characteristics of influenza virus at 25°C. Nature 213:612–614

86. Maassab HF (1969) Biological and immunologic characteristics of cold-adapted influenza virus. J Immunol 102:728–732

87. Spring SB, Maassab HF, Kendal AP, Murphy BR, Chanock RM (1977) Cold adapted variants of influenza A. II. Comparison of the genetic and biological properties of *ts* mutants and recombinants of the cold-adapted A/Ann Arbor/6/60 strain. Arch Virol 55:233–246

88. Wright PF, Okabe N, McKee KT Jr, Maasab HF, Karzon DT (1982) Cold-adapted recombinant influenza A virus vaccines in young seronegative children. J Infect Dis 146:71–79

89. Murphy BR, Tierney EL, Barbour BA, Yolken RH, Alling DW, Holley HP Jr, Mayner RE, Chanock RM (1980) Use of the enzyme-linked immunosorbent assay to detect serum antibody responses of volunteers who received attenuated influenza A virus vaccine. Infect Immun 29:342–347

90. Alexandrova GI, Smorodintsev AA (1965) Obtaining of an additionally attenuated vaccinating cryophilic influenza strain. Rev Roum Inframicrobiol 2:179

91. Alexieva RB, Petrova SM, Janceva BN (1971) Studies on some biological properties of vaccinal influenza strains cultivated at low temperatures. In: Gusic B (ed) Proceedings of the symposium on live influenza vaccine. Yugoslav Academy of Science and Arts, Zagreb, p 43

92. Zhilova GP, Alexandrova GI, Zykov MP, Smorodintsev AA (1977) Some problems with modern influenza prophylaxis with live vaccine. J Infect Dis 135:681–686

93. Morris CA, Freestone DS, Stealey VM, Oliver PR (1975) Recombinant WRL 105 strain live attenuated influenza vaccine. Immunogenicity, reactivity and transmissibility. Lancet 306:196–199

94. Schild GC, Oxford JS, de Jong JC (1983) Evidence for host-cell selection of influenza virus antigenic variants. Nature 303:706–709
95. Neiryncks S, Deroot T, Saelens X, Vanland Schoot P, Tou WM, Friers W (1999) A universal influenza A vaccine based on the extra cellular domain of the M2 protein. Nat Med 5:1157–1163
96. Rimmelzwaan GF, Baars M, van Beek R, van Amerongen G, Lövgren-Bengtsson K, Claas EC, Osterhaus AD (1997) Induction of protective immunity against influenza virus in a macaque model: comparison of conventional and ISCOM vaccines. J Gen Virol 78:757–765
97. Britain V (1989) Testament of youth: an autobiographical study of the years 1900–1925. Penguin, New York
98. Gibson HG, Bowman FB, Connor JI (1919) The etiology of influenza: a filterable virus as the cause (with some notes on the culture of the virus by the method of Noguchi). In: Studies of influenza in hospitals of the British armies in France, 1918, no. 36. HMSO, London, pp 19–36
99. Henle W, Henle G, Stokes J Jr (1943) Demonstration of the efficacy of vaccination against influenza type A by experimental infection of human beings. J Immunol 46:163
100. Bell JA, Ward TG, Kapikian AZ, Shelokov A, Reichelderfer TE, Huebner RJ (1957) Artificially induced Asian influenza in vaccinated and unvaccinated volunteers. J Am Med Assoc 165:1366–1373
101. Couch RB (1975) Assessment of immunity to influenza virus using artificial challenge of normal volunteers with influenza virus. Dev Biol Stand 28:295–306
102. Greenberg SB, Couch RB, Kasel JA (1973) Duration of immunity to type A influenza. Clin Res 21:600
103. Schulman JL, Khakpour M, Kilbourne ED (1968) Protective effects of specific immunity to viral neuraminidase on influenza virus infection of mice. J Virol 2:778–786
104. Beutner KR, Chow T, Rubi U, Strussenberg J, Clement J, Ogra PL (1979) Evaluation of a neuraminidase-specific influenza A virus vaccine in children. Antibody responses and effects on two successive outbreaks of natural infection. J Infect Dis 140:844–850
105. Slepushkin AN, Schild GC, Beare AS, Chinn S, Tyrrell DAJ (1971) Neuraminidase and resistance to vaccination with live influenza A2 Hong Kong vaccine. J Hyg (Camb) 69:571–578
106. Monto AS, Kendal AP (1973) Effect of neuraminidase antibody on Hong Kong influenza. Lancet 301:623–625
107. Larson HE, Tyrrell DAJ, Bowker CH, Potter CW, Schild GC (1978) Immunity to challenge in volunteers vaccinated with an inactivated current or earlier strain of influenza A (H3N2). J Hyg (Camb) 80:243–248
108. Fazekas de St. Groth S, Hannoun C (1973) Sélection par pression immunologique de mutants dominants du virus de la grippe A (Hong Kong). C R Hebd Seances Acad Sci 276:1917
109. Tyrrell D, Fielder M (2002) Cold wars: the fight against the common cold. Oxford University Press, Oxford
110. Fries L, Lambkin-Williams R, Gelder C, White G, Burt D, Lowell G, Oxford J (2004) FluInsure[TM], an inactivated trivalent influenza vaccine for intranasal administration, is protective in human challenge with A/Panama/2007/99 (H3N2) virus. In: Kawaoka Y (ed) Options for the control of influenza, V. International congress series, vol 1263. Elsevier, London, pp 661–665
111. Treanor JJ, Hayden FG (1998) Volunteer challenge studies. In: Nicholson KG, Webster RG, Hay AJ (eds) Textbook of influenza. Blackwell, Oxford
112. Jones S, Evans K, McElwaine-John H, Sharpe M, Oxford J, Lambkin-Williams R, Mant T, Nolan A, Zambon M (2008) DNA vaccination protects against an influenza challenge in a phase 1b double blind randomised placebo controlled clinical trial. Vaccine 27(18):2506–2512
113. Wilson N, Baker MG (2009) The emerging influenza pandemic: estimating the case fatality ratio. Euro Surveill 14:19255B
114. Garske T, Legrand J, Donnelly CA, Ward H, Cauchemez S, Fraser C, Ferguson NM, Ghani AC (2009) Assessing the severity of the novel A/H1N1 pandemic. BMJ 339:b2840
115. CDC. Novel H1N1 influenza vaccine. http://www.cdc.gov/h1n1flu/vaccination/public/vaccination_qa_pub.htm

Influenza and Influenza Vaccination in Children

Romina Libster and Kathryn M. Edwards

Abstract Ecological and active population-based surveillance studies have clearly shown the large burden of seasonal influenza disease in children, both in hospital and in outpatient settings. Mortality and encephalitis due to seasonal influenza have also been reported. The recent emergence of a novel H1N1 strain and its global spread have also had a major impact on children. Two influenza vaccines are licensed for use in children: trivalent inactivated and live-attenuated vaccines. Both have been shown to be efficacious for the prevention of clinical and laboratory-confirmed seasonal influenza. In recent comparative trials in young children, live-attenuated vaccines were shown to be more effective than trivalent inactivated vaccines for the prevention of laboratory-confirmed influenza. However, episodes of wheezing were increased in the youngest children receiving live-attenuated vaccine. Trivalent inactivated influenza vaccine has an excellent safety profile and has been mainly associated with minor local pain and tenderness at the injection site. Vaccine efficacy for the inactivated vaccine has been shown to be greater in older children. Vaccines for the prevention of the novel H1N1 strain have also been tested for safety and immunogenicity in children. The increased use of either inactivated or live influenza vaccines directed at seasonal and pandemic strains has the potential to reduce the influenza disease burden in children and to potentially extend herd protection to those who are unvaccinated.

R. Libster
Fundacion INFANT, Buenos Aires, Argentina
Department of Pediatrics, Vanderbilt Vaccine Research Program, Vanderbilt University School of Medicine, Nashville, TN 37232, USA

K.M. Edwards (✉)
Department of Pediatrics, Vanderbilt Vaccine Research Program, Vanderbilt University School of Medicine, Nashville, TN 37232, USA

G. Del Giudice and R. Rappuoli (eds.), *Influenza Vaccines for the Future*, 2nd edition,   149
Birkhäuser Advances in Infectious Diseases,
DOI 10.1007/978-3-0346-0279-2_7, © Springer Basel AG 2011

1   Introduction

Over the past several years, a number of ecological studies have demonstrated the excessive burden of influenza disease in children [1, 2]. Izurieta et al. [1] used local viral surveillance to define periods when the circulation of influenza viruses predominated over that of respiratory syncytial virus (RSV) and calculated rates of hospitalization for acute respiratory disease in children younger than 18 years of age enrolled in two large health maintenance organizations (HMOs). Among children without high-risk conditions, hospitalization rates in children younger than 2 years of age were 231 per 100,000 person-months in one HMO and 193 per 100,000 person-months in the other. In children 5–17 years of age, rates were 19 per 100,000 person-months in one HMO and 16 per 100,000 person-months in the other. Finally, among high-risk children 5–17 years of age, hospitalization rates were 386 per 100,000 person-months and 216 per 100,000 person-months in the two HMOs, respectively.

In another ecological study, Neuzil et al. [2] assessed the influenza burden in a large cohort of children less than 15 years of age enrolled in the Tennessee Medicaid program. Over a period of 19 years and a total of 2,035,143 person-years of observation, the average number of hospitalizations each year for cardiopulmonary conditions attributable to influenza was 10.4 per 1,000 children younger than 6 months of age, 5.0 per 1,000 for those 6–12 months, 1.9 per 1,000 for those 1–3 years, 0.9 per 1,000 for those 3–5 years, and 0.4 per 1,000 for those 5–15 years. In addition, for every 100 children there were an average of 6–15 outpatient visits and 3–9 courses of antibiotics attributable to influenza disease each year [2].

Rates observed in these ecological studies were confirmed through an active, prospective, population-based surveillance network [3–5]. Children younger than 5 years of age residing in three US counties were enrolled during hospitalizations or either outpatient or emergency department visits for acute respiratory tract infections or fever. Nasal and throat swabs were tested for influenza virus by viral culture and polymerase chain reaction assay, and epidemiological data were collected [5]. Combining data from four influenza seasons, the average annual hospitalization rates associated with influenza were 0.9 per 1,000 children (Table 1). The rates were 4.5 per 1,000 children less than 6 months of age, 0.9 per 1,000 children 6–23 months of age, and 0.3 per 1,000 children 24–59 months of age. The estimated burden of outpatient and emergency department visits associated with influenza was even greater and depended upon the severity of the influenza season (Table 2). During 2 years of outpatient surveillance, there were between 50 and 95 clinic visits and 6–27 emergency department visits per 1,000 children per year. Remarkably, only 28% of the hospitalized children with laboratory-confirmed influenza and only 17% of those seen in the outpatient settings with confirmed influenza were diagnosed with influenza by their treating physician. This is despite the availability of rapid diagnostic tests for the confirmation of influenza in young children [6–9]. Population-based estimates from other US studies have provided comparable rates using different study years, populations, and study methods [10–16].

Additional studies of influenza burden in children have also been conducted in other countries. Montes et al. [17] determined the incidence of virologically

Table 1 Rate of hospitalizations attributable to influenza per 1,000 children, according to age group and study year[a]

Age group	2000–2001	2001–2002	2002–2003	2003–2004	2000–2004
0–5 months					
Rate (95% CI)	2.4 (1.0–3.9)	4.3 (2.2–6.6)	2.3 (0.9–3.8)	7.2 (5.3–9.2)	4.5 (3.4–5.5)
6–23 months					
Rate (95% CI)	0.6 (0.2–1.2)	0.9 (0.4–1.3)	0.4 (0.1–0.7)	1.5 (1.0–2.1)	0.9 (0.7–1.2)
24–59 months					
Rate (95% CI)	0.2 (0.1–0.4)	0.3 (0.1–0.6)	0.04 (0.00–0.13)	0.6 (0.3–0.9)	0.3 (0.2–0.5)
0–59 months					
Rate (95% CI)	0.6 (0.3–0.8)	0.9 (0.6–1.2)	0.4 (0.2–0.6)	1.5 (1.2–1.9)	0.9 (0.8–1.1)

Modified from [5]

[a]Numbers are combined rates for three sites in the NVSN. CI denotes confidence interval. Counts were weighted for days of surveillance and proportion of eligible children enrolled

Table 2 Outpatient visits for acute respiratory tract infection or fever associated with confirmed influenza

Age group	Visits for acute respiratory tract infection or fever associated with confirmed influenza		Mean rate of visits for acute respiratory tract infection or fever, 1998–2002[a]	Estimated rate of visits attributable to influenza[b]	
	2002–2003	2003–2004	No./1,000 children (95% CI)	No./1,000 children (95% CI)	
	% (95% CI)			2002–2003	2003–2004
Outpatient clinics					
0–59 months	10.2 (7.5–13.6)	19.4 (16.0–23.1)	489 (387–591)	50 (35–71)	95 (72–125)
Emergency departments					
0–59 months	5.9 (3.7–8.9)	28.8 (25.0–32.7)	94 (78–110)	6 (4–9)	27 (22–33)

Modified from [5]

CI confidence interval

[a]The mean rate of visits for acute respiratory tract infection or fever per 1,000 children was calculated from the National Ambulatory Medical Care Survey/National Hospital Ambulatory Medical Care Survey

[b]Rates were calculated by multiplying the proportions of visits for acute respiratory tract infection or fever associated with confirmed influenza (columns 2 and 3) by the mean rate of visits for acute respiratory tract infections or fever, 1998–2002 (column 4)

confirmed influenza-related hospitalizations in children aged <5 years in southern Spain during three study years. Their average yearly hospitalization rates were 410 per 100,000 children less than 6 months of age, 80 per 100,000 children 6–11 months of age, 70 per 100,000 children 12–23 months of age, and 50 per 100,000 children aged 24–59 months. These rates are nearly identical to those reported by Poehling et al. [5]. In a retrospective, population-based study, Chiu et al. [18] determined the annual laboratory-confirmed influenza-associated hospitalization rates among children 15 years old or younger who lived in Hong Kong. The adjusted rates of hospitalization attributable to influenza were

2,800 per 100,000 children less than 1 year of age, 2,100 per 100,000 children 1–2 years of age, 900 per 100,000 children 2–5 years of age, 400 per 100,000 children 5–10 years of age, and nearly 100 per 100,000 children 10–15 years of age. These rates are considerably higher than those reported from either Spain or the USA [5, 17]. In a recent prospective population-based study, Chiu et al. [19] evaluated virologically confirmed hospitalization rates due to influenza virus infection in three consecutive seasons among children under 18 years old who lived in Hong Kong Island. Each year different viruses circulated; during 2003–2004 H3N2 predominated, during 2004–2005 86% of the viruses were H3N2, and during 2005–2006 94% of the viruses were H1N1. The highest rates of hospitalization for influenza A were seen in children <2 years of age and for influenza B in children aged 2–4 years. Hospitalization rates due to influenza A during the 2004–2005 season were 1,038 cases per 100,000 children aged <1 year. During the other two seasons, children 1 year of age had the highest hospitalization rates at 955 and 546 cases per 100,000 populations during 2003–2004 and 2005–2006 periods, respectively. Only 7% of the subjects had received influenza vaccination. Hospitalization rates reported in this study were lower than those described in the previous study from Hong Kong but still higher than those from other countries [20, 21]. Even with differences in rates, these studies highlight the important burden of influenza in young children.

Mortality associated with influenza also occurs in children. During the 2003–2004 influenza season in the USA, 153 pediatric influenza-associated deaths were reported to the Centers for Disease Control and Prevention (CDC) [22]. The median age of those who died was 3 years, 96 children were younger than 5 years old, and the highest mortality rate was noted in those less than 6 months of age. In terms of mortality, 47 of the children died outside a hospital setting, 45 died within 3 days of illness onset, and bacterial coinfections were identified in 24% of the children. Only 33% of the children had underlying medical conditions associated with increased influenza risk.

A recent paper highlights the role of bacterial superinfection in the mortality associated with influenza in children. One-hundred-sixty-six influenza-associated pediatric deaths were reported during the 3-year study period (2004–2007) with similar numbers of deaths during the first 2 years and increasing during the third year (47, 46, and 73). The percent with bacterial coinfection increased each year (6%, 15%, and 34%, respectively). The median number of days between onset of symptoms and death ranged from 3 to 4 days with 75% of deaths occurring within 7 days. *Staphylococcus aureus* was the most commonly identified bacterial pathogen; 60% (15 patients) of the isolates were methicillin resistant (MRSA), and 6 were methicillin susceptible (MSSA). The proportion of children with underlying high-risk conditions, including asthma, seizure disorders, and neuromuscular diseases, decreased from 55% in 2004–2005 period to 35% in 2006–2007. Most of the deaths occurred in previously healthy children without underlying medical condition [23].

Another severe complication of influenza is encephalopathy and has been described in Asian children, and less commonly in European and US children [24]. Influenza encephalitis has a fatality rate of nearly 30%, and nearly one third of the survivors are left with permanent disability. Influenza-associated encephalopathy occurs early in the influenza illness and is manifested by confusion, seizures, and progressive coma. Imaging studies show uniform cerebral edema with necrosis of the thalamus and other deep brain structures noted in 10–20% of victims. Elevated levels of proinflammatory cytokines have been measured in these patients and have been postulated to contribute to disease pathogenesis [24, 25].

2 Emergence of Pandemic H1N1 Virus

Triple-reassortant swine influenza A viruses, containing human, swine, and avian influenza genes, have been isolated from swine in the USA since 1998 [26, 27]. From 2005 to 2009, 12 cases of human infection with such viruses were reported in the USA [28]. Then, in April 2009, the CDC identified two cases of human infection with a novel swine-origin influenza A (H1N1) virus (S-OIV) characterized by a unique new combination of gene segments that had not been previously identified [29]. This virus rapidly spread throughout the world and by October 11, 2009, more than 399,232 laboratory-confirmed cases with over 4,735 deaths had been reported to World Health Organization (WHO). All these novel influenza H1N1 viral isolates were found to be antigenically and genetically similar to the A/California/7/2009-like pandemic H1N1 2009 virus [30].

Clinical characterization of 272 patients hospitalized with the novel H1N1 influenza virus in the USA from April 2009 to mid-June 2009 indicated that 25% of the patients were admitted to an intensive care unit, 7% died, and their median age was 26 years (range 1.3–57). Forty-five percent of the patients were children under the age of 18, 38% were between 18 and 49 years of age, and 5% were 65 years of age or older. Seventy-three percent of the patients had at least one underlying medical condition that included asthma; diabetes; heart, lung, or neurologic diseases; or pregnancy [31]. It was particularly striking that the proportion of children admitted with the pandemic H1N1 who had an underlying medical condition (60%) was higher than the proportion that was reported for children who were hospitalized with seasonal influenza (31–43%) [32, 33]. The morbidity and mortality associated with novel H1N1 2009 influenza virus infection appeared to be higher in patients between 5 and 59 years old, a pattern that is uncommon during seasonal influenza infections. Chowell et al. [34] reported a 87% of mortality with the novel H1N1 2009 season compared with 17% during previous seasonal periods in patients between 5 and 59 years old (Fig. 1).

3 The Role of Children in the Spread of Influenza Disease

During influenza infection, children shed higher titers of virus in the nasopharynx than adults and act as effective disseminators of infection [6, 35]. The impact of influenza in children was demonstrated in a study conducted in an elementary

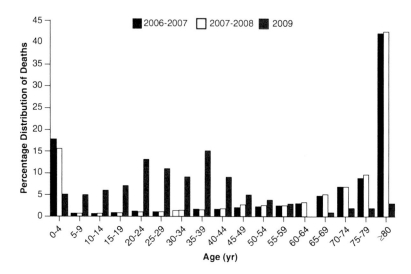

Fig. 1 Percentage distribution of deaths from severe pneumonia during the 2009 study period as compared with influenza seasons from 2006 to 2008, in Mexico, according to age group [34]. Copyright © [2009] Massachusetts Medical Society. All rights reserved

school, where illness episodes, school absenteeism, medication use, parental absenteeism from work, and the occurrence of secondary illnesses in other family members were assessed [36]. For every 100 school children enrolled during the 37 school days of the influenza season, there were 28 illness episodes and 63 missed school days attributable to influenza. In addition, for every 100 children followed, influenza accounted for an estimated 20 days of work missed by their parents and 22 secondary illness episodes among other family members. These findings support earlier observations made during an interpandemic influenza period in Houston in 1978 [37]. As can be seen in Fig. 2, school absenteeism in Houston preceded industrial absenteeism by several weeks, indicating that children have a central role in the transmission of influenza to older family members within a community.

4 Influenza Vaccination in Children

There are two seasonal influenza vaccines licensed for use in children, the trivalent inactivated vaccine (TIV) given by intramuscular injection and the trivalent live-attenuated influenza vaccine (LAIV) administered intranasally. TIV is licensed for use in all children 6 months of age and older, while LAIV is licensed for use in children, without a history of asthma, 2 years of age and older. Both of these vaccines have been studied in a number of safety, immunogenicity, and efficacy studies conducted in children of various ages. Because many other respiratory viruses mimic the symptoms of influenza, vaccine efficacy trials that use clinical

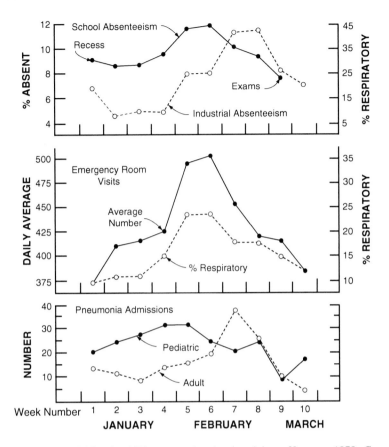

Fig. 2 Influenza morbidity in children precedes that in adults – Houston, 1978. Copyright © [1978] Massachusetts Medical Society. All rights reserved [37]

outcomes of influenza-like illness (ILI) generally have lower estimates of vaccine efficacy since they include a number of non-influenza cases. Efficacy trials that assess laboratory-confirmed influenza are regarded as the gold standards by which influenza vaccines are most appropriately judged. The results of such vaccine trials with seasonal vaccines are summarized in the next sections.

5 Efficacy of TIV

Although many pediatric studies of seasonal TIV have been conducted, a number of them have been of relatively small sample size and have used ILI as the efficacy outcomes of interest. Several reviews and meta-analyses of these trials provide a comprehensive assessment of the published literature [38–42]. Four reports discussing TIV efficacy are highlighted here [39, 42–44].

A comprehensive meta-analysis conducted by Manzoli et al. [42] evaluated all the published randomized clinical studies of TIV for the prevention of ILI and laboratory-confirmed influenza in healthy children and adolescents. Each trial was assessed for the quality of randomization, concealment of group allocation, and double blinding; studies judged to be inadequate were excluded. Data from nine randomized clinical studies of TIV using ILI as the study endpoint determined overall vaccine efficacy to be 45% [95% confidence interval (CI): 33–55%]. Data from 11 trials of TIV using laboratory-confirmed influenza as the study endpoint determined overall vaccine efficacy to be 62% (95% CI: 45–75%). TIV efficacy for both ILI and laboratory-confirmed influenza improved with increasing age of the children. These authors also attempted to determine vaccine efficacy for children less than 2 years of age but found only three studies of relatively small sample size [42]. One of these trials using ILI as the study endpoint showed a statistically significant vaccine efficacy, but two additional trials evaluating culture-confirmed influenza did not demonstrate significant vaccine efficacy [42]. Additionally, three studies that evaluated the impact of vaccine on acute otitis also showed no overall benefit of vaccine [42, 45, 46]. These authors concluded that in children younger than 2 years of age, "the scarcity of data available suggests that any conclusion should be avoided until further studies are published."

Zangwill and Belshe [39] also assessed the overall vaccine efficacy of TIV in another review and came to much the same conclusions. The results from five clinical studies of children <9 years of age receiving two doses of TIV and using laboratory-confirmed influenza as the efficacy criteria, showed a vaccine efficacy of 63% (95% CI: 45–70%). They also made several generalizations that mirror those of Manzoli et al.; protective efficacy increased with age of the child and the range of vaccine efficacy in children <5 years of age was broad and limited by the small sample size of the few existing studies.

From 1985 to 1990, a randomized, placebo-controlled comparative trial of inactivated and live vaccine for the prevention of laboratory-confirmed influenza A disease in individuals 1–65 years of age was conducted at Vanderbilt University [47]. Data from a subset of patients younger than 16 years were evaluated to determine TIV efficacy based on culture-positive illness and seroconversion [43]. During the 5 years of the study, 791 children younger than 16 years received 1,809 doses of inactivated vaccine, live vaccine, or placebo. In these children, inactivated vaccine was 91.4% and 77.3% efficacious in preventing symptomatic, culture-positive influenza A H1N1 and H3N2 illness, respectively. The efficacy of the inactivated vaccine using seroconversion for H1N1 and H3N2 serotypes was 67.1% and 65.5%, respectively. There were no statistically significant differences in vaccine efficacy between the inactivated vaccine and live vaccine for either study endpoint. The conclusion from that study was that inactivated vaccine was efficacious for the prevention of influenza disease in children 1–16 years old.

Finally, a recently published case–control study evaluated the effectiveness of TIV in 103 matched pairs of children less than 5 years of age over eight influenza seasons. Vaccine effectiveness (VE) for the prevention of laboratory-confirmed influenza among fully vaccinated children was 86% (95% CI: 29–97%) when compared with unvaccinated children. VE for partially vaccinated children was 73% (95% CI: 3–93%). The small sample size of the study, its retrospective nature, and the lack of underlying medical conditions were limitations [44].

6   Safety of TIV

Three large studies have assessed the safety of TIV in children and provide assurance that the vaccine is well tolerated in this age group [48–50]. Hambidge et al. [48] conducted a retrospective chart review of significant medically attended events at eight managed care organizations that participated in the CDC-funded Vaccine Safety Datalink (VSD). All children in this cohort who were 6–23 months of age and had received TIV between January 1991 and May 2003 were assessed. This represented 45,356 children who received a total of 69,359 TIV vaccinations. Any medically attended event associated with TIV was evaluated in four risk windows; 0–3, 1–14, 1–42, and 15–42 days after vaccination and compared with two control periods, one before vaccination and the other after the risk window. The results of this study indicate that there were very few medically attended events, none were serious, and none were significantly associated with the vaccine.

In another VSD study, France et al. [49] evaluated children aged younger than 18 who received TIV from January 1993 to December 1999. Risks of outpatient, emergency department, and inpatient visits during the 14 days after vaccination were compared with the risks of visits in two control periods. A total of 251,600 vaccination episodes were assessed. Study participants incurred 1,165, 230, and 489 different diagnoses during the 14 days after vaccination in the outpatient, emergency department, and inpatient settings, respectively. After medical record review of all of these diagnoses, only impetigo in nine children 6–23 months of age was significantly more common after vaccination when compared with the control periods. The conclusion of this large safety study was that TIV was well tolerated.

Finally, a recent study evaluated serious adverse events (SAEs) reported to the Vaccine Adverse Event Reporting System (VAERS), a passive surveillance system, after TIV in children 6–23 months of age. Two health care professionals independently reviewed all 104 SAEs reported to VAERS, including life-threatening illness, hospitalization, prolongation of hospitalization, congenital abnormality, or death in children 6–23 months of age vaccinated with TIV between 2003 and 2006. The two most frequent SAEs were fever and seizures. New onset asthma or asthma exacerbations were reported in only five patients, and causation was difficult to determine. Fifteen patients died from 1 to 14 days after vaccination, most of them were previously healthy children. One of them had myocarditis on autopsy. Despite the limitations of the passive surveillance and the retrospective nature of the study, the review did not identify previously unexpected SAEs and provided reassurance that TIV administration was generally safe [50].

7 Efficacy and Safety of LAIV

A number of studies have been published testing monovalent, bivalent, and trivalent experimental and manufacturing lot preparations of LAIV. One of the largest was a multicenter, double-blind, placebo-controlled trial of trivalent LAIV conducted in children 15–71 months old in the late 1990s [51]. In this pivotal study, 1,314 children were assigned to receive two doses of live-attenuated intranasal vaccine and 288 children were assigned to receive one dose of either live-attenuated vaccine or placebo. The strains included in the live-attenuated vaccine were antigenically equivalent to those in the contemporary TIV vaccine. Ill subjects were evaluated with viral cultures during the subsequent influenza season. A case of influenza was defined as illness associated with isolation of wild-type influenza virus from respiratory secretions. The intranasal vaccine was well tolerated with no SAEs reported. Among children who were initially seronegative, fourfold titer rises were noted in 61–96% of the subjects, depending on the influenza strain. Cases of influenza were significantly less common in the vaccine group than the placebo group, and vaccine efficacy against culture-confirmed influenza illness was 93% (95% CI: 88–96%). In addition, the one-dose LAIV regimen had 89% efficacy against culture-confirmed disease. Vaccines were well tolerated in this study.

To determine the safety of LAIV, a randomized, double-blind, placebo-controlled safety trial was conducted in nearly 10,000 healthy children 12 months to 17 years of age given live vaccine or placebo in a 2:1 randomization scheme [52]. Children <9 years of age received two doses of either vaccine or placebo with 28–42 days between doses. Enrolled children were followed for 42 days after each vaccination for all medically attended events. Acute respiratory tract events, systemic bacterial infections, acute gastrointestinal tract events, and rare events potentially associated with wild-type influenza were assessed, and none were found to be increased in the vaccine group. However, a statistically significant increase in the relative risk for reactive airway disease [4.06 (90% CI: 1.29–17.86)] was observed in children 18–35 months of age. Based on the high efficacy rates obtained in the Belshe et al. [51] study, but tempered by the safety concerns associated with wheezing in this large study, at that time LAIV was licensed for use in children over 5 years of age without a previous history of wheezing.

Given concerns over these reactive airway findings [52], another study was conducted directly comparing the efficacy and safety of LAIV with inactivated influenza vaccine in children 6–71 months of age with a history of recurrent respiratory tract infections [53]. Children were randomized to receive two doses of either LAIV ($n = 1,101$) or inactivated vaccine ($n = 1,086$) before the 2002–2003 influenza season. Participants were followed for culture-confirmed influenza illness and vaccine safety. Overall, there were 52.7% (95% CI: 21.6–72.2%) fewer cases of confirmed influenza caused by antigenically similar strains after LAIV than after TIV. There were no differences between the groups in the incidence of wheezing after vaccination.

To further compare the safety and efficacy of the LAIV and TIV in asthmatic children, Fleming et al. [54] randomized over 2,000 asthmatic children 6–17 years of age to either TIV or LAIV in an open-label study during the 2002–2003 influenza season. Participants were assessed for culture-confirmed influenza illness and vaccine safety. When the incidence of culture-confirmed influenza illness was compared between the two vaccine groups, the LAIV had significantly greater relative efficacy 34.7% (95% CI: 3.9–56.0%). No significant differences were noted between the two vaccine groups in the incidence of asthma exacerbations, mean peak expiratory flow rate findings, asthma symptom scores, or nighttime awakening scores. Runny nose and nasal congestion were more common in the recipients of LAIV, and more injection site reactions were noted after TIV.

Tam et al. [55] evaluated the efficacy and safety of LAIV against culture-confirmed influenza in a placebo-controlled trial during two influenza seasons in Asia. In year 1, 3,174 children 12–36 months of age were randomized to receive two doses of LAIV or placebo. In year 2, 2,947 subjects were again randomized to receive one dose of LAIV or placebo. Vaccine efficacy in year 1 was 72.9% (95% CI: 62.8–80.5%) against antigenically similar influenza subtypes and 70.1% (95% CI: 60.9–77.3%) against any strain. In year 2, LAIV was effective against antigenically similar (84.3%; 95% CI: 70.1–92.4%) and any (64.2%; 95% CI: 44.2–77.3%) influenza strains. No increase in wheezing episodes was noted in vaccine recipients in either study year.

In another comparative efficacy study, Belshe et al. [56] compared the safety and efficacy of LAIV and TIV in infants and young children during the 2004–2005 influenza season. Children 6–59 months of age, without a recent episode of wheezing illness or severe asthma, were randomly assigned in a 1:1 ratio to receive either LAIV or TIV in a double-blind manner. ILI was assessed with cultures and safety was carefully monitored. Overall, there were 54.9% fewer cases of culture-confirmed influenza in the LAIV recipients than in the TIV recipients (153 vs. 338 cases, $p < 0.001$). The better efficacy of live-attenuated vaccine was seen for both antigenically well-matched and drifted viruses. Among previously unvaccinated children, wheezing within 42 days of administration of dose one of LAIV was more common than with TIV. Rates of hospitalization for any cause during the 180 days after vaccination were higher among the recipients of LAIV who were 6–11 months of age (6.1%) than among the recipients of TIV (2.6%, $p = 0.002$). Based on these results, LAIV was licensed down to 2 years of age in children without a previous history of wheezing or asthma.

Belshe et al. recently summarized data from three efficacy trials of LAIV and focused on children 2–7 years of age [57]. Overall, the efficacy of LAIV when compared with placebo in seasons with matched strains varied from 69.2% (95% CI: 52.7, 80.4) to 94.6% (95% CI: 88.6, 97.5),in seasons with primarily mismatched strains was 87% (95% CI: 77.0, 92.6), and during late season epidemics was 73.8% (95% CI: 40.4, 89.4). Compared with TIV, LAIV recipients experienced 52.5% (95% CI: 26.7, 68.7) and 54.4% (95% CI: 41.8, 64.5) fewer cases of influenza illness caused by matched and mismatched strains, respectively. Events noted to be significantly increased after one dose of LAIV were runny nose/nasal congestion, muscle aches, decreased activity, and fever >100°F. Event rates after the second dose were generally lower than after the first dose. Hospitalizations and medically significant wheezing were not increased in these children. Similar findings were reported in another reanalysis of LAIV clinical trials recently published [58].

Finally, a large open-label, nonrandomized, community-based trial of a LAIV was conducted by Piedra et al. [59] and provides some of the most comprehensive LAIV safety data available. Medical records of all children who received LAIV were surveyed for SAEs and health care utilization 6 weeks after vaccination. In four study years, 18,780 doses of LAIV were administered to 11,096 children. A total of 4,529, 7,036, and 7,215 doses of LAIV-T were administered to children who were 18 months to 4 years, 5–9 years, and 10–18 years of age, respectively. During the four study years, 42 SAEs were identified, but none were attributed to LAIV-T. Compared with the prevaccination period, there were no increases in medically attended acute respiratory infections from 0 to 14 and 15 to 42 days after vaccination in children of all ages. A relative risk of 2.85 (95% CI: 1.01–8.03) for asthma events 15–42 days after vaccination was detected in children who were 18 months to 4 years of age during one study year, but was not significantly increased for the other 3 years [vaccine year 2, RR: 1.42 (95% CI: 0.59–3.42); vaccine year 3, RR: 0.47 (95% CI: 0.12–1.83); vaccine year 4, RR: 0.20 (95% CI: 0.03–1.54)]. They concluded that LAIV was safe in children [59].

8 H1N1 Vaccines

With the identification of the novel H1N1 strain, vaccine manufacturers rapidly began the process to produce, test, and license vaccine. On September 15, 2009, four influenza vaccine manufacturers received approval from the US Food and Drug Administration for influenza A (H1N1) 2009 monovalent influenza vaccines to be used in the prevention of influenza caused by the novel virus. Both live, attenuated, and inactivated influenza A (H1N1) 2009 monovalent vaccines were licensed, but none of the vaccines approved in the USA contained adjuvants. Children from 6 months to 9 years of age were recommended to receive two doses of the monovalent vaccine, while persons aged ≥ 10 were recommended to receive only one dose [60]. Groups recommended to receive the vaccine included pregnant women, household contacts of infants younger than 6 months, health care and emergency services personnel, individuals between 6 months and 24 years of age, and those aged 25 or older with underlying conditions that put them at high risk of complications from influenza [61].

Vaccine was also produced in a number of other countries, with the first vaccinations with the novel H1N1 vaccine occurring in China [62]. Two reports of the safety and immunogenicity of the novel H1N1 vaccine have recently appeared in the literature and more will likely appear in the next several months. In one trial conducted in Australia, two doses of an inactivated, split virus 2009

H1N1 vaccine were administered to healthy adults between the ages of 18 and 64 years. A total of 240 subjects, equally divided into two age groups (<50 and ≥50 years), were enrolled and underwent randomization to receive either 15 or 30 µg of hemagglutinin antigen by intramuscular injection. Antibody titers were measured using hemagglutination inhibition (HAI) and microneutralization assays at baseline and 21 days after the first vaccination. By day 21 after vaccination, antibody titers of 1:40 or more were observed in 96.7% of the subjects who received the 15-µg dose and in 93.3% of those who received the 30-µg dose. Local pain and tenderness were reported in 46.3% of subjects, and systemic symptoms were noted in 45.0% of subjects. Nearly all events were mild to moderate in intensity [63]. In another recently reported study 175 adults aged 18 to 50 received monovalent influenza A/California/2009 (H1N1) vaccine with and without MF-59 adjuvant. Subjects were randomly assigned to receive two intramuscular injections of vaccine containing 7.5 µg of hemagglutinin on day 0 in each arm or one injection on day 0 and the other on day 7, 14, or 21; two 3.75-µg doses of MF-59-adjuvanted vaccine, or 7.5 or 15 µg of nonadjuvanted vaccine, administered 21 days apart. Antibody responses were measured by HAI assay and a microneutralization assay. Preliminary data indicate that antibody titers, expressed as geometric means, were generally higher at day 14 among subjects who had received two 7.5-µg doses of the MF-59-adjuvanted vaccine than among those who received only one dose. Seroconversion rates after one dose of vaccine at day 21 were seen in 76% of the subjects by HI and in 92% by microneutralization, and after two doses in 88–92% and 92–96% of subjects, respectively. The most frequent local and systemic reactions were pain at the injection site and muscle aches, noted in 70% and 42% of subjects, respectively [64].

9   New Vaccine Approaches

Although both TIV and LAIV have been shown to be safe and effective in the prevention of influenza infections, the fact that two doses of vaccine are required in previously unimmunized young children, the need for annual reimmunization, and the lag time required for vaccine development and release are substantial limitations to the current vaccines. For these reasons, a number of new innovative influenza vaccine approaches have been devised and will be summarized in this section.

9.1   Adjuvants

Oil in water emulsion-based adjuvants have been shown to enhance the immunogenicity of a number of vaccines. One such adjuvant, MF-59, is already licensed in Europe and has been used in more than 45 million people [65]. Initially, MF-59 was combined with influenza vaccine and administered to the elderly, resulting in improved antibody levels when compared with standard TIV. In a recently published study, MF-59 has also shown to improve immune responses to influenza vaccine in young children [66]. This observer-blinded randomized study compared the immunogenicity, clinical tolerability, and safety of a MF-59-adjuvanted inactivated influenza subunit vaccine with standard TIV in unprimed healthy children between 6 and 36 months of age. Children were randomly assigned to receive two doses of either MF-59 adjuvanted vaccine ($n = 130$) or unadjuvanted split vaccine ($n = 139$). Then two subgroups of these children also received a booster dose 1 year later. HAI antibody titers were measured against influenza A and B strains included in the vaccines and against mismatched strains. Postvaccination HAI titers to all three vaccine strains were significantly higher with the adjuvanted vaccine ($p < 0.001$). In addition, adjuvanted vaccine induced significantly higher cross-reactivity against mismatched strains. After a single dose, 91% of the MF-95 group achieved seroprotection versus 49% ($p < 0.001$) receiving TIV alone. In the MF-59 group, 99% of the children developed seroprotective antibody to influenza B after the second dose when compared to only 33% in the control group receiving unadjuvanted vaccines ($p < 0.001$). This difference was even more pronounced in children between 6 and 11 months of age (100% vs. 12%, $p < 0.001$). Antibody titers remained significantly higher after 1 year in the MF-59 group. Clinical tolerability and safety were generally comparable between vaccine groups, although transient, mild solicited reactions were more frequent in the adjuvanted vaccine group. A response to a third dose of both vaccines was also evaluated in children from 16 to 48 months of age. Injection site pain was significantly higher in the older (≥ 3 years) recipients of the MF-59-adjuvanted vaccine when compared with recipients of the unadjuvanted product ($p < 0.01$). Yet, after both adjuvanted and unadjuvanted vaccines, reactions were of mild or moderate intensity and short duration. Children who received the adjuvanted vaccine during the previous season had higher antibody titers and seroprotection rates when compared with those who received unadjuvanted vaccine. Immune responses were significantly higher in the MF-59-adjuvanted group 3 weeks of the third dose of vaccine. Seroprotection rates after both adjuvanted and unadjuvanted vaccines were 100% for the two influenza A strains. However, seroconversion rates after the adjuvanted vaccine were 100% for influenza B compared with 68% after the unadjuvanted vaccine [67].

A recent review of 64 clinical trials of MF-59-adjuvanted influenza vaccine including 27,998 individuals aged 6 months to 100 years also showed reassuring safety data [68]. Solicited adverse events from 0 to 3 days after first vaccination were higher in the MF-59 group and were consistent with previous observations [69–73]. Hospitalization rates were lower in those who received MF-59-adjuvanted vaccine, but in the elderly, rates were comparable. In the overall analysis, 12.3 per 1,000 elderly subjects who received MF-59 and 14.0 per 1,000 in the control group died (adjusted RR 0.70, 95% CI: 0.54–0.91).

9.2   Cell Culture-Derived Vaccines

All currently licensed influenza vaccines in the USA are produced in embryonated hen's eggs, making rapid production of new vaccines problematic. In addition, a widespread epidemic of avian influenza could destroy the ability to produce such vaccines. The use of recombinant baculovirus to express foreign proteins in insect cells has been evaluated in several clinical trials [74]. A small study of this vaccine was recently reported in healthy children aged 6–59 months. Children were randomized into three groups; one group received two doses of TIV, another group received 22.5 µg of recombinant HA antigen, and the third group received 45 µg of recombinant HA antigen. In the younger children, the immunogenicity of TIV was significantly better than that of the recombinant antigen. Serologic responses to recombinant antigen were higher in the older children than the younger group but were still lower when compared with TIV. No serious vaccine-related adverse events occurred after either vaccine, and local and systemic reactions to both vaccines were generally similar. However, in the younger children, selected local and systemic symptoms were recorded significantly more frequently after the higher dose than the lower dose of the recombinant antigen [75].

10   Can Herd Immunity for Influenza be Achieved?

There are a number of highly contagious infections, such as measles and varicella, where immunization of a portion of the population confers protection to unimmunized individuals by decreasing the circulation of the pathogen, a concept called herd immunity. Several years ago, a study in Japan assessed the impact of influenza immunization of school children on influenza mortality in elderly persons and others at high risk [76]. From 1962 to 1987, Japanese school children were mandated to receive TIV, and most were vaccinated; in 1987 the laws were relaxed and in 1994 they were repealed. The study looked at influenza vaccination rates and death rates spanning this time period in Japan and compared them with data from the USA (Fig. 3).

After the vaccination program for school children was initiated in Japan, excess mortality rates dropped from values three to four times those in the USA to values similar to those in the USA. Routine vaccination of Japanese children was estimated to have prevented 37,000–49,000 deaths per year, or about one death for every 420 children vaccinated. As the vaccination mandate in Japan was relaxed, vaccination rates dropped and excess mortality rates increased. In contrast, excess mortality rates in the USA were nearly constant over the same period of time. The data from Japan suggested that vaccinating school children against influenza reduced influenza mortality among older persons, suggesting that herd immunity was occurring with influenza vaccine [76].

A similar study was recently reported from the USA, where school children were vaccinated with LAIV and its impact was assessed in their households and

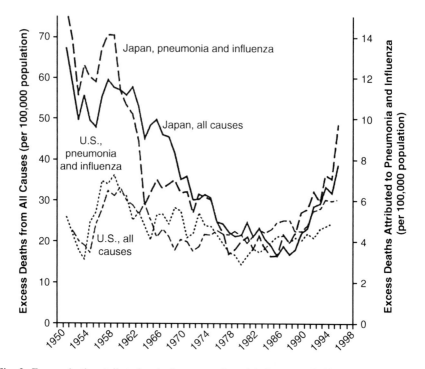

Fig. 3 Excess deaths attributed to both pneumonia and influenza and all causes, spanning the years when the Japanese school immunization program was dismantled [76]. Copyright © [2001] Massachusetts Medical Society. All rights reserved

community [77]. Eleven demographically similar clusters of elementary schools in four states were chosen. Within each cluster, one school was selected to receive vaccination (intervention school) and one or two schools in that cluster did not participate (control schools). During a predicted week of peak influenza activity in each state, all households with children in the intervention and control schools were asked about influenza vaccination and influenza-like illness. Persons living in intervention school households had significantly fewer influenza-like symptoms and outcomes during the recall week than those in control school households, even though they themselves might not have been immunized. This suggests that vaccinating children protects their unimmunized contacts, the essential mechanism of herd immunity.

11   Practical Implications for Influenza Vaccination of Children

For many years, all children with high-risk conditions associated with influenza have been recommended to receive annual influenza vaccination. These conditions include asthma or other chronic pulmonary diseases, significant cardiac disease, immunosuppressive disorders, human immunodeficiency virus infection, sickle cell anemia, long-term aspirin therapy, chronic renal disease, chronic metabolic disorders, and neurological disorders. Beginning with the 2006–2007 influenza season in the USA, all children between 6 months and 5 years of age were recommended to receive annual TIV to reduce the burden of both hospitalization and outpatient visits associated with influenza. With the universal recommendations for influenza vaccine in young children, studies were conducted to monitor vaccine uptake. Data from the National Immunization Survey measured vaccination rates in children 6–23 months of age 1 year after the universal influenza recommendations were issued. Although influenza immunization rates varied widely among the different states, overall 33.4% of children between 6 and 23 months of age received one dose of vaccine and 17.8% received two doses [78]. Continued assessments of influenza vaccination rates in this population are ongoing.

Considerable discussion then occurred surrounding the question whether routine influenza vaccination of all school children might reduce disease in both children and the community. A study conducted through the CDC-funded VSD addressed the simple question of whether two doses of TIV could be delivered to children <9 years of age who had not previously received vaccine [79]. A total of 125,928 children 6 months to 8 years of age were evaluated. Among children 6–23 months of age, a fairly high proportion of first time-vaccinated children also received a second vaccination, with rates of 44% in 2001–2002, 54% in 2002–2003, and 29% in 2003–2004 (a season with vaccine shortages). In contrast, among children 2–8 years of age, the corresponding rates were only 15%, 24%, and 12%. The fact that the majority of children who required two doses of vaccine did not receive them highlights some of the difficulties that are encountered in implementing universal vaccination of all school children in the primary care setting.

Might school-based vaccine delivery circumvent some of these problems? A recent report describing on-site administration of LAIV to all students in a large, metropolitan public school system demonstrated that large numbers of school children could be effectively immunized [80]. There were 53,420 students in the system; 56% of the elementary school students, 45% of the middle school students, and 30% of the high school students were immunized. This experience clearly highlights that a vaccination campaign in a large public school system can achieve relatively high coverage levels; however, considerable effort by the local health department was expended in the process. The results of this school-based immunization program were recently published and compared the impact of the program on disease burden in two Tennessee counties. The school-based immunization program was operative in Knox County but not in Davidson County. Twenty-two percent of Knox County children had laboratory-confirmed influenza infections,while 18% of Davidson County were positive ($p = 0.14$). More school-age than preschool-age children were influenza positive in both counties (27% vs. 14%, $p < 0.001$). Estimated influenza vaccine coverage in preschool-age children was comparable (36% for Knox County and 33% for Davidson County). In contrast, more Knox children aged 5–12 were vaccinated when compared with

Davidson County (44% vs. 12%, $p < 0.001$). Despite a school-based influenza campaign and universal vaccination recommendations, influenza was associated with a significant burden of illness in children. Influenza was responsible for a greater proportion of acute respiratory illness visits among school-age than pre-school-age children, supporting the ACIP recommendation to all children from 6 months to 18 years old. The data obtained from this study also show a direct benefit of the vaccination for school-age children but do not suggest any effect in younger children. Further studies are needed to better appreciate the impact of school-based vaccinations [81].

Given the evidence of the enormous burden of influenza infection in children and recognizing that vaccinations are an effective way to decrease morbidity and mortality, the Advisory Committee on Immunization Practices of the CDC first recommended annual influenza vaccination for children between 6 and 23 months of age in 2004. Then in 2006, the recommendations for universal influenza vacci-nation were expanded to 24–59 months of life. Finally, in 2008, all children aged 5–18 years were recommended for universal vaccination [82]. However, despite these strong recommendations, coverage levels remained suboptimal. During the 2008–2009 influenza season, average vaccine coverage with one or more vaccine doses in children aged 6–23 months was 47.8% (range 34.3–60.1%) and full vaccination coverage was 28.9% (range 19.8–39.7%). Among children aged 2–4 years vaccination coverage with one or more doses was 27.8% (range 17.3–38.1%) and full vaccination coverage was 21.8% (range 12.6–32.3%). Among children aged 5–10 years, vaccine coverage with one or more doses was 16.3% (range 9.4–23.7%) and full vaccination coverage was 12% (range 6.2–19.7%). Among children aged 11–12 years, 12.7% were fully vaccinated (range 6.6–18%), and among children aged 13–18 years, 9.1% (range 4.8–14.5%) were fully vaccinated.

Vaccination coverage rates in children vary widely among countries worldwide. A recently published population-based cross-sectional survey of 11 European countries reported influenza vaccination rates ranging from 4.2% in Ireland to 19% in Germany. Generally, most countries recommend influenza vaccination for all children older than 6 months with cardiac or renal diseases, diabetes, or immunocompromised conditions. However, since 2007 Austria and Finland have been the only European countries to also recommend universal influenza vaccina-tion for healthy children aged 6–23 monthsof age [83]. Lopez-de-Andres et al. [84] reported influenza vaccination rates in Spanish children of 6.8%, with higher coverage rates in children with high-risk conditions (asthma and/or diabetes). In Israel, the overall influenza vaccine coverage among children who visited the pediatric emergency room was 4.1%, with coverage in high-risk children of 6.5% and in children aged 6–24 months of 2.7% [85].

In summary, although influenza vaccination is the most effective method to prevent morbidity and mortality [86], coverage remains low. Only a few countries have a universal vaccination policy in children, while most recommend vaccination only in high-risk medical conditions. Greater efforts are needed to increase vaccine coverage among children worldwide.

12 Conclusion

Given the clear evidence that both live and inactivated influenza vaccines can prevent influenza disease, influenza vaccination should be offered to all children. The recent evidence of improved vaccine efficacy for LAIV in young children also suggests that it might be a better alternative to TIV in young children without a history of asthma. Also, given that influenza disease is so rarely specifically diagnosed [5] and that it can mimic other respiratory viral infections, it is imperative that laboratory-based surveillance be conducted to assess vaccine efficacy as influenza vaccine is utilized more broadly. The future for influenza prevention is bright, but continued attention to measuring vaccine effect is needed to sustain this effort.

References

1. Izurieta HS, Thompson WW, Kramarz P, Shay DK, Davis RL, DeStefano F, Black S, Shinefield H, Fukuda K (2000) Influenza and the rates of hospitalization for respiratory disease among infants and young children. N Engl J Med 342:232–239
2. Neuzil KM, Mellen BG, Wright PF, Mitchel EF Jr, Griffin MR (2000) The effect of influenza on hospitalizations, outpatient visits, and courses of antibiotics in children. N Engl J Med 342:225–231
3. Iwane MK, Edwards KM, Szilagyi PG, Walker FJ, Griffin MR, Weinberg GA, Coulen C, Poehling KA, Shone LP, Balter S et al (2004) New Vaccine Surveillance Network. Population-based surveillance for hospitalizations associated with respiratory syncytial virus, influenza virus, and parainfluenza viruses among young children. Pediatrics 113:1758–1764
4. Griffin MR, Walker FJ, Iwane MK, Weinberg GA, Staat MA, Erdman DD (2004) Epidemiology of respiratory infections in young children: insights from the New Vaccine Surveillance Network. Pediatr Infect Dis J 23(Suppl):S188–S192
5. Poehling KA, Edwards KM, Weinberg GA, Szilagyi P, Staat MA, Iwane MK, Bridges CB, Grijalva CG, Zhu Y, Bernstein DI et al (2006) New Vaccine Surveillance Network. The under-recognized burden of influenza in young children. N Engl J Med 355:31–40
6. Weinberg GA, Erdman DD, Edwards KM, Hall CB, Walker FJ, Griffin MR, Schwartz B, New Vaccine Surveillance Network Study Group (2004) Superiority of reverse-transcription polymerase chain reaction to conventional viral culture in the diagnosis of acute respiratory tract infections in children. J Infect Dis 189:706–710
7. Poehling KA, Griffin MR, Dittus RS, Tang YW, Holland K, Li H, Edwards KM (2002) Bedside diagnosis of influenza virus infections in hospitalized children. Pediatrics 110:83–88
8. Bonner AB, Monroe KW, Talley LI, Klasner AE, Kimberlin DW (2003) Impact of the rapid diagnosis of influenza on physician decision-making and patient management in the pediatric emergency department: results of a randomized, prospective, controlled trial. Pediatrics 112:363–367
9. Sharma V, Dowd MD, Slaughter AJ, Simon SD (2002) Effect of rapid diagnosis of influenza virus type A on the emergency department management of febrile infants and toddlers. Arch Pediatr Adolesc Med 156:41–43
10. Schrag SJ, Shay DK, Gershman K, Thomas A, Craig AS, Schaffner W, Harrison LH, Vugia D, Clogher P, Lynfield R et al (2006) Emerging Infections Program Respiratory Diseases Activity. Multistate surveillance for laboratory-confirmed, influenza-associated hospitalizations in children: 2003–2004. Pediatr Infect Dis J 25:395–400

11. Grijalva CG, Craig AS, Dupont WD, Bridges CB, Schrag SJ, Iwane MK, Schaffner W, Edwards KM, Griffin MR (2006) Estimating influenza hospitalizations among children. Emerg Infect Dis 12:103–109

12. Mullooly JP, Barker WH (1982) Impact of type A influenza on children: a retrospective study. Am J Public Health 72:1008–1016

13. Thompson WW, Shay DK, Weintraub E, Brammer L, Bridges CB, Cox NJ, Fukuda K (2004) Influenza-associated hospitalizations in the United States. JAMA 292:1333–1340

14. O'Brien MA, Uyeki TM, Shay DK, Thompson WW, Kleinman K, McAdam A, Yu XJ, Platt R, Lieu TA (2004) Incidence of outpatient visits and hospitalizations related to influenza in infants and young children. Pediatrics 113:585–593

15. Neuzil KM, Zhu Y, Griffin MR, Edwards KM, Thompson JM, Tollefson SJ, Wright PF (2002) Burden of interpandemic influenza in children younger than 5 years: a 25-year prospective study. J Infect Dis 185:147–152

16. Glezen WP, Greenberg SB, Atmar RL, Piedra PA, Couch RB (2000) Impact of respiratory virus infections on persons with chronic underlying conditions. JAMA 283:499–505

17. Montes M, Vicente D, Pérez-Yarza EG, Cilla G, Pérez-Trallero E (2005) Influenza-related hospitalisations among children aged less than 5 years old in the Basque Country, Spain: a 3-year study (July 2001–June 2004). Vaccine 23:4302–4306

18. Chiu SS, Lau YL, Chan KH, Wong WH, Peiris JS (2002) Influenza-related hospitalizations among children in Hong Kong. N Engl J Med 347:2097–2103

19. Chiu SS, Chan KH, Chen H, Young BW, Lim W, Wong WH, Lau YL, Peiris JS (2009) Virologically confirmed population-based burden of hospitalization caused by influenza A and B among children in Hong Kong. Clin Infect Dis 49(7):1016–1021

20. Forster J (2003) Influenza in children: the German perspective. Pediatr Infect Dis J 22: s215–s217

21. Schrag SJ, Shay DK, Gershman K, Thomas A, Craig AS, Schaffner W, Harrison LH, Vugia D, Clogher P, Lynfield R, Farley M, Zansky S, Uyeki T (2006) Emerging Infections Program Respiratory Diseases Activity. Multistate surveillance for laboratory-confirmed, influenza-associated hospitalizations in children: 2003–2004. Pediatr Infect Dis J 25(5):395–400

22. Bhat N, Wright JG, Broder KR, Murray EL, Greenberg ME, Glover MJ, Likos AM, Posey DL, Klimov A, Lindstrom SE et al (2005) Influenza Special Investigations Team. Influenza-associated deaths among children in the United States, 2003–2004. N Engl J Med 353: 2559–2567

23. Finelli L, Fiore A, Dhara R, Brammer L, Shay DK, Kamimoto L, Fry A, Hageman J, Gorwitz R, Bresee J, Uyeki T (2008) Influenza-associated pediatric mortality in the United States: increase of *Staphylococcus aureus* coinfection. Pediatrics 122(4):805–811

24. Morishima T, Togashi T, Yokota S, Okuno Y, Miyazaki C, Tashiro M, Okabe N (2002) Collaborative Study Group on influenza-associated encephalopathy in Japan. Encephalitis and encephalopathy associated with an influenza epidemic in Japan. Clin Infect Dis 35:512–517

25. Surtees R, DeSousa C (2006) Influenza virus associated encephalopathy. Arch Dis Child 91:455–456

26. Olsen CW (2002) The emergence of novel swine influenza viruses in North America. Virus Res 85:199–210

27. Vincent AL, Ma W, Lager KM, Janke BH, Richt JA (2008) Swine influenza viruses: a North American perspective. Adv Virus Res 72:127–154

28. Shinde V, Bridges CB, Uyeki TM et al (2009) Triple-reassortant swine influenza A (H1) in humans in the United States, 2005–2009. N Engl J Med 360:2616–2625

29. Centers for Disease Control and Prevention (CDC) (2009) Swine influenza A (H1N1) infection in two children – Southern California, March–April 2009. MMWR Morb Mortal Wkly Rep 58:400–402

30. http://www.who.int/csr/don/2009_10_16/en/index.html

31. Jain S, Kamimoto L, Bramley AM, Schmitz AM, Benoit SR, Louie J, Sugerman DE, Druckenmiller JK, Ritger KA, Chugh R, Jasuja S, Deutscher M, Chen S, Walker JD,

Duchin JS, Lett S, Soliva S, Wells EV, Swerdlow D, Uyeki TM, Fiore AE, Olsen SJ, Fry AM, Bridges CB, Finelli L, 2009 Pandemic Influenza A (H1N1) Virus Hospitalizations Investigation Team (2009) Hospitalized patients with 2009 H1N1 influenza in the United States, April–June 2009. N Engl J Med 361(20):1935–1944, PMID: 19815859

32. Schrag SJ, Shay DK, Gershman K et al (2006) Multistate surveillance for laboratory-confirmed, influenza-associated hospitalizations in children: 2003–2004. Pediatr Infect Dis J 25:395–400

33. Keren R, Zaoutis TE, Bridges CB et al (2005) Neurological and neuromuscular disease as a risk factor for respiratory failure in children hospitalized with influenza infection. JAMA 294:2188–2194

34. Chowell G, Bertozzi SM, Colchero MA, Lopez-Gatell H, Alpuche-Aranda C, Hernandez M, Miller MA (2009) Severe respiratory disease concurrent with the circulation of H1N1 influenza. N Engl J Med 361(7):674–679

35. Frank AL, Taber LH, Wells CR, Wells JM, Glezen WP, Paredes A (1981) Patterns of shedding of myxoviruses and paramyxoviruses in children. J Infect Dis 144:433–441

36. Neuzil KM, Hohlbein C, Zhu Y (2002) Illness among school children during influenza season: effect on school absenteeism, parental absenteeism from work, and secondary illness in families. Arch Pediatr Adolesc Med 156:986–991

37. Glezen WP, Couch RB (1978) Interpandemic influenza in the Houston area, 1974–76. N Engl J Med 298:587–592

38. Ruben FL (2004) Inactivated influenza virus vaccines in children. Clin Infect Dis 38:678–688

39. Zangwill KM, Belshe RB (2004) Safety and efficacy of trivalent inactivated influenza vaccine in young children: a summary for the new era of routine vaccination. Pediatr Infect Dis J 23:189–197

40. Jefferson T, Smith S, Demicheli V, Harnden A, Rivetti A, Di Pietrantonj C (2005) Assessment of the efficacy and effectiveness of influenza vaccines in healthy children: systematic review. Lancet 365:773–780

41. Negri E, Colombo C, Giordano L, Groth N, Apolone G, La Vecchia C (2005) Influenza vaccine in healthy children: a meta-analysis. Vaccine 23:2851–2861

42. Manzoli L, Schioppa F, Boccia A, Villari P (2007) The efficacy of influenza vaccine for healthy children: a meta-analysis evaluating potential sources of variation in efficacy estimates including study quality. Pediatr Infect Dis J 26:97–106

43. Neuzil KM, Dupont WD, Wright PF, Edwards KM (2001) Efficacy of inactivated and cold-adapted vaccines against influenza A infection, 1985 to 1990: the pediatric experience. Pediatr Infect Dis J 20:733–740

44. Joshi AY, Iyer VN, St Sauver JL, Jacobson RM, Boyce TG (2009) Effectiveness of inactivated influenza vaccine in children less than 5 years of age over multiple influenza seasons: a case–control study. Vaccine 27(33):4457–4461, Epub May 31, 2009

45. Hoberman A, Greenberg DP, Paradise JL, Rockette HE, Lave JR, Kearney DH, Colborn DK, Kurs-Lasky M, Haralam MA, Byers CJ et al (2003) Effectiveness of inactivated influenza vaccine in preventing acute otitis media in young children: a randomized controlled trial. JAMA 290:1608–1616

46. Clements DA, Langdon L, Bland C, Walter E (1995) Influenza A vaccine decreases the incidence of otitis media in 6- to 30-month-old children in day care. Arch Pediatr Adolesc Med 149:1113–1117

47. Edwards KM, Dupont WD, Westrich MK, Plummer WD Jr, Palmer PS, Wright PF (1994) A randomized controlled trial of cold-adapted and inactivated vaccines for the prevention of influenza A disease. J Infect Dis 169:68–76

48. Hambidge SJ, Glanz JM, France EK, McClure D, Xu S, Yamasaki K, Jackson L, Mullooly JP, Zangwill KM, Marcy SM et al (2006) Safety of trivalent inactivated influenza vaccine in children 6 to 23 months old. JAMA 296:1990–1997

49. France EK, Glanz JM, Xu S, Davis RL, Black SB, Shinefield HR, Zangwill KM, Marcy SM, Mullooly JP, Jackson LA, Chen R (2004) Safety of the trivalent inactivated influenza vaccine among children: a population-based study. Arch Pediatr Adolesc Med 158:1031–1036

50. Rosenberg M, Sparks R, McMahon A, Iskander J, Campbell JD, Edwards KM (2009) Serious adverse events rarely reported after trivalent inactivated influenza vaccine (TIV) in children 6–23 months of age. Vaccine 27(32):4278–4283

51. Belshe RB, Mendelman PM, Treanor J, King J, Gruber WC, Piedra P, Bernstein DI, Hayden FG, Kotloff K, Zangwill K et al (1998) The efficacy of live attenuated, cold-adapted, trivalent, intranasal influenzavirus vaccine in children. N Engl J Med 338:1405–1412

52. Bergen R, Black S, Shinefield H, Lewis E, Ray P, Hansen J, Walker R, Hessel C, Cordova J, Mendelman PM (2004) Safety of cold-adapted live attenuated influenza vaccine in a large cohort of children and adolescents. Pediatr Infect Dis J 23:138–144

53. Ashkenazi S, Vertruyen A, Arístegui J, Esposito S, McKeith DD, Klemola T, Biolek J, Kühr J, Bujnowski T, Desgrandchamps D et al (2006) Superior relative efficacy of live attenuated influenza vaccine compared with inactivated influenza vaccine in young children with recurrent respiratory tract infections. Pediatr Infect Dis J 25:870–879

54. Fleming DM, Crovari P, Wahn U, Klemola T, Schlesinger Y, Langussis A, Øymar K, Garcia ML, Krygier A, Costa H et al (2006) Comparison of the efficacy and safety of live attenuated cold-adapted influenza vaccine, trivalent, with trivalent inactivated influenza virus vaccine in children and adolescents with asthma. Pediatr Infect Dis J 25:860–869

55. Tam JS, Capeding MR, Lum LC, Chotpitayasunondh T, Jiang Z, Huang LM, Lee BW, Qian Y, Samakoses R, Lolekha S et al (2007) Efficacy and safety of a live attenuated, cold-adapted influenza vaccine, trivalent against culture-confirmed influenza in young children in Asia. Pediatr Infect Dis J 26:619–628

56. Belshe RB, Edwards KM, Vesikari T, Black SV, Walker RE, Hultquist M, Kemble G, Connor EM, CAIV-T Comparative Efficacy Study Group (2007) Live attenuated versus inactivated influenza vaccine in infants and young children. N Engl J Med 356:685–696

57. Belshe RB, Ambrose CS, Yi T (2008) Safety and efficacy of live attenuated influenza vaccine in children 2–7 years of age. Vaccine 26(Suppl 4):D10–D16

58. Rhorer J, Ambrose CS, Dickinson S, Hamilton H, Oleka NA, Malinoski FJ, Wittes J (2009) Efficacy of live attenuated influenza vaccine in children: a meta-analysis of nine randomized clinical trials. Vaccine 27(7):1101–1110

59. Piedra PA, Gaglani MJ, Riggs M, Herschler G, Fewlass C, Watts M, Kozinetz C, Hessel C, Glezen WP (2005) Live attenuated influenza vaccine, trivalent, is safe in healthy children 18 months to 4 years, 5 to 9 years, and 10 to 18 years of age in a community-based, nonrandomized, open-label trial. Pediatrics 116:e397–e407

60. Centers for Disease Control and Prevention (CDC) (2009) Update on influenza A (H1N1) 2009 monovalent vaccines. MMWR Morb Mortal Wkly Rep 58(39):1100–1101

61. Kuehn BM (2009) CDC names H1N1 vaccine priority groups. JAMA 302(11):1157–1158

62. Stone R (2009) Swine flu outbreak. China first to vaccinate against novel H1N1 virus. Science 325(5947):1482–1483

63. Greenberg ME, Lai MH, Hartel GF, Wichems CH, Gittleson C, Bennet J, Dawson G, Hu W, Leggio C, Washington D, Basser RL (2009) Response to a monovalent 2009 influenza A (H1N1) vaccine. N Engl J Med 361(25):2405–2413, PMID: 19745216

64. Clark TW, Pareek M, Hoschler K, Dillon H, Nicholson KG, Groth N, Stephenson I (2009) Trial of influenza A (H1N1) 2009 monovalent MF59-adjuvanted vaccine – preliminary report. N Engl J Med 361(25):2424–2435, PMID: 19745215

65. Rappuoli R, Del Giudice G, Nabel GJ, Osterhaus AD, Robinson R, Salisbury D, Stöhr K, Treanor JJ (2009) Public health. Rethinking influenza. Science 326(5949):50

66. Vesikari T, Pellegrini M, Karvonen A, Groth N, Borkowski A, O'hagan DT, Podda A (2009) Enhanced immunogenicity of seasonal influenza vaccines in young children using MF59 adjuvant. Pediatr Infect Dis J 28(7):563–571

67. Vesikari T, Groth N, Karvonen A, Borowski A, Pellegrini M (2009) MF59-adjuvanted influenza vaccine (FLUAD) in children: safety and immunogenicity following a second year seasonal vaccination. Vaccine 27(45):6291–6295

68. Pellegrini M, Nicolay U, Lindert K, Groth N, Della Cioppa G (2009) MF59-adjuvanted versus non-adjuvanted influenza vaccines: integrated analysis from a large safety database. Vaccine 27(49):6959–6965, PMID: 19751689

69. Treanor JJ, Campbell JD, Zangwill KM, Rowe T, Wolff M (2006) Safety and immunogenicity of an inactivated subvirion influenza A (H5N1) vaccine. N Engl J Med 24:1159–1169

70. Podda A (2001) The adjuvanted influenza vaccines with novel adjuvants: experience with the MF-59-adjuvanted vaccine. Vaccine 19:2673–2680

71. De Donato S, Granoff D, Minutello M, Lecchi G, Faccini M, Agnello M et al (1999) Safety and immunogenicity of MF-59-adjuvanted influenza vaccine in the elderly. Vaccine 17:3094–3101

72. Gasparini R, Pozzi T, Montomoli E, Fregapane E, Senatore F, Minutello M et al (2001) Increased immunogenicity of the MF59-adjuvanted influenza vaccine compared to a conventional subunit vaccine in elderly subjects. Eur J Epidemiol 17:135–140

73. Minutello M, Senatore F, Cecchinelli G, Bianchi M, Andreani T, Podda A et al (1999) Safety and immunogenicity of an inactivated subunit influenza virus vaccine combined with MF59 adjuvant emulsion in elderly subjects, immunized for the three consecutive influenza seasons. Vaccine 17:99–104

74. Holtz KM, Anderson DK, Cox MM (2003) Production of a recombinant influenza vaccine using baculovirus expression vector system. Bioprocess J 65:7312

75. King JC Jr, Cox MM, Reisinger K, Hedrick J, Graham I, Patriarca P (2009) Evaluation of the safety, reactogenicity and immunogenicity of FluBlok trivalent recombinant baculovirus-expressed hemagglutinin influenza vaccine administered intramuscularly to healthy children aged 6–59 months. Vaccine 27(47):6589–6594, PMID: 19716456

76. Reichert TA, Sugaya N, Fedson DS, Glezen WP, Simonsen L, Tashiro M (2001) The Japanese experience with vaccinating school children against influenza. N Engl J Med 344:889–896

77. King JC Jr, Stoddard JJ, Gaglani MJ, Moore KA, Magder L, McClure E, Rubin JD, Englund JA, Neuzil K, King JC et al (2006) Effectiveness of school-based influenza vaccination. N Engl J Med 355:2523–2532

78. http://www.cdc.gov/mmwr/preview/mmwrhtml/mm5539a1.htm

79. Jackson LA, Neuzil KM, Baggs J, Davis RL, Black S, Yamasaki KM, Belongia E, Zangwill KM, Mullooly J, Nordin J et al (2006) Compliance with the recommendations for 2 doses of trivalent inactivated influenza vaccine in children less than 9 years of age receiving influenza vaccine for the first time: a Vaccine Safety Datalink study. Pediatrics 118:2032–2037

80. Carpenter LR, Lott J, Lawson BM, Hall S, Craig AS, Schaffner W, Jones TF (2007) Mass distribution of free, intranasally administered influenza vaccine in a public school system. Pediatrics 120:e172–e178

81. Poehling KA, Talbot HK, Williams JV, Zhu Y, Lott J, Patterson L, Edwards KM, Griffin MR (2009) Impact of a school-based influenza immunization program on disease burden: comparison of two Tennessee counties. Vaccine 27(20):2695–2700

82. Centers for Disease Control and Prevention (CDC) (2009) Influenza vaccination coverage among children and adults – United States, 2008–09 influenza season. MMWR Morb Mortal Wkly Rep 58(39):1091–1095

83. Blank P, Schwenkglenks M, Szucs T (2009) Vaccination coverage rates in eleven European countries during two consecutive influenza seasons. J Infect 58:446–458

84. Lopez-de-Andres A, Hernandez-Barrera V, Carrasco-Garrido P, Gil-de-Miguel A, Jimenez-Garcia R (2009) Influenza vaccination coverage among Spanish children, 2006. Public Health 123(7):465–469

85. Stein M, Yossepowitch O, Somekh E (2005) Influenza vaccine coverage in paediatric population from central Israel. J Infect 50(5):382–385

86. CDC (2008) Prevention and control of influenza: recommendations of the Advisory Committee on immunization Practices (ACIP). MMWR Recomm Rep 57(RR07):1–60

The Immune Response to Influenza A Viruses

Justine D. Mintern, Carole Guillonneau, Stephen J. Turner,
and Peter C. Doherty

Abstract The influenza A viruses are dangerous pathogens with the potential to provoke devastating disease. The challenge for the medical research community is to design preventive measures and therapeutic interventions that will limit the severe consequences of pandemic influenza A virus infections. Vaccines have long been available, but there is considerable scope for improvement as they target only the prevailing influenza A virus strains, do not give broad immunity, and work poorly in the elderly, the target group that is most at risk of fatal disease. Improved vaccines will only emerge if the development strategy is based on a firm understanding of the host immune response to the virus. Here, we summarize the research to date that details immune mechanisms participating in the control and elimination of influenza A viruses.

1 Introduction

The influenza viruses are *Orthomyxoviruses* with an eight-segmented, negative-sense, single-stranded RNA genome. There are three types: influenza A, B, and C. The influenza A viruses that cause the most serious problems in humans are the subject of this review. These pathogens are classified according to their two major surface glycoproteins: hemagglutinin (HA or H) and neuraminidase (NA or N). Infecting both mammalian and avian species, the highly contagious influenza A

J.D. Mintern (✉), C. Guillonneau, and S.J. Turner
Department of Microbiology and Immunology, The University of Melbourne, Parkville, VIC 3010, Australia
e-mail: mintern@wehi.edu.au

P.C. Doherty
Department of Microbiology and Immunology, The University of Melbourne, Parkville, VIC 3010, Australia
Department of Immunology, St Jude Children's Research Hospital, Memphis, TN 38105, USA

G. Del Giudice and R. Rappuoli (eds.), *Influenza Vaccines for the Future*, 2nd edition,
Birkhäuser Advances in Infectious Diseases,
DOI 10.1007/978-3-0346-0279-2_8, © Springer Basel AG 2011

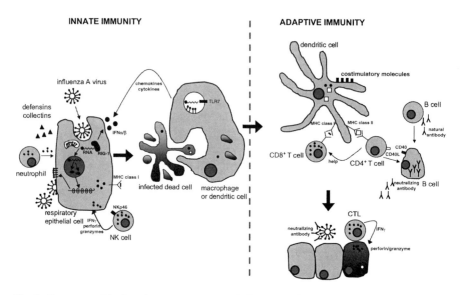

Fig. 1 Summary of the host immune response to influenza A virus

viruses are responsible for widespread morbidity and mortality [1]. In mammals, infection is established in the upper and lower respiratory tracts, provoking an illness that is associated with fever, myalgia, congestion, pharyngitis, and, in severe cases, pneumonia. Early on, some of the very virulent influenza A viruses can induce a "cytokine shock" syndrome mediated via the innate immune response pathway. Fortunately, infection also elicits potent adaptive immunity and long-term memory, though the virus can mutate readily, allowing strains with variant HA molecules to cause successive pandemics. The current killed or subunit vaccines induce effective antibody responses in normal adults, though they do not promote a virus-specific CD8$^+$ T-cell response and memory and they are poorly immunogenic in those who are even marginally immunologically compromised. The major task for immunologists interested in the problem that influenza virus poses is to develop better vaccines. Most of our detailed knowledge about immunity to the influenza A viruses is derived from the murine model that allows rigorous analysis due to the availability of an extensive panel of defined analytical reagents. Here, we provide a comprehensive summary of a large body of research examining the immune mechanisms that act to control influenza A virus infection (Fig. 1). This information should provide a useful basis for the informed design of novel, next generation influenza A virus vaccines.

2 Detection of Influenza A Virus

Invading influenza A viruses are detected in the host environment by "pattern recognition receptors" (PRRs) [2]. Previously, the molecular target was considered to be double-stranded viral RNA (dsRNA) recognized by the PRR, toll-like receptor 3 (TLR3) [3, 4]. A role for TLR3 was questioned, however, given that the concentration of dsDNA is unlikely to be sufficient to signal TLR3 [5]. It is now considered that influenza A virus infection does not generate dsRNA at all [6]. Instead, the influenza A virus polymerase generates single-stranded RNA (ssRNA) with an uncapped 5'-phosphate that serves as the molecular signature identified by the immune system [6]. The cytoplasmic RNA helicase, RIG-1 [6, 7], but not MDA5 [6, 8], is responsible for influenza A virus recognition, which occurs independently of viral replication [7]. In addition to RIG-1, TLR7 is implicated in influenza A virus detection. Expressed in the endosomal compartments of plasmacytoid dendritic cells (DCs) and B cells, TLR7 detects influenza A virus ssRNA [9, 10]. The participation of multiple PRRs in the surveillance of influenza A virus may reflect cell type-specific roles [11]. Influenza A virus infection also activates NOD-like receptor-associated inflammasomes that are critical for the processing and release of IL-1β [12–14]. Once influenza A virus is recognized, PRRs initiate multiple signaling cascades that facilitate both innate and adaptive immunity to enable viral eradication.

3   Innate Immunity and the Influenza A Viruses

Innate immunity directed against influenza A virus provides an immediate and rapid response to the pathogen. The pulmonary infiltrate of innate immune cells is comprised mainly of natural killer (NK) cells, neutrophils, and macrophages. The NK cell represents the major innate response element and is detected in the infected lung as early as 48 h following influenza A virus infection [15, 16]. Protection is thought to be mediated by both cytokine production (IFN-γ and TNF-α) and direct cytotoxicity of virus-infected cells [17]. Influenza A virus-infected cells are recognized by NKp46 [18] and NKp44 [19] interaction with HA. The critical role for this pathway in influenza control is illustrated by the fatal infection that occurs in mice that lack NKp46 [20]. Together with NK cells, neutrophils also contribute to influenza A virus clearance through the secretion of an array of proinflammatory molecules that serve to limit viral replication [21–23]. Finally, alveolar macrophages (AMs) are also present in the innate pulmonary infiltrate, although they form only a small contribution early, they are recruited in large numbers later by the T-cell response. AMs represent the major phagocytic cell type resident in the lung [24], acting to scavenge influenza A virus-derived antigen [25]. In addition, AMs secrete proinflammatory cytokines including tumor necrosis factor (TNF)-α, interleukin (IL-1)-β, IL-6, and interferon (IFN)-α/β [26, 27] together with the chemokines macrophage inflammatory protein (MIP)-1α, monocyte chemotactic protein (MCP)-1, RANTES, and IFN-inducible protein (IP)-10 [21, 26, 28–30]. The magnitude and duration of the potent AM inflammatory response are negatively regulated via CD200R/CD200 [31]. The AM can also modulate adaptive T-cell immunity to influenza A viruses [32]. Present in the lung during active viral replication, AMs are fully susceptible to influenza A virus infection [26]. Unlike in epithelial cells, however, the infection is nonproductive with little, if any, virion release [26, 33], though it does lead to subsequent apoptosis [33]. Depletion of macrophages during influenza A virus infection results in elevated viral titers and increased morbidity and mortality [21]. In contrast, macrophages can elicit damage to the infected respiratory tissue [34]. Therefore, multiple immune cell types participate in the immediate innate response to influenza A viruses.

The pulmonary infiltrate releases a torrent of innate immune molecules that are considered to limit influenza A virus infection. A long list of cytokines and chemokines are potentially involved. A major player is type I IFN, representing the most potent cytokine attack against the virus [35]. So potent is the IFN response that the influenza A viruses encode a protein (NS2) to disable this pathway (described in Sect. 6). Nasal and pulmonary IFN-α and -β rise rapidly following influenza A virus infection [36] and act to directly limit viral replication and induce further cytokines and/or chemokine secretion that enhances recruitment and activation of multiple immune cell types. Type I IFN serves to enhance macrophage function, promote antigen presentation by antigen-presenting cells (APCs), and modulate adaptive immunity. The importance of this pathway is exemplified by the severe pulmonary disease that develops following influenza A virus infection of mice with disrupted type I IFN signaling [37, 38]. Plasmacytoid DCs are the major producers of type 1 IFN in response to many viruses, including influenza A virus [39–42]. Other cytokines implicated in influenza A virus immunity include TNF-α [43], IL-6 [44, 45], IL-1 [46], IL-18 [47], and IL-12 [48, 49]. In contrast, mice that lack functional IFN-γ can efficiently clear influenza A viruses, suggesting only a minor or redundant role for IFN-γ in the response [50–52]. Chemokines with defined roles in influenza A virus immunity include MIP-1α [53] and CCR5 [54], as illustrated by the elevated disease burden following infection of the chemokine-deficient mice. Finally, while cytokines and chemokines are important in the immune control of influenza A virus infections, their contribution can be detrimental as they elicit potentially fatal "cytokine shock" [55]. Recent studies dramatically illustrate the devastating impact of increased inflammatory infiltrates on viral-induced pathology. In animal models, infection with the reconstructed 1918 influenza A virus promotes massive inflammatory infiltrates with significantly higher levels of cytokines (IFN-γ, TNF-α, IL-1, IL-6, IL-12, IL-18, and granulocyte-colony-stimulating factor) and chemokines (MIP-2, MIP-1α/β, MCP-1) [21, 56–58]. Therefore, particularly early on, potent inflammatory antiviral activity may be dangerous, rather than protective, to the host due to the deleterious impact on lung pathology.

Collectins are collagen-like lectins that participate in innate immunity to viral pathogens [59]. Collectin family members, the surfactant proteins A (SP-A), and SP-D, are constitutively present in the fluids that line the respiratory tract [60]. Together with the mannan-binding lectin (MBL), SP-A and SP-D contribute to influenza A virus clearance via a number of mechanisms. Hemagglutination and viral infectivity are inhibited by SP-A [61, 62], SP-D [61, 63], and MBL [61, 64, 65]. In addition, complement-mediated lysis of influenza A virus-infected cells is enhanced by MBL [66], while SP-A and SP-D promote the binding and uptake of influenza A viruses by neutrophils [61, 67] and SP-A promotes opsonization and phagocytosis of influenza A virus by the AM population [68]. The sensitivity of different influenza A viral strains to collectin-mediated defense correlates with the degree of glycosylation of the HA glycoprotein [66, 69].

Defensins are cationic peptides produced by both leukocytes and epithelial cells. Defensins can exert direct microbial activity or promote immunity by acting as chemotactic agents. Examples of defensin-mediated anti-influenza A virus activity include retrocyclin-2 (o-defensin) and human β defensin 3 inhibition of HA-mediated membrane fusion [70]. The human neutrophil peptide (HNP) 1 (α-defensin) directly inactivates influenza A virus [65, 71].

4   Humoral Immunity and the Influenza A Viruses

Humoral immunity provides host defense through B lymphocyte secretion of antibody. Protective antibodies target antigenic structures exposed on the pathogen surface. Antibody-mediated immunity contributes to defense against the influenza A viruses [72–75] but is not always essential for optimal viral clearance [76, 77]. In any case, the influenza A viruses elicit a diverse spectrum of antiviral antibody responses. Natural antibodies present the first line of antibody-mediated defense [78]. These are low-affinity antibodies that restrict early virus dissemination [78] and promote the recruitment of viral antigen to the secondary lymphoid organs [79]. Natural antibodies reduce the overall load of influenza A virus and, as such, are required for optimal specific IgG antibody responses [75, 80]. Secretion of natural antibodies requires the transcriptional repressor Blimp-1: mice with Blimp-1-deficient B cells are more susceptible to influenza A virus infection [81]. Although natural antibodies are involved in the primary response to influenza A viruses, they are not required for optimal protection from secondary challenge [82]. Furthermore, while natural antibodies clearly display antiviral properties, effective virus clearance requires the induction of neutralizing antibody. Such neutralizing antibodies can be rapidly induced and possess high affinity (or avidity) for viral antigen. Mostly, virus neutralization is thought to be optimally achieved via antibody-mediated interference with viral binding to the host receptors required for cell entry or egress. Consequently, the influenza virus HA is heavily targeted by neutralizing antibodies [83, 84]. Crystallographic examination of HA in complex with neutralizing antibodies shows that antibody binding can occur at the same site as host receptor binding [85] or in distal regions where receptor binding is obstructed by steric hindrance [86]. Anti-HA neutralizing antibodies can also interfere with HA-mediated membrane fusion [87]. Similar to HA, NA is also targeted by neutralizing antibodies [88]. Neutralizing antibodies represent the major target of current influenza A virus vaccine strategies. While most neutralizing antibody strategies target HA or NA [89], the matrix protein 2 (M2) represents an interesting potential vaccine candidate [90]. M2 is a transmembrane protein expressed at the infected cell surface [91], but in contrast to HA and NA, is highly conserved among influenza A virus strains. Unfortunately thus far, M2-targeted vaccine strategies have elicited only weak immunity that does not protect mice from lethal challenge [92].

CD4$^+$ T-helper cells contribute to humoral immunity by promoting B-cell differentiation into immunoglobulin class-switched, antibody-secreting cells. In most studies, the production of anti-influenza A virus antibody is CD4$^+$ T-cell dependent [74, 93–95], although exceptions are reported [73, 74]. Classically, CD4$^+$ T-cell help involves (1) the recognition of viral antigen and (2) the delivery of an activation signal to the B cell via the TNFR family member, CD40. Mice deficient in CD40 generate significantly impaired influenza A virus-specific antibody responses [93, 96]. Of interest, CD4$^+$ T cells can help B lymphocytes by noncognate interactions that do not require specific influenza A virus antigen recognition [93].

5   T-Cell Immunity and the Influenza A Viruses

5.1   Dendritic Cells

DCs enable pathogen-derived antigens to be presented in a context that facilitates successful T-cell immunity [97]. Specialized in antigen presentation, the DCs facilitate (1) the acquisition of antigen, (2) processing and presentation of antigenic peptides in the context of host major histocompatibility complex (MHC) molecules, and (3) the provision of costimulatory signals. Immunity to influenza A virus infection requires DCs for both primary [98] and secondary T-cell responses [99, 100]. Many DC subsets are involved including the CCR2-dependent "inflammatory" DCs [101, 102], while plasmacytoid DCs are dispensable for influenza A virus clearance [103]. DC can control the magnitude of influenza A virus-specific T-cell immunity via FasL-mediated apoptosis [104]. In the respiratory tract, an extensive network of DC populations is present both in the lung [105] and in the draining lymph node [106]. Furthermore, pulmonary infection recruits additional DC populations into the lung [107–109]. To acquire influenza A virus antigen, DC may simply be directly infected with the virus. Infection induces the maturational changes (upregulation of costimulatory molecules and MHC class II) that are necessary for DC stimulation of T cells [110–112]. Infection can result in the expression of influenza NA at the DC surface, with NA-mediated removal of sialic acids serving to both enhance and inhibit DC function depending on the multiplicity of infection [113, 114]. DCs can also acquire influenza A virus-derived antigen released following the apoptotic lysis of infected respiratory cells [115, 116]. Once antigen is acquired, lung DCs migrate to the lymph node that drains the respiratory tract [107, 117, 118]. Migration occurs early after infection (24–48 h), and then the DCs display a refractory state to further inflammatory stimuli [107]. The lymph node also contains a resident DC set that has no direct access to the airways. Despite this, these resident DCs can also present influenza A virus-derived antigen [117]. Therefore, antigen transfer between the resident and migratory lung DC subsets must occur [119, 120]. Most experiments indicate that MHC class I presentation of influenza A virus-derived antigen in the lung draining lymph node ceases beyond 12–14 days [121, 122], although recently it has been suggested that antigen presentation can occur for up to 2 months following infection [123]. MHC class II presentation is also reported to persist for as long as 4 weeks after infection [124]. This is surprising given that infectious virus is cleared by day 10 [125]. Therefore, it has been postulated that the respiratory lymph node DCs can serve as a reservoir for antigen, with a depot being maintained well beyond the clearance of pathogen from the infected respiratory tissue [123, 126]. This, however, remains a contentious issue as the presence of an influenza A virus antigen depot was not detected in a separate independent study [127].

5.2   Costimulation

The participation of DCs in adaptive immunity is critical due to the rich array of costimulatory molecules expressed at the cell surface. A growing list of costimu-latory molecules has been identified, most of which belong to either CD28/B7 [128] or TNFR [129] families. Costimulation serves to enhance the antigen-specific signals that are delivered through the T-cell receptor (TCR). As such, costimulation is required for optimal T-cell immunity in many viral infections [130]. The major pathway of costimulation is via the CD28/B7 interaction that plays an important role in influenza A virus immunity. This signal contributes to the generation of influenza A virus-specific T-cell immunity at multiple levels. For $CD8^+$ T cells, CD28/B7 contributes to expansion [131–133], cytotoxicity, and/or effector cyto-kine production [131, 134, 135], recruitment to the infected airways [134], and survival [135]. In contrast, the hierarchy of T-cell response magnitude to individual influenza A virus-derived epitopes (a phenomenon termed immunodominance [136, 137]) is not altered in the absence of CD28/B7 signaling [138]. Mice deficient in CD28/B7 also display impaired influenza-specific neutralizing antibody responses [133]. While CD28/B7 plays a prominent part early in response to influenza A virus infection, 41BB/41BBL is important for sustained $CD8^+$ T-cell expansion and is critical for optimal recall responses [131, 133, 139]. Effective $CD4^+$ T-cell immunity during influenza A virus infection also requires CD28/B7 [133], OX40/OX40L [140], and ICOS/ICOSL [141]-mediated costimulation. The accumulation of T cells in influenza A virus-infected lungs depends on CD27/CD70 signaling [132, 142]. This is due to its impact on T-cell survival and/or migration to the infected respiratory tract [132]. Together, multiple costimulatory signals are delivered via the DCs to promote optimal adaptive immunity and, in turn, influenza A virus elimination.

5.3 CD8$^+$ T Cells

Effector CD8$^+$ T cells, also known as cytolytic T lymphocytes (CTLs), are important in the normal clearance of influenza A viruses [143]. Mice deficient in CD8$^+$
T cells show delayed influenza A virus clearance, though they eventually control
infection with all but the most virulent viruses [144]. The influenza A virus-specific
CD8$^+$ T-cell response has been extensively characterized utilizing murine models
of infection, particularly with the HKx31 (H3N2) and PR/8 (H1N1) influenza A
viruses. CD8$^+$ T cells are primed, are activated, and expand in the lung draining
lymph nodes during the first week or so after primary infection [121, 145]. Activated CD8$^+$ T cells then traffic to the respiratory airways and the infected lung to
mediate viral clearance [146]. The trafficking [147] and retention of CD8$^+$ T cells in
the lung [148] are dependent on LFA-1 expression. At the site of infection, CD8$^+$
T cells target virus-infected cells that express peptide derived from influenza A
virus protein associated with major histocompatibility complex class I (MHC I). An
array of epitopes is recognized in the C57BL/6 (B6) mouse model, with the
dominant epitopes (in terms of response magnitude) seen by CD8$^+$ T cells being
provided by the viral polymerase A (PA$_{224-233}$) [149] and nucleoprotein (NP$_{366-374}$)
[150, 151]. Subdominant epitopes are derived from the basic polymerase subunit 1
(PB1$_{703-711}$) [152], the mitochondrial protein PB1-F2$_{62-70}$ [152, 153], nonstructural
protein 2 (NS2$_{114-121}$) [151], and matrix protein 1 (M1$_{128-135}$) [154]. In the absence
of the dominant epitopes, subdominant epitope-specific CD8$^+$ T cells account for a
compensatory response, although a slight delay in viral clearance is observed [155,
156]. Depending on the experimental model, 30–90% of CD8$^+$ T cells recovered
from the respiratory tract are influenza A virus specific at the peak of the primary
response, illustrating their enrichment in the pneumonic lung [137, 151, 152, 157].
Epitope-specific CD8$^+$ T cells can be found widely dispersed throughout various
body organs, including the lung, spleen, bone marrow, blood, liver, and nondraining
lymph nodes [157, 158]. Once their target antigen is recognized, CD8$^+$ T cells exert
multiple effector functions. Cytokines such as IFNγ, TNF-α, and IL-2 are secreted
by influenza A virus-specific CD8$^+$ T cells [159]. In addition, CD8$^+$ T cells mediate
direct cytolysis of influenza A virus-infected target cells by the exocytosis of
cytolytic granules that contain perforin and granzymes [160–163] and/or through
the expression of Fas-ligand (FasL) [164–166]. CD8$^+$ T cells also exert regulation
of the inflammatory process via the production of IL-10 [167].

Following influenza A virus clearance, virus-specific CD8$^+$ T cells decrease in
number until a plateau is reached approximately 2 months following infection [122,
157]. After primary infection, the codominant D^{b}NP$_{366-374}$ and D^{b}PA$_{224-233}$-spe-
cific CD8$^+$ T-cell populations contract at the same rate [157] to memory pools that
are approximately equivalent in number and represent 10% of the population at the
peak of the response [168]. Influenza A virus-specific CD8$^+$ T cells persist as a
stable population for the life of a laboratory mouse [157, 169, 170]. Retention of
memory CD8$^+$ T cells in nonlymphoid tissue, such as the lung, is mediated by
T-cell expression of VLA-1 [171]. Secondary challenge recruits the memory CD8$^+$

T cells that expand in the lymph nodes and promote viral clearance approximately 2 days earlier than after primary infection [157]. During secondary infection, the NP$_{366-374}$ CD8$^+$ T-cell population is clearly dominant representing up to 80% of the virus-specific CTL responses [122, 137, 151, 152]. This dominance is maintained in the memory populations that persist following the peak of the secondary response (day 8) [122]. The skewed immunodominance hierarchy observed in secondary versus primary influenza A virus infection was initially thought to be largely a consequence of differential antigen presentation [172], though it is now considered that T-cell precursor frequency and antigen dose are likely to be important determining variables [173].

5.4   CD4$^+$ T Cells

Virus-specific CD4$^+$ T cells are important participants in influenza immunity [174, 175]. Although, acting alone, these cells do not normally eliminate virus [176], they exert distinct roles in both humoral immunity (as discussed) and CD8$^+$ T-cell responses. A vigorous, heterogenous CD4$^+$ T-cell response is elicited following influenza A virus infection [175]. Again, the process of clonal expansion and differentiations is initiated in the lung draining lymph node, with the peak response in the respiratory airways occurring 6–7 days following infection [175]. This is dominated by producers of the Th1 cytokines, such as IL-2, IFN-γ, and TNF-α [177]. CD4$^+$ T cells also secrete IL-10 contributing to the regulation of the inflammatory response [167]. Following influenza A virus clearance, CD4$^+$ T cells demonstrate increased contraction in the respiratory tract compared with influenza A virus-specific CD8$^+$ T cells [178, 179]. A major role for CD4$^+$ T cells is the provision of "help" for optimal CD8$^+$ T-cell immunity. Although CD4$^+$ T cells are not required for primary influenza-specific CD8$^+$ T-cell responses, presumably due to the direct activation of DC by viral infection [180–182], they are critical for the optimal establishment of CD8$^+$ T-cell memory. The absence of CD4$^+$ T cells during primary influenza A virus infections leads to a significant reduction in the size and magnitude of the secondary response and impaired viral clearance [77, 180]. Activation of CD4$^+$ T cells requires antigen-specific signaling via TCR recognition of antigens presented in the context of MHC class II molecules. Until recently, the spectrum of influenza A virus CD4$^+$ T-cell epitopes was much less well characterized than the panel known for the CD8$^+$ subset. Recently however, 20–30 peptides were identified for the influenza-specific CD4$^+$ T-cell response in C57BL/6 mice, with the majority being derived from the NP and HA proteins [183]. There is some evidence that influenza MHC class II epitopes are persisting for a substantial interval after the virus has been cleared from the host [124]. Overall, the adaptive immune response to the influenza A viruses involves complex interactions between a spectrum of functionally different cell types and their secretions.

6   Influenza A Virus Escape

The major influenza A virus escape mechanism rests in the inherent genetic variation of these RNA viruses, combined with the selective pressure exerted by HA-specific neutralizing antibody [184–186]. This process is known as "antigenic drift." Lacking proof reading capacity, the influenza A virus RNA polymerase promotes the accumulation of nucleotide point mutations. Such mutations generate approximately 3.5 amino acid substitutions per year [187]. Circulating viral subtypes are then selected where substitutions have occurred and maintain viral fitness [188] but abrogate immune recognition. For example, virus escape mutants are poorly recognized by neutralizing antibody due to (1) introduced steric interference with antibody binding [85], (2) virus conformational changes that render antibody binding energetically unfavorable [86], or (3) the introduction of new oligosaccharide attachment sites to surface glycoproteins that obscure antibody binding [189, 190]. Retention of amino acid substitutions at the HA membrane distal surface, an area targeted by antibodies, is favored over those buried within the protein [83]. Virus-specific CTL immunity can also be targeted by antigenic drift [191]. Here, viruses are selected with mutations that interfere with epitope binding to MHC class I or with epitopes that are no longer recognized by the TCR. Both $NP_{388-391}$ [192, 193] and $NP_{418-426}$ [194, 195] CTL peptides have shown evidence of antigenic drift. Hypervariability within a CTL epitope correlates with the functional avidity of the TCR [196]. Such antigenic drift can function to limit cross-protective immunity against multiple influenza A virus strains and, as a consequence, contribute to seasonal epidemics.

While antigenic drift represents a subtle mode of immune escape, influenza A viruses can also undergo major antigenic variation to outmaneuver the immune system. This takes place by "antigenic shift," where infection of the same cell with two distinct influenza A virus strains allows reassortment of the viral genomic segments, generating a new hybrid influenza A virus. Reassortment can occur following infection with different species-adapted viruses. For example, pigs can be infected with both human and avian influenza A viruses. Simultaneous infection may thereby generate a reassortment virus where the "human" pathogen acquires an "avian" virus HA or NA gene. In this case, for the HA and NA in particular, there would be no prevailing immunity in the human population, leading to the possibility of a human pandemic [197, 198]. Such antigenic shift involving avian and human strains has been implicated in two of the influenza A virus pandemics that have occurred in the twentieth century; the 1957 H2N2 [199, 200] and 1968 H3N2 [187, 200] infections. Of interest, the influenza A virus that provoked the 1918 pandemic did not arise through antigenic shift. Instead the 1918 H1N1 virus, which was responsible for millions of deaths worldwide, is believed to be an entirely avian viral strain that mutated in a way that allowed it to infect humans [201, 202].

The nonstructural protein 1 (NS1) encoded by influenza A virus provides a mode of immune escape that does not require manipulation of the genome. NS1 inhibits the host cell IFNα/β response [203, 204], a major pathway of immune defense against the virus (as discussed). Type 1 IFN induction is antagonized by NS1-mediated suppression of IFN-induced proteins dsRNA-activated protein kinase, $2'-5'$-oligo (A) synthetase [205–207], the transcription factors NFκB [208], and the IFN regulatory factor-3 [209]. Containing an RNA-binding domain at its N-terminus [208], it was previously considered that NS1 sequestered influenza A virus dsRNA [210]. Instead, NS1 forms a complex with RIG-1, the cellular sensor of influenza A virus uncapped ssRNA [6]. Therefore, NS1 acts to disable the host mechanism for detection of viral-derived RNA and the induction of the IFN response. Influenza A viruses lacking the NS1 protein are good vaccine candidates as the absence of this immunomodulatory protein greatly enhances the immunogenicity of the virus [211].

7   Heterotypic Influenza A Virus Immunity

Heterotypic immunity in this system is defined by cross-reactive, protective responses between serologically different (HA-distinct) influenza A viruses. It would obviously be advantageous if, for example, prior infection with a human influenza A virus could generate immune memory that provides at least some resistance to a highly pathogenic avian virus that suddenly adapted to transmit between people [212, 213]. Clearly, promoting heterotypic immunity is a desirable strategy for influenza A virus vaccine development. Described many decades ago [214], heterotypic immunity has now been shown for many influenza A virus combinations [215–218]. At least in mice, heterotypic immunity can both be long lasting and provide protection against otherwise lethal virus challenge. The best understood component of such responses is CTL immunity directed at generally conserved, internal viral proteins [215, 217, 218]. However, there is also evidence for the retention of a measure of heterotypic immunity in mice lacking CD8$^+$ T cells [216, 219]. In addition to the CD8$^+$ T effectors, CD4$^+$ T cells, nonneutralizing IgA antibody, NKT cells, and $\gamma\delta$ T cells have all been considered as possible players [217]. Immunization with a low dose of a cold-adapted, attenuated influenza A virus provides one vaccination strategy that has the potential to induce at least some degree of long-term, heterotypic immunity [220]. The promotion of such responses is clearly a worthwhile focus for future vaccination strategies.

8   Influenza A Virus Immunity and Vaccination

Ultimately, studies of the immune response to influenza A virus aim to provide the foundation for strategies that will combat influenza-mediated disease. Vaccination is the major weapon to enable reduced morbidity, mortality, and economic damage associated with widespread influenza A virus infection. The 2009 HINI pandemic highlights the urgency of developing safe and effective vaccines to emerging influenza A virus strains. H5N1 avian influenza A virus is another immediate concern. H5N1 is a highly pathogenic virus that possesses the capacity to provoke a debilitating pandemic of greater severity than that of H1NI. As such, much effort has been employed to design a suitable H5N1 vaccine. Eliciting high titer neutralizing antibody is a major priority of any vaccination strategy, although cell-mediated immunity is also considered important. Cell-mediated immunity is powerful in that it has the potential to provide universal protection against divergent viral strains [221, 222]. Many vaccine formulations have been tested to date, but the most widely utilized platform is the inactivated, attenuated H5N1 virus (whole virion, subvirion, or surface antigen). Studies indicate that two doses of this vaccine, together with an adjuvant such as MF59, elicit cross-protective immunogenic responses in healthy subjects [223–225]. Mechanisms underlying protection include the expansion of antigen-specific CD4$^+$ T cells, which serves as a reliable correlate of vaccine protection [226]. H5N1 vaccination studies provide valuable lessons that are currently being harnessed for a swift and rapid response to the 2009 HINI pandemic.

9 Conclusion

The influenza A viruses pose intriguing challenges for vaccine design [227]. Moving beyond the currently available products will depend on exploiting our understanding of immune defense mechanisms against this important and potentially very dangerous group of human pathogens. Here, we have briefly summarized a current view of how these viruses are controlled by elements of both innate and adaptive host response, together with the escape strategies that influenza A viruses exploit to survive in nature and to maintain transmission at the species level. An ideal vaccine could be thought to induce high levels of neutralizing antibody and CTL memory. This might optimally be achieved by promoting more effective DC vaccination, perhaps via the pathway of driving the innate response in ways that enhance T-cell immunity. An important caveat is, though, that much of our understanding of (particularly) the innate and T-cell responses to the influenza A viruses is based on mouse experiments. As we go forward to develop vaccine candidates, it is important that the analysis of influenza virus cell-mediated immunity, in particular, should be greatly extended in human subjects.

References

1. Lewis DB (2006) Avian flu to human influenza. Annu Rev Med 57:139–154
2. Janeway CA Jr, Medzhitov R (2002) Innate immune recognition. Annu Rev Immunol 20:197–216

3. Le Goffic R, Balloy V, Lagranderie M, Alexopoulou L, Escriou N, Flavell R, Chignard M, Si-Tahar M (2006) Detrimental contribution of the toll-like receptor (TLR)3 to influenza A virus-induced acute pneumonia. PLoS Pathog 2:e53

4. Guillot L, Le Goffic R, Bloch S, Escriou N, Akira S, Chignard M, Si-Tahar M (2005) Involvement of toll-like receptor 3 in the immune response of lung epithelial cells to double-stranded RNA and influenza A virus. J Biol Chem 280:5571–5580

5. Edelmann KH, Richardson-Burns S, Alexopoulou L, Tyler KL, Flavell RA, Oldstone MB (2004) Does toll-like receptor 3 play a biological role in virus infections? Virology 322:231–238

6. Pichlmair A, Schulz O, Tan CP, Naslund TI, Liljestrom P, Weber F, Reis e Sousa C (2006) RIG-I-mediated antiviral responses to single-stranded RNA bearing 5′-phosphates. Science 314:997–1001

7. Hornung V, Ellegast J, Kim S, Brzozka K, Jung A, Kato H, Poeck H, Akira S, Conzelmann KK, Schlee M, Endres S, Hartmann G (2006) 5′-Triphosphate RNA is the ligand for RIG-I. Science 314:994–997

8. Kato H, Sato S, Yoneyama M, Yamamoto M, Uematsu S, Matsui K, Tsujimura T, Takeda K, Fujita T, Takeuchi O, Akira S (2005) Cell type-specific involvement of RIG-I in antiviral response. Immunity 23:19–28

9. Diebold SS, Kaisho T, Hemmi H, Akira S, Reis e Sousa C (2004) Innate antiviral responses by means of TLR7-mediated recognition of single-stranded RNA. Science 303:1529–1531

10. Lund JM, Alexopoulou L, Sato A, Karow M, Adams NC, Gale NW, Iwasaki A, Flavell RA (2004) Recognition of single-stranded RNA viruses by toll-like receptor 7. Proc Natl Acad Sci USA 101:5598–5603

11. Kato H, Takeuchi O, Sato S, Yoneyama M, Yamamoto M, Matsui K, Uematsu S, Jung A, Kawai T, Ishii KJ, Yamaguchi O, Otsu K, Tsujimura T, Koh CS, Reis e Sousa C, Matsuura Y, Fujita T, Akira S (2006) Differential roles of MDA5 and RIG-I helicases in the recognition of RNA viruses. Nature 441:101–105

12. Ichinohe T, Lee HK, Ogura Y, Flavell R, Iwasaki A (2009) Inflammasome recognition of influenza virus is essential for adaptive immune responses. J Exp Med 206:79–87

13. Allen IC, Scull MA, Moore CB, Holl EK, McElvania-TeKippe E, Taxman DJ, Guthrie EH, Pickles RJ, Ting JP (2009) The NLRP3 inflammasome mediates in vivo innate immunity to influenza A virus through recognition of viral RNA. Immunity 30:556–565

14. Thomas PG, Dash P, Aldridge JR Jr, Ellebedy AH, Reynolds C, Funk AJ, Martin WJ, Lamkanfi M, Webby RJ, Boyd KL, Doherty PC, Kanneganti TD (2009) The intracellular sensor NLRP3 mediates key innate and healing responses to influenza A virus via the regulation of caspase-1. Immunity 30:566–575

15. Leung KN, Ada GL (1981) Induction of natural killer cells during murine influenza virus infection. Immunobiology 160:352–366

16. Stein-Streilein J, Bennett M, Mann D, Kumar V (1983) Natural killer cells in mouse lung: surface phenotype, target preference, and response to local influenza virus infection. J Immunol 131:2699–2704

17. Biron CA, Nguyen KB, Pien GC, Cousens LP, Salazar-Mather TP (1999) Natural killer cells in antiviral defense: function and regulation by innate cytokines. Annu Rev Immunol 17:189–220

18. Mandelboim O, Lieberman N, Lev M, Paul L, Arnon TI, Bushkin Y, Davis DM, Strominger JL, Yewdell JW, Porgador A (2001) Recognition of haemagglutinins on virus-infected cells by NKp46 activates lysis by human NK cells. Nature 409:1055–1060

19. Arnon TI, Lev M, Katz G, Chernobrov Y, Porgador A, Mandelboim O (2001) Recognition of viral hemagglutinins by NKp44 but not by NKp30. Eur J Immunol 31:2680–2689

20. Gazit R, Gruda R, Elboim M, Arnon TI, Katz G, Achdout H, Hanna J, Qimron U, Landau G, Greenbaum E, Zakay-Rones Z, Porgador A, Mandelboim O (2006) Lethal influenza infection in the absence of the natural killer cell receptor gene Ncr1. Nat Immunol 7:517–523

21. Tumpey TM, Garcia-Sastre A, Taubenberger JK, Palese P, Swayne DE, Pantin-Jackwood MJ, Schultz-Cherry S, Solorzano A, Van Rooijen N, Katz JM, Basler CF (2005) Pathogenicity of influenza viruses with genes from the 1918 pandemic virus: functional roles of alveolar macrophages and neutrophils in limiting virus replication and mortality in mice. J Virol 79:14933–14944

22. Fujisawa H (2001) Inhibitory role of neutrophils on influenza virus multiplication in the lungs of mice. Microbiol Immunol 45:679–688

23. Ratcliffe DR, Nolin SL, Cramer EB (1988) Neutrophil interaction with influenza-infected epithelial cells. Blood 72:142–149

24. Sibille Y, Reynolds HY (1990) Macrophages and polymorphonuclear neutrophils in lung defense and injury. Am Rev Respir Dis 141:471–501

25. Fujimoto I, Pan J, Takizawa T, Nakanishi Y (2000) Virus clearance through apoptosis-dependent phagocytosis of influenza A virus-infected cells by macrophages. J Virol 74:3399–3403

26. Hofmann P, Sprenger H, Kaufmann A, Bender A, Hasse C, Nain M, Gemsa D (1997) Susceptibility of mononuclear phagocytes to influenza A virus infection and possible role in the antiviral response. J Leukoc Biol 61:408–414

27. Gong JH, Sprenger H, Hinder F, Bender A, Schmidt A, Horch S, Nain M, Gemsa D (1991) Influenza A virus infection of macrophages. Enhanced tumor necrosis factor-alpha (TNF-alpha) gene expression and lipopolysaccharide-triggered TNF-alpha release. J Immunol 147:3507–3513

28. Kaufmann A, Salentin R, Meyer RG, Bussfeld D, Pauligk C, Fesq H, Hofmann P, Nain M, Gemsa D, Sprenger H (2001) Defense against influenza A virus infection: essential role of the chemokine system. Immunobiology 204:603–613

29. Sprenger H, Meyer RG, Kaufmann A, Bussfeld D, Rischkowsky E, Gemsa D (1996) Selective induction of monocyte and not neutrophil-attracting chemokines after influenza A virus infection. J Exp Med 184:1191–1196

30. Bussfeld D, Kaufmann A, Meyer RG, Gemsa D, Sprenger H (1998) Differential mononuclear leukocyte attracting chemokine production after stimulation with active and inactivated influenza A virus. Cell Immunol 186:1–7

31. Snelgrove RJ, Goulding J, Didierlaurent AM, Lyonga D, Vekaria S, Edwards L, Gwyer E, Sedgwick JD, Barclay AN, Hussell T (2008) A critical function for CD200 in lung immune homeostasis and the severity of influenza infection. Nat Immunol 9:1074–1083

32. Wijburg OL, DiNatale S, Vadolas J, van Rooijen N, Strugnell RA (1997) Alveolar macrophages regulate the induction of primary cytotoxic T-lymphocyte responses during influenza virus infection. J Virol 71:9450–9457

33. Fesq H, Bacher M, Nain M, Gemsa D (1994) Programmed cell death (apoptosis) in human monocytes infected by influenza A virus. Immunobiology 190:175–182

34. Herold S, Steinmueller M, von Wulffen W, Cakarova L, Pinto R, Pleschka S, Mack M, Kuziel WA, Corazza N, Brunner T, Seeger W, Lohmeyer J (2008) Lung epithelial apoptosis in influenza virus pneumonia: the role of macrophage-expressed TNF-related apoptosis-inducing ligand. J Exp Med 205:3065–3077

35. Theofilopoulos AN, Baccala R, Beutler B, Kono DH (2005) Type I interferons (alpha/beta) in immunity and autoimmunity. Annu Rev Immunol 23:307–336

36. Wyde PR, Wilson MR, Cate TR (1982) Interferon production by leukocytes infiltrating the lungs of mice during primary influenza virus infection. Infect Immun 38:1249–1255

37. Durbin JE, Fernandez-Sesma A, Lee CK, Rao TD, Frey AB, Moran TM, Vukmanovic S, Garcia-Sastre A, Levy DE (2000) Type I IFN modulates innate and specific antiviral immunity. J Immunol 164:4220–4228

38. Garcia-Sastre A, Durbin RK, Zheng H, Palese P, Gertner R, Levy DE, Durbin JE (1998) The role of interferon in influenza virus tissue tropism. J Virol 72:8550–8558

39. Cella M, Jarrossay D, Facchetti F, Alebardi O, Nakajima H, Lanzavecchia A, Colonna M (1999) Plasmacytoid monocytes migrate to inflamed lymph nodes and produce large amounts of type I interferon. Nat Med 5:919–923

40. Nakano H, Yanagita M, Gunn MD (2001) CD11c(+)B220(+)Gr-1(+) cells in mouse lymph nodes and spleen display characteristics of plasmacytoid dendritic cells. J Exp Med 194:1171–1178

41. Bruno L, Seidl T, Lanzavecchia A (2001) Mouse pre-immunocytes as non-proliferating multipotent precursors of macrophages, interferon-producing cells, CD8alpha(+) and CD8alpha(−) dendritic cells. Eur J Immunol 31:3403–3412

42. O'Keeffe M, Hochrein H, Vremec D, Caminschi I, Miller JL, Anders EM, Wu L, Lahoud MH, Henri S, Scott B, Hertzog P, Tatarczuch L, Shortman K (2002) Mouse plasmacytoid cells: long-lived cells, heterogeneous in surface phenotype and function, that differentiate into CD8(+) dendritic cells only after microbial stimulus. J Exp Med 196:1307–1319

43. Seo SH, Webster RG (2002) Tumor necrosis factor alpha exerts powerful anti-influenza virus effects in lung epithelial cells. J Virol 76:1071–1076

44. Jego G, Palucka AK, Blanck JP, Chalouni C, Pascual V, Banchereau J (2003) Plasmacytoid dendritic cells induce plasma cell differentiation through type I interferon and interleukin 6. Immunity 19:225–234

45. Lee SW, Youn JW, Seong BL, Sung YC (1999) IL-6 induces long-term protective immunity against a lethal challenge of influenza virus. Vaccine 17:490–496

46. Schmitz N, Kurrer M, Bachmann MF, Kopf M (2005) Interleukin-1 is responsible for acute lung immunopathology but increases survival of respiratory influenza virus infection. J Virol 79:6441–6448

47. Denton AE, Doherty PC, Turner SJ, La Gruta NL (2007) IL-18, but not IL-12, is required for optimal cytokine production by influenza virus-specific CD8(+) T cells. Eur J Immunol 37(2):368–375

48. Bhardwaj N, Seder RA, Reddy A, Feldman MV (1996) IL-12 in conjunction with dendritic cells enhances antiviral CD8+ CTL responses in vitro. J Clin Invest 98:715–722

49. Monteiro JM, Harvey C, Trinchieri G (1998) Role of interleukin-12 in primary influenza virus infection. J Virol 72:4825–4831

50. Nguyen HH, van Ginkel FW, Vu HL, Novak MJ, McGhee JR, Mestecky J (2000) Gamma interferon is not required for mucosal cytotoxic T-lymphocyte responses or heterosubtypic immunity to influenza A virus infection in mice. J Virol 74:5495–5501

51. Bot A, Bot S, Bona CA (1998) Protective role of gamma interferon during the recall response to influenza virus. J Virol 72:6637–6645

52. Baumgarth N, Kelso A (1996) In vivo blockade of gamma interferon affects the influenza virus-induced humoral and the local cellular immune response in lung tissue. J Virol 70:4411–4418

53. Cook DN, Beck MA, Coffman TM, Kirby SL, Sheridan JF, Pragnell IB, Smithies O (1995) Requirement of MIP-1 alpha for an inflammatory response to viral infection. Science 269:1583–1585

54. Dawson TC, Beck MA, Kuziel WA, Henderson F, Maeda N (2000) Contrasting effects of CCR5 and CCR2 deficiency in the pulmonary inflammatory response to influenza A virus. Am J Pathol 156:1951–1959

55. La Gruta NL, Kedzierska K, Stambas J, Doherty PC (2007) A question of self-preservation: immunopathology in influenza virus infection. Immunol Cell Biol 85(2):85–92

56. Kobasa D, Jones SM, Shinya K, Kash JC, Copps J, Ebihara H, Hatta Y, Kim JH, Halfmann P, Hatta M, Feldmann F, Alimonti JB, Fernando L, Li Y, Katze MG, Feldmann H, Kawaoka Y (2007) Aberrant innate immune response in lethal infection of macaques with the 1918 influenza virus. Nature 445:319–323

57. Kobasa D, Takada A, Shinya K, Hatta M, Halfmann P, Theriault S, Suzuki H, Nishimura H, Mitamura K, Sugaya N, Usui T, Murata T, Maeda Y, Watanabe S, Suresh M, Suzuki T,

Suzuki Y, Feldmann H, Kawaoka Y (2004) Enhanced virulence of influenza A viruses with the haemagglutinin of the 1918 pandemic virus. Nature 431:703–707

58. Kash JC, Tumpey TM, Proll SC, Carter V, Perwitasari O, Thomas MJ, Basler CF, Palese P, Taubenberger JK, Garcia-Sastre A, Swayne DE, Katze MG (2006) Genomic analysis of increased host immune and cell death responses induced by 1918 influenza virus. Nature 443:578–581

59. Holmskov U, Thiel S, Jensenius JC (2003) Collections and ficolins: humoral lectins of the innate immune defense. Annu Rev Immunol 21:547–578

60. Crouch E, Hartshorn K, Ofek I (2000) Collectins and pulmonary innate immunity. Immunol Rev 173:52–65

61. Hartshorn KL, White MR, Shepherd V, Reid K, Jensenius JC, Crouch EC (1997) Mechanisms of anti-influenza activity of surfactant proteins A and D: comparison with serum collectins. Am J Physiol 273:L1156–L1166

62. Benne CA, Kraaijeveld CA, van Strijp JA, Brouwer E, Harmsen M, Verhoef J, van Golde LM, van Iwaarden JF (1995) Interactions of surfactant protein A with influenza A viruses: binding and neutralization. J Infect Dis 171:335–341

63. Hartshorn K, Chang D, Rust K, White M, Heuser J, Crouch E (1996) Interactions of recombinant human pulmonary surfactant protein D and SP-D multimers with influenza A. Am J Physiol 271:L753–L762

64. Hartshorn KL, Sastry K, White MR, Anders EM, Super M, Ezekowitz RA, Tauber AI (1993) Human mannose-binding protein functions as an opsonin for influenza A viruses. J Clin Invest 91:1414–1420

65. Daher KA, Selsted ME, Lehrer RI (1986) Direct inactivation of viruses by human granulocyte defensins. J Virol 60:1068–1074

66. Reading PC, Hartley CA, Ezekowitz RA, Anders EM (1995) A serum mannose-binding lectin mediates complement-dependent lysis of influenza virus-infected cells. Biochem Biophys Res Commun 217:1128–1136

67. Hartshorn KL, Reid KB, White MR, Jensenius JC, Morris SM, Tauber AI, Crouch E (1996) Neutrophil deactivation by influenza A viruses: mechanisms of protection after viral opsonization with collectins and hemagglutination-inhibiting antibodies. Blood 87:3450–3461

68. Benne CA, Benaissa-Trouw B, van Strijp JA, Kraaijeveld CA, van Iwaarden JF (1997) Surfactant protein A, but not surfactant protein D, is an opsonin for influenza A virus phagocytosis by rat alveolar macrophages. Eur J Immunol 27:886–890

69. Hartley CA, Reading PC, Ward AC, Anders EM (1997) Changes in the hemagglutinin molecule of influenza type A (H3N2) virus associated with increased virulence for mice. Arch Virol 142:75–88

70. Leikina E, Delanoe-Ayari H, Melikov K, Cho MS, Chen A, Waring AJ, Wang W, Xie Y, Loo JA, Lehrer RI, Chernomordik LV (2005) Carbohydrate-binding molecules inhibit viral fusion and entry by crosslinking membrane glycoproteins. Nat Immunol 6:995–1001

71. Doss M, White MR, Tecle T, Gantz D, Crouch EC, Jung G, Ruchala P, Waring AJ, Lehrer RI, Hartshorn KL (2009) Interactions of alpha-, beta-, and theta-defensins with influenza A virus and surfactant protein D. J Immunol 182:7878–7887

72. Graham MB, Braciale TJ (1997) Resistance to and recovery from lethal influenza virus infection in B lymphocyte-deficient mice. J Exp Med 186:2063–2068

73. Lee BO, Rangel-Moreno J, Moyron-Quiroz JE, Hartson L, Makris M, Sprague F, Lund FE, Randall TD (2005) CD4 T cell-independent antibody response promotes resolution of primary influenza infection and helps to prevent reinfection. J Immunol 175:5827–5838

74. Mozdzanowska K, Furchner M, Zharikova D, Feng J, Gerhard W (2005) Roles of CD4+ T-cell-independent and -dependent antibody responses in the control of influenza virus infection: evidence for noncognate CD4+ T-cell activities that enhance the therapeutic activity of antiviral antibodies. J Virol 79:5943–5951

75. Kopf M, Brombacher F, Bachmann MF (2002) Role of IgM antibodies versus B cells in influenza virus-specific immunity. Eur J Immunol 32:2229–2236

76. Topham DJ, Tripp RA, Hamilton-Easton AM, Sarawar SR, Doherty PC (1996) Quantitative analysis of the influenza virus-specific CD4+ T cell memory in the absence of B cells and Ig. J Immunol 157:2947–2952
77. Riberdy JM, Christensen JP, Branum K, Doherty PC (2000) Diminished primary and secondary influenza virus-specific CD8(+) T-cell responses in CD4-depleted Ig(-/-) mice. J Virol 74:9762–9765
78. Ochsenbein AF, Zinkernagel RM (2000) Natural antibodies and complement link innate and acquired immunity. Immunol Today 21:624–630
79. Ochsenbein AF, Pinschewer DD, Odermatt B, Ciurea A, Hengartner H, Zinkernagel RM (2000) Correlation of T cell independence of antibody responses with antigen dose reaching secondary lymphoid organs: implications for splenectomized patients and vaccine design. J Immunol 164:6296–6302
80. Baumgarth N, Herman OC, Jager GC, Brown LE, Herzenberg LA, Chen J (2000) B-1 and B-2 cell-derived immunoglobulin M antibodies are nonredundant components of the protective response to influenza virus infection. J Exp Med 192:271–280
81. Savitsky D, Calame K (2006) B-1 B lymphocytes require Blimp-1 for immunoglobulin secretion. J Exp Med 203:2305–2314
82. Harada Y, Muramatsu M, Shibata T, Honjo T, Kuroda K (2003) Unmutated immunoglobulin M can protect mice from death by influenza virus infection. J Exp Med 197:1779–1785
83. Skehel JJ, Wiley DC (2000) Receptor binding and membrane fusion in virus entry: the influenza hemagglutinin. Annu Rev Biochem 69:531–569
84. Kwong PD, Wilson IA (2009) HIV-1 and influenza antibodies: seeing antigens in new ways. Nat Immunol 10:573–578
85. Bizebard T, Gigant B, Rigolet P, Rasmussen B, Diat O, Bosecke P, Wharton SA, Skehel JJ, Knossow M (1995) Structure of influenza virus haemagglutinin complexed with a neutralizing antibody. Nature 376:92–94
86. Fleury D, Barrere B, Bizebard T, Daniels RS, Skehel JJ, Knossow M (1999) A complex of influenza hemagglutinin with a neutralizing antibody that binds outside the virus receptor binding site. Nat Struct Biol 6:530–534
87. Ekiert DC, Bhabha G, Elsliger MA, Friesen RH, Jongeneelen M, Throsby M, Goudsmit J, Wilson IA (2009) Antibody recognition of a highly conserved influenza virus epitope. Science 324:246–251
88. Murphy BR, Kasel JA, Chanock RM (1972) Association of serum anti-neuraminidase antibody with resistance to influenza in man. N Engl J Med 286:1329–1332
89. Belshe RB, Gruber WC, Mendelman PM, Cho I, Reisinger K, Block SL, Wittes J, Iacuzio D, Piedra P, Treanor J, King J, Kotloff K, Bernstein DI, Hayden FG, Zangwill K, Yan L, Wolff M (2000) Efficacy of vaccination with live attenuated, cold-adapted, trivalent, intranasal influenza virus vaccine against a variant (A/Sydney) not contained in the vaccine. J Pediatr 136:168–175
90. Neirynck S, Deroo T, Saelens X, Vanlandschoot P, Jou WM, Fiers W (1999) A universal influenza A vaccine based on the extracellular domain of the M2 protein. Nat Med 5:1157–1163
91. Lamb RA, Zebedee SL, Richardson CD (1985) Influenza virus M2 protein is an integral membrane protein expressed on the infected-cell surface. Cell 40:627–633
92. Jegerlehner A, Schmitz N, Storni T, Bachmann MF (2004) Influenza A vaccine based on the extracellular domain of M2: weak protection mediated via antibody-dependent NK cell activity. J Immunol 172:5598–5605
93. Sangster MY, Riberdy JM, Gonzalez M, Topham DJ, Baumgarth N, Doherty PC (2003) An early CD4+ T cell-dependent immunoglobulin A response to influenza infection in the absence of key cognate T-B interactions. J Exp Med 198:1011–1021
94. Scherle PA, Palladino G, Gerhard W (1992) Mice can recover from pulmonary influenza virus infection in the absence of class I-restricted cytotoxic T cells. J Immunol 148:212–217

95. Topham DJ, Tripp RA, Sarawar SR, Sangster MY, Doherty PC (1996) Immune CD4+ T cells promote the clearance of influenza virus from major histocompatibility complex class II -/- respiratory epithelium. J Virol 70:1288–1291

96. Lee BO, Moyron-Quiroz J, Rangel-Moreno J, Kusser KL, Hartson L, Sprague F, Lund FE, Randall TD (2003) CD40, but not CD154, expression on B cells is necessary for optimal primary B cell responses. J Immunol 171:5707–5717

97. Shortman K, Liu YJ (2002) Mouse and human dendritic cell subtypes. Nat Rev Immunol 2:151–161

98. Sigal LJ, Rock KL (2000) Bone marrow-derived antigen-presenting cells are required for the generation of cytotoxic T lymphocyte responses to viruses and use transporter associated with antigen presentation (TAP)-dependent and -independent pathways of antigen presentation. J Exp Med 192:1143–1150

99. Belz GT, Wilson NS, Smith CM, Mount AM, Carbone FR, Heath WR (2006) Bone marrow-derived cells expand memory CD8+ T cells in response to viral infections of the lung and skin. Eur J Immunol 36:327–335

100. Zammit DJ, Cauley LS, Pham QM, Lefrancois L (2005) Dendritic cells maximize the memory CD8 T cell response to infection. Immunity 22:561–570

101. Aldridge JR Jr, Moseley CE, Boltz DA, Negovetich NJ, Reynolds C, Franks J, Brown SA, Doherty PC, Webster RG, Thomas PG (2009) TNF/iNOS-producing dendritic cells are the necessary evil of lethal influenza virus infection. Proc Natl Acad Sci USA 106:5306–5311

102. Nakano H, Lin KL, Yanagita M, Charbonneau C, Cook DN, Kakiuchi T, Gunn MD (2009) Blood-derived inflammatory dendritic cells in lymph nodes stimulate acute T helper type 1 immune responses. Nat Immunol 10:394–402

103. GeurtsvanKessel CH, Willart MA, van Rijt LS, Muskens F, Kool M, Baas C, Thielemans K, Bennett C, Clausen BE, Hoogsteden HC, Osterhaus AD, Rimmelzwaan GF, Lambrecht BN (2008) Clearance of influenza virus from the lung depends on migratory langerin+CD11b-but not plasmacytoid dendritic cells. J Exp Med 205:1621–1634

104. Legge KL, Braciale TJ (2005) Lymph node dendritic cells control CD8+ T cell responses through regulated FasL expression. Immunity 23:649–659

105. Sung SS, Fu SM, Rose CE Jr, Gaskin F, Ju ST, Beaty SR (2006) A major lung CD103 (alphaE)-beta7 integrin-positive epithelial dendritic cell population expressing langerin and tight junction proteins. J Immunol 176:2161–2172

106. Henri S, Vremec D, Kamath A, Waithman J, Williams S, Benoist C, Burnham K, Saeland S, Handman E, Shortman K (2001) The dendritic cell populations of mouse lymph nodes. J Immunol 167:741–748

107. Legge KL, Braciale TJ (2003) Accelerated migration of respiratory dendritic cells to the regional lymph nodes is limited to the early phase of pulmonary infection. Immunity 18:265–277

108. McWilliam AS, Napoli S, Marsh AM, Pemper FL, Nelson DJ, Pimm CL, Stumbles PA, Wells TN, Holt PG (1996) Dendritic cells are recruited into the airway epithelium during the inflammatory response to a broad spectrum of stimuli. J Exp Med 184:2429–2432

109. Yamamoto N, Suzuki S, Shirai A, Suzuki M, Nakazawa M, Nagashima Y, Okubo T (2000) Dendritic cells are associated with augmentation of antigen sensitization by influenza A virus infection in mice. Eur J Immunol 30:316–326

110. Nonacs R, Humborg C, Tam JP, Steinman RM (1992) Mechanisms of mouse spleen dendritic cell function in the generation of influenza-specific, cytolytic T lymphocytes. J Exp Med 176:519–529

111. Macatonia SE, Taylor PM, Knight SC, Askonas BA (1989) Primary stimulation by dendritic cells induces antiviral proliferative and cytotoxic T cell responses in vitro. J Exp Med 169:1255–1264

112. Bhardwaj N, Bender A, Gonzalez N, Bui LK, Garrett MC, Steinman RM (1994) Influenza virus-infected dendritic cells stimulate strong proliferative and cytolytic responses from human CD8+ T cells. J Clin Invest 94:797–807

113. Oh S, Eichelberger MC (1999) Influenza virus neuraminidase alters allogeneic T cell proliferation. Virology 264:427–435

114. Oh S, McCaffery JM, Eichelberger MC (2000) Dose-dependent changes in influenza virus-infected dendritic cells result in increased allogeneic T-cell proliferation at low, but not high, doses of virus. J Virol 74:5460–5469

115. Albert ML, Sauter B, Bhardwaj N (1998) Dendritic cells acquire antigen from apoptotic cells and induce class I-restricted CTLs. Nature 392:86–89

116. Wilson NS, Behrens GM, Lundie RJ, Smith CM, Waithman J, Young L, Forehan SP, Mount A, Steptoe RJ, Shortman KD, de Koning-Ward TF, Belz GT, Carbone FR, Crabb BS, Heath WR, Villadangos JA (2006) Systemic activation of dendritic cells by toll-like receptor ligands or malaria infection impairs cross-presentation and antiviral immunity. Nat Immunol 7:165–172

117. Belz GT, Smith CM, Kleinert L, Reading P, Brooks A, Shortman K, Carbone FR, Heath WR (2004) Distinct migrating and nonmigrating dendritic cell populations are involved in MHC class I-restricted antigen presentation after lung infection with virus. Proc Natl Acad Sci USA 101:8670–8675

118. Vermaelen KY, Carro-Muino I, Lambrecht BN, Pauwels RA (2001) Specific migratory dendritic cells rapidly transport antigen from the airways to the thoracic lymph nodes. J Exp Med 193:51–60

119. Carbone FR, Belz GT, Heath WR (2004) Transfer of antigen between migrating and lymph node-resident DCs in peripheral T-cell tolerance and immunity. Trends Immunol 25:655–658

120. Randolph GJ (2006) Migratory dendritic cells: sometimes simply ferries? Immunity 25:15–18

121. Lawrence CW, Braciale TJ (2004) Activation, differentiation, and migration of naive virus-specific CD8+ T cells during pulmonary influenza virus infection. J Immunol 173:1209–1218

122. Flynn KJ, Riberdy JM, Christensen JP, Altman JD, Doherty PC (1999) In vivo proliferation of naive and memory influenza-specific CD8(+) T cells. Proc Natl Acad Sci USA 96:8597–8602

123. Zammit DJ, Turner DL, Klonowski KD, Lefrancois L, Cauley LS (2006) Residual antigen presentation after influenza virus infection affects CD8 T cell activation and migration. Immunity 24:439–449

124. Jelley-Gibbs DM, Brown DM, Dibble JP, Haynes L, Eaton SM, Swain SL (2005) Unexpected prolonged presentation of influenza antigens promotes CD4 T cell memory generation. J Exp Med 202:697–706

125. Doherty PC, Christensen JP (2000) Accessing complexity: the dynamics of virus-specific T cell responses. Annu Rev Immunol 18:561–592

126. Julia V, Hessel EM, Malherbe L, Glaichenhaus N, O'Garra A, Coffman RL (2002) A restricted subset of dendritic cells captures airborne antigens and remains able to activate specific T cells long after antigen exposure. Immunity 16:271–283

127. Mintern JD, Bedoui S, Davey GM, Moffat JM, Doherty PC, Turner SJ (2009) Transience of MHC Class I-restricted antigen presentation after influenza A virus infection. Proc Natl Acad Sci USA 106:6724–6729

128. Sharpe AH, Freeman GJ (2002) The B7-CD28 superfamily. Nat Rev Immunol 2:116–126

129. Croft M (2003) Co-stimulatory members of the TNFR family: keys to effective T-cell immunity? Nat Rev Immunol 3:609–620

130. Bertram EM, Dawicki W, Watts TH (2004) Role of T cell costimulation in anti-viral immunity. Semin Immunol 16:185–196

131. Halstead ES, Mueller YM, Altman JD, Katsikis PD (2002) In vivo stimulation of CD137 broadens primary antiviral CD8+ T cell responses. Nat Immunol 3:536–541

132. Hendriks J, Xiao Y, Borst J (2003) CD27 promotes survival of activated T cells and complements CD28 in generation and establishment of the effector T cell pool. J Exp Med 198:1369–1380

133. Bertram EM, Lau P, Watts TH (2002) Temporal segregation of 4-1BB versus CD28-mediated costimulation: 4-1BB ligand influences T cell numbers late in the primary response and regulates the size of the T cell memory response following influenza infection. J Immunol 168:3777–3785

134. Lumsden JM, Roberts JM, Harris NL, Peach RJ, Ronchese F (2000) Differential requirement for CD80 and CD80/CD86-dependent costimulation in the lung immune response to an influenza virus infection. J Immunol 164:79–85

135. Liu Y, Wenger RH, Zhao M, Nielsen PJ (1997) Distinct costimulatory molecules are required for the induction of effector and memory cytotoxic T lymphocytes. J Exp Med 185:251–262

136. Yewdell JW, Del Val M (2004) Immunodominance in TCD8+ responses to viruses: cell biology, cellular immunology, and mathematical models. Immunity 21:149–153

137. Belz GT, Stevenson PG, Doherty PC (2000) Contemporary analysis of MHC-related immunodominance hierarchies in the CD8+ T cell response to influenza A viruses. J Immunol 165:2404–2409

138. Chen W, Bennink JR, Morton PA, Yewdell JW (2002) Mice deficient in perforin, CD4+ T cells, or CD28-mediated signaling maintain the typical immunodominance hierarchies of CD8+ T-cell responses to influenza virus. J Virol 76:10332–10337

139. DeBenedette MA, Wen T, Bachmann MF, Ohashi PS, Barber BH, Stocking KL, Peschon JJ, Watts TH (1999) Analysis of 4-1BB ligand (4-1BBL)-deficient mice and of mice lacking both 4-1BBL and CD28 reveals a role for 4-1BBL in skin allograft rejection and in the cytotoxic T cell response to influenza virus. J Immunol 163:4833–4841

140. Kopf M, Ruedl C, Schmitz N, Gallimore A, Lefrang K, Ecabert B, Odermatt B, Bachmann MF (1999) OX40-deficient mice are defective in Th cell proliferation but are competent in generating B cell and CTL Responses after virus infection. Immunity 11:699–708

141. Bertram EM, Tafuri A, Shahinian A, Chan VS, Hunziker L, Recher M, Ohashi PS, Mak TW, Watts TH (2002) Role of ICOS versus CD28 in antiviral immunity. Eur J Immunol 32:3376–3385

142. Hendriks J, Gravestein LA, Tesselaar K, van Lier RA, Schumacher TN, Borst J (2000) CD27 is required for generation and long-term maintenance of T cell immunity. Nat Immunol 1:433–440

143. Doherty PC, Topham DJ, Tripp RA, Cardin RD, Brooks JW, Stevenson PG (1997) Effector CD4+ and CD8+ T-cell mechanisms in the control of respiratory virus infections. Immunol Rev 159:105–117

144. Bender BS, Croghan T, Zhang L, Small PA Jr (1992) Transgenic mice lacking class I major histocompatibility complex-restricted T cells have delayed viral clearance and increased mortality after influenza virus challenge. J Exp Med 175:1143–1145

145. Tripp RA, Sarawar SR, Doherty PC (1995) Characteristics of the influenza virus-specific CD8+ T cell response in mice homozygous for disruption of the H-2IAb gene. J Immunol 155:2955–2959

146. Cerwenka A, Morgan TM, Dutton RW (1999) Naive, effector, and memory CD8 T cells in protection against pulmonary influenza virus infection: homing properties rather than initial frequencies are crucial. J Immunol 163:5535–5543

147. Galkina E, Thatte J, Dabak V, Williams MB, Ley K, Braciale TJ (2005) Preferential migration of effector CD8+ T cells into the interstitium of the normal lung. J Clin Invest 115:3473–3483

148. Thatte J, Dabak V, Williams MB, Braciale TJ, Ley K (2003) LFA-1 is required for retention of effector CD8 T cells in mouse lungs. Blood 101:4916–4922

149. Belz GT, Xie W, Altman JD, Doherty PC (2000) A previously unrecognized H-2D(b)-restricted peptide prominent in the primary influenza A virus-specific CD8(+) T-cell response is much less apparent following secondary challenge. J Virol 74:3486–3493

150. Townsend AR, Rothbard J, Gotch FM, Bahadur G, Wraith D, McMichael AJ (1986) The epitopes of influenza nucleoprotein recognized by cytotoxic T lymphocytes can be defined with short synthetic peptides. Cell 44:959–968

151. Flynn KJ, Belz GT, Altman JD, Ahmed R, Woodland DL, Doherty PC (1998) Virus-specific CD8+ T cells in primary and secondary influenza pneumonia. Immunity 8:683–691

152. Belz GT, Xie W, Doherty PC (2001) Diversity of epitope and cytokine profiles for primary and secondary influenza a virus-specific CD8+ T cell responses. J Immunol 166:4627–4633

153. Chen W, Calvo PA, Malide D, Gibbs J, Schubert U, Bacik I, Basta S, O'Neill R, Schickli J, Palese P, Henklein P, Bennink JR, Yewdell JW (2001) A novel influenza A virus mitochondrial protein that induces cell death. Nat Med 7:1306–1312

154. Vitiello A, Yuan L, Chesnut RW, Sidney J, Southwood S, Farness P, Jackson MR, Peterson PA, Sette A (1996) Immunodominance analysis of CTL responses to influenza PR8 virus reveals two new dominant and subdominant Kb-restricted epitopes. J Immunol 157:5555–5562

155. Andreansky SS, Stambas J, Thomas PG, Xie W, Webby RJ, Doherty PC (2005) Consequences of immunodominant epitope deletion for minor influenza virus-specific CD8+-T-cell responses. J Virol 79:4329–4339

156. Webby RJ, Andreansky S, Stambas J, Rehg JE, Webster RG, Doherty PC, Turner SJ (2003) Protection and compensation in the influenza virus-specific CD8+ T cell response. Proc Natl Acad Sci USA 100:7235–7240

157. Marshall DR, Turner SJ, Belz GT, Wingo S, Andreansky S, Sangster MY, Riberdy JM, Liu T, Tan M, Doherty PC (2001) Measuring the diaspora for virus-specific CD8+ T cells. Proc Natl Acad Sci USA 98:6313–6318

158. Turner SJ, Diaz G, Cross R, Doherty PC (2003) Analysis of clonotype distribution and persistence for an influenza virus-specific CD8+ T cell response. Immunity 18:549–559

159. La Gruta NL, Turner SJ, Doherty PC (2004) Hierarchies in cytokine expression profiles for acute and resolving influenza virus-specific CD8+ T cell responses: correlation of cytokine profile and TCR avidity. J Immunol 172:5553–5560

160. Johnson BJ, Costelloe EO, Fitzpatrick DR, Haanen JB, Schumacher TN, Brown LE, Kelso A (2003) Single-cell perforin and granzyme expression reveals the anatomical localization of effector CD8+ T cells in influenza virus-infected mice. Proc Natl Acad Sci USA 100:2657–2662

161. Liu B, Mori I, Hossain MJ, Dong L, Chen Z, Kimura Y (2003) Local immune responses to influenza virus infection in mice with a targeted disruption of perforin gene. Microb Pathog 34:161–167

162. Mintern JD, Guillonneau C, Carbone FR, Doherty PC, Turner SJ (2007) Cutting edge: tissue-resident memory CTL down-regulate cytolytic molecule expression following virus clearance. J Immunol 179:7220–7224

163. Moffat JM, Gebhardt T, Doherty PC, Turner SJ, Mintern JD (2009) Granzyme A expression reveals distinct cytolytic CTL subsets following influenza A virus infection. Eur J Immunol 39:1203–1210

164. Topham DJ, Tripp RA, Doherty PC (1997) CD8+ T cells clear influenza virus by perforin or Fas-dependent processes. J Immunol 159:5197–5200

165. Price GE, Huang L, Ou R, Zhang M, Moskophidis D (2005) Perforin and Fas cytolytic pathways coordinately shape the selection and diversity of CD8+-T-cell escape variants of influenza virus. J Virol 79:8545–8559

166. Fujimoto I, Takizawa T, Ohba Y, Nakanishi Y (1998) Co-expression of Fas and Fas-ligand on the surface of influenza virus-infected cells. Cell Death Differ 5:426–431

167. Sun J, Madan R, Karp CL, Braciale TJ (2009) Effector T cells control lung inflammation during acute influenza virus infection by producing IL-10. Nat Med 15:277–284

168. Kedzierska K, La Gruta NL, Turner SJ, Doherty PC (2006) Establishment and recall of CD8+ T-cell memory in a model of localized transient infection. Immunol Rev 211:133–145

169. Hogan RJ, Usherwood EJ, Zhong W, Roberts AA, Dutton RW, Harmsen AG, Woodland DL (2001) Activated antigen-specific CD8+ T cells persist in the lungs following recovery from respiratory virus infections. J Immunol 166:1813–1822

170. Wiley JA, Hogan RJ, Woodland DL, Harmsen AG (2001) Antigen-specific CD8(+) T cells persist in the upper respiratory tract following influenza virus infection. J Immunol 167:3293–3299

171. Ray SJ, Franki SN, Pierce RH, Dimitrova S, Koteliansky V, Sprague AG, Doherty PC, de Fougerolles AR, Topham DJ (2004) The collagen binding alpha1beta1 integrin VLA-1 regulates CD8 T cell-mediated immune protection against heterologous influenza infection. Immunity 20:167–179

172. Crowe SR, Turner SJ, Miller SC, Roberts AD, Rappolo RA, Doherty PC, Ely KH, Woodland DL (2003) Differential antigen presentation regulates the changing patterns of CD8+ T cell immunodominance in primary and secondary influenza virus infections. J Exp Med 198:399–410

173. La Gruta NL, Kedzierska K, Pang K, Webby R, Davenport M, Chen W, Turner SJ, Doherty PC (2006) A virus-specific CD8+ T cell immunodominance hierarchy determined by antigen dose and precursor frequencies. Proc Natl Acad Sci USA 103:994–999

174. Brown DM, Roman E, Swain SL (2004) CD4 T cell responses to influenza infection. Semin Immunol 16:171–177

175. Roman E, Miller E, Harmsen A, Wiley J, Von Andrian UH, Huston G, Swain SL (2002) CD4 effector T cell subsets in the response to influenza: heterogeneity, migration, and function. J Exp Med 196:957–968

176. Mozdzanowska K, Furchner M, Maiese K, Gerhard W (1997) CD4+ T cells are ineffective in clearing a pulmonary infection with influenza type A virus in the absence of B cells. Virology 239:217–225

177. Graham MB, Braciale VL, Braciale TJ (1994) Influenza virus-specific CD4+ T helper type 2 T lymphocytes do not promote recovery from experimental virus infection. J Exp Med 180:1273–1282

178. Powell TJ, Brown DM, Hollenbaugh JA, Charbonneau T, Kemp RA, Swain SL, Dutton RW (2004) CD8+ T cells responding to influenza infection reach and persist at higher numbers than CD4+ T cells independently of precursor frequency. Clin Immunol 113:89–100

179. Homann D, Teyton L, Oldstone MB (2001) Differential regulation of antiviral T-cell immunity results in stable CD8+ but declining CD4+ T-cell memory. Nat Med 7:913–919

180. Belz GT, Wodarz D, Diaz G, Nowak MA, Doherty PC (2002) Compromised influenza virus-specific CD8(+)-T-cell memory in CD4(+)-T-cell-deficient mice. J Virol 76:12388–12393

181. Bender A, Bui LK, Feldman MA, Larsson M, Bhardwaj N (1995) Inactivated influenza virus, when presented on dendritic cells, elicits human CD8+ cytolytic T cell responses. J Exp Med 182:1663–1671

182. Larsson M, Messmer D, Somersan S, Fonteneau JF, Donahoe SM, Lee M, Dunbar PR, Cerundolo V, Julkunen I, Nixon DF, Bhardwaj N (2000) Requirement of mature dendritic cells for efficient activation of influenza A-specific memory CD8+ T cells. J Immunol 165:1182–1190

183. Crowe SR, Miller SC, Brown DM, Adams PS, Dutton RW, Harmsen AG, Lund FE, Randall TD, Swain SL, Woodland DL (2006) Uneven distribution of MHC class II epitopes within the influenza virus. Vaccine 24:457–467

184. Palese P, Young JF (1982) Variation of influenza A, B, and C viruses. Science 215:1468–1474

185. Yewdell JW, Webster RG, Gerhard WU (1979) Antigenic variation in three distinct determinants of an influenza type A haemagglutinin molecule. Nature 279:246–248

186. Webster RG, Laver WG, Air GM, Schild GC (1982) Molecular mechanisms of variation in influenza viruses. Nature 296:115–121

187. Bean WJ, Schell M, Katz J, Kawaoka Y, Naeve C, Gorman O, Webster RG (1992) Evolution of the H3 influenza virus hemagglutinin from human and nonhuman hosts. J Virol 66:1129–1138

188. Berkhoff EG, de Wit E, Geelhoed-Mieras MM, Boon AC, Symons J, Fouchier RA, Osterhaus AD, Rimmelzwaan GF (2005) Functional constraints of influenza A virus epitopes limit escape from cytotoxic T lymphocytes. J Virol 79:11239–11246

189. Wiley DC, Wilson IA, Skehel JJ (1981) Structural identification of the antibody-binding sites of Hong Kong influenza haemagglutinin and their involvement in antigenic variation. Nature 289:373–378

190. Caton AJ, Brownlee GG, Yewdell JW, Gerhard W (1982) The antigenic structure of the influenza virus A/PR/8/34 hemagglutinin (H1 subtype). Cell 31:417–427

191. Price GE, Ou R, Jiang H, Huang L, Moskophidis D (2000) Viral escape by selection of cytotoxic T cell-resistant variants in influenza A virus pneumonia. J Exp Med 191:1853–1867

192. Rimmelzwaan GF, Boon AC, Voeten JT, Berkhoff EG, Fouchier RA, Osterhaus AD (2004) Sequence variation in the influenza A virus nucleoprotein associated with escape from cytotoxic T lymphocytes. Virus Res 103:97–100

193. Voeten JT, Bestebroer TM, Nieuwkoop NJ, Fouchier RA, Osterhaus AD, Rimmelzwaan GF (2000) Antigenic drift in the influenza A virus (H3N2) nucleoprotein and escape from recognition by cytotoxic T lymphocytes. J Virol 74:6800–6807

194. Boon AC, de Mutsert G, Graus YM, Fouchier RA, Sintnicolaas K, Osterhaus AD, Rimmelzwaan GF (2002) Sequence variation in a newly identified HLA-B35-restricted epitope in the influenza A virus nucleoprotein associated with escape from cytotoxic T lymphocytes. J Virol 76:2567–2572

195. Boon AC, de Mutsert G, van Baarle D, Smith DJ, Lapedes AS, Fouchier RA, Sintnicolaas K, Osterhaus AD, Rimmelzwaan GF (2004) Recognition of homo- and heterosubtypic variants of influenza A viruses by human CD8+ T lymphocytes. J Immunol 172:2453–2460

196. Boon AC, de Mutsert G, Fouchier RA, Osterhaus AD, Rimmelzwaan GF (2006) The hypervariable immunodominant NP418-426 epitope from the influenza A virus nucleoprotein is recognized by cytotoxic T lymphocytes with high functional avidity. J Virol 80:6024–6032

197. Webby RJ, Webster RG (2003) Are we ready for pandemic influenza? Science 302:1519–1522

198. Cox NJ, Subbarao K (2000) Global epidemiology of influenza: past and present. Annu Rev Med 51:407–421

199. Schafer JR, Kawaoka Y, Bean WJ, Suss J, Senne D, Webster RG (1993) Origin of the pandemic 1957 H2 influenza A virus and the persistence of its possible progenitors in the avian reservoir. Virology 194:781–788

200. Kawaoka Y, Krauss S, Webster RG (1989) Avian-to-human transmission of the PB1 gene of influenza A viruses in the 1957 and 1968 pandemics. J Virol 63:4603–4608

201. Reid AH, Taubenberger JK, Fanning TG (2004) Evidence of an absence: the genetic origins of the 1918 pandemic influenza virus. Nat Rev Microbiol 2:909–914

202. Taubenberger JK, Reid AH, Lourens RM, Wang R, Jin G, Fanning TG (2005) Characterization of the 1918 influenza virus polymerase genes. Nature 437:889–893

203. Garcia-Sastre A, Egorov A, Matassov D, Brandt S, Levy DE, Durbin JE, Palese P, Muster T (1998) Influenza A virus lacking the NS1 gene replicates in interferon-deficient systems. Virology 252:324–330

204. Diebold SS, Montoya M, Unger H, Alexopoulou L, Roy P, Haswell LE, Al-Shamkhani A, Flavell R, Borrow P, Reis e Sousa C (2003) Viral infection switches non-plasmacytoid dendritic cells into high interferon producers. Nature 424:324–328

205. Bergmann M, Garcia-Sastre A, Carnero E, Pehamberger H, Wolff K, Palese P, Muster T (2000) Influenza virus NS1 protein counteracts PKR-mediated inhibition of replication. J Virol 74:6203–6206

206. Li S, Min JY, Krug RM, Sen GC (2006) Binding of the influenza A virus NS1 protein to PKR mediates the inhibition of its activation by either PACT or double-stranded RNA. Virology 349:13–21

207. Min JY, Krug RM (2006) The primary function of RNA binding by the influenza A virus NS1 protein in infected cells: inhibiting the 2′–5′ oligo (A) synthetase/RNase L pathway. Proc Natl Acad Sci USA 103:7100–7105

208. Wang X, Li M, Zheng H, Muster T, Palese P, Beg AA, Garcia-Sastre A (2000) Influenza A virus NS1 protein prevents activation of NF-kappaB and induction of alpha/beta interferon. J Virol 74:11566–11573

209. Talon J, Horvath CM, Polley R, Basler CF, Muster T, Palese P, Garcia-Sastre A (2000) Activation of interferon regulatory factor 3 is inhibited by the influenza A virus NS1 protein. J Virol 74:7989–7996

210. Garcia-Sastre A (2001) Inhibition of interferon-mediated antiviral responses by influenza A viruses and other negative-strand RNA viruses. Virology 279:375–384

211. Ferko B, Stasakova J, Romanova J, Kittel C, Sereinig S, Katinger H, Egorov A (2004) Immunogenicity and protection efficacy of replication-deficient influenza A viruses with altered NS1 genes. J Virol 78:13037–13045

212. Jameson J, Cruz J, Terajima M, Ennis FA (1999) Human CD8+ and CD4+ T lymphocyte memory to influenza A viruses of swine and avian species. J Immunol 162:7578–7583

213. Jameson J, Cruz J, Ennis FA (1998) Human cytotoxic T-lymphocyte repertoire to influenza A viruses. J Virol 72:8682–8689

214. Schulman JL, Kilbourne ED (1965) Induction of partial specific heterotypic immunity in mice by A single infection with influenza A virus. J Bacteriol 89:170–174

215. Kreijtz JH, Bodewes R, van Amerongen G, Kuiken T, Fouchier RA, Osterhaus AD, Rimmelzwaan GF (2007) Primary influenza A virus infection induces cross-protective immunity against a lethal infection with a heterosubtypic virus strain in mice. Vaccine 25:612–620

216. Epstein SL, Lo CY, Misplon JA, Lawson CM, Hendrickson BA, Max EE, Subbarao K (1997) Mechanisms of heterosubtypic immunity to lethal influenza A virus infection in fully immunocompetent, T cell-depleted, beta2-microglobulin-deficient, and J chain-deficient mice. J Immunol 158:1222–1230

217. Benton KA, Misplon JA, Lo CY, Brutkiewicz RR, Prasad SA, Epstein SL (2001) Hetero-subtypic immunity to influenza A virus in mice lacking IgA, all Ig, NKT cells, or gamma delta T cells. J Immunol 166:7437–7445

218. Nguyen HH, Moldoveanu Z, Novak MJ, van Ginkel FW, Ban E, Kiyono H, McGhee JR, Mestecky J (1999) Heterosubtypic immunity to lethal influenza A virus infection is asso-ciated with virus-specific CD8(+) cytotoxic T lymphocyte responses induced in mucosa-associated tissues. Virology 254:50–60

219. Nguyen HH, van Ginkel FW, Vu HL, McGhee JR, Mestecky J (2001) Heterosubtypic immunity to influenza A virus infection requires B cells but not CD8+ cytotoxic T lympho-cytes. J Infect Dis 183:368–376

220. Powell TJ, Strutt T, Reome J, Hollenbaugh JA, Roberts AD, Woodland DL, Swain SL, Dutton RW (2007) Priming with cold-adapted influenza a does not prevent infection but elicits long-lived protection against supralethal challenge with heterosubtypic virus. J Immunol 178:1030–1038

221. Brown LE, Kelso A (2009) Prospects for an influenza vaccine that induces cross-protective cytotoxic T lymphocytes. Immunol Cell Biol 87:300–308

222. Thomas PG, Keating R, Hulse-Post DJ, Doherty PC (2006) Cell-mediated protection in influenza infection. Emerg Infect Dis 12:48–54

223. Lin J, Zhang J, Dong X, Fang H, Chen J, Su N, Gao Q, Zhang Z, Liu Y, Wang Z, Yang M, Sun R, Li C, Lin S, Ji M, Liu Y, Wang X, Wood J, Feng Z, Wang Y, Yin W (2006) Safety and immunogenicity of an inactivated adjuvanted whole-virion influenza A (H5N1) vaccine: a phase I randomised controlled trial. Lancet 368:991–997

224. Leroux-Roels I, Borkowski A, Vanwolleghem T, Drame M, Clement F, Hons E, Devaster JM, Leroux-Roels G (2007) Antigen sparing and cross-reactive immunity with an adjuvanted

rH5N1 prototype pandemic influenza vaccine: a randomised controlled trial. Lancet 370:580–589

225. Galli G, Hancock K, Hoschler K, DeVos J, Praus M, Bardelli M, Malzone C, Castellino F, Gentile C, McNally T, Del Giudice G, Banzhoff A, Brauer V, Montomoli E, Zambon M, Katz J, Nicholson K, Stephenson I (2009) Fast rise of broadly cross-reactive antibodies after boosting long-lived human memory B cells primed by an MF59 adjuvanted prepandemic vaccine. Proc Natl Acad Sci USA 106:7962–7967

226. Galli G, Medini D, Borgogni E, Zedda L, Bardelli M, Malzone C, Nuti S, Tavarini S, Sammicheli C, Hilbert AK, Brauer V, Banzhoff A, Rappuoli R, Del Giudice G, Castellino F (2009) Adjuvanted H5N1 vaccine induces early CD4+ T cell response that predicts long-term persistence of protective antibody levels. Proc Natl Acad Sci USA 106:3877–3882

227. Doherty PC, Turner SJ, Webby RG, Thomas PG (2006) Influenza and the challenge for immunology. Nat Immunol 7:449–455

Correlates of Protection Against Influenza

Emanuele Montomoli, Barbara Capecchi, and Katja Hoschler

Abstract Correlates of protection against influenza viruses have not been fully defined, but it is widely believed that protection against influenza can be conferred by serum hemagglutinin (HA) antibodies. The immune responses to injected influenza vaccines are routinely assessed by titrating serological HA antibodies. It is generally accepted that neutralizing and HA antibodies, as well as antibodies to neuraminidase, can be detected in serum 3–4 weeks post primary infection or vaccination. Serological assays commonly used to quantify antibodies specific for influenza viruses include hemagglutination inhibition (HI), single radial hemolysis (SRH), microneutralization (MN), ELISA and Western blot, of which, historically, HI and SRH are the most widely applied methods, the latter being increasingly replaced by MN. Each method used for antibody titration has different characteristics, and the validity index and specific use (seroepidemiology, serodiagnosis, response to vaccination, etc.) have to be considered while selecting the most suitable assay. Recently, ELISA tests have been improved, thanks to the elucidation of the structure of HA and the availability of this protein after recombinant expression. While the amount of data collected by conventional assays (HI and SRH) has permitted a fairly good optimization, serological measures are used to characterize the number of antibodies before and after vaccination. HI is the assay used most frequently for influenza antibody titration; however, it has low sensitivity in detecting responses to avian viruses in mammalian sera and alternative serological tests are needed. SRH utilizes a complement-mediated hemolysis reaction to

E. Montomoli (✉)
Department of Physiopathology, Experimental Medicine and Public Health, Laboratory of Molecular Epidemiology, University of Siena, Via Aldo Moro 3, 53100 Siena, Italy
e-mail: montomoli@unisi.it

B. Capecchi,
Novartis Vaccines and Diagnostics, Via Fiorentina 1, 53100 Siena, Italy

K. Hoschler
Health Protection Agency, Specialist and Reference Microbiology Division, ERNVL, Influenza Unit, Centre for Infections, 61 Colindale Avenue, London, UK

G. Del Giudice and R. Rappuoli (eds.), *Influenza Vaccines for the Future*, 2nd edition,
Birkhäuser Advances in Infectious Diseases,
DOI 10.1007/978-3-0346-0279-2_9, © Springer Basel AG 2011

measure the amount of antibody produced. This test appears to be as sensitive as the MN assay. HI and SRH assays are not functional tests for measuring immunity to influenza and suffer from several technical drawbacks. Improvements in these assays will be a further step in the preparation of new influenza vaccines, particularly for cell-derived products. Additional immunological assessments, such as cell-mediated immunity and the role of neuraminidase, need to be explored to give better insight into the overall effects of vaccination.

1 Introduction

The Centers for Disease Control and Prevention estimate that between 114,000 and 146,000 individuals are hospitalized each year because of influenza. Although the exact tabulations of illnesses and complications attributable to influenza virus infection are not available, the preceding estimates indicate that morbidity and mortality caused by influenza are major health problems.

The influenza virus belongs to the family of *Orthomyxoviridae* and is classified into three different types A, B and C on the basis of different epitopes, which are the antigenic differences in their respective nucleocapsids. Influenza A and B are the two types of influenza viruses that cause epidemic human disease. Influenza A viruses are further categorized into subtypes, e.g., H1N1, H2N2, and H3N2 on the basis of two surface antigens: hemagglutinin (HA) and neuraminidase (NA). Influenza B viruses are not categorized into subtypes. Since 1977, influenza A (H3N2 and H1N1 subtypes) and B viruses have been in global co-circulation. These types are further separated into groups on the basis of their antigenic characteristics. New influenza virus variants result from frequent antigenic change (i.e., antigenic drift) arising from point mutations that occur during viral replication. Influenza B viruses undergo antigenic drift less rapidly than influenza A viruses.

Influenza is characterized by the occurrence of frequent unpredictable epidemics, and much less frequent worldwide pandemics. Epidemics arise because different strains of influenza are constantly generated through antigenic drift, and individuals become less or not at all protected in some years. A pandemic is responsible for higher morbidity and mortality than an epidemic because it affects a larger proportion of the population. The burden of epidemics, however, is cumulatively greater than that of pandemics. A worldwide pandemic is caused by the spread of a new influenza subtype arising from an antigenic shift [1]. When such a subtype enters the population, it is likely that antibodies against previously circulating strains do not provide adequate protection; this lack of protective immunity means that the new virus can easily infect exposed individuals.

If such a virus demonstrates the additional ability to transmit efficiently from person to person, the result is a global outbreak of disease that affects a high percentage of individuals in a short period of time and is likely to cause substantially increased morbidity and mortality in all countries of the world. Over 50 million people are estimated to have succumbed to the most devastating influenza pandemic in 1918, the so-called "Spanish flu." The "Asian flu" of 1957 has been responsible for about 70,000 deaths in the USA only.

Past findings have identified the H2, H5, H6, H7 and H9 subtypes of the influenza A virus as the subtypes that are most likely to be transmitted to humans, thus presenting a potential pandemic threat [2]. However, the current ongoing pandemic, which was declared by the World Health Organization (WHO) in May 2009, is caused by the reassortment of the swine classical H1N1 with the PB1, HA, and NA segments from a human H3N2 strain, and a triple reassortment of swine classical H1N1, with the PB1, HA, and NA segments of a human H3N2 strain and the PB2 and PA segments of the avian lineage [3, 4, 5]. However, the subtypes mentioned before still pose a significant pandemic threat.

Influenza viruses cause disease across all age groups. Rates of infection are highest among young children, but rates of serious illness and death are highest among persons aged ≥ 65 years and persons of any age who have medical conditions that place them at increased risk for complications from influenza [6]. Studies on the morbidity and mortality associated with influenza suggest that hospitalization rates for adults, with medical conditions that place them at high risk for influenza, often increase fivefold during epidemics, leading to an average 172,000 excess hospitalizations during each epidemic in the USA [7]. This has important economic consequences, with the annual productivity loss estimated to be more than US$760 million and hospitalization costs to be in excess of US$300 million, for each epidemic in the USA alone [8]. The total economic impact is considerable, and in industrialized countries, total estimated costs (direct and indirect) may reach approximately US$10–60 million per million population.

In an avian influenza virus, the HA, characteristically, has glutamine at position 226 and glycine at position 228 (human viruses have leucine at 226 and serine at 228), which form a narrow receptor binding pocket that preferentially binds to host cell receptors containing sialyloligosaccharides (SA) terminated by an N-acetyl sialic acid linked to a galactose via an $\alpha 2,3$ linkage (the major form in the avian trachea and intestine).

While a correlate of protection has not been fully defined, challenge studies in human volunteers indicate that protection against influenza can be conferred by serum antibodies. The immune responses to injectable influenza vaccines are routinely assessed using serological HA antibody measurements. It is generally accepted that neutralizing HA antibodies, as well as antibodies to neuraminidase, can be detected in serum approximately 1–2 weeks after primary infection and peak at 3–4 weeks post infection [9].

2 Influenza Vaccines and Criteria for Licensure

Influenza surveillance information, regarding the presence of influenza viruses in the community as well as diagnostic testing, can aid clinical judgment and guide treatment decisions. Several commercial, rapid, diagnostic tests are available that can be used by laboratories in outpatient settings to detect influenza viruses in a few minutes. These rapid tests differ in the types of influenza viruses they can detect.

Antiviral drugs are an adjunct to vaccination for the control and prevention of influenza; however, these agents are not a substitute for vaccination. Four currently licensed antiviral agents against influenza are available in the USA: amantadine, rimantadine, zanamivir, and oseltamivir. With seasonal influenza, the decision to prescribe an antiviral drug for the prevention or treatment of influenza must be based on the certainty, or the high probability, that a person has been or will be exposed to the virus, or on a diagnosis of influenza, and is usually only recommended for patients with underlying medical conditions that increase the risk of complications from an influenza infection. However, in the initial phase of the current pandemic, antiviral drugs have been widely used in a healthy population in the attempt to prevent or delay the onset of a full pandemic.

Amantadine and rimantadine are chemically related antiviral drugs that are active against influenza A viruses only. Amantadine was approved in 1966 for prophylaxis of influenza A/H2N2 infection and was, later, also approved for the treatment and prophylaxis of influenza type A virus infections among adults and children aged ≥1 year. Rimantadine was approved in 1993 for the treatment and prophylaxis of infection among adults, and for prophylaxis among children. Neither antiviral drug has been used widely due to their narrow spectrum of activity, the rapid onset of resistance, and the related adverse effects [10]. Zanamivir and oseltamivir are NA inhibitors that are active against both influenza A and B viruses. The site of enzyme activity of the influenza NA is highly conserved between different types, subtypes and strains of influenza, and has, therefore, emerged as the target of this new class of antiviral agents that are effective in prevention and treatment. In the US, both drugs were approved in 1999 for the treatment of uncomplicated influenza infections. Zanamivir was approved for the treatment of patients aged ≥7 years, and oseltamivir was approved for treatment of patients aged ≥1 year and for prophylaxis in persons of age ≥13 years. These antiviral drugs are only effective if started soon after the onset of disease.

Influenza vaccination is the primary method for preventing influenza and its severe complications. Vaccination is associated with a reduction in influenza-related respiratory illness and physician visits at all ages, in hospitalizations and deaths among high-risk persons, otitis media among children, and work absenteeism among adults. Vaccination with inactivated influenza virus currently represents the most important measure for reducing the impact of influenza.

The two types of vaccines which are currently licensed are inactivated vaccine and attenuated vaccine. Inactivated vaccines, which are generally administered parenterally are produced by the propagation of the virus in embryonated hen's eggs. The vaccine is available containing whole, split (chemically disrupted) and subunit (purified surface glycoproteins) virus. To enhance the immunogenicity of purified subunit antigens, several new adjuvants have been promoted [11, 12, 13]. The deve lopment of cell culture-based vaccines is an attractive alternative approach to the use of hen's eggs and is a potentially faster mechanism, as strains do not need to be egg-adapted prior to production [14, 15].

Live attenuated vaccines that can be administered by nasal spray have been licensed in the USA in 2003 and might soon be widely available in the rest of the world. This vaccine type has been shown to be as efficacious as the inactivated vaccine [16, 17]. Live influenza vaccines elicit systemic and local mucosal immune responses that include stimulating secretory immunoglobulin IgA in the respiratory tract, which is a portal for the virus.

Rapid and early diagnosis of influenza virus infection is an important activity that aids in the surveillance of circulating strains and enables the early vaccination or prophylactic treatment of high-risk groups. Laboratory diagnosis of influenza is generally made by detection of the virus or its genome in respiratory secretions by virus culture or molecular methods (e.g. RT-PCR).

The influenza virus surface glycoprotein, HA, is a major antigenic determinant in the production of virus-neutralizing antibodies generated during an infection or immunization. Serological assays commonly used to quantify antibodies specific for influenza viruses include hemagglutination inhibition (HI), single radial hemolysis (SRH), virus microneutralization (MN), ELISA and Western blot; the most widely used assays are HI and MN.

Serological measures are used to characterize the amount of antibody before and after vaccination, and to compare the seroresponse in subjects with different treatment regimens or other characteristics (dose, age, etc.). Measures that are most frequently used are geometric mean titer (GMT), seroconversion, significant titer increase, and seroprotection rate. Considerable discrepancies were found in the use of serological measures in several studies [18].

Three criteria need to be fulfilled (and at least one of the assessments should meet the indicated requirements) for the yearly vaccine registration in the European Union (CPMP/BWP/214/96) [19]. A tabular presentation of these criteria is provided in Table 1. As there are currently no criteria for the licensure of pandemic vaccines, the serological results were analyzed using the CPMP criteria required for the annual registration of seasonal vaccines. According to guideline CPMP/VEG/4717/03 in the dossier on the structure and content for pandemic influenza vaccine marketing authorization application [20], it is anticipated that mock-up pandemic vaccines should at least be able to elicit sufficient immunological responses to meet all three of the current standards set for existing vaccines in adults or older adults, as defined in CPMP/BWP/214/96. All sera should be assayed for anti-HA

Table 1 Serological criteria to meet CPMP/BWP/214/96 requirements by age group [19]

Test	HI	SRH	CPMP/BWP/214/96 criterion	
			Age group	
			18–60 years	>61 years
Geometric mean ratio (pre- to post-vaccination)			>2.5	>2
Seroprotection	Titer ≥ 40	≥ 25 mm^2	>70% of subjects	>60% of subjects
Seroconversion or significant increase	Negative at pre-vacc and post-vacc titer ≥ 40	negative at pre-vacc and post-vacc titer ≥ 25 mm^2	>40% of subjects	>30% of subjects

antibody against the prototype strains by HI or SRH tests. In the interpretation of HI and SRH immunogenicity results, criteria established from the Committee for Medical Products for Human use (CHMP) are to be taken into consideration.

Vaccines against strains with pandemic potential need to follow the Guidance in the dossier on structure and content for influenza vaccines derived from strains with a pandemic potential for use outside of the core dossier of marketing authorization (CHMP/VWP/263499/2006). This guideline addresses the content of marketing authorization applications for inactivated avian influenza vaccines produced in eggs or in cell cultures. The recommendations include the same three criteria as the seasonal vaccines (i.e., seroconversion rate, seroprotection, and significant increase in GMT) as defined in CHMP/BWP/214/96, with seroconversion rates being the most important. This guideline also is valid for vaccines containing, or derived from, influenza strains with a high pandemic potential from other animals (e.g., pig) or those of non-H1/H3 human origin.

As discussed for the European Region, licensure of "new" seasonal and pandemic influenza vaccines in the USA needs to be supported by the submission of clinical data. Immunogenicity bridging studies should be conducted to compare the immune response observed in the clinical endpoint efficacy study with that elicited in other populations. Suitable endpoints may be the hemagglutination inhibition antibody responses to each viral strain included in the vaccine. Studies should be powered to assess the primary endpoints, GMT, and rates of seroconversion, which is defined as the percentage of subjects with either a pre-vaccination HI titer < 1:10 and post-vaccination HI titer ≥ 1:40 or a pre-vaccination HI titer ≥ 1:10 and a minimum fourfold increase in post-vaccination HI antibody titer.

Identification of an immune correlate of protection during the course of a clinical endpoint efficacy study may facilitate the design and interpretation of such bridging studies.

The same criteria should be adopted for licensure of pandemic influenza vaccines. The hemagglutination inhibition (HI) antibody assay has been used to assess seasonal vaccine activity and may be appropriate for the evaluation of the pandemic influenza vaccine. Appropriate endpoints may include: (1) the percentage of subjects achieving an HI antibody titer ≥ 1:40, and (2) rates of seroconversion, defined as the percentage of subjects with either a pre-vaccination HI titer < 1:10 and a post-vaccination HI titer ≥ 1:40 or a pre-vaccination HI titer ≥ 1:10 and a minimum fourfold increase in post-vaccination HI antibody titer. In a pre-pandemic setting it is likely that most subjects will not have been exposed to the pandemic influenza viral antigens. Therefore, it is possible that vaccinated subjects may reach both suggested endpoints. Thus, for studies that enroll subjects who are immunologically naïve to the pandemic antigen, one HI antibody assay endpoint, such as the percentage of subjects achieving an HI antibody titer ≥ 1:40, may be considered.

Other endpoints and the corresponding immunologic assays, such as the microneutralization assay, might also be used to support the approval of a pandemic influenza vaccine.

For both seasonal and pandemic vaccines, the same EMEA criteria, reported in Table 1, may support an accelerated approval in the USA.

3   *In Vitro* Assays to Assess Protective Antibody Levels

3.1   *Hemagglutination Inhibition*

HI is easy to use, can be rapidly performed with little equipment and the reagents are cheap and widely available: reasons which have contributed to its status as the most employed assay for serum influenza antibody titration. It is one of the "classic" methods in influenza serology and was standardized in 1942 [21]. Although HI is not a functional test for measuring immunity to influenza, it is the most commonly used reference method for the assessment of anti-HA antibody levels and the results are usually concordant with the MN test when used with seasonal influenza A strains. The method can be performed using an inactivated virus and is based on the ability of influenza virus to agglutinate red blood cells (RBCs) and the inhibition of this agglutination by anti-HA antibodies. The biological relevance of the assay is rooted in the fact that agglutination is mediated by the receptor binding site of the HA, which is also a common target for neutralizing antibodies, and therefore, quantification of these antibodies by HI is both a correlate of protection and a widely accepted surrogate marker for prediction of vaccine efficacy and the immunity to infection from specific strains.

The HI antibody titer is expressed as the reciprocal of the highest serum dilution showing complete inhibition of hemagglutination using four or eight viral hemagglutination units (Fig. 1). The HI reaction is as immunity to influenza strain specific. The rationale for the use of this assay to predict immunity is derived from observations in human challenge experiments. There is a positive linear correlation between pre-challenge antibody titers and percentage protection and also an inverse relationship between antibody titers and disease severity, and virus shedding [22, 23, 24]. Most results indicate that after vaccination (inactivated vaccine) HI titers between 30 and 40 confer protection from infection in 50% of the subjects [25], while higher antibody titers, in the range of 120 to 160, protect around 90% of the subjects [24]. While HI titers > 40 are considered as the threshold beyond which serious illness is unlikely to occur, a fourfold increase in HI titer between samples taken before and after vaccination is the minimum increase considered necessary for classification as seroconversion. These findings are reflected in current CPMP recommendations for the immunogenicity of influenza vaccines [19] (Table 1).

However, despite the establishment of this assay as the standard technique for the measurement of antibodies to an influenza virus, the HI test sometimes suffers from technical drawbacks and it has been shown in several international comparative studies that both HI and MN can show considerable variability between laboratories. HI assays are influenced by the binding avidity of the virus and by the species of the RBCs used. Another important source of variation in the HI test seems to be the difference in the sensitivity of RBCs from individual animals of the same species.

Furthermore, an antibody to NA can sterically block the access to the HA of the virus by the RBC receptors and thereby inhibit hemagglutination. One of the minor disadvantages of the assay is the fact that sera have to be treated with

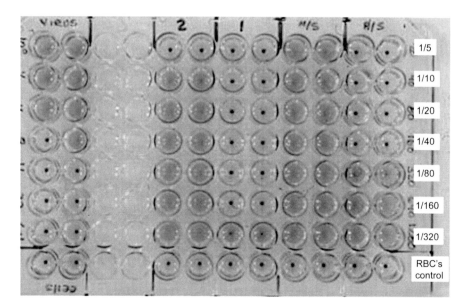

Fig. 1 Basics of horse RBC HI assay. V-bottomed 96-well plates; RDE-treated sera; 1:5 starting dilution; serial twofold dilutions in final volume of 25 µl; virus; A/Vietnam/1203/2004 (H5N1) BPL-inactivated; 4 HAU in 25 µl volume; horse RBCs; collected in citrate dextrose acid (ACD) solution; washed and standardized to 1% v/v in PBS/0.5% BSA; added to assay in 50 µl volume; 60 min incubation time (at room temperature) to allow horse RBC to settle

receptor-destroying enzyme (RDE), prior to testing, in order to remove nonspecific inhibitors as these may bind virus, and could therefore interfere with the assay. For H1 and H3 strains, this test has the advantage of good sensitivity and shows good reproducibility, but as mentioned earlier, the variability in results between laboratories can be significant. On the other hand, the HI shows relatively low sensitivity for antibody to influenza B virus [26]. It is possible to increase assay sensitivity by ether treatment of the virus, but this can also be a source of variability, and can potentially reduce the strain specificity of the test [27, 28 29].

It has been shown in various studies that the sensitivity of the HI was apparently too low to detect responses to avian viruses in mammalian sera, so that alternative serological tests were needed [30] or modifications had to be introduced to the standard protocol – one of which is the use of subunit HA rather than intact virus and this has been shown to improve assay sensitivity [31].

It is now understood that the reason for the observed low sensitivity and the resulting underestimation of antibody levels to avian and other viruses with pandemic potential arises from the altered binding specificity with respect to their cellular sialic acid receptor between human and non-human influenza viruses.

The receptor specificity of influenza viruses correlates with their ability to agglutinate RBCs from different species. Human viruses preferentially bind to oligosaccharides containing *N*-acetylneuraminic acid α2,6-galactose (NeuAc α2,6Gal), while avian and equine influenza strains bind to NeuAc α2,3Gal. Many animal species, including the horse and cow, have high amount of NeuAc α2,3Gal receptors but virtually no NeuAc α2,6Gal receptors in erythrocytes. Chicken RBCs have less NeuAc α2,6Gal and more NeuAc α2,3Gal, turkey RBCs have more NeuAc α2,6Gal than chicken RBCs [21]. Therefore, seasonal H1 and H3 influenza viruses preferentially agglutinate chicken, or turkey, but not horse or cow RBCs, whereas avian viruses preferentially agglutinate RBCs from the horse or cow.

In accordance with these hypotheses, the sensitivity of the HI assay is largely determined by the type of erythrocytes used, and the measurement of HI titers against avian viruses has been significantly improved by use of horse erythrocytes. Turkey erythrocytes could be responsible for the relative insensitivity of HI in the detection of H5 antibody [32, 33]. The main problem with horse HI is a horse-to-horse variation, and specificity is reduced with increasing age of the erythrocytes, therefore it is recommended to use blood within 1 week after collection.

3.2   Single Radial Hemolysis

SRH was developed in 1975 [34]. It is routinely used for the detection of influenza-specific and rubella IgG antibodies. The test utilizes antibody diffusion in agar gel to measure the antibody content of the test sera. Complement-mediated hemolysis, induced by influenza antigen-antibody complexes, produces easily discernible zones, the sizes of which are proportionate to concentrations of specific antibody in the sera (Fig. 2) [35, 36]. Antibody responses to natural infections and vaccinations are readily detected by this method. Advantages of this assay are that sera do not need to be pre-treated to inactivate nonspecific inhibitors, sera can be analyzed without dilution, only a pre-incubation of samples at 56°C for 30 min is needed to inactivate the complement, the test is easily standardized, and it may be more sensitive than HI, particularly for the pandemic H5 strains. This test appears to be as sensitive as the MN assay [37].

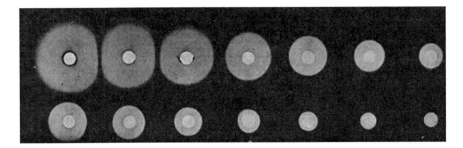

Fig. 2 Single radial hemolysis reactions of human serum in agarose gel immunoplates containing guinea-pig complement (final concentration 1:30) and chicken erythrocytes (1%) treated with A/Port Chalmers/1/73 (H3N2) virus. The clear areas represent zones of lysed erythrocytes produced by antibody to hemagglutinin. The wells in the *top row* contain serial, twofold dilutions (1:1 to 1:64) of a potent human serum having an HI titer of 1:2,560 with A/Port Chalmers/1/73 (H3N2) virus. The *bottom row* contains similar dilutions of a serum with an HI titer of 1:256 [34]

SRH is usually performed in PVC immunoplates, which are prepared using sheep RBCs for H1 or H3 antibody detection. The amount of live or inactivated whole influenza virus used to sensitize the RBCs is 2,000 UE/ml in a 10% RBC suspension, and 5 μl of heated-inactivated serum is added to wells in SRH plates. After incubation for 18 h at 4 °C and 3 h at 37°C, the halos of hemolysis are measured and areas are calculated. Areas of hemolysis equal or higher than 25 mm^2 are considered seroprotective. In the case of H5 strains, better results were obtained using turkey erythrocytes.

The SRH test works well with inactivated viruses, so serology of H5N1 can be safely analyzed at a biosafety level 2 containment facility. Although this test can detect H5N1 antibody, it cross-reacts with nonspecific antibodies that are present in both human and rabbit sera. Therefore, a preliminary screen for cross-reactivity and confirmatory tests with an alternative technique are recommended.

SRH is suitable for screening a large number of samples. This feature has made the test useful for rapid screening of antibodies against newly detected influenza variants, making it valuable for large-scale sero-epidemiological studies. The test allows smaller differences in antibody level to be detected than is possible by conventional HI tests. It has a good correlation with the MN assay for pandemic H5 strains and is EMEA approved.

When a comparison between the HI and SRH tests for seasonal strains was made (Fig. 3), a close correlation between the antibody potencies measured by both test systems was observed. Serum samples with high HI titers (1:1,256–1:5,120) gave zone diameters of 9–11 mm (hemolysis area 64–95 mm^2). Of the 15 samples shown in Fig. 3 and HI tests (titer <1:10), ten were also negative by SRH [34]. Completely different results were achieved when comparing HI with SRH and MN for H5 strains (Fig. 4).

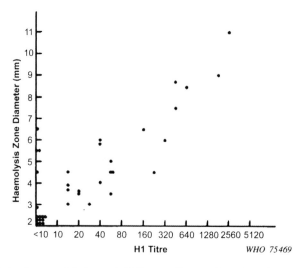

Fig. 3 Correlation between HI titer and SRH zone diameter for A/Port Chalmers/1/73 (H3N2) virus in 37 human serum samples [34]

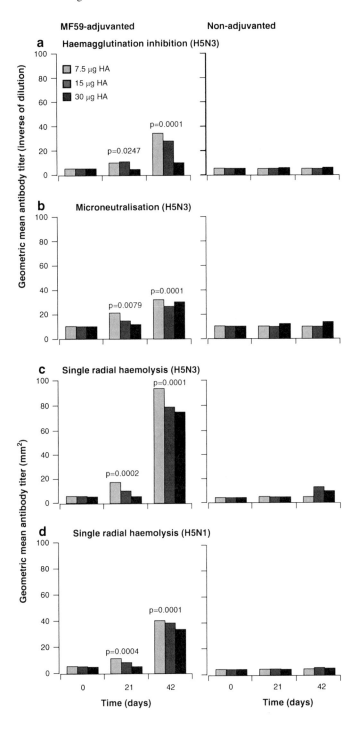

3.3   Microneutralization

The MN assay (or virus neutralization assay, VN) requires infectious virus as the antigen. The advantages of the MN assay are that it is more sensitive than HI, more specific, and is suitable for automation [39]. This assay detects functional anti-HA antibodies, which are highly specific for the subtype in question. Moreover, this assay can be developed quickly upon recognition of a novel virus and can be used even before suitable recombinant or purified viral proteins become available for use in other assays. In addition, the MN protocol can be optimized to also measure responses against other envelope glycoproteins, i.e., NA; this is in contrast to the HI assay that only measures responses against the HA component. The MN test seems to be advantageous when antibody levels are low, with negligible or negative titers in the HI test under the same conditions [40, 41].

MN tests have not been used widely in serological studies of seasonal influenza because they are lengthy (overnight test at minimum) compared to HI and SRH, and therefore were not as practical for screening large numbers of samples [31]. However, their use has rapidly increased over the past years when they were used for the analysis of strains with pandemic potential, even though the need for live virus results in the necessity of a higher containment facility when working with these strains. At the front line of an outbreak, especially in resource-limited regions, biosafety level 3, or higher, laboratory facilities are not always available. This safety issue has been overcome by the routine availability of reverse genetics (RG) viruses as challenge virus [42, 43] and has helped to allow wide implementation of this assay format. When assays are required to test for antibodies against highly pathogenic (HP) viruses – e.g., H5N1 – RG viruses are often used to circumvent the need for high containment and thus improve the safety of the assay. RG viruses are created by reassorting the genes for the HA and NA of an HP virus with the remaining genes of a low pathogenic (LP) virus (e.g., A/Puerto Rico/8/34), usually in a 12 plasmid rescue system [44]. This results in a LP virus with the backbone of the LP donor and the surface proteins of the HP virus and can serve as the input virus in neutralizations and other serological assays.

For the MN assay, a recognized correlate of protection does not yet exist; however, a fourfold increase in titer after vaccination has been used in the literature to assess immune responses to H5 viral antigens by MN [45, 46, 47]. Moreover, titers >80 are used as surrogate for the description of exposure in a population as these seem to have been indicative of infection with H5N1 during the outbreak of H5 in 1997 [31, 47]. While there is discrepancy between these two cut-off values, various considerations have to be kept in mind: these titers are used to describe two distinct circumstances (vaccination versus infection), proper correlates of protection have

Fig. 4 Geometric mean titers of antibody for MF59-adjuvanted and conventional surface-antigen H5N3 vaccine before and after two and three doses of vaccine. (**a**) Hemagglutination inhibition test using H5N3 antigen. (**b**) Microneutralization using H5N3 antigen. (**c**) Single radial hemolysis using H5N3 antigen. (**d**) Single radial hemolysis using H5N1 antigen [38]

yet to be defined and lastly, it is also known that MN titers determined by different laboratories may vary, so that no absolute titers can be specified at the moment.

The use of a standard – created from pooled plasma of subjects immunized with clade 1 A/Vietnam/1194/2004 vaccine (whole virus formulation) – with a known neutralization titer has been proposed and established by WHO as an International Standard for antibodies to A/Vietnam/1194/2004 for the immunogenicity assessment of H5N1 vaccines.

Based on the sensitivity and specificity of the analysis described, the MN assay is now routinely used as part of sero-epidemiological investigations in outbreaks of avian influenza to detect antibody against H5 and H7 viruses [48, 49]. MN assays are usually performed in 96-well plates, where sera (heat-inactivated for 30 min at 56 °C) are tested by mixing with an equal volume of influenza virus at $100\times$ TCID50. After incubation of serum and virus to allow neutralization, susceptible cells are added or the mixture is transferred onto a cell monolayer to allow infection. After incubation for at least 14 h, the cells are fixed and the presence of viral protein is detected by ELISA using a monoclonal antibody to the influenza nucleoprotein. Alternatively, remaining infectivity can be detected in the culture medium using simple hemagglutination of animal RBCs [50] or by inspection of the cytopathic effect of the infection on the cell monolayer.

Figure 5 shows the relationship between HI and MN antibody titers against seasonal strains. The MN antibody titers are slightly higher than the HI titers against the homologous strains. This figure also demonstrates that correlation is dependent on the strain analyzed. A decrease in correlation between MN and HI was also observed during the detection of antibodies that may be present in sera from individuals receiving the H5N1 Vietnam vaccine (Fig. 6).

3.4 Pseudotyped Assays

Pseudotyped viral particles have been extensively used to express glycoproteins from a wide variety of high containment viral pathogens. With the pseudotype system, only the envelope of interest, such as HA from influenza virus, is required, with no possibility of recombination or virus escape. The particles undergo abortive replication and do not give rise to replication-competent progeny. Pseudotypes are generated by co-transfection of different plasmids: the envelope gene construct, which contains the the retroviral gag-pol construct encoding the structural proteins (expressed from *gag*) and the enzymatic proteins (expressed from *pol*) responsible for processing the structural proteins and ensuring the integration of a transfer gene, and additionally the *rev* gene is included for efficient processing; the transfer gene construct: this is only carrying the packaging signal to ensure the efficient incorporation of the marker gene into the particles and to regulate its expression once the gene is integrated. The most common marker genes used encode for the GFP, β-gal or luciferase. For influenza, the neuraminidase is also required for the release of HA-pseudotyped particles from the surface of the producer cells.

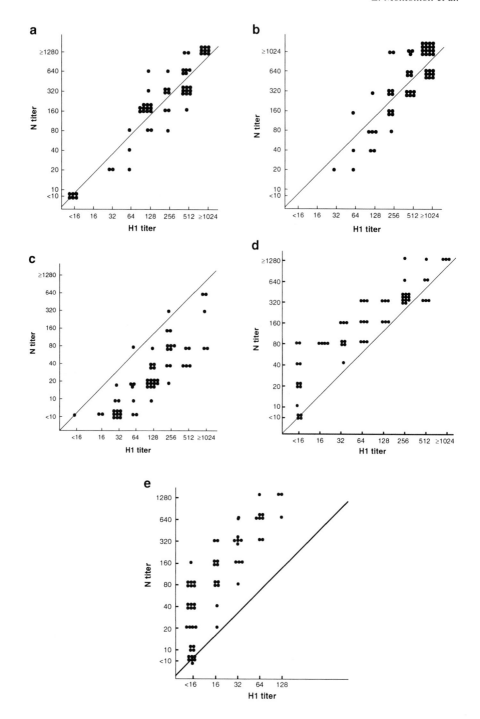

It can be either supplied from an exogenous font, or encoded by a plasmid. These pseudotypes, collected from the culture supernatant of the producer cells, encode the reporter gene and bear the envelope glycoprotein on the surface. The transfer of the marker gene to target cells depends on the function of the envelope protein, therefore, the titer of neutralizing antibodies against the envelope can be measured by a reduction in marker gene transfer [53] (Fig. 7).

Pseudoparticles expressing H5 from H5N1 influenza virus have been amply produced and used to set up a pseudotype neutralization assay (PPN) [54, 55, 56, 57]. As pseudoparticles are unable to replicate, the PPN assay has the great advantage of being able to be carried out at BSL2, and therefore is compatible with the containment level available in most laboratories. Moreover, by using luciferase as the reporter gene, it is promptly adaptable to high-throughput formats to evaluate immunogenicity of several pandemic vaccine formulations; and in principle, it can be easily standardized among different laboratories. Finally, the PPN assay allows a rapid and easy assessment of cross-neutralizing response by using pseudotypes bearing hemagglutinins (HA) from different clades without the need to have access to multiple clinical isolates which are often difficult to obtain and handle.

This assay has been recently used to determine the neutralizing antibody responses to influenza H5N1 in vaccinated individuals. Neutralizing titers measured by the PPN assay are significantly correlated to those obtained by the other well established serological methods, such as MN, HI or SRH described above (Fig. 8). The PPN assay is being proposed as a valid alternative to the classical serological methods, in particular, for the evaluation of cross-reactive antibody response to highly pathogenic wild-type influenza viruses, with the ultimate goal of introducing it into routine use [57]. Of course, this will require an international effort to standardize and validate the assay.

Further development of β-Gal-based pseudotype assays will allow wider application as an ELISA-type assay that can be performed in laboratories without specialized equipment.

In addition, these pseudotypes are readily exploitable for the development of novel entry and exit assays which can be used to screen new antiviral compounds. For example, the requirement of NA for the release of HA-pesudotyped particles was exploited during the development of a high-throughput assay to evaluate neuraminidase inhibitors [58].

Fig. 5 Relationship between HI and neutralizing antibody (MN) titers against A/Yamagata/120/86 (H1N1) (**a**), A/Fukuoka/C29/85 (H3N2) (**b**), A/Shisen/2/87 (H3N2) (**c**), B/Nagasaki/1/87 (**d**) and B/Osaka/152/88 (**e**). The serum antibodies were measured by MN and HI tests. When they were titrated against vaccine strains (A; B; D) differences between the MN and HI titers were small, when they were titrated against heterologous strains the differences were large. The possibility that non-infectious virus particles consume neutralizing antibodies and thus result in lower actual titers, as suspected in the neutralization tests with heterologous strains, seems unlikely because the ratio between infectivity and the hemagglutination titer to A/Shisen/2/87 (H3N2) was not lower than that of other strains. Therefore, these observations seem to indicate that the neutralization test is more specific than the HI test [51]

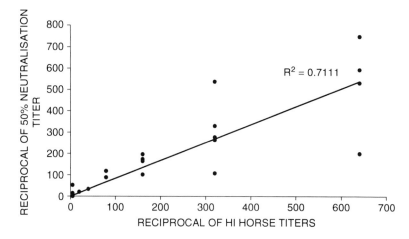

Fig. 6 Correlation between microneutralization (MN) and horse RBCs HI titers using samples from subjects vaccinated with A/Duck/Singapore/97 (H5N3) vaccine (MF59 adjuvated; $n = 48$), (R = correlation coefficient) [52]

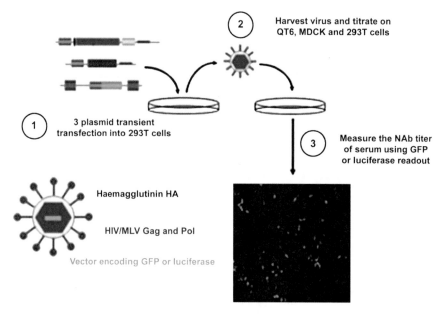

Fig. 7 H5N1 HA retroviral pseudotypes. MLV(HA) and HIV(HA) pseudotype construction and neutralization assay for influenza A H5N1 [54]

3.5 Other Assays

Antibodies against viral neuraminidase can inhibit its enzyme activity [59, 60] and reduce the number of viral particles that are released after infection and replication.

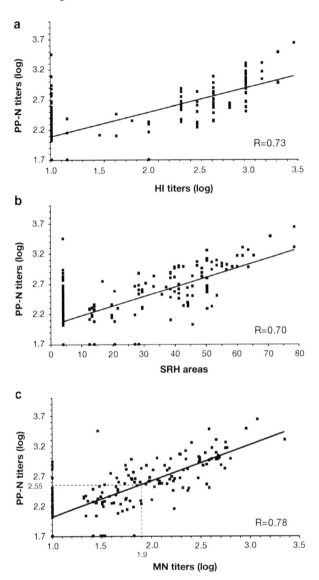

Fig. 8 Comparison of PPN with HI, SRH and MN titers. Scatterplots showing the correlation of antibody logarithmic titers measured by PPN versus HI (**a**), SRH (**b**) and MN (**c**) assays performed against the vaccine strain A/Vietnam/1194/2004 (H5N1). Total number of sera assayed is 226. Graphs show the linear regression fitted to the data by using *Excel*. Pearson's correlation analysis was used to assess the correlation coefficient (R) between PPN $\log_{10}$ titers and MN, HI $\log_{10}$ titers or SRH areas. In panel **c**, *vertical dashed line* indicate the value of MN $\log_{10}$ titer = 1.9 (corresponding to a titer of 1:80), the proposed threshold of protective antibodies, *horizontal dashed line* indicate the corresponding value of PPN $\log_{10}$ titer = 2.55 (corresponding to a titer of 1:357) [57]

These inhibiting antibodies do not neutralize infectivity: they are important determinants for the course of disease [61], and it has been shown in mice that they have the potential to protect against lethal doses of highly pathogenic avian influenza viruses. Antibodies against NA were found to decrease viral replication in the lungs and reduce disease severity upon subsequent challenge [62]. Furthermore, it has been observed that antibodies against a subtype matched NA might be partially cross-protective in the event of a pandemic with a new HA subtype [63, 64]. This partial protection of cross-reactive NA antibodies has since been demonstrated in mice using the H5N1 virus [65]. Additionally, the seasonal influenza vaccine may boost cross-reactive immunity to H5N1 which might be mediated by N1 and involve either cellular or humoral responses [66]. Consequently, hopes for cross-protection by anti-N1 antibodies in a pandemic caused by H5N1 have raised an increasing interest in workable NI assay protocols over the past few years.

Assays for the quantitation of NA inhibition antibodies have been recently adapted from the standard [67] to the 96-well format (Sandbulte and Eichelberger, personal communication) in which neuraminidase activity is determined by colorimetric analysis of sialic acid release from fetuin, which is the substrate.

To date no correlate of the protection afforded by NA inhibiting antibodies has been determined, although NI titers as low as 4 have been associated with resistance to viral replication and illness in human challenge studies [68]. The NI assay is not widely used yet and it requires the generation of RG viruses with matched NA and mismatched HA subtype to avoid interference of anti-HA antibodies in sera [69].

ELISA-type assays are generally a good tool to determine antibody titers to various pathogens in serum and bodily fluids, and are often commercially available. This assay type is suitable for screening a large number of samples and therefore offers the possibility of automation, but the assay has no correlate of protection. Moreover, their diagnostic value in influenza serology is limited, as shown in various experiments to determine titers against influenza of the H1 [70, 71] and H5 [31, 72] subtypes. The reason for this is the prevalence of anti-HA antibodies in the human population. Influenza infection is usually first experienced in childhood and re-infection occurs over a lifetime. As a result, the majority of adults show some degree of antibody reactivity with HA of currently circulating strains. This leads to nonspecific reactivity of adult human sera in ELISA which measures all antibodies, not just the functional antibodies, and therefore, most likely results from cross-reactive epitopes common to HAs of different influenza subtypes [53]. Use of HA1 instead of full length HA can improve specificity, but not completely prevent cross-reactivity, so results determined using this assay format might be misleading in the prediction of immunity.

However, when combined with Western blotting (WB), ELISA shows improved specificity and retains improved sensitivity when compared with the MN assay and WB combination [31, 52, 73].

WB is useful only for confirmation. This technique is too laboratory intensive to be considered a diagnostic test for screening several thousand sera, and is generally

Table 2 Sensitivities and specificities of serologic assays for detection of antibodies to H5N1 virus

Age group[a]	Parameter[b]	Values (%)				
		Individual serologic tests[c]			Combination of tests[d]	
		N	E	W	N-W	E-W
Child	Sensitivity ($n = 8$)	88	100	100	88	100
	Specificity ($n = 24$)	100	92	83	100	100
Adult	Sensitivity ($n = 85$)	80	80	80	80	80
	Specificity ($n = 85$)	93	62[e,f]	85	96	84[e]

[a]Serum samples from individuals from 1 to 14 years of age (child) or from individuals 18 to 59 years of age (adult)

[b]Sensitivity, number of H5N1 virus-infected patients testing positive for antibody divided by the total number of patients with confirmed H5N1 infections tested. Specificity, number of control age-matched sera tested minus the number of control sera testing positive for antibody divided by total number of control sera

[c]Tests for determination of H5N1 virus positivity (N microneutralization test, W Western blotting, E ELISA)

[d]Combination of tests. Microneutralization test with A/Hong Kong/156/97 virus followed by western blot confirmation with rHA of A/Hong Kong/156/97 virus (N-W) and ELISA with rHA of A/Hong Kong/156/97 virus followed by western blot confirmation with rHA of A/Hong Kong/156/97 virus (E-W)

[e]Number of samples, 50

[f]Statistical analysis for positive association between test result and known status of samples was not significant (Fisher exact test; $P = 0.067$). All other assays were significant ($P = 0.003$) [31]

used as a secondary serological test to confirm other EMEA-accepted serological tests, i.e., MN assay or ELISA.

The sensitivities of WB and ELISA are generally higher than MN, and specificities of MN are higher than WB and ELISA, particularly for pandemic strains (Table 2). To determine whether the MN assay and/or ELISA could be used to detect H5-specific antibody in single serum samples, the relative sensitivities and specificities of the assays were compared (Table 2). The test using the MN assay was notably superior to that of ELISA. When combined with WB, each test improved in specificity; however, the maximum sensitivity and specificity were still achieved by a combination of the MN assay and theWB [31]. The ELISA and WB tests detected antibodies of lower avidity and/or quantity than that required for detection by the MN assay.

There are two realistic options for the rapid development of neutralization assays to make them more widely applicable for pandemic strains: the use of reverse genetics to engineer a safer, attenuated virus by deletion of the polybasic cleavage site in HA, as is done for the development of inactivated vaccines for pandemic influenza [74, 75], or the construction of viral pseudotypes bearing the influenza HA glycoproteins as surrogate viruses for use in neutralization assays. The first option has its inherent problems, i.e., the issue of possible reversion to the wild-type virus via genetic reassortment [76].

4   Conclusion

The traditional HI method, performed using turkey RBCs, offers a simple and speedy assay format for the detection of human antibodies to currently circulating seasonal influenza strains, particularly H1 and H3; this assay may be useful for large serosurveys as an initial screening tool. Detection of antibodies to avian influenza viruses in mammalian species, including humans, using this traditional HI method, has generally failed, even in cases where infection was confirmed by virus isolation; however, change of RBC species from turkey to horse has increased the assay sensitivity to the level of the MN assay. The HI assay is considered the "golden standard" in influenza serology and is less difficult and less time consuming to perform than the MN or SRH assays.

SRH assay requires only a small amount of whole inactivated virus, thus there is no need to adapt the virus to rapid growth. This feature has made this test useful for rapid screening of antibodies against newly detected influenza variants. It can be valuable in large-scale sero-epidemiological studies of new influenza virus variants. SRH has a big disadvantage as it can detect antibodies to internal virus antigens in addition to those antibodies directed against surface glycoproteins, and may lack specificity for the detection of antibodies to HA. However, interpretation of results is complicated as the relationship between HI titer and the hemolytic area obtained may not be easy to determine.

The MN assay is specific and more sensitive than the HI assay, and is suitable for automation. However, the assay time is generally longer than that of the HI. Furthermore, live, titrated virus (by TCID50 determination or plaque assay) is required, and thus, a high containment laboratory is needed when testing pandemic strains.

An ELISA test specific for HA antibodies requires highly purified antigen, which can be difficult to obtain in sufficient amounts and also suffers from reduced specificity compared to assays measuring functional antibodies.

The PPN is as sensitive as horse erythrocyte HI or MN for the detection of antibodies against H5N1. It is safer, and can be applied in a high-throughput format for human and animal influenza surveillance and for the evaluation of vaccines. To achieve maximum sensitivity in serological assays, the selection of virus isolated from the same influenza outbreak, or the use of an antigenically equivalent strain is required for an optimal antigenic match. Competent molecular virology laboratories could produce HA pseudotype virus within 2–3 weeks of the availability of viral RNA. Further studies are underway, making use of a panel of H5 retroviral pseudotypes with HA components derived from H5N1 viruses involved in the recent human and avian outbreaks.

Development of such assays will be an important further step in preparation of new influenza vaccines, not only for pre-pandemic and pandemic products, but also for cell-derived vaccines. Additional immunological assessments such as cell-mediated immunity and NA inhibition need to be explored to give more insight into the overall effects of vaccination.

References

1. Treanor J (2004) Influenza vaccine – outmaneuvering antigenic shift and drift. N Engl J Med 350:218–220
2. Webby RJ, Webster RG (2003) Are we ready for pandemic influenza? Science 302:1519–1522
3. Webby RJ, Swenson SL, Krauss SL, Gerrish PJ, Goyal SM et al (2000) Evolution of swine H3N2 influenza viruses in the United States. J Virol 74(18):8243–8251
4. Zhou NN, Senne DA, Landgraf JS, Swenson SL, Erickson G et al (1999) Genetic reassortment of avian, swine, and human influenza A viruses in American pigs. J Virol 73(10):8851–8856
5. Vincent AL, Ma W, Lager KM, Janke BH, Richt JA (2008) Swine influenza viruses: a North American perspective. Adv Virus Res 72:127–154
6. Barker WH, Mullooly JP (1980) Impact of epidemic type A influenza in a defined adult population. Am J Epidemiol 112:798–811
7. Barker WH (1986) Excess pneumonia and influenza-associated hospitalization during influenza epidemics in the United States, 1970–78. Am J Public Health 76:761–765
8. Dorrell L, Hassan I, Marshall S, Chakraverty P, Ong E (1997) Clinical and serological responses to an inactivated influenza vaccine in adults with HIV infection, diabetes, obstructive airways disease, elderly adults and healthy volunteers. Int J STD AIDS 8:776–779
9. Künzel W, Glathe H, Engelmann H, Van Hoecke C (1996) Kinetics of humoral antibody response to trivalent inactivated split influenza vaccine in subjects previously vaccinated or vaccinated for the first time. Vaccine 12:1108–1110
10. Stephenson I, Nicholson KG (2001) Influenza: vaccination and treatment. Eur Respir J 17:1282–1293
11. Betts RF, Treanor JJ (2000) Approaches to improved influenza vaccination. Vaccine 18 (16):1690–1695
12. Podda A (2001) The adjuvanted influenza vaccines with novel adjuvants: experience with the MF59-adjuvanted vaccine. Vaccine 19(17–19):2673–2680
13. Podda A, Del Giudice G (2003) MF59-adjuvanted vaccines: increased immunogenicity with an optimal safety profile. Expert Rev Vaccines 2(2):197–203
14. Groth N, Montomoli E, Gentile C, Manini I, Bugarini R, Podda A (2009) Safety, tolerability and immunogenicity of a mammalian cell-culture-derived influenza vaccine: a sequential phase I and phase II clinical trial. Vaccine 27(5):786–791
15. Ehrlich HJ, Müller M, Fritsch S, Zeitlinger M, Berezuk G, Löw-Baselli A, van der Velden MV, Pöllabauer EM, Maritsch F, Pavlova BG, Tambyah PA, Oh HM, Montomoli E, Kistner O, Noel Barrett P (2009) A cell culture (vero)-derived H5N1 whole-virus vaccine induces cross-reactive memory responses. J Infect Dis 200(7):1113–1118
16. Couch RB, Kasel JA, Glezen WP, Cate TR, Six HR, Taber LH, Frank AL, Greenberg SB, Zahradnik JM, Keitel WA (1986) Influenza: its control in persons and populations. J Infect Dis 153:431–440
17. Beyer WE, Palache AM, de Jong JC, Osterhaus AD (2002) Cold-adapted live influenza vaccine versus inactivated vaccine: systemic vaccine reactions, local and systemic antibody response, and vaccine efficacy. A meta-analysis. Vaccine 20:1340–1353
18. Beyer WE, Palache AM, Lüchters G, Nauta J, Osterhaus AD (2004) Seroprotection rate, mean fold increase, seroconversion rate: which parameter adequately expresses seroresponse to influenza vaccination? Virus Res 103:125–132
19. EMEA (1996) Note for guidance on harmonisation of requirements for influenza vaccines (CPMP/BWP/214/96)
20. EMEA (2003) Guideline on dossier structure and content for pandemic influenza vaccine marketing authorisation application (CPMP/VEG/4717/03)
21. Hirst GK (1941) The agglutination of red cells by allantoic fluid of chick embryos infected with influenza virus. Science 94:22–23

22. Goodeve AC, Jennings R, Potter CW (1983) The use of the single radial haemolysis test for assessing antibody response and protective antibody levels in an influenza B vaccine study. J Biol Stand 11:289–296

23. Al-Khayatt R, Jennings R, Potter CW (1984) Interpretation of responses and protective levels of antibody against attenuated influenza A viruses using single radial haemolysis. J Hyg (Lond) 93:301–312

24. Hannoun C, Megas F, Piercy J (2004) Immunogenicity and protective efficacy of influenza vaccination. Virus Res 103:133–138

25. Hobson D, Curry RL, Beare AS, Ward-Gardner A (1972) The role of serum haemagglutination-inhibiting antibody in protection against challenge infection with influenza A2 and B viruses. J Hyg (Lond) 70:767–777

26. Wright PF, Bryant JD, Karzon DT (1980) Comparison of influenza B/Hong Kong virus infections among infants, children, and young adults. J Infect Dis 141:430–435

27. Monto AS, Maassab HF (1981) Ether treatment of type B influenza virus antigen for the haemagglutination inhibition test. J Clin Microbiol 13:54–57

28. Kendal AP, Cate TR (1983) Increased sensitivity and reduced specificity of haemagglutination inhibition tests with ether-treated influenza B/Singapore/222/79. J Clin Microbiol 18:930–934

29. Palmer DF, Coleman MT, Dowdle WR, Schild GC (1975) Advanced laboratory techniques for influenza diagnosis. Immunology series, procedural guide 6. US Department of Health, Education, and Welfare; Public Health Service; Center for Disease Control, 34

30. Hinshaw VS, Webster RG, Easterday BC, Bean WJ Jr (1981) Replication of avian influenza A viruses in mammals. Infect Immun 34:354–361

31. Rowe T, Abernathy RA, Hu-Primmer J, Thompson WW, Lu X, Lim W, Fukuda K, Cox NJ, Katz JM (1999) Detection of antibody to avian influenza A (H5N1) virus in human serum by using a combination of serologic assays. J Clin Microbiol 37:937–943

32. Stephenson I, Wood JM, Nicholson KG, Zambon MC (2003) Sialic acid receptor specificity on erythrocytes affects detection of antibody to avian influenza haemagglutinin. J Med Virol 70:391–398

33. Profeta ML, Palladino G (1986) Serological evidence of human infections with avian influenza viruses. Arch Virol 90:355–360

34. Schild GC, Pereira MS, Chakraverty P (1975) Single-radial-hemolysis: a new method for the assay of antibody to influenza haemagglutinin. Applications for diagnosis and seroepidemiologic surveillance of influenza. Bull World Health Organ 52:43–50

35. Mumford J, Wood J (1993) WHO/OIE meeting: consultation on newly emerging strains of equine influenza. 18–19 May 1992, Animal Health Trust, Newmarket, Suffolk, UK. Vaccine 11:1172–1175

36. Wood JM, Gaines-Das RE, Taylor J, Chakraverty P (1994) Comparison of influenza serological techniques by international collaborative study. Vaccine 12:167–174

37. Wood JM, Melzack D, Newman RW, Major DL, Zambon M, Nicholson KG, Podda A (2001) A single radial haemolysis assay for antibody to H5 haemagglutinin. Int Congr Ser 1219:761–766

38. Nicholson KG, Colegate AE, Podda A, Stephenson I, Wood J, Ypma E, Zambon MC (2001) Safety and antigenicity of non-adjuvanted and MF59-adjuvanted influenza A/Duck/Singapore/97 (H5N3) vaccine: a randomised trial of two potential vaccines against H5N1 influenza. Lancet 357:1937–1943

39. Gross PA, Barry DW, D'Esopo N (1976) Influenza immunization in chronic bronchitis: local and systemic immune response. Am Rev Respir Dis 114:305–313

40. Benne CA, Harmsen M, de Jong JC, Kraaijeveld CA (1994) Neutralization enzyme immunoassay for influenza virus. J Clin Microbiol 32:987–990

41. Harmon MW, Rota PA, Walls HH, Kendal AP (1988) Antibody response in humans to influenza virus type B host cell-derived variants after vaccination with standard (egg-derived) vaccine or natural infection. J Clin Microbiol 26:333–337

42. Fodor E, Devenish L, Engelhardt OG et al (1999) Rescue of influenza A virus from recombinant DNA. J Virol 73:9679–9682
43. Subbarao K, Chen H, Swayne D et al (2003) Evaluation of a genetically modified reassortant H5N1 influenza A virus vaccine candidate generated by plasmid-based reverse genetics. Virology 305:192–200
44. Neumann G, Fujii K, Kino Y, Kawaoka Y (2005) An improved reverse genetics system for influenza A virus generation and its implications for vaccine production. Proc Natl Acad Sci USA 102:16825–16829
45. Treanor JJ, Campbell JD, Zangwill KM, Rowe T, Wolff M (2006) Safety and immunogenicity of an inactivated subvirion influenza A (H5N1) vaccine. N Engl J Med 354:1343–1351
46. Bresson JL, Perronne C, Launay O, Gerdil C, Saville M, Wood J, Höschler K, Zambon MC (2006) Safety and immunogenicity of an inactivated split-virion influenza A/Vietnam/1194/2004 (H5N1) vaccine: phase I randomised trial. Lancet 367:1657–1664
47. Katz JM, Lim W, Bridges CB et al (1999) Antibody response in individuals infected with avian influenza A (H5N1) viruses and detection of anti-H5 antibody among household and social contacts. J Infect Dis 180:1763–1770
48. Kayali G, Setterquist SF, Capuano AW, Myers KP, Gill JS, Gray GC (2008) Testing human sera for antibodies against avian influenza viruses: horse RBC haemagglutination inhibition vs. microneutralization assays. J Clin Virol 43(1):73–78
49. Myers KP, Setterquist SF, Capuano AW, Gray GC (2007) Infection due to 3 avian influenza subtypes in United States veterinarians. Clin Infect Dis 45(1):4–9
50. Frank AL, Puck J, Hughes BJ, Cate TR (1980) Microneutralization test for influenza A and B and parainfluenza 1 and 2 viruses that uses continuous cell lines and fresh serum enhancement. J Clin Microbiol 12:426–432
51. Okuno Y, Tanaka K, Baba K, Maeda A, Kunita N, Ueda S (1990) Rapid focus reduction neutralization test of influenza A and B viruses in microtitre system. J Clin Microbiol 28:1308–1313
52. Stephenson I, Wood JM, Nicholson KG, Charlett A, Zambon MC (2004) Detection of anti-H5 responses in human sera by HI using horse erythrocytes following MF59-adjuvanted influenza A/Duck/Singapore/97 vaccine. Virus Res 103:91–95
53. Govorkova EA, Smirnov Yu A (1997) Cross-protection of mice immunized with different influenza A (H2) strains and challenged with viruses of the same HA subtype. Acta Virol 41:251–257
54. Temperton NJ, Hoschler K, Major D, Nicolson C, Manvell R, Hien VM, Ha DQ, de Jong M, Zambon MC, Takeuchi Y, Weiss RA (2007) A sensitive retroviral pseudotype assay for influenza H5N1-neutralizing antibodies. Influenza Other Respi Viruses 1:105–112
55. Kong WP, Hood C, Yang ZY, Wei CJ, Xu L, García-Sastre A, Tumpey TM, Nabel GJ (2006) Protective immunity to lethal challenge of the 1918 pandemic influenza virus by vaccination. Proc Natl Acad Sci USA 103(43):15987–15991
56. Wang W, Butler EN, Veguilla V, Vassell R, Thomas JT, Moos M Jr, Ye Z, Hancock K, Weiss CD (2008) Establishment of retroviral pseudotypes with influenza haemagglutinins from H1, H3, and H5 subtypes for sensitive and specific detection of neutralizing antibodies. J Virol Methods 153(2):111–119
57. Alberini I, Del Tordello B, Fasolo A, Temperton NJ, Galli G, Gentile C, Montomoli E, Hilbert AK, Banzhoff A, Del Giudice G, Donnelly J, Rappuoli R, Capecchi B (2009) Pseudoparticle neutralization is a reliable assay to measure immunity and cross-reactivity to H5N1 influenza viruses. Vaccine 27(43):5998–6003
58. Su CY, Wang SY, Shie JJ, Jeng KS, Temperton NJ, Fang JM, Wong CH, Cheng YS (2008) In vitro evaluation of neuraminidase inhibitors using the neuraminidase-dependent release assay of haemagglutinin-pseudotyped viruses. Antivir Res 79:199–205
59. Qiu M, Fang F, Chen Y et al (2006) Protection against avian influenza H9N2 virus challenge by immunization with haemagglutinin- or neuraminidase-expressing DNA in BALB/c mice. Biochem Biophys Res Commun 343:1124–1131

60. Johansson BE, Bucher DJ, Kilbourne ED (1989) Purified influenza virus haemagglutinin and neuraminidase are equivalent in stimulation of antibody response but induce contrasting types of immunity to infection. J Virol 63:1239–1246

61. Kilbourne ED, Pokorny BA, Johansson B et al (2004) Protection of mice with recombinant influenza virus neuraminidase. J Infect Dis 189:459–461

62. Chen Z, Kadowaki S, Hagiwara Y et al (2000) Cross-protection against a lethal influenza virus infection by DNA vaccine to neuraminidase. Vaccine 18:3214–3222

63. Monto AS, Kendal AP (1973) Effect of neuraminidase antibody on Hong Kong influenza. Lancet 1:623–625

64. Tamura M, Webster RG, Ennis FA (1994) Subtype cross-reactive, infection-enhancing antibody responses to influenza A viruses. J Virol 68:3499–3504

65. Sandbulte MR, Jimenez GS, Boon AC et al (2007) Cross-reactive neuraminidase antibodies afford partial protection against H5N1 in mice and are present in unexposed humans. PLoS Med 4:e59

66. Gioia C, Castilletti C, Tempestilli M et al (2008) Cross-subtype immunity against avian influenza in persons recently vaccinated for influenza. Emerg Infect Dis 14:121–128

67. Webster RG, Campbell CH (1972) An inhibition test for identifying the neuraminidase antigen on influenza viruses. Avian Dis 16:1057–1066

68. Clements ML, Betts RF, Tierney EL, Murphy BR (1986) Serum and nasal wash antibodies associated with resistance to experimental challenge with influenza A wild-type virus. J Clin Microbiol 24:157–160

69. Schulman JL, Kilbourne ED (1969) Independent variation in nature of haemagglutinin and neuraminidase antigens of influenza virus: distinctiveness of haemagglutinin antigen of Hong Kong-68 virus. Proc Natl Acad Sci USA 63:326–333

70. Burlington DB, Clements ML, Meiklejohn G, Phelan M, Murphy BR (1983) Haemagglutinin-specific antibody responses in immunoglobulin G, A, and M isotypes as measured by enzyme-linked immunosorbent assay after primary or secondary infection of humans with influenza A virus. Infect Immun 41:540–545

71. Murphy BR, Phelan MA, Nelson DL et al (1981) Haemagglutinin-specific enzyme-linked immunosorbent assay for antibodies to influenza A and B viruses. J Clin Microbiol 13:554–560

72. Stelzer-Braid S, Wong B, Robertson P et al (2008) A commercial ELISA detects high levels of human H5 antibody but cross-reacts with influenza A antibodies. J Clin Virol 43:241–243

73. Doller G, Schuy W, Tjhen KY, Stekeler B, Gerth HJ (1992) Direct detection of influenza virus antigen in nasopharyngeal specimens by direct enzyme immunoassay in comparison with quantitating virus shedding. J Clin Microbiol 30:866–869

74. Wood JM, Robertson JS (2004) From lethal virus to life-saving vaccine: developing inactivated vaccines for pandemic influenza. Nat Rev Microbiol 2:842–847

75. Hoffmann E, Krauss S, Perez D, Webby R, Webster RG (2002) Eight-plasmid system for rapid generation of influenza virus vaccines. Vaccine 20:3165–3170

76. Temperton NJ, Wright E (2009) Retroviral pseudotypes. Encyclopedia of life sciences (ELS). Wiley, Chichester. doi:10.1002/9780470015902.a0021549

The Role of Animal Models In Influenza Vaccine Research

Catherine J. Luke and Kanta Subbarao

Abstract A major challenge for research on influenza vaccines is the selection of an appropriate animal model that accurately reflects the disease and the protective immune response to influenza infection in humans. Vaccines for seasonal influenza have been available for decades and there is a wealth of data available on the immune response to these vaccines in humans, with well-established correlates of protection for inactivated influenza virus vaccines. Many of the seminal studies on vaccines for epidemic influenza have been conducted in human subjects. Studies in humans are performed less frequently now than they were in the past. Therefore, as the quest for improved influenza vaccines continues, it is important to consider the use of animal models for the evaluation of influenza vaccines, and a major challenge is the selection of an appropriate animal model that accurately reflects the disease and the protective immune response to influenza infection in humans.

The emergence of highly pathogenic H5N1 avian influenza (AI) viruses and the threat of a pandemic caused by AI viruses of this or another subtype has resulted in a resurgence of interest in influenza vaccine research. The development of vaccines for pandemic influenza presents a unique set of obstacles, not the least of which is that the demonstration of efficacy in humans is not possible. As the correlates of protection from pandemic influenza are not known, we rely on extrapolation of the lessons from seasonal influenza vaccines and on data from the evaluation of pandemic influenza vaccines in animal models to guide our decisions on vaccines for use in humans. The features and contributions of commonly used animal models for influenza vaccine research are discussed. The recent emergence of the pandemic 2009 H1N1 influenza virus underscores the unpredictable nature of influenza viruses and the importance of pandemic preparedness.

C.J. Luke (✉) and K. Subbarao
Laboratory of Infectious Diseases, National Institute of Allergy and Infectious Diseases, National Institutes of Health, Bethesda, MD 20892, USA
e-mail: cluke@niaid.nih.gov

G. Del Giudice and R. Rappuoli (eds.), *Influenza Vaccines for the Future*, 2nd edition,   223
Birkhäuser Advances in Infectious Diseases,
DOI 10.1007/978-3-0346-0279-2_10, © Springer Basel AG 2011

1 Influenza Viruses

Influenza is a negative-sense, single-stranded RNA virus belonging to the family *Orthomyxoviridae*. *Orthomyxoviridae* consist of four genera: influenza A, influenza B, influenza C and Thogoto viruses. The proteins of influenza A viruses are encoded by genes on eight RNA segments. Influenza A viruses are widely distributed in nature and can infect a wide variety of birds and mammals, including humans. Influenza A virus subtypes are classified on the basis of the antigenicity of their surface glycoproteins, hemagglutinin (HA) and neuraminidase (NA) [1, 2] into 16 HA subtypes and 9 NA subtypes, and all of these subtypes have been found to infect birds [2, 3]. Waterfowl and shorebirds are the natural reservoirs of AI viruses.

In their natural hosts, most AI infections are not associated with clinical disease, and the viruses are generally thought to be in evolutionary stasis [4]. In humans, relatively few subtypes of influenza A viruses have caused sustained outbreaks of disease; viruses bearing H1, H2, and H3 HA genes and N1 and N2 NA genes have circulated in the human population during the twentieth century. H1N1 viruses appeared in 1918 and circulated until 1957, when they were replaced by H2N2 viruses. These, in turn, were replaced in 1968 by H3N2 viruses, which continue to circulate. In 1977, H1N1 viruses reappeared and have continued to co-circulate with the H3N2 viruses. Influenza A and B viruses continue to cause epidemics in humans each winter.

In addition to the seasonal influenza epidemics, the potential also exists for an influenza pandemic at any time. A pandemic occurs when an influenza strain with a novel HA subtype (with or without a novel NA subtype) appears and spreads in a susceptible human population. In the twentieth century, influenza pandemics occurred in 1918, 1957, 1968, and most recently, in 2009, with the emergence of the swine-origin pandemic H1N1 influenza virus. The influenza pandemic of 1918 was associated with severe morbidity and significant mortality but the pandemics of 1957 and 1968 were milder [5]. To date, disease caused by the pandemic 2009 H1N1 influenza virus does not appear to be more severe than disease seen with epidemic influenza, but there is a significant difference in the age groups most affected, with the majority of cases of pandemic 2009 H1N1 infection and hospitalization occurring in children and young adults and the highest mortality occurring in adults aged 24–49 years.

AI viruses in their natural reservoir in waterfowl and shorebirds are one source from which novel HA and NA subtypes are introduced into the human population. An influenza virus with a novel HA and/or NA can be introduced into the human population by direct spread from either wild birds or domestic poultry, as was seen when an H5N1 AI virus infected humans in 1997 [6]. Alternatively, avian and human influenza viruses can reassort, generating a virus that can efficiently spread in humans, as happened in the 1957 H2N2 and 1968 H3N2 pandemics [7]. The pandemic 2009 H1N1 influenza virus is derived from influenza viruses that were circulating in pigs rather than birds. This virus is a reassortant, bearing gene segments that were originally derived from avian, human and classical swine influenza viruses [8].

Influenza A viruses also infect and cause disease in a wide variety of mammalian species, including swine, horses, ferrets, mink, dogs, seals, and whales. The currently circulating highly pathogenic AI H5N1 viruses that emerged in Asia in 2003 can also infect and cause lethal infection in felids, including tigers, leopards and domestic cats [9, 10].

Although several animal species can be infected with influenza A viruses naturally and experimentally, an ideal animal model for studying infection and immunity to human influenza has not been identified. Several animal species are permissive to infection with influenza A and B viruses to varying degrees and some exhibit clinical signs of illness and pathological changes in the respiratory tract that are similar to those seen in human influenza. In this chapter, we discuss the main features of the animal models used for the evaluation of influenza vaccines, their advantages and disadvantages, and their contribution to research on vaccines against influenza in humans. We also discuss the role of animal models in the development of vaccines against pandemic influenza. Veterinary vaccines for swine, equine, avian and canine influenza can be evaluated in their natural hosts and are not discussed.

2 Influenza Vaccines

Vaccines have been available for epidemic or "seasonal" influenza since the 1940s. Inactivated influenza virus vaccines are still largely the same as they were when first developed. They are still generally produced in embryonated hen's eggs. There has been much investment recently in the development of cell-based influenza vaccines, of which at least two are licensed in Europe and several others are in development in Europe and the United States. Live attenuated influenza vaccines were first licensed in the United States in 2003, and are currently approved for annual use in healthy individuals between 2 and 49 years of age.

A serum hemagglutination inhibiting (HAI) antibody titer of 1:32 or 1:40 or greater is associated with protection from seasonal influenza [11–13], and this is used as a measure to predict the protective efficacy of the seasonal inactivated influenza virus vaccines. The correlates of protection for live attenuated vaccines are less clear-cut. These vaccines elicit systemic and mucosal immune responses and mucosal antibodies in the respiratory tract are believed to play a major role in the protection afforded by these vaccines [14–16].

Antigenic drift describes the gradual change in the antigenicity of an influenza virus which allows the virus to escape neutralization by antibodies that have already been induced by prior infection or immunization with previously circulating strains. Antigenic drift results from point mutations in and around antibody-combining sites in the HA and NA proteins. Influenza virus vaccines are unusual in that one or more of the components of the trivalent vaccine formulation may have to be changed annually to keep pace with antigenic drift of the virus, but as long as a licensed manufacturing process is used, the change in composition of the vaccine is considered a strain change and is not treated as a new vaccine. Approval of seasonal influenza vaccines for use in humans requires limited testing in animals, and an evaluation of immunogenicity in humans is required in Europe but not in the United States.

In recent years, a resurgence of interest in the improvement of seasonal influenza vaccines, and the looming threat of a possible influenza pandemic have spurred efforts to develop vaccines that could thwart the spread of an emerging pandemic virus. Extensive pre-clinical characterization of these new vaccines in animals will be necessary. Many researchers are engaged in efforts aimed at developing "universal" influenza vaccines that can protect against both epidemic and pandemic strains by targeting the more conserved antigens of the virus, such as nucleoprotein (NP) or the matrix protein (M), thus eliminating the need for having to constantly update the composition of the annual seasonal influenza vaccine. The immune responses to candidate universal vaccines are entirely different from those elicited by the currently licensed seasonal inactivated influenza virus vaccines, where protective immunity is based mainly on neutralizing antibodies produced against the HA protein. Animal models in which different types of immune responses can be evaluated are needed.

One of the major challenges faced during the development of pandemic influenza vaccines is that the correlates of protection from AI viruses of pandemic potential are not known. Efficacy of these novel influenza vaccines cannot be established in humans, so estimates of efficacy are based on the information gleaned from challenge studies in animals.

3 Animal Models for Influenza

Despite the diversity of mammalian species infected by influenza viruses in nature only a few species are amenable to study in the laboratory. Tables 1–3 and the following sections summarize the features of the most commonly used small animal models in the study of influenza, and their respective utilities in the evaluation of influenza vaccines are summarized in Table 4. Commonly used laboratory animal species may not be fully permissive for infection with wild-type, non-adapted isolates of influenza viruses, and can vary in susceptibility to infection with specific virus strains and subtypes. Other variables that can influence the outcome of infection are the use of anesthesia, route of virus administration and the volume of inoculum used.

3.1 Rodent Models

Rodent models of infectious diseases are attractive for a number of scientific and practical reasons. They are small and relatively inexpensive to purchase and house. Many inbred strains are available and a battery of immunological reagents are available for some species.

Table 1 The use of the mouse model for the evaluation of vaccines against influenza

Influenza virus subtypes tested	Findings	References
Human influenza H1N1, H3N2, H2N2	Human influenza virus isolates require adaptation to cause illness (lethality) in mice Infection under anesthesia results in viral pneumonia Clinical signs include ruffled fur, hunching, labored breathing, unsteady gait, hypothermia, and weight loss Inflammation is observed in the respiratory tract	[18, 31, 32, 181]
Reconstructed 1918 H1N1 pandemic virus	Causes illness in mice and replicates efficiently in the respiratory tract without prior adaptation Up to 13% loss of body weight is observed Lethal to mice with an MDT[a] of 4.5 days No extrapulmonary spread observed Necrotizing bronchitis and bronchiolitis, and moderate to severe peribronchial and alveolar edema present	[138]
HPAI[b] H5N1	Most isolates cause severe illness and death without prior adaptation Replicate efficiently in the respiratory tract without prior adaptation Cause significant weight loss Most isolates are lethal in mice with a MDT of 6-8 days Some isolates are detected in extrapulmonary sites including the brain Variable virulence in mice is observed with isolates from Hong Kong from 1997, and 2003–2004, and viruses isolated from Europe and South America	[100–104, 133, 148]
H7	HP[c] and LP[d] isolates replicate efficiently in respiratory tract of mice without prior adaptation, with some viruses causing weight loss and death Extrapulmonary spread to the brain and spleen observed following intranasal infection with some isolates Histopathologic observations following intranasal infections with human isolates include necrosis and inflammation throughout the respiratory tract, but no lesions in the brain, heart, spleen, liver or kidneys Histopathological lesions are observed following intranasal infection with HP avian isolates	[115–119]
H9N2	Replicate efficiently in lungs of mice without prior adaptation Conflicting reports of lethality in mice Adaptation by passage in mouse lungs results in increased virulence Replication in brain reported following intranasal infection with non-adapted and mouse-adapted viruses	[132–137, 182, 183]
H6	Varying replication efficiency in respiratory tract depending on isolate Replication more efficient in the lower respiratory tract than in the upper respiratory tract	[132, 145]

(*continued*)

Table 1 (continued)

Influenza virus subtypes tested	Findings	References
	A/teal/W312/HK/97 (H6N1) and A/quail/HK/1721-30/99 (H6N1) lethal for mice when administered at high titers	
	Significant weight loss (average 24%) is observed in infected mice	
Pandemic 2009 H1N1	Replicate efficiently in the upper and lower respiratory tract	[171, 172]
	Some isolates cause weight loss and are lethal when administered at high titers	

[a]*MDT* mean time to death
[b]*HPAI* highly pathogenic avian influenza
[c]*HP* highly pathogenic
[d]*LP* low pathogenicity

Table 2 The use of the ferret model for the evaluation of vaccines against influenza

Influenza virus subtypes tested	Findings	References
Human influenza H1N1, H3N2, H2N2 viruses	Efficient replication of non-adapted isolates in respiratory tract	[17, 147, 184, 185]
	Isolated report of the presence of an H3N2 human influenza virus in the brain	
	Signs of illness include fever, sneezing, rhinorrhea, and weight loss	
	Mild inflammatory changes are observed upon histopathological examination of lungs of infected animals	
Reconstructed 1918 H1N1 pandemic virus	Replication to high titers in respiratory tract	[156]
	Severe disease observed including lethargy, anorexia, severe weight loss and high fever	
	Infection is lethal in 2/3 of inoculated animals; death occurs by Day 11	
	Virus is not detected in brain or heart	
	Necrotizing bronchiolitis, and moderate to severe alveolitis with edema observed upon histopathological examination	
HPAI[a] H5N1	Efficient replication in respiratory tract and evidence of extrapulmonary spread to brain, spleen and intestines	[104, 147, 149]
	Most isolates cause severe disease, including fever, rhinitis, sneezing, severe lethargy, hind limb paresis and diarrhea	
	Many isolates cause lethal infection in ferrets	
	Histopathologic observations include inflammatory changes in the lungs (bronchioloitis, bronchitis, interstitial pneumonia) and inflammation in the brain	

(*continued*)

Table 2 (continued)

Influenza virus subtypes tested	Findings	References
H6	Replicate to varying levels in the respiratory tract, with lower titers of virus in the upper respiratory tract than in the lungs	[145, 155]
	Transient weight loss observed with Eurasian and North American isolates	
	Transient elevation in body temperature	
H7	Replicate to varying levels in the respiratory tract	[115, 116]
	Some HP H7 viruses replicate in the brain. A/NL/219/2003 (H7N7) HPAI causes severe disease with neurologic symptoms and mortality	
	HP H7 viruses generally replicate to higher titers in the lungs than LP H7 viruses and duration of replication is longer	
Pandemic 2009 H1N1	Viruses replicate efficiently in the upper and lower respiratory tract	[171, 172, 176]
	Viruses replicate to higher titers in the lungs than seasonal H1N1 influenza viruses	
	Some isolates caused signs of illness (weight loss, fever), severe illness and death	
	One isolate was detected in rectal swabs	
AI[b] subtypes H1N1, H2N1, H6N2, H2N2, H2N3, H3N2, H10N7, H3N6, H7N7, seal H7N7 isolate	Efficient replication in the upper respiratory tract No signs of illness with any of these isolates	[155]

[a]*HPAI* highly pathogenic avian influenza
[b]*AI* avian influenza

Table 3 The use of the hamster model for the evaluation of vaccines against influenza

Influenza virus subtypes tested	Findings	References
Human influenza H3N2	Non-adapted isolates replicate in the upper and lower respiratory tract. No clinical signs of infection are observed	[35–37, 161]
HPAI[a] H5N1	Non-adapted A/HK/483/97 (H5N1) resulted in lethal infection with deaths of all inoculated animals by Day 6 post-inoculation	[161]
	Virus is detectable in the lungs and brain	
H9N2	Non-adapted A/HK/1073/99 (H9N2) replicates to high titers in the lungs but is not detected in the brain	[161]
	Infection is not lethal	
H9N5	Non-adapted A/dk/HK/702/79 (H9N5) replicates efficiently in the lungs	[161]
	Infection is not lethal	

[a]*HPAI* highly pathogenic avian influenza

Table 4 Comparison of the utility of commonly used animal models in the evaluation of influenza vaccines

Species	Utility in vaccine evaluation
Mouse	Determination of level of replication of live attenuated vaccine candidates in comparison to wild-type viruses
	Evaluation of antibody responses to vaccination by HAI[a] assay, Nt Ab[b] assay, ELISA[c]
	Evaluation of cellular immune responses to vaccination
	Evaluation of vaccine efficacy and effects of adjuvants
	General safety test for manufactured candidate vaccine
Ferret	Determination of level of replication of live attenuated vaccine candidates in comparison to wild-type viruses
	Evaluation of antibody responses to vaccination by HAI assay, Nt Ab assay, ELISA
	Limited evaluation of cellular immune responses to vaccination
	Evaluation of vaccine efficacy and effects of adjuvants
	Toxicology studies
Hamster	Determination of level of replication of temperature-sensitive live attenuated vaccine candidates
	Evaluation of vaccine immunogenicity by HAI assay, Nt Ab assay and ELISA
	Evaluation of vaccine efficacy

[a]*HAI* hemagglutination inhibition
[b]*NtAb* neutralizing antibody
[c]*ELISA* enzyme linked immunosorbent assay

3.2 Mice

Mice have been used for influenza vaccine research from the earliest days of the study of influenza virus biology. Shortly after the first human influenza virus was isolated from ferrets in 1933 by Wilson Smith and colleagues at the National Institute for Medical Research in London [17], it was discovered that human influenza viruses would cause disease in mice only if they were first adapted to the species by serial passages in the lungs [18]. This was subsequently found to be true for all human influenza virus isolates. One of the most commonly used human influenza viruses in mice is influenza A/Puerto Rico/8/34 (PR8), an H1N1 virus with a complex passage history, including several passages in ferrets, and hundreds of passages in eggs and mice (CB Smith, CDC, Atlanta, GA, personal communication). This virus is well adapted to mice and causes a lethal infection. The need for adaptation through serial passage of human influenza viruses is one of the major drawbacks of using mice in influenza research, because many mutations can arise during adaptation to the murine host; [19–22] these can alter their replication kinetics, and can result in the ability of the virus to escape the host innate immune responses [23].

Influenza viruses that cause disease and are lethal for mice provide a useful endpoint for vaccine efficacy studies. Depending on the strain of virus used, mice may become lethargic, anorexic, develop ruffled fur, and may also exhibit neurological symptoms of infection, in addition to weight loss, which is often the primary objective measure of the severity of infection. Body temperature is not a useful measurement in mice because hypothermia can occur following infection with mouse-adapted viruses. Irrespective of whether an influenza virus induces morbidity or mortality in mice, the level of replication of influenza viruses in the lungs is the most informative endpoint for efficacy studies in mice since even a modest reduction in titer of infectious virus in the lungs can be associated with survival from lethal infection [24, 25]. Mice immunized with influenza viruses or vaccines develop serum HAI and neutralizing antibodies, the titers of which correlate with protection from subsequent challenge. Studies by Virelizier [26] demonstrated that antibody alone could protect against influenza infection in mice. Passive transfer of immune serum to naive mice resulted in a reduction in the replication of virus in the lungs and protected the recipient mice from lethal influenza pneumonitis, but did not prevent tracheitis or replication of virus in the upper respiratory tract [27]. The observation that passively transferred serum antibodies can reduce pulmonary viral replication but not viral replication in the upper respiratory tract is not unique to influenza A. Similar observations have been reported with influenza C virus [28], respiratory syncytial virus (RSV) [29] and severe acute respiratory syndrome-associated coronavirus (SARS-CoV) infections [30]. Measuring the amount of virus in various tissues in cases where high levels of serum antibody are present, for example, when vaccines are administered with adjuvant, should be done by quantitative molecular methods to rule out the possibility of ex vivo neutralization by serum antibody during tissue preparation. Such ex vivo neutralization has been shown to account for a reduction of up to 300-fold in detectable virus in the lungs of mice that had undergone passive transfer of immune serum against SARS-CoV [30]. The use of nasal and bronchiolar wash samples, instead of tissue homogenates, for viral quantitation was also employed as a solution to this issue [28].

The level of anesthesia can influence the outcome of influenza infection in mice. Mice infected under anesthesia develop pneumonia, while infection is limited to the upper respiratory tract when awake mice are infected [31, 32]. The volume of inoculum administered intranasally also influences the extent to which virus is distributed in the respiratory tract [32]. Immunologically, the lack of a functional Mx gene in standard laboratory strains of mice is a disadvantage of this model for studies in which the innate immune response to infection is important [33, 34]. However, the ready availability of mice, their relatively low cost, the available variety of genetic backgrounds and targeted genetic defects, and the immunological reagents available still make the mouse an attractive and heavily utilized animal model for studies on influenza.

3.3 Hamsters

Influenza virus infection of hamsters with non-adapted human influenza viruses does not result in clinical disease, but the virus replicates to high titers in the nasal turbinates and lungs following an intranasal infection [35–37]. As with mice, the hamster represents a readily available small animal model that can be used for pre-clinical evaluation of candidate vaccines, but it has not been as extensively used as mice for studies of inactivated influenza virus vaccines. The body temperature of Golden Syrian hamsters is about 39°C, while that of mice is 37°C. Thus, hamsters have been used for the evaluation of live attenuated temperature-sensitive vaccines with shut-off temperatures $\geq$38.8°C [38].

3.4 Guinea Pigs

Guinea pigs can be infected with non-adapted human influenza viruses, although the amount of virus needed to infect guinea pigs is about ten times more than the amount needed to infect hamsters or ferrets [39]. Infection of guinea pigs with A/England/42/72 (H3N2) did not result in febrile illness or other clinical signs of influenza infection. The virus was isolated from the nasal washes of animals infected with influenza A/England/42/72 (H3N2), A/Hong Kong/1/1968 (H3N2) or A/FM/1/47 (H1N1) viruses, but titers of virus shed in the nasal secretions were not as high as those observed following experimental infection of ferrets. Infection of guinea pigs with influenza A/HK/1/68 (H3N2) virus resulted in pneumonia, which developed slowly and was reversible. This model was used to study the effects of environmental pollutants or drugs on the respiratory tract [40]. Lowen and colleagues [41] reported that guinea pigs of the Hartley strain are highly susceptible to non-adapted influenza A/Panama/2007/99 (H3N2) virus. Intranasal infection resulted in virus replication in the nose and lungs, with higher titers of virus being recovered from the lungs. The virus could be recovered from the upper respiratory tract for up to 9 days post inoculation, whereas shedding declined to undetectable levels in the lungs by day 5. Virus replication was not associated with any effects on body temperature or weight of the animals, and no other clinical signs of illness were observed.

3.5 Rats

Common laboratory strains of rat are described as "semi-permissive" for influenza infection, and infant rats are of some utility in the evaluation of live attenuated influenza vaccines, but they have not been used extensively to study influenza infection [42–44].

The cotton rat (*Sigmodon hispidus*) has been used in the laboratory as a model for several infectious diseases (reviewed in [45]). In particular, the cotton rat model was extensively used in the development of therapeutic antibody treatments for RSV and has provided much useful information for vaccine development against this pathogen. Sadowski and co-workers reported that intranasal administration of human influenza virus to lightly anesthetized, outbred young adult cotton rats resulted in virus replication in the respiratory tract, the production of pulmonary lesions and a strong immune response [46]. In recent years, there has been some renewed interest in the cotton rat as a laboratory animal model for human influenza virus infection. Species-specific reagents that permit more detailed analysis of viral pathogenesis and immune responses in this species have been developed [45] and inbred cotton rats are now available. The advantages of this model include the fact that cotton rats can be infected by non-adapted human influenza viruses, inbred animals are available, the virus replicates in the upper and lower respiratory tract, some clinical parameters can be measured, and viral infection results in histopathological changes in the lungs that are similar to those seen during natural infection of humans [47]. To date only a limited number of human influenza viruses have been evaluated in cotton rats.

3.6   Ferrets

Ferrets are exquisitely susceptible to infection with human influenza viruses. The initial isolation of a human influenza virus by Smith and colleagues was from ferrets [17]. The ferret model of influenza has remained the same since this fortuitous discovery, and, in the opinion of many researchers, the ferret remains the ideal small animal model for influenza research. Ferrets can be infected with non-adapted human influenza virus isolates. Influenza virus infection in ferrets is primarily an upper respiratory tract infection, and infected ferrets exhibit clinical signs of infection similar to those seen during human influenza including fever, rhinitis and sneezing. The disadvantages of the ferret as a model for studying influenza vaccines include expense, special housing requirements, a limited number of suppliers, difficulties in obtaining animals that are seronegative for influenza virus, their exquisite sensitivity to other respiratory pathogens and ease of acquiring infection from their handlers, and the lack of species-specific reagents, although this last point does not present an obstacle for the evaluation of HAI and neutralizing antibody responses. In addition, the high body temperature of ferrets (average temperature of 38.8°C) may limit their utility in the evaluation temperature-sensitive live attenuated influenza vaccines.

3.7   Non-Human Primates

Non-human primates have not been used extensively for influenza vaccine research. From a practical standpoint, these animals are expensive and they have not proven to be the best model for the study of vaccines for influenza. Old World and New World species of monkeys have been evaluated as models of human influenza infection. It was determined early in the days of the study of influenza virus biology that non-human primate species were not as susceptible to human influenza viruses as their human relatives. Burnet reported in 1941 [48] that clinical signs of infection were only apparent in cynomolgus macaques when they were infected via the intratracheal route as opposed to the intranasal route. Interestingly, mortality was observed in animals inoculated with the "W.S. Egg" strain, which was a mouse-adapted human influenza virus that had been passaged in eggs. Burnet reported that pathological changes consistent with those seen in human influenza infection were observed in the lungs of infected monkeys. The observation that intratracheal infection in monkeys might be required to achieve clinical signs of infection was supported by studies conducted by Saslaw and colleagues [49] in Rhesus macaques. Intratracheal infection of Rhesus macaques with a lung filtrate from mice infected with mouse-adapted A/PR/8/34 (H1N1) virus resulted in clinical signs of illness on day 2 post infection (p.i.), which resolved by day 4 p.i., whereas no signs of illness were apparent in monkeys inoculated with the same virus preparation intranasally, although both groups of animals showed hematological and serological evidence of infection.

Cynomolgus macaques were explored as a model for the evaluation of the immunogenicity and efficacy of an immunostimulating complex (ISCOM) influenza vaccine by Rimmelzwaan and colleagues [50]. Cynomolgus macaques inoculated intratracheally with the human influenza A/Netherlands/18/94 (H3N2) virus did not develop clinical signs of illness but virus could be recovered from lung lavage, nasal swabs and pharyngeal swab samples. Histopathological examinations were not performed.

Pigtailed macaques (*Macaca nemestrina*) were infected with a recombinant human influenza A/Texas/91 (H1N1) virus following virus administration via the trachea, tonsils and conjunctiva [51]. The animals exhibited clinical signs of infection, including loss of appetite, weight loss, nasal discharge and moderate fever, and histopathological observations that were consistent with progressive pneumonia. Virus was recovered from lung tissue at day 4 p.i. but not at day 7 p.i.

New World monkeys – including squirrel and cebus monkeys – have been evaluated as models for influenza vaccine studies. Murphy et al. [52] demonstrated that adult squirrel monkeys could be infected with intratracheally administered human influenza viruses. Mild illness that manifested as afebrile coryza was seen and, although radiographic evidence of pneumonia was not observed, the animals shed virus from the respiratory tract. Further studies evaluated the ability of AI viruses to replicate and cause illness in this species [53]. Different viruses caused varying degrees of clinical illness; some influenza viruses were completely attenuated in squirrel monkeys, while others replicated efficiently and caused clinical signs which were of a severity similar to that seen in human H3N2 influenza infection. Squirrel monkeys were employed to evaluate the level of attenuation of avian/human influenza virus reassortants, in a study comparing the replication of reassortants in chimpanzees and human volunteers [54]; the findings in squirrel monkeys were not predictive of the level of attenuation of the reassortant viruses in humans.

Cebus apella and *Cebus albifrons* monkeys were evaluated as models for influenza infection by Grizzard et al. [55]. The monkeys were inoculated either intranasally or intratracheally with two human influenza A viruses: A/Victoria/75

(H3N2) and A/New Jersey/76 (H1N1). All animals that received the A/Victoria/75 (H3N2) strain developed clinical signs of illness, and showed evidence of infection by either virus shedding or serology. Radiographic evidence of pulmonary disease was only seen in animals inoculated intratracheally with A/Victoria/75 (H3N2). Eight of ten animals inoculated intratracheally with the A/New Jersey/76 (H1N1) virus had mild upper respiratory tract illness, but only one of ten animals shed virus. However, all of these animals seroconverted. Histopathological evidence of inflammation in the lungs and trachea was seen in animals inoculated intratracheally with either strain, although the lesions in the animals that received A/Victoria/75 (H3N2) were more severe.

Chimpanzees are considered to be a valuable animal model to study infections of humans because of their close evolutionary relationship with the human species. However, the use of chimpanzees as animal models in research is logistically difficult. They are extremely expensive animals that require long-term care and stringent isolation since they are susceptible to several human pathogens. Chimpanzees have been used for some studies of influenza [54, 56, 57]. Influenza A and B viruses replicated to high titer in seronegative chimpanzees, but viral replication was not associated with illness. The advantages of studying influenza in this species include the fact that chimpanzees have the same body temperature as humans, their lower respiratory tract can be repeatedly sampled safely, they display permissiveness for vectored vaccines, similar to humans (for example, vaccinia-based vaccines), they are evolutionarily close to humans, and this may mean that similar host-range restrictions for replication of viruses may be present which could facilitate the selection of live attenuated candidate vaccines for testing in humans.

There is renewed interest in the use of non-human primates for evaluation of vaccines for pandemic influenza (see *Vaccines for pandemic influenza* below).

4 Animal Models in Influenza Vaccine Research

The three general areas of vaccine research and development in which animal models are utilized are the evaluation of vaccine safety, immunogenicity and efficacy. The following sections describe the use of animal models in each of these aspects of the pre-clinical evaluation of influenza vaccines.

4.1 Safety

In the early days of clinical testing of live attenuated vaccines against seasonal influenza, it was recognized that an animal model that could predict the attenuation of these vaccines would allow more rapid progression to immunogenicity and efficacy testing. Ideally, systematic comparisons of the behavior of attenuated virus vaccine candidates in animal models and in humans are needed to achieve this end. Researchers began to address this question in the late 1970s and early 1980s, and the infant rat was extensively investigated as a model to predict the restriction in the replication of live attenuated influenza vaccines in humans [42–44]. In general, attenuation in the infant rat model correlated with attenuation in humans, although there were exceptions. Other species evaluated for this purpose include mice, hamsters, ferrets, and chimpanzees.

Although vaccine safety can only be fully assessed when a vaccine is administered to human subjects, regulatory authorities usually recommend standard tests for pre-clinical evaluation of the safety of new canidiate vaccines. The primary safety concern for inactivated influenza virus vaccines is reactogenicity, and for live attenuated influenza vaccines, it is their level of attenuation and their genetic stability. Standard toxicology tests on new vaccine candidates are often performed in rabbits, although current WHO guidelines for nonclinical evaluation of vaccines recommend that toxicology studies be performed in an animal species that most closely reflects the immune response to the vaccine in humans, or is "sensitive to the biological effects of the vaccine", and use the same dose and route of administration to be studied in clinical trials [58]. The design and results of such studies should be reviewed with special attention to experimental details such as the route of administration, volume and quantity of virus in the inoculum, and whether or not anesthesia was used, particularly for live attenuated vaccines, because each of these factors can influence the outcome. Toxicity following administration of very high doses of live influenza virus to animals via a variety of routes has been reported in the literature. For example, administration of 10^9 EID_{50} of influenza virus administered intranasally resulted in complete pulmonary consolidation and death in mice, and this pathology occurred despite restricted replication of virus in lung tissue [59]. Henle and Henle [60] reported inflammation in the gut, damage to the liver and spleen, and death in mice given high doses of influenza virus intraperitoneally. Similar findings were observed in rats, rabbits and guinea pigs. Lung inflammation was observed in ferrets administered high titer live attenuated influenza viruses intranasally [61] and systemic signs of illness were reported in human volunteers who received attenuated influenza viruses at doses that exceeded 10^7 $TCID_{50}$ [62–64]. In these studies, signs of clinical illness, including fever and other systemic signs, appeared within 48 h of administration of the virus, which is more rapid, in general, than the appearance of symptoms associated with a productive influenza virus infection. The systemic symptoms did not correlate with the titer of virus shed in respiratory secretions, or with the occurrence of respiratory symptoms. The occurrence of systemic illness in humans following administration of high doses of influenza virus in the absence of high levels of virus replication may be explained by the innate immune response to an abortive infection of epithelial cells.

The current procedures for marketing approval of vaccines for seasonal influenza do not involve extensive safety testing in animals. In the US, a standard general safety test, which is designed to detect extraneous toxic components in the vaccine preparation, is usually performed with the final drug product in mice and guinea pigs [65]. This test is performed for both inactivated and live attenuated virus vaccines. For inactivated influenza virus vaccines, the vaccine can be administered via either the subcutaneous or intraperitoneal route for the guinea pig test, whereas only intraperitoneal route can be used for other types of vaccine. The vaccine formulation must also be certified to be free of endotoxin.

New vaccine candidates or novel preparations (including vaccines prepared by currently licensed methodologies that are now formulated with adjuvant) require extensive pre-clinical safety testing. In addition to tests such as repeat dose toxicology testing and general safety testing, some tests would be appropriate depending on the specific type of vaccine, e.g., demonstration of attenuation of live attenuated vaccines compared to the wild-type parent virus in more than one animal species [16, 66, 67], and biodistribution studies for plasmid-based vaccines [68–72].

Ferrets have been used to assess the attenuation of cold-adapted live attenuated vaccines against influenza [73]. These studies showed that cold-adapted 6-2 reassortant vaccine viruses generated from human influenza viruses failed to replicate in the lower respiratory tract of ferrets. Since ferrets are a good model for influenza infection in humans, they can also be used in toxicological studies of influenza vaccines.

The attenuation phenotype of several live attenuated influenza vaccine candidates was evaluated using the hamster model [35, 37]. For the small number of temperature-sensitive, cold-adapted reassortant influenza viruses tested in hamsters, and later in humans, there was a general correlation between the level of replication in hamsters and humans. However, in studies with AI/human influenza virus reassortants, the findings in hamsters did not accurately predict the level of attenuation of the viruses in humans [74]. Such data are important because they demonstrate that the genetic determinants of attenuation of influenza viruses are different in different species.

Non-human primate species have not been extensively used in studies the safety of influenza vaccines. Chimpanzees were used in several studies to evaluate the level of attenuation and the safety of candidate live attenuated vaccines [74]. Regulatory authorities in Europe require neurovirulence testing of live attenuated influenza vaccines and inactivated vaccines that are to be administered intranasally [75]. Since influenza viruses are not central nervous system pathogens in humans, the wisdom of such requirements, which were designed to determine the safety of live attenuated vaccines for truly neurotropic viruses such as poliovirus, can be questioned. The neonatal rat was recently proposed as a model for the study of neurovirulence of intranasally administered influenza vaccines [76], and a few influenza strains have been evaluated in this model. Some viruses replicated in the brain following intranasal administration, but pronounced lesions or dramatic behavioral changes were not demonstrated in infected animals.

4.2 Immunogenicity

The vast majority of studies conducted in animals for influenza vaccine research are those that evaluate the immune response to candidate vaccines. Although it is clear that the immune responses to vaccines in animals are not often identical to and may not be directly predictive of those seen in humans, the first step in the proof-of-principle for a new vaccine is to establish its immunogenicity in animals before proceeding to clinical evaluation. The immune responses measured in the animal model should be relevant to the desired response in humans. Such studies may provide useful information regarding the regimen and routes of vaccination and can guide the design of clinical trials.

4.3   Strain-Specific Immunity Directed Against the HA

It is well established that the primary correlate of protection for inactivated whole-virus or subunit influenza vaccines administered parenterally is serum antibody directed against the HA protein. Most studies that are conducted to evaluate immune responses to influenza vaccines are done in mice and ferrets. The measurement of antibody responses in animal models is very straightforward, since HAI and neutralizing antibody assays do not require species-specific reagents. Limited studies have been conducted to evaluate the guinea pig as a model to study immunity to influenza virus. Phair and colleagues [39] demonstrated that infection of guinea pigs with unadapted human influenza viruses resulted in resistance to challenge with a homologous virus, and that passive transfer of hyperimmune serum to naive guinea pigs also conferred protection against infection. However, the levels of HAI antibody detected in the serum following infection were lower than those observed in ferrets or hamsters, and infected guinea pigs did not produce detectable levels of local antibody in their nasal secretions. In addition, high levels of nonspecific inhibitors of hemagglutination were present in guinea pig sera, making measurement of specific HAI antibodies problematic [39]. Phair et al. did, however, demonstrate that guinea pigs exhibited a delayed-type hypersensitivity response to influenza infection which resembled that seen in humans, although this response did not appear to be involved in resistance to infection.

Humoral immune responses to the HA of human influenza viruses and vaccines have been studied extensively in ferrets. Early studies determined that naive ferrets were not protected against influenza infection by vaccination with killed virus [77]. These observations were confirmed in later studies using formalin-inactivated vaccines [78]. However, killed vaccine, administered with an adjuvant to naive ferrets, provided partial protection against infection [79]. Thus, immune responses, in the ferret, to vaccination with inactivated virus vaccines against human influenza viruses do not appear to be identical to those seen in humans, since humans do not generally require an adjuvant to achieve protective levels of HAI antibodies. In contrast to the findings with inactivated influenza viruses, immunization with live influenza virus resulted in protection against subsequent challenge [78]. An explanation for this difference may be that in ferrets, influenza infection is primarily an upper respiratory tract infection, and adjuvant is required to elicit higher levels of serum antibody needed to restrict replication of virus in the upper respiratory tract. Several studies have demonstrated that higher levels of serum antibody are required to provide protection against respiratory viruses in the nose of animals than in the lungs [27–29].

4.4 Heterosubtypic Immunity

In recent years, particularly since the emergence of the highly pathogenic H5N1 viruses in Asia in 2003, and the challenges in developing H5N1 vaccines, there has been a resurgence of interest in heterosubtypic immunity – the ability of an immune response elicited by a particular influenza A virus to protect against an influenza A virus of a different subtype. Heterosubtypic immunity against influenza has been demonstrated in a number of studies in mice but the precise mechanism of this immunity is not clear [24, 80–82]. Previously, it was thought that this phenomenon was mediated by cellular immune responses, but recent studies suggest that antibodies are the primary mediators of heterosubtypic immunity [82] and that the diversity of the antibody repertoire is important [83].

Heterosubtypic immunity has also been observed in ferrets [84, 85], although there was some debate as to the length of time for which such immunity persists. McLaren and Potter [84] reported that it did not persist beyond 10 weeks after vaccination, but in another study, protection against infection with a heterosubtypic virus was observed 18 months following immunization [86]. In both cases, hetero-subtypic immunity did not prevent infection but it did limit virus replication following challenge.

The utility of the cotton rat model in addressing the issue of heterosubtypic immunity was explored [87]. The endpoints in this study were respiratory rate, virus replication in the lungs and nasal tissues, and pulmonary histopathology. A statistically significant reduction in respiratory rate was seen following challenge with A/Wuhan/359/95 (H3N2) in cotton rats that had been immunized with either the homologous virus or with a virus of a different subtype, A/PR/8/34 (H1N1), 4 weeks earlier, compared to non-immunized animals. This reduction in respiratory rate correlated with a statistically significant reduction in virus titers in the lungs and nasal tissues in immunized animals. Cotton rats that were immunized with the heterosubtypic A/PR/8/34 (H1N1) virus had the same extent of alveolitis, interstitial pneumonia and airway debris as non-immune, infected animals, and, like the cotton rats that were immunized with homologous virus, they had more severe early peribronchiolitis than was observed during primary infection. This peribronchiolitis could be indicative of a memory response in the heterosubtypic immune animals. However, the heterosubtypic-immune cotton rats had less bronchiolar epithelial damage than those animals immunized with homologous virus.

The role of heterosubtypic immunity through prior exposure or vaccination in humans, although inferred from retrospective analysis of data from influenza pandemics [88], is extremely complex and cannot be readily determined. Studies in young infants and children on the effect of pre-existing immunity on replication and immunogenicity of heterosubtypic attenuated influenza viruses suggested that heterosubtypic immunity in humans is weak [89].

4.5   Immune Responses to Other Influenza Proteins

An approach that is being explored in the development of novel vaccines for influenza is that of universal influenza vaccines that target the conserved proteins of the virus – NP, M1, and M2. A number of modalities, such as NP and M DNA vaccines [90–92], baculovirus-expressed recombinant M2 protein [93], M2 peptides [94] and recombinant M2 protein incorporated into hepatitis B core antigen [95–97], have been tested in mice, and prevent death but not illness following challenge with a heterologous virus. In the case of candidate universal vaccines for influenza, new animal models and assays that can measure antibody and cellular responses to viral antigens other than the HA and NA are needed . As the immune responses to these conserved antigens are not well characterized in humans, at present, it is not clear whether these responses are accurately reflected in animal models. Undoubtedly, more information will be obtained in this area in the future as candidate universal vaccines are evaluated in clinical trials.

Recently, there has also been interest in the role of immune responses to the NA component of seasonal vaccines in protection against related subtypes of influenza, including potential pandemic strains [98]. Antibodies to the NA protein can modulate the severity of influenza illness [99] but the NA content of inactivated influenza virus vaccines is not currently standardized.

4.6   Efficacy

Animal models are also used to evaluate the efficacy of new candidate influenza vaccines. The most commonly used animal models for such studies are mice and ferrets. In mice and ferrets, it has been established that antibodies against the HA can prevent infection or ameliorate disease following challenge with influenza virus. Reduction in virus titer in the lower respiratory tract following a challenge correlates with protection, so quantitative virology is the most relevant measure of vaccine efficacy for vaccines designed to generate antibody responses to HA. Additional endpoints such as morbidity, mortality and pathological findings may provide supporting evidence of protection from infection and disease. Although demonstration of vaccine efficacy in an animal model is not an absolute requirement in the pre-clinical evaluation of a candidate vaccine from a regulatory standpoint, it does provide evidence that immune responses to the vaccine are biologically relevant.

5   Vaccines for Pandemic Influenza

The direct transmission of HPAI H5N1, H7N7 and low pathogenicity AI (LPAI) H9N2 viruses from birds to humans, associated in many cases with severe morbidity and mortality, has raised concerns about the emergence of one of these viruses as a pandemic virus and has, therefore, prompted efforts to develop vaccines against AI viruses of pandemic potential. Evaluation and characterization of a suitable animal model for these other influenza virus subtypes is a critical step in the development of such vaccines.

6   Animal Models

In the following section we describe the features of the animal models that have been developed to study AI viruses, and their contributions to the evaluation of pandemic vaccines. In addition, the predictive value of the various animal models in the evaluation of safety and immunogenicity of several live attenuated pandemic influenza candidate vaccines, that have been evaluated in clinical trials, will be discussed.

6.1   Mice

Mice have been used in pre-clinical studies of inactivated and live attenuated pandemic influenza virus vaccines. Reports in the scientific literature that describe characterization of the replication, pathogenicity and the immune response of AI viruses in mice focus on viruses of the H5, H6, H7, and H9 subtypes.

6.1.1   H5N1 Viruses and Vaccines

Several studies demonstrated that H5N1 viruses that were isolated from human cases in Hong Kong in 1997 cause disease and death in mice without prior adaptation [100–102]. These viruses varied for their ability to cause disease and death in BALB/c mice and generally fell into two distinct groups – those that were highly virulent, and those with low virulence for mice – and one virus (A/HK/156/ 97) was of intermediate virulence in two of the studies [101, 102]; however, Gao et al. [100] found this isolate to be one of the most highly virulent in this model. The 50% lethal dose of H5N1 viruses that were highly virulent for mice was 10–1,000 times lower than that of low virulence strains, they replicated to titers that were up to 1,000 times higher in the lungs of mice early in the course of infection, and they replicated in extrapulmonary sites, including the brain. Viral antigen was observed by immunohistochemistry in the lungs of mice infected with A/HK/483/97 (H5N1), a highly virulent strain, and A/HK/486/97 (H5N1), a less virulent strain, and was associated with necrotic bronchi. Viral antigen was also observed, in both glial cells and neurons, in the brain of mice infected with the highly virulent influenza A/HK/483/97 (H5N1) virus, a finding also reported by Gao et al. [100]. In addition, Gao et al. reported the presence of viral antigen in cardiac myofibers of mice infected with the highly virulent influenza A/HK/483/97 (H5N1) virus. The ability of the H5N1 viruses to replicate and cause disease and death in mice did not correlate with their ability to kill chickens [102], and the relevance of replication of these viruses in extrapulmonary sites in mice to the disease in humans is not clear, although a general correlation between the level of virulence in mice and the severity and outcome of disease in humans was observed with 11 of 15 viruses evaluated [101]. Dybing and colleagues [103] reported that infection of mice with highly pathogenic H5 AI viruses that were isolated from Scotland [influenza A/ck/Scotland/59 (H5N1)], Italy [influenza A/ck/Italy/1485-330/97 (H5N2)], Queretaro [influenza A/ck/Queretaro/7653-20/95 (H5N1)] and England [influenza A/tk/England/91 (H5N1)], caused little or no disease in BALB/c mice. HPAI H5N1 influenza viruses isolated from humans in Asia in 2004 caused weight loss, ruffled fur, listlessness and pronounced leukopenia, and were lethal in mice without prior adaptation, and replicated outside the respiratory tract [104]. In the same study, HPAI H5N1 viruses isolated from birds, and a single human isolate, were less virulent for mice.

Lu et al. [102] used the BALB/c mouse model to evaluate the immunogenicity and efficacy of a vaccine against H5N1 influenza, based on an antigenically related non-pathogenic AI virus, A/duck/Singapore-Q/F119-3/97 (H5N3). They found that two doses of inactivated vaccine were required to elicit HAI antibody responses of a magnitude that would be protective in human influenza in the majority of vaccinated animals, and that the addition of an alum adjuvant resulted in higher levels of HAI antibody and a greater seroconversion rate. These findings generally agreed with the observations made in humans when a similar vaccine was tested in clinical studies: two doses of vaccine were necessary to achieve acceptable levels of antibody, and the addition of adjuvant, in this case MF59 (instead of alum used in the studies in mice), increased the magnitude of the antibody response as well as the seroconversion rate [105–107]. The efficacy of this vaccine in mice was determined by measuring the level of virus replication in the lungs and protection against lethal challenge with an H5N1 isolate that was highly virulent for mice.

The efficacy of several different H5N1 virus vaccines has been evaluated in mice and in all cases, the vaccines were found to be immunogenic and protective in mice (reviewed in [108]). When tested in Phase I studies in humans, inactivated H5N1 virus vaccines were found to be suboptimally immunogenic, requiring high doses [109, 110] to elicit neutralizing and HAI antibody responses. The administration of whole virion vaccines and inactivated virus vaccines with adjuvant increased the immunogenicity in mice and in humans [109, 111]. It is unclear whether data obtained in mice with pandemic influenza vaccines are predictive of vaccine immunogenicity in humans since pre-clinical data for the specific vaccine formulations that have been tested in humans to date have not been reported.

Cold-adapted live attenuated vaccine candidates against H5N1 AI viruses have been evaluated in pre-clinical studies in mice [66]. H5N1 vaccine candidates, bearing the modified HA and the NA from various HPAI H5N1 human isolates and the six internal protein genes from the A/Ann Arbor/6/60 cold-adapted (*ca*) donor virus, were restricted in replication in the lungs of mice compared to the corresponding wild-type virus, were found be immunogenic and conferred protection against challenge with homologous and heterologous wild-type viruses, although two doses of the vaccine virus were required to fully protect mice against replication of homologous wild-type viruses in the lungs [66]. Another live attenuated cold-adapted candidate H5N1 vaccine, a 7:1 reassortant virus which derived the HA from the low pathogenicity virus A/duck/Potsdam/86 (H5N2) and the remaining genes from the A/Leningrad/17/57 (H2N2) cold-adapted virus that is the donor virus for the seasonal live attenuated vaccine used in Russia, was evaluated in mice [112–114]. Only a single dose of this live vaccine virus was evaluated, but similar findings were reported: it was restricted in replication in the respiratory tract of mice and was immunogenic. The ability of the live attenuated H5N2 virus to elicit local IgA antibody responses in nasal washes has been demonstrated in mice. Modest levels of neutralizing antibodies were detected 6 weeks after a single dose of the H5N2 live attenuated vaccine, and the vaccine conferred protection against lethal challenge with a wild-type HPAI H5N1 virus. The predictive value of the mouse model for the evaluation of the safety and immunogenicity of these vaccines is discussed below in the section entitled "Clinical Evaluation of Live Attenuated Candidate Vaccines for Pandemic Influenza".

6.1.2 H7 Viruses and Vaccines

Representative low pathogenicity and highly pathogenic H7 AI viruses from both the Eurasian and North American lineages replicated in mice without prior adaptation [115, 116]. Highly pathogenic H7 viruses demonstrated extrapulmonary spread to the spleen and brain, as has been observed with HPAI H5N1 isolates, although H7 viruses were detected in the brain earlier during infection (day 1 p.i. for H7 and day 4 for H5) [116]. de Wit et al. [117] reported that intranasal infection of mice with the non-adapted HPAI A/Netherlands/219/2003 H7N7 virus, that was isolated from a fatal human case, resulted in severe illness, as indicated by weight loss, lethargy, ruffled fur, and lethality. The rate of loss in body weight was similar over a range of doses of virus between 3×10^3 and 3×10^6 EID_{50}. The virus was detected in the spleen, liver, kidneys and brain, as well as in the lungs of mice. This model was used for the evaluation of the immunogenicity and efficacy of candidate H7 influenza vaccines [117]. A single dose of an ISCOM vaccine and two doses of a subunit vaccine failed to protect mice against lethal infection with the A/NL/219/2003 (H7N7) virus, with one exception. Mice vaccinated with two doses of 1 µg or 5 µg ISCOM vaccine exhibited a small temporary loss in body weight but otherwise appeared healthy after challenge. Vaccination with two doses of the ISCOM vaccine resulted in at least a 1,000-fold reduction in virus replication in the lungs, and near-complete reduction of extrapulmonary replication of the challenge virus. However, in all vaccinated mice, virus was still present in the lungs at high titers.

Munster et al. [118] reported that the human HPAI H7N7 viruses A/NL/219/ 2003 and A/NL/33/2003 both caused lethal infection in mice when administered intranasally at a high dose (dose not specified). At a dose of 5×10^2 TCID$_{50}$, influenza A/NL/219/2003 virus, which was isolated from a fatal human case, resulted in loss of body weight, ruffled fur, lethargy, and respiratory problems from day 2 p.i. and infected mice were euthanized on day 5 p.i., whereas in mice that were infected intranasally with 5×10^2 TCID$_{50}$ of influenza A/NL/33/2003 virus, isolated from a human with conjunctivitis in the same outbreak, no signs of illness or loss in body weight were observed up to day 7 p.i. The influenza A/NL/ 219/2003 virus replicated to a titer that was more than 1,000-fold higher compared to the titer in the lungs of mice infected with the influenza A/NL/33/2003 virus, and it was isolated from the brain, spleen, liver, and kidney of all infected animals. Influenza A/NL/33/2003 virus was isolated from the brain of only one out of three mice, and was not detected in the other organs examined. Histopathological findings in all mice infected with influenza A/NL/219/2003 virus included necrosis and inflammation throughout the respiratory tract that was pronounced in the trachea and became progressively milder in the bronchi, bronchioles, and alveoli. In contrast, lesions in the respiratory tract were only observed in one out of four mice infected with the influenza A/NL/33/2003 virus, and were characterized as mild to moderate cell necrosis, with neutrophil infiltrates in the trachea, bronchi, and bronchioles. Lesions were not observed upon histopathological examination of the brain, heart, spleen, liver or kidneys of mice infected with either virus. Viral antigen expression was limited to the tissues of the respiratory tract in mice infected with either virus, but was more abundant in mice infected with the influenza A/NL/ 219/2003 virus. Rigoni and colleagues [119] reported that HPAI H7N1 viruses isolated from chickens and ostriches could infect and replicate in mice without adaptation, and were associated with disease signs of varying severity. Bronchitis, tracheitis, alveolitis, and brain lesions were observed in mice infected with three HPAI H7N1 influenza viruses. However, the influenza A/ostrich/2332/00 virus caused more severe lesions and spread more rapidly in the lungs and brain than the other two viruses (influenza A/ostrich/984/00 and influenza A/ck/5093/99) [119].

Low pathogenicity H7 viruses replicated to high titers in the upper and lower respiratory tract of mice, but were not lethal, even at high doses. Immunogenicity of these viruses was also evaluated in mice [116].

Several reassortant viruses, bearing the HA or NA genes from H7 avian influenza viruses and the internal protein genes from A/PR/8/34 (H1N1) [PR8], have been described. Jadhao et al. [120] and Pappas et al. [121] reported the evaluation of the egg-based PR8 reassortant H7 influenza virus vaccines in mice. An H7N7-PR8 reassortant was generated which derived its HA from the low pathogenicity A/ mallard/Netherlands/12/2000 (H7N3) virus and its NA from the low pathogenicity A/mallard/Netherlands/2/2000 (H10N7) virus. An H7N2-PR8 vaccine was generated with the HA and NA from the low pathogenicity A/turkey/Virginia/4529/02

(H7N2) virus. Mice immunized with two doses of the formalin-inactivated H7N7-PR8 or H7N2-PR8 vaccines, with or without alum, mounted a serum HAI antibody response that increased after the second vaccination. Antibody responses were generally higher when the vaccine was administered with an adjuvant. Mice that received two doses of the vaccine were protected from lethal challenge with highly pathogenic H7 influenza viruses. Evaluation of these vaccines in clinical trials is planned.

The immunogenicity and efficacy of a cell-based H7N1 avian influenza split virion vaccine, derived from the HPAI A/chicken/Italy/13474/99 (H7N1) virus, have been studied in mice [122]. Low titers of HAI antibodies were detected in the sera after two doses of 12 or 20 μg of HA. Titers were generally higher if vaccine was administered with an adjuvant. Vaccinated mice shed significantly less virus than unvaccinated animals following intranasal challenge with the HPAI A/ck/Italy/13474/99 (H7N1) virus and were protected from both disease and weight loss. Vaccination also conferred significant protection against lethal challenge. The same vaccine was tested in a Phase I clinical trial in sixty healthy adults [123]. Two doses were administered, with or without adjuvant. Serum HAI and neutralizing antibody titers, after two doses, were low, but were higher in the individuals who received an adjuvant (21 vs. 50% for the 12 μg dose and 23 vs. 62% for the 24 μg dose). Antibody secreting cells were also detected in those individuals with detectable HAI or neutralizing antibody titers, which were associated with IL-2 production.

The mouse model has also been employed for the evaluation of the attenuation, immunogenicity and protective efficacy of a candidate cold-adapted, live attenuated, influenza vaccine of the H7N3 subtype [124]. The HA and NA genes of this vaccine virus were derived from the low pathogenicity AI virus A/chicken/British Columbia/CN-6/2004 (H7N3), and its six internal protein genes were from the A/Ann Arbor/6/60 *ca* virus that is the backbone of FluMist®. In mice, the vaccine virus did not cause weight loss, and was restricted in replication in the lower respiratory tract compared to the low pathogenicity wild-type parent virus and an antigenically related HPAI H7N3 wild-type virus, and it appeared to have delayed replication kinetics in the upper respiratory tract compared to the wild-type parent virus. A single dose of the H7N3 *ca* vaccine virus was immunogenic in mice and provided complete protection against lethality and pulmonary replication following challenge with H7 influenza viruses of the North American lineage. Two doses of vaccine were required to confer protection against H7 influenza viruses of the Eurasian lineage.

6.1.3 H9 Viruses and Vaccines

Human infections with H9N2 AI viruses were first reported in 1999 [125, 126] and, although the illness in the infected individuals was relatively mild, there is still concern over the pandemic potential of H9 viruses because viruses of this subtype are highly prevalent in birds [127–131]. The pathogenicity of human and avian H9 influenza viruses in mice has been studied by several laboratories, with a view of establishing an animal model that can be used to study strategies for prevention of pandemic influenza, including vaccines and antiviral drugs. Some H9 influenza viruses replicate in the respiratory tract of mice without prior adaptation [128, 132–134], but serial passage of the A/quail/Hong Kong/G1/97 (H9N2) virus in mice resulted in an increase in the virulence and in the extrapulmonary spread and lethality of this virus in intranasally infected mice [132, 133]. Data from different laboratories that have used the same H9N2 virus to infect mice are not consistent. Some of the factors that can influence the outcome of infection are anesthesia, dose, volume and route of virus administration, and passage history. It is difficult to compare studies when complete information is not provided. For example, in studies reported by Lu et al. [134], the human influenza A/Hong Kong/1073/99 (H9N2) virus replicated efficiently in the lungs of mice but failed to cause death or signs of disease, significant weight loss or to spread to extrapulmonary sites. However, Leneva et al. [132] reported that infection of mice with this virus resulted in 40% mortality and significant weight loss in the surviving mice. In these discordant studies, mice were anesthetized with CO_2 [134] or with metofane [132], were infected by the same route using virus that had been propagated in embryonated eggs, at approximately the same dose (10^6 EID_{50}), but inoculum volumes used were not stated in either study, so it is not clear why this virus was lethal in one study and not in the other. Similarly, a lethal challenge of mice with the human influenza A/Hong Kong/1073/99 (H9N2) virus was reported as part of a study to determine the efficacy of an M2 liposome vaccine [135], although this virus did not cause disease or lethality in the hands of other investigators [134, 136, 137] . All laboratories delivered virus intranasally to anesthetized mice. However, in the study reported by Ernst et al. [135], mice were anesthetized intraperitoneally with ketamine/xylazine, whereas in the other two studies, inhalational anesthesia was used, which may have resulted in a lighter state of anesthesia.

The mouse model has been used to evaluate the level of attenuation and the protective efficacy of a candidate cold-adapted, live attenuated, H9N2 vaccine bearing the HA and NA from the influenza A/ckHK/G9/97 (H9N2) virus and the internal protein genes from the influenza A/Ann Arbor/6/60 cold-adapted virus [136]. The H9N2 live attenuated vaccine was restricted in replication and protected mice from challenge with homologous and heterologous wild-type H9N2 influenza viruses.

6.1.4 1918 H1N1 Pandemic Virus

Like the highly pathogenic H5N1 AI viruses, the fully reconstructed recombinant 1918 H1N1 pandemic influenza virus was highly lethal in mice without prior adaptation [138]. The mean time to death in mice infected intranasally was 4.5 days. However, in contrast to the highly pathogenic H5N1 influenza viruses, this virus was not detected in extrapulmonary tissues. Histopathological findings included necrotizing bronchitis and bronchiolitis, moderate to severe alveolitis and severe peribronchial and alveolar edema.

The mouse model appears to be potentially useful for the evaluation of pandemic influenza vaccines. Most AI viruses studied in mice, to date, can replicate without adaptation, although the outcome of infection with some AI viruses is clearly different, depending not only on the particular virus being studied but also on the laboratory in which the studies were conducted. It is important that AI viruses continue to be evaluated in mice, using standardized inoculation procedures and doses and with the measurement of the same endpoints so that the utility of this model can be maximized for the evaluation of pandemic influenza vaccines.

6.1.5   H6 Viruses and Vaccines

Although most of the pandemic influenza vaccine development efforts have focused on the subtypes of AI that have caused infections in humans, namely H5N1, H7, and H9 viruses, in theory, AI viruses of all subtypes have the potential to cause pandemics and therefore it is prudent to develop animal models to study the pathogenicity of these viruses and to evaluate experimental vaccines that may be needed in the future. There is concern regarding the pandemic potential of H6 AI viruses, since these viruses are highly prevalent in many avian species around the world [139–143]; they have a high propensity to reassort, and an H6N1 virus, A/teal/W312/Hong Kong/97, has been implicated as the donor of the internal protein genes of the H5N1 AI viruses that emerged in 1997 [140, 144]. In addition, there is serological evidence of human infections with H6 AI viruses in China [143].

The replication, pathogenicity and immunogenicity of several H6 AI viruses have been studied in mice [145]. Fourteen temporally and antigenically diverse H6 AI viruses of various NA subtypes, from both the Eurasian and North American lineages, were evaluated in BALB/c mice. Following intranasal inoculation of 10^5 $TCID_{50}$ of virus, replication of varying efficiency was observed in the respiratory tract of mice. Eleven of the 14 viruses replicated in the lower respiratory tract, ten in the upper respiratory tract; only one of the viruses failed to replicate to detectable levels in mice. Higher titers of the viruses were observed in the lungs of mice compared to the nasal turbinates. Two viruses from Hong Kong, A/teal/W312/HK/97 (H6N1) and A/quail/HK/1721-30/99 (H6N1) caused significant weight loss, illness, and death in mice, but their replication appeared to be limited to the respiratory tract. H6 AI viruses that replicated well in the lungs elicited high neutralizing antibody titers in infected mice, but the immunogenicity of H6 viruses did not correlate with their efficiency of replication in the respiratory tract. The cross-reactivity of the neutralizing antibodies was not an accurate predictor of protection. Live attenuated, cold-adapted candidate vaccines were generated from three of the H6 AI viruses studied [146]. Immunogenicity and efficacy of the candidate vaccines were evaluated in mice. A single intranasal dose of each vaccine virus elicited serum neutralizing and HAI antibody, and fully protected mice against replication of the wild-type parent H6 AI virus in the lower respiratory tract. Cross-reactive antibody titers against heterologous H6 viruses were significantly lower than against the homologous parent virus. A second dose of vaccine in mice boosted the antibody titers, and improved cross-protection against the heterologous H6 AI viruses. As had been seen in the initial studies in mice, the level of neutralizing antibody elicited by the H6 candidate vaccines was a poor predictor of their ability to cross-protect against antigenically distinct H6 AI viruses. A candidate A/teal/HK/97 (H6N1) cold-adapted vaccine elicited the broadest cross-protective response, and this vaccine virus is currently undergoing evaluation in human clinical trials.

6.2   Ferrets

6.2.1   H5N1 Viruses and Vaccines

The ability of a limited number of AI subtypes to replicate and cause disease in ferrets has been investigated, and not surprisingly, the behavior of H5 subtype viruses has been the most studied. Zitzow and colleagues [147] demonstrated that two H5N1 influenza viruses isolated from human cases of infection in Hong Kong in 1997 were capable of replication not only in the respiratory tract, but also in the brain, spleen and intestines of ferrets. Virus replication was associated with clinical signs of disease such as severe lethargy, sneezing, rhinitis, hind limb paresis and, in some cases, diarrhea, and some H5N1 viruses were lethal to ferrets. However, the hierarchy in the severity of disease seen with the different H5N1 1997 isolates upon infection of mice, was not observed in ferrets: influenza A/HK/483/97 and A/HK/486/97 were equally virulent after intranasal infection of ferrets, whereas the A/HK/483/97 virus was more virulent in mice than the A/HK/486/97 virus was in several studies [100–102, 148]. As with mice, the significance, with respect to humans, of disease signs and the extrapulmonary replication of H5N1 viruses in ferrets is not clear, particularly since, in the same study, Zitzow et al. reported the isolation of a human H3N2 influenza virus from the brain of ferrets following intranasal infection. Similar studies have been conducted using human and avian H5N1 viruses isolated in 2004–2005 [104, 149]. Govorkova et al. [149] evaluated four human H5N1 influenza isolates and nine avian H5N1 isolates from Asia from 2004. A wide spectrum of infectivity, severity of disease and lethality was observed in ferrets inoculated with these viruses. The H5N1 viruses isolated from humans and two of the avian isolates caused severe disease in ferrets with some lethality. However, it is difficult to draw general conclusions regarding the behavior of these viruses in this model because of the small numbers of animals used (only two animals per group for all but one of the viruses tested), and the variability in infectivity of the viruses examined. For example, although the influenza A/Vietnam/3046/2004 virus caused severe disease in two out of two the inoculated ferrets, it was lethal in only one animal, and virus was only recovered from the nasal washes. In contrast, the influenza A/Vietnam/3062/2004 virus, which was also lethal in one out of two ferrets inoculated, was recovered from the lungs, brain, spleen, and intestine of these animals. Similarly, Maines et al. [104] evaluated H5N1 isolates from Asia from 2004 using the ferret model. Although the viruses used in this study were different from those used by Govorkova et al. (with the exception of A/Vietnam/ 1203/2004), similar findings were reported: the human isolates caused severe disease, with some lethality, in ferrets. Again, small numbers of animals were used (three per group for most of the isolates tested) and some variability in infectivity and severity of disease was observed. In the study conducted by Zitzow et al., gross pathological changes observed in ferrets infected with highly virulent HPAI H5N1 viruses included focal areas of redness in the lungs, consolidation of the lungs and rare discoloration of the liver, petechiae on the liver and lesions on the intestines and kidneys [147]. Maines et al. [104] reported the presence of hemorrhage in the adipose tissue surrounding the liver, kidney and bladder in two-thirds of infected ferrets. Histopathological findings in the lungs of infected ferrets included acute bronchiolitis, bronchopneumonia, interstitial pneumonia with suppurative exudates in the bronchi, bronchioles and adjacent alveolar spaces, prominent epithelial necrosis and marked intraalveolar edema by day 3 p.i., and bronchitis, bronchiolitis and pneumonia observed on days 6–7 p.i. [104, 147, 149]. Inflammatory changes were also evident in the brain of ferrets infected with highly virulent HPAI H5N1 viruses at days 5–6 p.i., including in the glial nodules with perivascular infiltration of lymphocytes and polymorphonuclear leukocytes in the brain parenchyma, neuronophagia and lymphocytic infiltrates in the choroid plexus [147, 149]. Viral antigen was observed by immunohistochemistry in neurons in the same areas of the brain as the inflammation [104]. Govorkova et al. [149] reported histopathological changes in the liver, including diffuse vacuolization of the hepatocellular cytoplasm, mononuclear infiltrates, periportal hemorrhage, and hepatocellular necrosis. Generally, the viruses isolated from avian species caused less severe disease than those isolated from humans.

The number of ferrets inoculated with each virus was small and ferrets are an outbred species, so the significance of the variability in data such as virus replication and clinical illness are difficult to interpret. Until the scientific community has more experience with the behavior of AI viruses in animal models, it would be prudent to compare new isolates with well-characterized strains and to study these pathogens in more than one model.

The ferret model has also been used to evaluate the efficacy of several experimental inactivated [150, 151] and live attenuated [66, 112, 114] vaccines against H5N1 influenza. Inactivated H5N1 vaccines were immunogenic and protective in the ferret model [150, 151]. However, inactivated H5N1 vaccines that were tested in clinical trials were suboptimally immunogenic [109, 110]. The attenuation of cold-adapted live attenuated H5N1 vaccines has been demonstrated in ferrets. These vaccine candidates were also immunogenic and protective against challenge with homologous and heterologous H5N1 wild-type viruses in ferrets [66]. Protection from lethal H5N1 infection and the level of replication of the challenge virus in the lungs and other tissues are the endpoints used for evaluation of efficacy in this model. Van Riel et al. [152] demonstrated that the pattern of attachment of H5N1 influenza human isolates in the respiratory tract of ferrets was similar to that seen in the human respiratory tract; the virus attached predominantly to type II pneumocytes, alveolar macrophages and nonciliated cuboidal epithelial cells of the terminal bronchioles in the lower respiratory tract and became progressively rarer more proximally, i.e., towards the trachea. This pattern of H5N1 virus attachment, predominantly in the lower respiratory tract, is thought to be related to the distribution of α-2,3 sialic acid receptors [153]. However, other investigators found that H5N1 influenza viruses were able to infect ex vivo cultures of the human upper respiratory tract, i.e., nasopharyngeal, adenoid and tonsillar tissues, despite the lack of α-2,3 sialic acid receptors in these tissues [154]. The tropism of H5N1 influenza viruses in the respiratory tract of humans and other species remains equivocal and further studies, in which a number of different isolates are evaluated in larger numbers of animals, are needed.

6.2.2 H7 Viruses and Vaccines

The behavior of AI viruses of the H7 subtype has been studied in ferrets. Human isolates of highly pathogenic H7N7 influenza viruses replicated to higher titers in the upper and lower respiratory tract of ferrets than low pathogenicity H7N2 influenza viruses isolated from humans. The H7N7 viruses also replicated in nonrespiratory tissues [115]. The H7N7 isolate A/NL/219/2003 caused severe illness, including significant weight loss, caused neurological symptoms and was lethal in 2 out of 3 ferrets inoculated. Another highly pathogenic H7N7 AI virus, A/NL/230/2003, and the low pathogenicity H7N2 viruses evaluated in this study, did not cause severe disease and were not lethal in this model. Joseph et al. [116] demonstrated that the pattern of antigenic relatedness of H7 subtype AI viruses, determined using post-infection ferret sera, was similar to that observed in mice. The ferret model was used to evaluate attenuation, immunogenicity and efficacy of the H7N3 *ca* live attenuated vaccine virus [124]. The vaccine virus was restricted in replication in the upper respiratory tract of ferrets and did not replicate to detectable levels in the lungs or in the brain. Neutralizing antibodies were detected in the sera of ferrets immunized with a single dose of the H7N3 *ca* vaccine 4 weeks after immunization, and a second dose of vaccine provided a boost in the antibody response. Two doses of vaccine significantly reduced the replication of homologous and heterologous highly pathogenic H7 influenza viruses in the lungs of ferrets and prevented their spread to the brain and the olfactory bulb.

Ferrets immunized with an inactivated vaccine derived from an H7N1-PR8 reassortant based on HPAI A/chicken/Italy/13474/99 (H7N1), with alum adjuvant, mounted a serum HAI (GMT 76) and neutralizing antibody (range 42–200) response after two 24 μg doses of vaccine [122]. Cross-reactive HAI titers against heterologous Eurasian and North American H7 viruses were detectable but low (titer 8–160). Vaccination of ferrets resulted in reduced signs of illness, shedding of virus from the upper and lower respiratory tract and systemic spread following challenge with HPAI A/chicken/Italy/13474/99 (H7N1).

6.2.3 Other AI Subtypes

There are few reports in the scientific literature that describe the replication and clinical signs resulting from infection of ferrets with other AI subtypes. Hinshaw et al. [155] demonstrated that AI viruses of the H2, H3, H6, H7, and H10 subtypes, as well as an H7N7 virus isolated from a seal, replicated in the upper respiratory tract of ferrets, but elicited low or undetectable levels of antibody. None of these AI isolates tested caused any signs of disease in infected ferrets. Replication, pathogenesis, and immunogenicity of AI viruses of the H6 subtype were evaluated in the ferret model. Following evaluation in the mouse model of infection, four AI viruses of the H6 subtype that replicated to varying degrees in mice were studied in ferrets [145]. As in mice, the viruses replicated to lower titers in the upper respiratory tract than in the lungs, although the difference in titers was much less than in mice (~10-fold lower titers in ferrets vs. 10–1,000-fold difference seen in mice). All four viruses replicated to a peak titer of about 10^7 $TCID_{50}/g$ in ferret lungs, although the peak titer occurred at different timepoints post-infection. Transient weight loss and fever were observed in ferrets infected with the A/teal/HK/97 and A/quail/HK/99 viruses that were lethal in mice, but also in ferrets that received the influenza A/mallard/Alberta/85 (H6N2) virus, which caused no signs of illness in mice. Ferrets infected with influenza A/duck/HK/77 (H6N9) did not exhibit weight loss or fever, but, unlike mock-infected ferrets, they failed to gain weight during the period of observation. Antibody responses elicited by an infection in ferrets generally correlated with those seen in mice, but, as in the mouse model, the antibody responses did not correlate with virus replication. In the ferret model, live attenuated, cold-adapted H6 AI candidate vaccine viruses were attenuated compared to the corresponding wild-type H6 virus. None of the vaccine viruses caused signs of illness in ferrets, nor did they replicate in the lungs. A single intranasal dose of the vaccine viruses elicited serum neutralizing and HAI antibodies in ferrets, and, as in mice, conferred complete protection in the lower respiratory tract following wild-type virus challenge. The levels of neutralizing antibody induced in ferrets by these vaccine viruses did not accurately predict the outcome of challenge with heterologous H6 viruses. The H6 AI viruses generally behaved in a similar fashion in ferrets and in mice, but species-specific differences in the cross-reactive antibody responses were observed.

6.2.4 1918 H1N1 Pandemic Virus

The reconstructed 1918 H1N1 influenza virus replicated to high titers in the upper respiratory tract of ferrets following intranasal inoculation [156]. All inoculated ferrets exhibited severe signs of disease that included lethargy, anorexia, sneezing, rhinorrhea, severe weight loss and high fever from day 2 p.i., and two out of three animals succumbed to infection by day 11. Unlike the highly pathogenic H5N1 viruses in ferrets, viral replication was not detected in tissues outside the respiratory tract. Necrotizing bronchiolitis, moderate to severe alveolitis and edema were observed in the lungs of infected ferrets on day 3 p.i. The presence of viral antigen in the upper and lower portions of the bronchi, bronchial and bronchiolar epithelium and in the hyperplasic epithelium within the alveoli was observed.

6.3 Cats

There are few reports in the literature on influenza infection in cats. In studies conducted by Paniker and Nair in the 1970s [157, 158], intranasal infection of anesthetized cats with influenza A/Hong/Kong/1968 (H3N2) virus freshly isolated from human cases or laboratory- and egg-adapted isolates did not result in clinical signs of influenza but virus was recovered from pharyngeal secretions, and infection induced HAI antibodies and was transmitted to contact animals. Infected cats did not display clinical signs of influenza. Hinshaw and colleagues [155] later demonstrated that intranasally administered H7N7 and H7N3 AI viruses replicated in the upper respiratory tract of cats without clinical signs of disease, and the cats developed HAI antibodies after infection.

6.3.1 H5N1 AI Viruses

There was little interest in influenza infection and immunity in cats until the recent re-emergence of highly pathogenic avian H5N1 viruses in Asia, when it was reported that a number of big cats, namely tigers and leopards in the zoos in Thailand, became infected with HPAI H5N1 viruses, apparently after they were fed infected chicken carcasses [9]. Infection in many of these felids was fatal, and later, anecdotal reports of H5N1 infection in domestic cats in areas where there were outbreaks of H5N1 infection in avian populations contributed to a surge in interest in H5N1 influenza in cats. The pattern of attachment of a human H5N1 influenza virus to respiratory tract tissues of a cat was similar to that seen with human tissue [152].

Experimental infection of European short haired cats with an H5N1 virus isolated from a human in Vietnam in 2004 resulted in clinical disease, virus replication in respiratory and extra-pulmonary tissues, and pathological changes consistent with H5N1 infections in humans [10, 159]. Clinical signs, including significant elevation in body temperature, decreased activity, conjunctivitis and labored breathing were seen in cats experimentally infected intratracheally or by feeding on infected chicks [10]. Similar disease symptoms were observed in sentinel cats that became infected from being housed with cats that had been infected intratracheally. Illness in contact cats became apparent about 3 days later than in the cats infected via the intratracheal route. Peak viral titers in the throat swabs of the intratracheally infected cats were $\sim 10^{4.5}$ TCID$_{50}$/ml, whereas the peak titers observed in nasal swabs ranged from $10^{2.5}$ to $10^{5.0}$ TCID$_{50}$/ml [159]. The virus was also recovered from rectal swabs of cats infected by feeding on infected chicks, but the titers of virus in these samples varied widely. In addition, cats infected through feeding had lesions in the intestines. In animals infected intratracheally or by feeding, the virus was also recovered from extra-pulmonary tissues, most often from the brain, liver, kidney and heart. Infected sentinel cats did not have detectable virus in tissues outside the respiratory tract; however, pathological changes were observed in the adrenal glands in one of the two sentinel cats infected in this manner. These studies demonstrated that HPAI H5N1 viruses are capable of extrapulmonary spread in cats, and can cause severe disease and even death in animals infected intratracheally or by feeding on infected bird carcasses. These observations also raise the possibility that the gastrointestinal tract may serve as a source for HPAI infection in cats.

Karaca et al. [160] studied the immunogenicity of a fowlpox-based H5 vaccine in cats. HAI antibodies were detected in serum of cats following a single subcutaneous dose of the vaccine, and a significant boost in antibody titers was observed following a second vaccination.

It remains to be seen if cats will be used extensively in the evaluation of vaccines against pandemic influenza.

6.4 Hamsters

6.4.1 H9 Viruses and Vaccines

Saito and colleagues conducted a study to evaluate the replication and pathogenicity of influenza viruses of various subtypes in Syrian hamsters [161]. The influenza A/HK/1073/99 (H9N2) virus replicated to high titers in the lungs, but was not lethal to hamsters and was not detected in the brain. The HPAI H5N1 influenza A/HK/483/97 virus, that was highly virulent in mice, was also lethal in hamsters, with all animals succumbing to infection by day 6 p.i., and, as in mice, virus was recovered from the brain of the infected hamsters. Avian H9N2 and H9N5 isolates could replicate in the lungs of hamsters, but did so to lower titers compared to human isolates. The human H9N2 virus elicited low levels of neutralizing antibody in infected hamsters, whereas the avian H9N2 isolate did not elicit detectable neutralizing antibody. The behavior of this limited number of AI isolates in the Syrian hamster model suggests that the effects of this viruses may be similar to that observed in mice, and further evaluation of this model for evaluating the efficacy of pandemic influenza vaccines is warranted.

6.4.2 Non-Human Primates

There is renewed interest in the use of non-human primates for immunogenicity studies of pandemic vaccines; this is based on the presumption that immune responses in these animals, which have a closer evolutionary relationship to humans, may be more predictive of the responses in humans than in smaller animals like mice and ferrets. To date, few data are available on the serological responses of non-human primates to AI virus vaccines.

6.4.3 H5N1 AI Viruses

The use of cynomolgus macaques as a model for influenza virus infection in humans was revisited following the emergence of the highly pathogenic H5N1 AI viruses in 1997 [162]. The initial human H5N1 influenza isolate, A/Hong Kong/156/1997, isolated from a fatal case of influenza in a child [6], was inoculated at multiple sites, including the trachea, tonsils and conjunctiva. Three of four animals developed fever within 2 days, and one showed signs of anorexia and acute respiratory distress. High titers of virus were recovered from lungs on day 4 p.i., and the virus was also isolated from the trachea, tracheobronchial lymph nodes and the heart. The virus was not recovered from these tissues on day 7 p.i. The virus was also recovered from bronchioalveolar lavage from 2 out of 2 animals on days 3 and 5 p.i.; from pharyngeal swabs of two animals on day 5 p.i., and from nasal swabs of one animal on days 3 and 7. Viral RNA was detected by RT-PCR in the brains of two animals on day 4 p.i., and in the spleen of all four animals tested on day 7 p.i. Pathological changes in the lungs of infected animals included pulmonary consolidation, necrotizing broncho-interstitial pneumonia and flooding of alveoli with edema fluid, fibrin, erythrocytes, cell debris, macrophages, and neutrophils, and inflammatory changes were seen in multiple organs [163].

Infection of Rhesus macaques with avian H5N1 isolates, reported by Chen et al. [164], indicated that results of intranasal inoculation varied depending on the influenza virus isolate used. Clinical signs of infection, including elevation in body temperature, anorexia and increased respiratory rate were observed in macaques inoculated with the following H5N1 viruses: A/bar-headed goose/Qinghai/1/2005, A/great cormorant/Qinghai/3/2005 and A/duck/Guangxi/35/2001. Pathological changes were seen in the lungs of all of the infected animals, but were more pronounced in the monkeys inoculated with the duck isolate. However, the only virus to be re-isolated from infected animals was A/duck/Guangxi/35/2001, and this virus was isolated from respiratory tract secretions and tissues and also from the spleen, liver and the heart.

The Rhesus macaque model has been used to evaluate the immunogenicity and efficacy of a candidate live attenuated cold-adapted H5N1 vaccine [113]. The vaccine virus derives its HA and NA from the clade 2.3 A/Anhui/2/2005 (H5N1) virus and the six internal protein genes from the cold-adapted A/AA/6/60 ca virus. The multibasic cleavage site in the HA gene was removed. Animals were inoculated intranasally with 10^7 EID_{50} of the vaccine virus on days 0 and 28. Serum antibodies in the vaccinated macaques were detected by ELISA 2 weeks following the first dose, with an apparent boost after the second dose. Four weeks after the first dose of vaccine all animals had detectable levels of neutralizing antibodies in the serum. After the second dose of vaccine, HAI and neutralizing antibodies were detected in the sera of all the vaccinated animals. HAI and neutralizing titers against a heterologous H5N1 virus were two to fourfold lower than against the homologous virus. T cell responses, measured by IFN-γ ELISPOT, were detected following the second dose of the vaccine. Three weeks after the second vaccination, the macaques were challenged intratracheally with 10^6 EID_{50} of either the parent A/Anhui/2/2005

(H5N1) virus, or the A/bar headed goose/Qinghai/1/2005 (H5N1) virus. Control animals had symptoms of illness including anorexia, fever and loss of appetite from day 1 post-challenge. Four control animals were euthanized on day 3 post-challenge and the remaining 4 animals gradually recovered. None of the vaccinated animals exhibited any clinical signs of illness. Pathological changes in the lungs of the unvaccinated control animals were more severe than in the vaccinated animals, and viral antigen was only detected in cells of the control animals. Virus was not isolated from any organs of vaccinated animals, whereas high titers of virus were detected in the respiratory tissues of the control animals.

Rudenko et al. described the evaluation of the safety, immunogenicity and protective efficacy of a live attenuated cold-adapted vaccine virus, which is based on the low pathogenicity A/duck/Potsdam/86 (H5N2) virus, as a candidate vaccine against H5N1 in Java macaques [165]. This vaccine candidate is a 7:1 reassortant, and derives its HA from A/duck/Potsdam/86 (H5N2) and its NA and internal protein genes from the donor virus for the live attenuated influenza vaccines used in Russia, A/Leningrad/17/57 (H2N2). Monkeys vaccinated with two doses of the H5N2 cold-adapted vaccine virus, 21 days apart, did not exhibit signs of illness, and virus was recovered from 2 out of 4 animals, at titers between $10^{1.2}$ and $10^{4.2}$ EID_{50}/ml between days 3 and 5 after the first dose. The dose of vaccine virus used was not reported. The H5N2 cold-adapted vaccine virus elicited only modest HAI responses in the vaccinated macaques. The animals were challenged with the HPAI A/chicken/Kurgan/02/2005 (H5N1) isolate. Vaccinated animals developed a fever, but it was of a lower grade and of a shorter duration than that observed in the control animals, and shedding of challenge virus occurred in vaccinated animals, but it was for a shorter duration than in the control animals. These data suggest that there may be a small protective effect of the H5N2 cold-adapted vaccine virus against heterologous H5 virus challenge, however, the numbers of animals used was small and the immune responses that were observed were not consistent between animals. The H5N2 cold-adapted vaccine has been evaluated in Phase 1 and Phase 2 clinical trials in small numbers of volunteers [166]. The vaccine was evaluated at two different dose levels, and two doses were found to be safe and immunogenic in 47–55% of subjects. HAI antibodies were detected in the serum of the vaccine recipients and IgA antibodies were detected by ELISA in nasal wash samples. Qualitatively, these responses were similar to those observed in the mouse model. The level of replication of the vaccine virus in humans was not reported.

6.4.4 1918 H1N1 Pandemic Virus

Cynomolgus macaques were evaluated as a model for studying the reconstructed 1918 H1N1 pandemic influenza virus [167]. Monkeys were infected by multiple routes – intratracheally, orally, on the tonsils and conjunctiva – based on the earlier studies with HPAI H5N1 influenza viruses in this species [162]. Animals infected with the reconstructed 1918 virus had severe clinical illness, high levels of virus replication in the respiratory tract and severe pathological changes in the lungs compared to control animals infected with a recombinant human H1N1 influenza virus, A/Kawasaki/173/01 [167].

There may be a place for non-human primates as models for the evaluation of pandemic influenza vaccines, but the currently available data are not sufficient to support the use of these animals for immunogenicity or efficacy studies. Further studies are needed to characterize AI infection and the immune responses to AI viruses and vaccines in these species.

Clinical Evaluation of Live Attenuated Candidate Vaccines for Pandemic Influenza

The development of live attenuated vaccines against influenza viruses with pandemic potential has rapidly progressed from pre-clinical evaluation to early stage clinical testing in recent years. Data from both mouse and ferret models suggested that vaccine viruses of H5, H6, H7, and H9 subtypes, though restricted in replication in the respiratory tract compared to wild-type viruses, elicited serum antibody responses and were protective against both lethal challenge and pulmonary and extra-pulmonary replication following wild-type virus challenge. In addition, cross-protection against heterologous wild-type viruses was observed to varying degrees. Studies in non-human primates also showed that live attenuated H5 influenza virus vaccines could replicate in the respiratory tract and elicit serum HAI responses [113, 165].

In clinical trials involving small numbers of healthy adults, live attenuated cold-adapted H5N1 vaccine candidates, based on the clade 1 viruses A/Vietnam/1203/2004 and A/Hong Kong/213/2003, were found to be highly restricted in replication and poorly immunogenic [168]. A live vaccine virus, based on the A/Vietnam/1203/04 (H5N1) virus, when administered in two doses at $10^{7.5}$ TCID$_{50}$ per dose, failed to elicit neutralizing antibody in the serum of vaccinees and elicited serum HAI antibody in only 10% of the study subjects. Serum IgA and nasal wash IgA responses were detected in 52% and 19% of subjects, respectively; serum or local IgA responses had not been measured in ferret studies. Although the underlying reasons have not yet been elucidated, the poor predictive value of the mouse and ferret models with respect to replication and immunogenicity of these particular vaccine candidates was unexpected.

Rudenko et al. reported that a live attenuated cold-adapted vaccine, with the HA from a low pathogenicity avian H5N2 virus, elicited serum HAI and neutralizing antibodies in about 50% of volunteers after two doses, and resulted in the production of local IgA in the respiratory tract in 65% of vaccinees [166]. The level of shedding of the vaccine virus in volunteers was not determined. The reasons for the superior immunogenicity of the H5N2 *ca* vaccine virus, compared to the H5N1 vaccines based on the A/AA/6/60 *ca* in these small clinical studies, are not fully understood. It is possible that the donor cold-adapted H2N2 virus used to generate the H5N2 *ca* vaccine virus is less attenuated than the A/AA/6/60 *ca* donor virus, resulting in a vaccine virus that replicates more efficiently, however, this cannot be confirmed since replication of the H5N2 *ca* vaccine virus in humans was not reported. Given the poor predictive value of the mouse and ferret models with respect to the replication and immunogenicity of clade 1 live attenuated H5N1 vaccines in humans [66, 168], it will be interesting to see how the A/Anhui/2/2005 cold-adapted vaccine described by Fan et al. [113] behaves in human clinical trials.

The H7N3 *ca* vaccine virus was highly restricted in replication in Phase I clinical studies but elicited an immune response in over 90% of subjects [169]. However, serum IgA, and not HAI or neutralizing antibody, was the most frequently observed indication of immunogenicity of this vaccine in humans, with a serum IgA response being detected in 71% of subjects, and 62% and 48% of the subjects developing a fourfold or greater rise in HAI or neutralizing antibody, respectively. Studies in mice and ferrets did not accurately predict such restricted replication of the vaccine virus and it is difficult to determine the predictive value of the mouse and ferret studies in terms of immunogenicity, since the number of human subjects in whom the vaccine was evaluated was small, and serum IgA responses were not studied in animals.

An H9N2 *ca* vaccine was evaluated in a Phase I clinical trial in humans [170]. Despite being highly restricted in replication, the vaccine virus was immunogenic in all subjects in at least one assay (HAI or neutralization assay). Again, the degree of restriction of replication of the vaccine virus in humans was not predicted by studies in mice.

Pandemic 2009 H1N1 Influenza Vaccines

The emergence of the pandemic 2009 H1N1 influenza virus prompted a rapid response from the research community and vaccine manufacturers to develop a vaccine against the emerging virus. Inactivated and live attenuated vaccines, based on the A/California/07/2009 (H1N1) virus, were produced using the same manufacturing process and regulatory infrastructure as for seasonal influenza vaccines in order to make vaccine available expeditiously. The pandemic 2009 H1N1 vaccines were evaluated in limited clinical trials to support licensure but extensive pre-clinical testing in animals was not performed. Inactivated and live attenuated vaccines for the novel H1N1 influenza virus were licensed in September 2009 in the US. Animal models, however, will be needed for the continued study of the pandemic 2009 H1N1 influenza virus and for the evaluation of alternative approaches to develop vaccines against this pathogen. Several laboratories have reported studies of pandemic 2009 H1N1 influenza virus isolates in laboratory animals. These studies are summarized in the following section.

6.5 Mice

Pandemic 2009 H1N1 influenza viruses isolated from humans replicated efficiently in the respiratory tract of BALB/c mice without prior adaptation [171, 172]. However, differences in the severity of disease caused by the A/California/04/2009 (H1N1) isolate were reported by the two laboratories. Maines et al. reported that the A/California/04/2009 (H1N1) isolate was highly infectious in the mouse model, with a 50% mouse infectious dose (MID_{50}) of between $10^{0.5}$ and $10^{1.5}$

plaque forming units (PFU), but this virus and two other isolates (A/Texas/15/2009 and A/Mexico/4108/2009) were not lethal in mice; whereas Itoh and co-workers reported that the A/California/04/2009 (H1N1) virus was lethal in mice at an LD_{50} of $10^{5.8}$ PFU, and that mortality was also observed in mice infected with another isolate, WSLH34939 (LD_{50} of $10^{4.5}$ PFU). In both studies, weight loss was observed in mice following infection with A/California/04/2009, but the disease was far more severe in the study conducted by Itoh et al. In both studies, the peak virus titer in the lungs at day 3 post-infection was similar (between $10^{5.8}$ and $10^{7.8}$ PFU), and virus replication was restricted to the respiratory tract. Higher levels of virus replication and more severe pathological changes were observed in the lungs of mice infected with the pandemic 2009 H1N1 influenza virus compared to those inoculated with a recent seasonal human H1N1 influenza virus. Prominent bronchitis and alveolitis, with positive staining for viral antigen were observed on day 3 p.i. in mice infected with A/California/04/2009, with signs of regeneration present by day 6 p.i. [171]. Mice that were inoculated with the seasonal H1N1 virus had progressed to bronchitis and peribronchitis by day 6 p.i., but there was much less extensive staining for virus antigen in the tissues from these animals.

An inactivated split-virion vaccine for pandemic 2009 H1N1 influenza, administered with or without adjuvant (MF59), was evaluated in mice [173]. These studies suggested that a single dose of vaccine required an adjuvant to elicit a serum HAI response that was predictive of protection. Interim data from human clinical studies demonstrated that a single dose of 7.5 µg of HA, administered with MF59, did indeed elicit serum antibody responses that were predictive of protection, according to the criteria for the licensure of seasonal influenza vaccines [174]. However, data from recipients of the same vaccine administered without adjuvant were not reported [174]. Interim data from clinical trials of an inactivated pandemic 2009 H1N1 influenza vaccine suggest that, surprisingly, a single dose of unadjuvanted vaccine is sufficiently immunogenic to meet the criteria established for the licensure of seasonal influenza vaccines [175]. The preliminary data from clinical trials of the pandemic 2009 H1N1 vaccine show evidence of immunologic priming to the novel H1N1 virus. The complex previous immunologic experience of humans with influenza viruses, either by prior infection or vaccination, cannot be emulated easily in experimental animals.

6.6   Ferrets

The replication and virulence of pandemic 2009 H1N1 viruses, compared to recent seasonal H1N1 human influenza viruses, were evaluated in ferrets [171, 172, 176]. Beyond a mild level of inactivity, overt clinical signs of influenza were not observed in ferrets inoculated with A/California/04/2009 [171, 172]. More pronounced clinical features, some resulting in euthanasia, were observed in ferrets inoculated with virus isolates from Texas and Mexico [172], but not in animals that received pandemic 2009 H1N1 virus isolates from Wisconsin, the Netherlands and

Japan [171]. The clinical symptoms seen in ferrets did not reflect the severity of infection in the patients from whom the viruses were isolated.

In general, the pandemic 2009 H1N1 viruses and recent seasonal H1N1 influenza viruses replicated efficiently and to similar levels in the upper respiratory tract of ferrets, but the pandemic 2009 H1N1 viruses achieved higher titers in the lungs [171, 172, 176]. The virus was also detected in rectal swabs and tissue samples taken from the intestinal tract of infected ferrets [172]. There have been sporadic reports of gastrointestinal symptoms in human cases of pandemic 2009 H1N1 influenza infection [177], but this does not appear to be common, and the significance of this observation, with respect to pathogenesis of infection with these viruses, is not clear.

Pathologic changes were observed in the respiratory tract of ferrets inoculated with either seasonal or pandemic H1N1 viruses, but the changes were more extensive and more severe in ferrets infected with the pandemic 2009 H1N1 viruses. Itoh et al. reported similar levels of viral antigen in the nasal mucosa of animals that received either seasonal or pandemic virus. The lungs of ferrets inoculated with the seasonal H1N1 virus A appeared mostly normal, whereas A/California/04/2009-infected ferrets had more severe bronchopneumonia with prominent expression of viral antigen in the peribronchial glands and in a few alveolar cells. Similarly, ferrets inoculated with a pandemic H1N1 isolate from the Netherlands had mild to moderate, multi-focal, necrotizing rhinitis, tracheitis, bronchitis and bronchiolitis on day 3 p.i. with viral antigen observed in many cells in the nasal cavity, trachea, bronchus, and bronchioles while the pathologic changes and the presence of viral antigen were limited to the upper respiratory tract and were less extensive, respectively, in ferrets inoculated with seasonal H1N1 influenza [176]. By day 7 p.i., most of the virus-infected cells had been cleared from the respiratory tract of ferrets inoculated with either the seasonal or pandemic H1N1 virus. The pandemic 2009 H1N1 influenza viruses were also found to efficiently transmit via direct contact and respiratory droplets. In summary, the pandemic 2009 H1N1 viruses replicated more efficiently than seasonal H1N1 influenza viruses in the lower respiratory tract of ferrets. This increased level of replication was associated with more severe pathologic changes in the lower respiratory tract, but did not generally result in more severe clinical illness.

6.7   Non-Human Primates

As described above, there have been several reports of the use of non-human primate species as models for studies of influenza infection and for the evaluation of experimental influenza vaccines. To date, there is only one report of infection of non-human primates with pandemic 2009 H1N1 influenza viruses. Itoh and colleagues studied the infection of cynomolgus macaques with A/California/04/2009 (H1N1) virus [171]. As with the previous studies of avian H5N1 influenza viruses in this model, multiple routes of inoculation were used to establish infection:

animals were inoculated with a total dose of $10^{7.4}$PFU via the intratracheal, intranasal, ocular and oral routes. Macaques inoculated with the A/California/04/2009 (H1N1) virus experienced a greater increase in body temperature than animals that received a recent seasonal H1N1 virus, but they exhibited no other clinical signs of infection. The pandemic 2009 H1N1 virus replicated more efficiently in both the upper and lower respiratory tracts of macaques, achieving titers of between $10^{4.3}$ and $10^{6.9}$PFU in the lungs on day 3 p.i. High titers of virus ($>10^5$PFU) were still detected in the oro/nasopharynx, tonsil and bronchi of one animal on day 7 p.i., but it had been cleared from the other respiratory tissues.

Pathologic changes were observed in animals inoculated with either pandemic or seasonal H1N1 influenza viruses, but these lesions were more severe in the animals that received the pandemic 2009 H1N1 virus. On day 3 p.i., an edematous exudate and inflammatory infiltrates in the alveolar spaces with severe thickening of the alveolar walls were observed. Cells which appeared to be type I pneumocytes, that were positive for viral antigen, were distributed in the inflammatory lesions, and many type II pneumocytes were also positive for virus antigen. A thickening of the alveolar wall was also observed in large sections of lungs from monkeys infected with the seasonal H1N1 influenza virus, with prominent inflammatory cells in the alveolar wall. However, cells staining positive for viral antigen were sparse, and were only type I, not type II, pneumocytes. By day 7 p.i., the lung pathology remained more severe in the animals that received the pandemic virus than in those infected with the seasonal influenza virus, and many antigen-positive cells were still visible; however, regenerative changes were also evident.

7 Correlates of Protection from AI Viruses and Regulatory Concerns

Despite the fact that the correlates of protection from AI virus infections in humans are not known, the criteria for licensing pandemic influenza vaccines are based on the previous experience with vaccines against seasonal influenza. In Europe and the United States, regulatory authorities have published guidances for vaccine manufacturers that attempt to balance the need for expedited approval of pandemic influenza vaccines with the requirements for the demonstration of safety and immunogenicity of candidate vaccines.

In the United States, for example, a guidance for vaccine manufacturers, published in 2007 [178], states that licensure of both inactivated and live attenuated vaccines for pandemic influenza should be based on the percent of subjects achieving an HAI antibody titer of 1:40 or greater, and on the rate of seroconversion, which is defined as a fourfold or greater rise in post-vaccination HAI antibody titer. This could be particularly problematic for live attenuated AI vaccines, since experience with seasonal live influenza vaccines indicates that serum antibody levels do not correlate with the efficacy of such vaccines. Results from clinical trials conducted so far with live attenuated AI vaccines suggest that the measurement of immune responses other than serum HAI and neutralizing antibody, for example, serum IgA levels in pre-clinical studies, may be of value. Efficacy studies in animal models, although not an absolute requirement, may at least provide evidence that biologically relevant immune responses are elicited by candidate vaccines.

This guidance is intended to allow for rapid marketing approval of pandemic influenza vaccines that are produced using manufacturing processes that are already validated for seasonal influenza vaccines so that the licensure of the pandemic vaccine is essentially a strain change. Such approval requires much more limited testing of the candidate vaccines in animal models. In the European Union, manufacturers are required to submit information on the production and pre-clinical testing of a "mock-up" pandemic vaccine. In the event of a pandemic, a vaccine made in the same way as the mock-up vaccine, but based on the nascent pandemic virus, will be produced and will be subject to limited pre-clinical characterization, including immunogenicity studies in animals on at least one batch of the product [179]. Efficacy studies of the actual pandemic vaccine formulation in animals are not required. However, extensive pre-clinical testing of the vaccine candidate is required for new vaccine modalities and formulations, including formulations of approved vaccines with adjuvants.

In the US, a regulatory mechanism was introduced under what is commonly referred to as the "animal rule" [180] for obtaining marketing approval of vaccines for which efficacy studies in healthy human volunteers are either unethical or not feasible. This regulation stipulates that, in cases where efficacy of vaccines in humans cannot be definitively determined, marketing approval for a vaccine may be granted based on "adequate and well-controlled animal studies", provided the basis for vaccine efficacy is reasonably well understood, and that the animal responds to the vaccine in a manner that is predictive of the response in humans. Studies in more than one animal species would typically be required, unless a single animal model is available that can faithfully predict the efficacy of a vaccine in humans. It is unclear, at this time, whether this rule will eventually be applied to vaccines for pandemic influenza. In any event, it is critical that the predictive value of the available animal models for immunogenicity and efficacy of pandemic influenza vaccines be determined systematically using the same vaccine formulations that are progressing into clinical studies.

8 Conclusion

Although several animal species support the replication of human and AI viruses, a survey of the literature leads to the conclusion that there is no single ideal animal model for the evaluation of influenza vaccines. Some animal models are more suitable than others in predicting the attenuation of live virus vaccines, or more closely reflect the human immune response to vaccines. Animal models certainly play a crucial role in the evaluation of influenza vaccines, but the limitations of the models must be taken into account when decisions regarding which vaccine candidates should move forward into clinical trials are made.

The evaluation of vaccines for pandemic influenza presents additional challenges, in that, the correlates of protection from AI viruses are not known, and so there may be a greater need for reliance on data from animal studies for these vaccines. It is critical that the behavior of AI viruses with pandemic potential be characterized in a range of animal models. Even from limited observations, it is clear that replication of AI viruses and their ability to cause disease in animals depends on the host species, and is subtype and even strain specific. To date, the level of replication and the immunogenicity of live attenuated AI candidate vaccine viruses seen in animal models have not accurately predicted the behavior of these vaccine viruses in humans. Therefore, pre-clinical safety, immunogenicity and efficacy data from animal studies must be carefully considered in the evaluation of pandemic influenza vaccines.

Acknowledgments We thank Brian Murphy for critical review of this manuscript. This research was supported in part by the Intramural Research Program of the NIAID, NIH.

References

1. Palese P, Shaw ML (2007) Orthomyxoviridae: the viruses and their replication. In: Knipe DM, Howley PM, Griffin DE, Lamb RA, Martin MA, Roizman B, Straus SE (eds) Fields virology, 5th edn. Lippincott Williams & Wilkins, Philadelphia, pp 1647–1689
2. Wright PF, Neumann G, Kawaoka Y (2007) Orthomyxoviruses. In: Knipe DM, Howley PM, Griffin DE, Lamb RA, Martin MA, Roizman B, Straus SE (eds) Fields virology, 5th edn. Lippincott Williams & Wilkins, Philadelphia, pp 1691–1740
3. Fouchier RA, Munster V, Wallensten A, Bestebroer TM, Herfst S, Smith D, Rimmelzwaan GF, Olsen B, Osterhaus AD (2005) Characterization of a novel influenza A virus hemagglutinin subtype (H16) obtained from black-headed gulls. J Virol 79:2814–2822
4. Webster RG, Bean WJ, Gorman OT, Chambers TM, Kawaoka Y (1992) Evolution and ecology of influenza A viruses. Microbiol Rev 56:152–179
5. Cox NJ, Subbarao K (2000) Global epidemiology of influenza: past and present. Annu Rev Med 51:407–421
6. Subbarao K, Klimov A, Katz J, Regnery H, Lim W, Hall H, Perdue M, Swayne D, Bender C, Huang J et al (1998) Characterization of an avian influenza A (H5N1) virus isolated from a child with a fatal respiratory illness. Science 279:393–396
7. Kawaoka Y, Krauss S, Webster RG (1989) Avian-to-human transmission of the PB1 gene of influenza A viruses in the 1957 and 1968 pandemics. J Virol 63:4603–4608
8. Garten RJ, Davis CT, Russell CA, Shu B, Lindstrom S, Balish A, Sessions WM, Xu X, Skepner E, Deyde V et al (2009) Antigenic and genetic characteristics of swine-origin 2009 A(H1N1) influenza viruses circulating in humans. Science 325:197–201
9. Keawcharoen J, Oraveerakul K, Kuiken T, Fouchier RA, Amonsin A, Payungporn S, Noppornpanth S, Wattanodorn S, Theambooniers A, Tantilertcharoen R et al (2004) Avian influenza H5N1 in tigers and leopards. Emerg Infect Dis 10:2189–2191
10. Kuiken T, Rimmelzwaan G, van Riel D, van Amerongen G, Baars M, Fouchier R, Osterhaus A (2004) Avian H5N1 influenza in cats. Science 306:241

11. Couch RB, Kasel JA (1983) Immunity to influenza in man. Annu Rev Microbiol 37:529–549
12. Hobson D, Curry RL, Beare AS, Ward-Gardner A (1972) The role of serum hemagglutina-tion-inhibiting antibody in protection against challenge infection with influenza A2 and B viruses. J Hyg 70:767–777
13. Treanor J, Wright PF (2003) Immune correlates of protection against influenza in the human challenge model. In: Brown F, Haaheim LR, Schild GC (eds) Laboratory correlates of immunity to influenza – a reassessment. Karger, Basel, pp 97–104
14. Clements ML, Murphy BR (1986) Development and persistence of local and systemic antibody responses in adults given live attenuated or inactivated influenza A virus vaccine. J Clin Microbiol 23:66–72
15. Gorse GJ, O'Connor TZ, Newman FK, Mandava MD, Mendelman PM, Wittes J, Peduzzi PN (2004) Immunity to influenza in older adults with chronic obstructive pulmonary disease. J Infect Dis 190:11–19
16. Murphy BR, Coelingh K (2002) Principles underlying the development and use of live attenuated cold-adapted influenza A and B virus vaccines. Viral Immunol 15:295–323
17. Smith W, Andrewes CH, Laidlaw PP (1933) A virus obtained from influenza patients. Lancet 222:66–68
18. Andrewes CH, Laidlaw PP, Smith W (1934) The susceptibility of mice to the viruses of human and swine influenza. Lancet 224:859–862
19. Brown EG (1990) Increased virulence of a mouse-adapted variant of Influenza A/FM/1/47 virus is controlled by mutations in genome segments 4, 5, 7 and 8. J Virol 64:4523–4533
20. Brown EG, Liu H, Chang Kit L, Baird S, Nesrallah M (2001) Pattern of mutation in the genome of influenza A virus on adaptation to increased virulence in the mouse lung: identification of functional themes. Proc Natl Acad Sci USA 98:6883–6888
21. Smeenk CA, Brown EG (1994) The Influenza virus variant A/FM/1/47-MA possesses single amino acid replacements in the hemagglutinin, controlling virulence, and in the matrix protein, controlling virulence as well as growth. J Virol 68:530–534
22. Smeenk CA, Wright KE, Burns BF, Thaker AJ, Brown EG (1996) Mutations in the hemagglutinin and matrix genes of a virulent influenza virus variant, A/FM/1/47-MA, control different stages in pathogenesis. Virus Res 44:79–95
23. Grimm D, Staeheli P, Hufbauer M, Koerner I, Martinez-Sobrido L, Solorzano A, Garcia-Sastre A, Haller O, Kochs G (2007) Replication fitness determines high virulence of influenza A virus in mice carrying functional Mx resistance gene. Proc Natl Acad Sci USA 104:6806–6811
24. Epstein SL, Lo CY, Misplon JA, Lawson CM, Hendrickson BA, Max EE, Subbarao K (1997) Mechanisms of heterosubtypic immunity to lethal influenza A virus infection in fully immunocompetent, T cell-depleted, ß2-microglobulin-deficient, and J chain-deficient mice. J Immunol 158:1222–1230
25. Tumpey TM, Szretter KJ, Van Hoeven N, Katz JM, Kochs G, Haller O, Garcia-Sastre A, Staeheli P (2007) The Mx1 gene protects mice against pandemic 1918 and highly lethal human H5N1 influenza viruses. J Virol 81:10818–10821
26. Virelizier J (1975) Host defenses against influenza virus: the role of anti-hemagglutinin antibody. J Immunol 115:434–439
27. Ramphal R, Cogliano RC, Shands JWJ, Small PAJ (1979) Serum antibody prevents lethal murine influenza pneumonitis but not tracheitis. Infect Immun 25:992–997
28. Takiguchi K, Sugawara K, Hongo S, Nishimura H, Kitame F, Nakamura K (1992) Protective effect of serum antibody on respiratory infection of influenza C virus in rats. Arch Virol 122:1–11
29. Prince GA, Horswood RL, Chanock RM (1985) Quantitative aspects of passive immunity to respiratory syncytial virus infection in infant cotton rats. J Virol 55:517–520
30. Subbarao K, McAuliffe J, Vogel L, Fahle G, Fischer S, Tatti K, Packard M, Shieh WJ, Zaki S, Murphy B (2004) Prior infection and passive transfer of neutralizing antibody

prevent replication of severe acute respiratory syndrome coronavirus in the respiratory tract of mice. J Virol 78:3572–3577

31. Iida T, Bang FB (1963) Infection of the upper respiratory tract of mice with influenza virus. Am J Hyg 77:169–176

32. Yetter RA, Lehrer S, Ramphal R, Small PAJ (1980) Outcome of influenza infection: effect of site of initial infection and heterotypic immunity. Infect Immun 29:654–662

33. Staeheli P, Grob R, Meier E, Sutcliffe JG, Haller O (1988) Influenza virus-susceptible mice carry Mx genes with a large deletion or a nonsense mutation. Mol Cell Biol 8:4518–4523

34. Staeheli P, Haller O, Boll W, Lindenmann J, Weissmann C (1986) Mx protein: constitutive expression in 3T3 cells transformed with cloned Mx cDNA confers selective resistance to influenza virus. Cell 44:147–158

35. Abou-Donia H, Jennings R, Potter CW (1980) Growth of influenza A viruses in hamsters. Arch Virol 65:99–107

36. Heath AW, Addison C, Ali M, Teale D, Potter CW (1983) In vivo and in vitro hamster models in the assessment of virulence of recombinant influenza viruses. Antiviral Res 3:241–252

37. Murphy BR, Wood FT, Massicot JG, Chanock RM (1978) Temperature-sensitive mutants of influenza virus. XVI. Transfer of the two *ts* lesions present in the Udorn/72-*ts*-1A2 donor virus to the Victoria/3/75 wild-type virus. Virology 88:244–251

38. Subbarao EK, Kawaoka Y, Murphy BR (1993) Rescue of an influenza A virus wild-type PB2 gene and a mutant derivative bearing a site-specific temperature-sensitive and attenuating mutation. J Virol 67:7223–7228

39. Phair JP, Kauffman CA, Jennings R, Potter CW (1979) Influenza virus infection of the guinea pig: immune response and resistance. Med Microbiol Immunol 165:241–254

40. Azoulay-Dupuis E, Lambre CR, Soler P, Moreau J, Thibon M (1984) Lung alterations in guinea-pigs infected with influenza virus. J Comp Path 94:273–283

41. Lowen AC, Mubareka S, Tumpey TM, Garcia-Sastre A, Palese P (2006) The guinea pig as a transmission model for human influenza viruses. Proc Natl Acad Sci USA 103:9988–9992

42. Ali M, Maassab HF, Jennings R, Potter CW (1982) Infant rat model of attenuation for recombinant influenza viruses prepared from cold-adapted attenuated A/Ann/Arbor/6/60. Infect Immun 38:610–619

43. Mahmud MIA, Jennings R, Potter CW (1979) The infant rat as a model for assessment of the attenuation of human influenza viruses. J Med Microbiol 12:43–54

44. Teh C, Jennings R, Potter CW (1980) Influenza virus infection of newborn rats: virulence of recombinant strains prepared from influenza virus strain A/Okuda/57. J Med Microbiol 13:297–306

45. Niewiesk S, Prince G (2002) Diversifying animal models: the use of hispid cotton rats (*Sigmodon hispidus*) in infectious diseases. Lab Anim 36:357–372

46. Sadowski W, Wilczynski J, Semkow R, Tulimowska M, Krus S, Kantoch M (1987) The cotton rat (*Sigmodon hispidus*) as an experimental model for studying viruses in respiratory tract infections. II. Influenza viruses types A and B. Med Dosw Mikrobiol 39:43–55

47. Ottolini MG, Blanco JC, Eichelberger MC, Porter DD, Pletneva L, Richardson JY, Prince GA (2005) The cotton rat provides a useful small-animal model for the study of influenza virus pathogenesis. J Gen Virol 86:2823–2830

48. Burnet FM (1941) Influenza virus "A" infections of cynomolgus monkeys. Aust J Exp Biol Med Sci 19:281–290

49. Saslaw S, Wilson HE, Doan CA, Woolpert OC, Schwab JL (1946) Reactions of monkeys to experimentally induced influenza virus A infection. An analysis of the relative roles of humoral and cellular immunity under conditions of optimal or deficient nutrition. J Exp Med 84:113–125

50. Rimmelzwaan GF, Baars M, van Beek R, Van Amerongen G, Lovgren-Bengtsson K, Claas ECJ, Osterhaus ADME (1997) Induction of protective immunty against influenza

virus in a macaque model: comparison of conventional and ISCOM vaccines. J Gen Virol 78:757–765

51. Baskin CR, Garcia-Sastre A, Tumpey TM, Bielefeldt-Ohmann H, Carter VS, Nistal-Villan E, Katze MG (2004) Integration of clinical data, pathology, and cDNA microarrays in influenza virus-infected pigtailed macaques (*Macaca nemestrina*). J Virol 78:10420–10432

52. Murphy BR, Lewis Sly D, Hosier NT, London WT, Chanock RM (1980) Evaluation of three strains of influenza A virus in humans and in owl, cebus and squirrel monkeys. Infect Immun 28:688–691

53. Murphy BR, Hinshaw VS, Lewis Sly D, London WT, Hosier NT, Wood FT, Webster RG, Chanock RM (1982) Virulence of avian influenza A viruses for squirrel monkeys. Infect Immun 37:1119–1126

54. Snyder MH, Clements ML, Herrington D, London WT, Tierney EL, Murphy BR (1986) Comparison by studies in squirrel monkeys, chimpanzees, and adult humans of avian-human influenza A virus reassortants derived from different avian influenza virus donors. J Clin Microbiol 24:467–469

55. Grizzard MB, London WT, Sly DL, Murphy BR, James WD, Parnell WP, Chanock RM (1978) Experimental production of respiratory tract disease in cebus monkeys after intra-tracheal or intranasal infection with influenza A/Victoria/3/75 or influenza A/New Jersey/76 virus. Infect Immun 21:201–205

56. Murphy BR, Hall SL, Crowe J, Collins PL, Subbarao EK, Connors M, London WT, Chanock RM (1992) The use of chimpanzees in respiratory virus research. In: Erwin J, Landon JC (eds) Chimpanzee conservation and public health: environments for the future. Diagnon/Bioqual, Rockville, MD

57. Snyder MH, London WT, Tierney EL, Maassab HF, Murphy BR (1986) Restricted replication of a cold-adapted reassortant influenza A virus in the lower respiratory tract of chimpanzees. J Infect Dis 154:370–371

58. WHO (2005) WHO guidelines on nonclinical evaluation of vaccines. World Health Organization, Geneva. Annex 1

59. Sugg JY (1949) An Influenza virus pneumonia of mice that is nontransferable by serial passage. J Bacteriol 57:399–403

60. Henle W, Henle G (1946) Studies on the toxicity of influenza viruses. II. The effect of intra-abdominal and intravenous injection of influenza viruses. J Exp Med 84:639–661

61. Jin H, Manetz S, Leininger J, Luke C, Subbarao K, Murphy B, Kemble G, Coelingh KL (2007) Toxicological evaluation of live attenuated, cold-adapted H5N1 vaccines in ferrets. Vaccine 25:8664–8672

62. Betts RF, Douglas GRJ, Maassab HF, DeBorde DC, Clements ML, Murphy BR (1988) Analysis of virus and host factors in a study of A/Peking/2/79 (H3N2) cold-adapted vaccine recombinant in which vaccine-associated illness occurred in normal volunteers. J Med Virol 26:175–183

63. Murphy BR, Holley HP, Berquist EJ, Levine MM, Spring SB, Maassab HF, Kendal AP, Chanock RM (1979) Cold-adapted variants of influenza A virus: evaluation in adult sero-negative volunteers of A/Scotland/840/74 and A/Victoria/3/75 cold-adapted recombinants derived from the cold-adapted A/Ann Arbor/6/60 strain. Infect Immun 23:253–259

64. Okuno Y, Nakamura K, Yamamura T, Takahashi M, Toyoshima K, Kunita N, Sugai T, Fujita T (1960) Studies on attenuation of influenza virus. Proc Jpn Acad 36:299–303

65. Center for Biologics Evaluation and Research (CBER) 21 CFR PART 610 General biological products standards, Rockville, MD, USA, (Food and Drug Administration)

66. Suguitan AL Jr, McAuliffe J, Mills KL, Jin H, Duke G, Lu B, Luke CJ, Murphy B, Swayne DE, Kemble G et al (2006) Live, attenuated influenza A H5N1 candidate vaccines provide broad cross-protection in mice and ferrets. PLoS Med 3:e360

67. WHO (2003) Production of pilot lots of inactivated influenza vaccines from reassortants derived from avian influenza viruses. Interim biosafety risk assessment. World Health Organization, Geneva

68. Gonin P, Gaillard C (2002) Gene transfer vector biodistribution: pivotal safety studies in clinical gene therapy development. Gene Ther 11:S98–S108

69. Leamy VL, Martin T, Mahajan R, Vilalta A, Rusalov D, Hartikka J, Bozoukova V, Hall KD, Morrow J, Rolland AP et al (2006) Comparison of rabbit and mouse models for persistence analysis of plasmid-based vaccines. Hum Vaccin 2:113–118

70. Ledwith BJ, Manam S, Troilo PJ, Barnum AB, Pauley CJ, Griffiths TG, Harper LB, Schock HB, Zhang H, Faris JE et al (2002) Plasmid DNA vaccines: assay for integration into host genomic DNA. Dev Biol (Basel) 104:33–43

71. Manam S, Ledwith BJ, Barnum AB, Troilo PJ, Pauley CJ, Harper LB, Griffiths TG, Niu Z, Denisova L, Follmer TT et al (2000) Plasmid DNA vaccines: tissue distribution and effects of DNA sequence, adjuvants and delivery method on integration into host DNA. Intervirology 43:273–281

72. Winegar RA, Monforte JA, Suing KD, O'Loughlin KG, Rudd CJ, Macgregor JT (1996) Determination of tissue distribution of an intramuscular plasmid vaccine using PCR and in situ DNA hybridization. Hum Gene Ther 7:2185–2194

73. Maassab HF, Kendal AP, Abrams GD, Monto AS (1982) Evaluation of a cold-recombinant influenza virus vaccine in ferrets. J Infect Dis 146:780–790

74. Murphy BR, Sly DL, Tierney EL, Hosier NT, Massicot JG, London WT, Chanock RM, Webster RG, Hinshaw VS (1982) Reassortant virus derived from avian and human influenza A viruses is attenuated and immunogenic in monkeys. Science 218:1330–1332

75. EMEA (2003) Points to consider on the development of live attenuated Influenza vaccines, European Agency for the Evaluation of Medicinal Products (EMEA), London, CPMP/BWP/2289/01

76. Rubin SA, Liu D, Pletnikov M, McCullers JA, Ye Z, Levandowski RA, Johannessen J, Carbone KM (2004) Wild-type and attenuated influenza virus infection of the neonatal rat brain. J Neurovirol 10:305–314

77. Smith W, Andrewes CH, Laidlaw PP (1935) Influenza: experiments on the immunization of ferrets and mice. Brit J Exp Pathol 16:291–302

78. Potter CW, Oxford JS, Shore SL, McLaren C, Stuart-Harris CH (1972) Immunity to influenza in ferrets. I. Response to live and killed virus. Brit J Exp Pathol 53:153–167

79. Potter CW, Shore SL, McLaren C, Stuart-Harris CH (1972) Immunity to influenza in ferrets. 2. Influence of adjuvants on immunization. Brit J Exp Pathol 53:168–179

80. Benton KA, Misplon JA, Lo CY, Brutkiewicz RR, Prasad SA, Epstein SL (2001) Heterosubtypic immunity to Influenza A virus in mice lacking IgA, all Ig, NKT cells or γδ T cells. J Immunol 166:7437–7445

81. Epstein SL, Lo CY, Misplon JA, Bennink JR (1998) Mechanism of protective immunity against influenza virus infection in mice without antibodies. J Immunol 160:322–327

82. Nguyen HH, van Ginkel FW, Vu HL, McGhee JR, Mestecky J (2001) Heterosubtypic immunity to Influenza A virus infection requires B cells but not CD8+ cytotoxic T lymphocytes. J Infect Dis 183:368–376

83. Nguyen HH, Zemlin M, Ivanov II, Andrasi J, Zemlin C, Vu HL, Schelonka R, Schroeder HWJ, Mestecky J (2007) Heterosubtypic immunity to influenza A virus infection requires a properly diversified antibody repertoire. J Virol 81:9331–9338

84. McLaren C, Potter CW (1974) Immunity to influenza in ferrets. VII. Effect of previous infection with heterotypic and heterologous influenza viruses on the response of ferrets to inactivated influenza virus vaccines. J Hyg 72:91–100

85. McLaren C, Potter CW, Jennings R (1974) Immunity to influenza in ferrets. X. Intranasal immunization of ferrets with inactivated influenza A virus vaccines. Infect Immun 9:985–990

86. Yetter RA, Barber WH, Small PAJ (1980) Heterotypic immunity to influenza in ferrets. Infect Immun 29:650–653

87. Straight TM, Ottolini MG, Prince GA, Eichelberger MC (2006) Evidence of a cross-protective immune response to influenza A in the cotton rat model. Vaccine 24:6264–6271

88. Epstein SL (2006) Prior H1N1 influenza infection and susceptibility of Cleveland family study participants during the H2N2 pandemic of 1957. J Infect Dis 193:49–53

89. Steinhoff MC, Fries LF, Karron RA, Clements ML, Murphy BR (1993) Effect of heterosubtypic immunity on infection with attenuated influenza A virus vaccines in young children. J Clin Microbiol 31:836–838

90. Epstein SL, Tumpey TM, Misplon JA, Lo CY, Cooper LA, Subbarao K, Renshaw M, Sambhara S, Katz JM (2002) DNA vaccine expressing conserved influenza virus proteins protective against H5N1 challenge infection in mice. Emerg Infect Dis 8:796–801

91. Okuda K, Ihata A, Watabe S, Okada E, Yamakawa T, Hamajima K, Yang J, Ishii N, Nakazawa M, Okuda K et al (2001) Protective immunity against influenza A virus induced by immunization with DNA plasmid containing influenza M gene. Vaccine 19:3681–3691

92. Ulmer J, Donnelly J, Parker S, Rhodes G, Felgner P, Dwarki V, Gromkowski S, Deck R, DeWitt C, Friedman A et al (1993) Heterologous protection against influenza by injection of DNA encoding a viral protein. Science 259:1745–1749

93. Slepushkin VA, Katz JM, Black RA, Gamble WC, Rota PA, Cox NJ (1995) Protection of mice against influenza A virus challenge by vaccination with baculovirus-expressed M2 protein. Vaccine 13:1399–1402

94. Fan J, Liang X, Horton M, Perry HC, Citron MP, Heidecker GJ, Fu TM, Joyce J, Przysiecki CT, Keller PM et al (2004) Preclinical study of influenza virus A M2 peptide conjugate in mice, ferrets, and rhesus monkeys. Vaccine 22:2993–3003

95. DeFilette M, Friers W, Martens W, Birkett A, Ramne A, Lowenadler B, Lycke N, Jou WM, Saelens X (2006) Improved design and intranasal delivery of an M2e-based human influenza A vaccine. Vaccine 24:6597–6601

96. DeFilette M, Min Jou W, Birkett A, Lyons K, Schultz B, Tonkyro A, Resch S, Friers W (2005) Universal influenza A vaccine: optimization of M2-based constructs. Virology 337:149–161

97. Neirynck S, Deroo T, Saelens X, Vanlandschoot P, Jou WM, Friers W (1999) A universal influenza A vaccine based on the extracellular domain of the M2 protein. Nat Med 5:1157–1163

98. Sandbulte MR, Jimenez GS, Boon AC, Smith LR, Treanor JJ, Webby RJ (2007) Crossreactive neuraminidase antibodies afford partial protection against H5N1 in mice and are present in unexposed humans. PLoS Med 4:e59

99. Murphy BR, Kasel JA, Chanock RM (1972) Association of serum anti-neuraminidase antibody with resistance to influenza in man. N Engl J Med 286:1329–1332

100. Gao P, Watanabe S, Ito T, Goto H, Wells K, McGregor M, Cooley AJ, Kawaoka Y (1999) Biological heterogeneity, including systemic replication in mice, of H5N1 influenza A virus isolates from humans in Hong Kong. J Virol 73:3184–3189

101. Katz JM, Lu X, Tumpey TM, Smith CB, Shaw MW, Subbarao K (2000) Molecular correlates of influenza A H5N1 virus pathogenesis in mice. J Virol 74:10807–10810

102. Lu X, Tumpey TM, Morken T, Zaki SR, Cox NJ, Katz JM (1999) A mouse model for the evaluation of pathogenesis and immunity to influenza A (H5N1) viruses isolated from humans. J Virol 73:5903–5911

103. Dybing JK, Schultz-Cherry S, Swayne DE, Suarez DL, Perdue ML (2000) Distinct pathogenesis of Hong Kong-origin H5N1 viruses in mice compared to that of other highly pathogenic H5 avian influenza viruses. J Virol 74:1443–1450

104. Maines TR, Lu XH, Erb SM, Edwards L, Guarner J, Greer PW, Nguyen DC, Szretter KJ, Chen LM, Thawatsupha P et al (2005) Avian influenza (H5N1) viruses isolated from humans in Asia in 2004 exhibit increased virulence in mammals. J Virol 79:11788–11800

105. Nicholson KG, Colegate AE, Podda A, Stephenson I, Wood J, Ypma E, Zambon MC (2001) Safety and antigenicity of non-adjuvanted and MF59-adjuvanted influenza A/Duck/

Singapore/97 (H5N3) vaccine: a randomised trial of two potential vaccines against H5N1 influenza. Lancet 357:1937–1943

106. Stephenson I, Nicholson KG, Colegate A, Podda A, Wood J, Ypma E, Zambon M (2003) Boosting immunity to influenza H5N1 with MF59-adjuvanted H5N3 A/Duck/Singapore/97 vaccine in a primed human population. Vaccine 21:1687–1693

107. Stephenson I, Nicholson KG, Gluck R, Mischler R, Newman RW, Palache AM, Verlander NQ, Warburton F, Wood JM, Zambon MC (2003) Safety and antigenicity of whole virus and subunit influenza A/Hong Kong/1073/99 (H9N2) vaccine in healthy adults: phase I randomised trial. Lancet 362:1959–1966

108. Subbarao K, Luke CJ (2007) H5N1 viruses and vaccines. PLoS Pathog 3:e40

109. Bresson JL, Perronne C, Launay O, Gerdil C, Saville M, Wood J, Hoschler K, Zambon MC (2006) Safety and immunogenicity of an inactivated split-virion influenza A/Vietnam/1194/2004 (H5N1) vaccine: phase I randomised trial. Lancet 367:1657–1664

110. Treanor JJ, Campbell JD, Zangwill KM, Rowe T, Wolff M (2006) Safety and immunogenicity of an inactivated subvirion influenza A (H5N1) vaccine. N Engl J Med 354:1343–1351

111. Lin J, Zhang J, Dong X, Fang H, Chen J, Su N, Gao Q, Zhang Z, Liu Y, Wang Z et al (2006) Safety and immunogenicity of an inactivated adjuvanted whole-virion influenza A (H5N1) vaccine: a phase I randomised controlled trial. Lancet 368:991–997

112. Desheva JA, Lu XH, Rekstin AR, Rudenko LG, Swayne DE, Cox NJ, Katz JM, Klimov AI (2006) Characterization of an influenza A H5N2 reassortant as a candidate for live-attenuated and inactivated vaccines against highly pathogenic H5N1 viruses with pandemic potential. Vaccine 24:6859–6866

113. Fan S, Gao Y, Shinya K, Li CK, Li Y, Shi J, Jiang Y, Suo Y, Tong T, Zhong G et al (2009) Immunogenicity and protective efficacy of a live attenuated H5N1 vaccine in nonhuman primates. PLoS Pathog 5:e1000409

114. Lu X, Edwards LE, Desheva JA, Nguyen DC, Rekstin AR, Stephenson I, Szretter KJ, Cox NJ, Rudenko LG, Klimov A et al (2006) Cross-protective immunity in mice induced by live-attenuated or inactivated vaccines against highly pathogenic influenza A (H5N1) viruses. Vaccine 24:6588–6593

115. Belser JA, Lu X, Maines TR, Smith C, Li Y, Donis RO, Katz JM, Tumpey TM (2007) Pathogenesis of avian influenza (H7) virus infection in mice and ferrets: enhanced virulence of Eurasian H7N7 viruses isolated from humans. J Virol 81:11139–11147

116. Joseph T, McAuliffe J, Lu B, Jin H, Kemble G, Subbarao K (2007) Evaluation of replication and pathogenicity of avian influenza a H7 subtype viruses in a mouse model. J Virol 81:10558–10566

117. de Wit E, Munster V, Spronken MIJ, Bestebroer TM, Baas C, Beyer WEP, Rimmelzwaan GF, Osterhaus ADME, Fouchier RAM (2005) Protection of mice against lethal infection with highly pathogenic H7N7 influenza A virus by using a recombinant low-pathogenicity vaccine strain. J Virol 79:12401–12407

118. Munster VJ, de Wit E, van Riel D, Beyer WEP, Rimmelzwaan GF, Osterhaus ADME, Kuiken T, Fouchier RAM (2007) The molecular basis of the pathogenicity of the Dutch highly pathogenic human influenza A H7N7 viruses. J Infect Dis 196:258–265

119. Rigoni M, Shinya K, Toffan A, Milani A, Bettini F, Kawaoka Y, Cattoli G, Capua I (2007) Pneumo- and neurotropism of avian origin Italian highly pathogenic avian influenza H7N1 isolates in experimentally infected mice. Virology 364:28–35

120. Jadhao SJ, Achenbach J, Swayne DE, Donis R, Cox N, Matsuoka Y (2008) Development of Eurasian H7N7/PR8 high growth reassortant virus for clinical evaluation as an inactivated pandemic influenza vaccine. Vaccine 26:1742–1750

121. Pappas C, Matsuoka Y, Swayne DE, Donis RO (2007) Development and evaluation of an Influenza virus subtype H7N2 vaccine candidate for pandemic preparedness. Clin Vaccine Immunol 14:1425–1432

122. Cox RJ, Major D, Hauge S, Madhun AS, Brokstad KA, Kuhne M, Smith J, Vogel FR, Zambon M, Haaheim LR et al (2009) A cell-based H7N1 split influenza virion vaccine

confers protection in mouse and ferret challenge models. Influenza Other Respi Viruses 3:107–117

123. Cox RJ, Madhun AS, Hauge S, Sjursen H, Major D, Kuhne M, Hoschler K, Saville M, Vogel FR, Barclay W et al (2009) A phase I clinical trial of a PER.C6 cell grown influenza H7 virus vaccine. Vaccine 27:1889–1897

124. Joseph T, McAuliffe J, Lu B, Vogel L, Swayne D, Jin H, Kemble G, Subbarao K (2008) A live attenuated cold-adapted influenza A H7N3 virus vaccine provides protection against homologous and heterologous H7 viruses in mice and ferrets. Virology 378:123–132

125. Guo Y, Li J, Cheng X (1999) Discovery of men infected by avian influenza A (H9N2) virus. Chinese. Zhonghua Shi Yan He Lin Chuang Bing Du Xue Za Zhi 13:105–108

126. Peiris M, Yuen KY, Leung CW, Chan KH, Ip PLS, Lai RWM, Orr WK, Shortridge KF (1999) Human infection with influenza H9N2. Lancet 354:916–917

127. Guan Y, Shortridge KF, Krauss S, Webster RG (1999) Molecular characterization of H9N2 influenza viruses: were they the donors of the "internal" genes of H5N1 viruses in Hong Kong ? Proc Natl Acad Sci USA 96:9363–9367

128. Guo YJ, Krauss S, Senne DA, Mo IP, Lo KS, Xiong XP, Norwood M, Shortridge KF, Webster RG, Guan Y (2000) Characterization of the pathogenicity of members of the newly established H9N2 influenza virus lineages in Asia. Virology 267:279–288

129. Shortridge KF (1999) Poultry and the influenza H5N1 outbreak in Hong Kong, 1997: abridged chronology and virus isolation. Vaccine 17(Suppl 1):S26–S29

130. Xu KM, Li KS, Smith GJ, Li JW, Tai H, Zhang JX, Webster RG, Peiris JS, Chen H, Guan Y (2007) Evolution and molecular epidemiology of H9N2 influenza A viruses from quail in southern China, 2000–2005. J Virol 81:2635–2645

131. Xu KM, Smith GJ, Bahl J, Duan L, Tai H, Vijaykrishna D, Wang J, Zhang JX, Li KS, Fan XH et al (2007) The genesis and evolution of H9N2 influenza viruses in poultry from southern China, 2000 to 2005. J Virol 81:10389–10441

132. Leneva IA, Goloubeva O, Fenton RJ, Tisdale M, Webster RG (2001) Efficacy of zanamivir against avian influenza A viruses that possess genes encoding H5N1 internal proteins and are pathogenic in mammals. Antimicrob Agents Chemother 45:1216–1224

133. Leneva IA, Roberts N, Govorkova EA, Goloubeva OG, Webster RG (2000) The neuraminidase inhibitor GS4104 (oseltamivir phosphate) is efficacious against A/Hong Kong/156/97 (H5N1) and A/Hong Kong/1074/99 (H9N2) influenza viruses. Antiviral Res 48:101–115

134. Lu X, Renshaw M, Tumpey TM, Kelly GD, Hu-Primmer J, Katz JM (2001) Immunity to influenza A H9N2 viruses induced by infection and vaccination. J Virol 75:4896–4901

135. Ernst WA, Kim HJ, Tumpey TM, Jansen AD, Tai W, Cramer DV, Adler-Moore JP, Fujii G (2006) Protection against H1, H5, H6 and H9 influenza A infection with liposomal matrix 2 epitope vaccines. Vaccine 24:5158–5168

136. Chen H, Matsuoka Y, Swayne D, Chen Q, Cox NJ, Murphy BR, Subbarao K (2003) Generation and characterization of a cold-adapted influenza A H9N2 reassortant as a live pandemic influenza virus vaccine candidate. Vaccine 21:4430–4436

137. Chen H, Subbarao K, Swayne D, Chen Q, Lu X, Katz J, Cox N, Matsuoka Y (2003) Generation and evaluation of a high-growth reassortant H9N2 influenza A virus as a pandemic vaccine candidate. Vaccine 21:1974–1979

138. Tumpey TM, Basler CF, Aguilar PV, Zeng H, Solorzano A, Swayne DE, Cox NJ, Katz JM, Taubenberger JK, Palese P et al (2005) Characterization of the reconstructed 1918 Spanish influenza pandemic virus. Science 310:77–80

139. Cheung CL, Vijaykrishna D, Smith GJ, Fan XH, Zhang JX, Bahl J, Duan L, Huang K, Tai H, Wang J et al (2007) Establishment of influenza A virus (H6N1) in minor poultry species in southern China. J Virol 81:10402–10412

140. Chin PS, Hoffmann E, Webby R, Webster RG, Guan Y, Peiris M, Shortridge KF (2002) Molecular evolution of H6 influenza viruses from poultry in Southeastern China: prevalence of H6N1 influenza viruses possessing seven A/Hong Kong/156/97 (H5N1)-like genes in poultry. J Virol 76:507–516

141. Myers KP, Setterquist SF, Capuano AW, Gray GC (2007) Infection due to 3 avian influenza subtypes in United States veterinarians. Clin Infect Dis 45:4–9
142. Shortridge KF (1982) Avian influenza A viruses of southern China and Hong Kong: ecological aspects and implications for man. Bull World Health Organ 60:129–135
143. Shortridge KF (1992) Pandemic influenza: a zoonosis? Semin Respir Infect 7:11–25
144. Hoffmann E, Stech J, Leneva I, Krauss S, Scholtissek C, Chin PS, Peiris M, Shortridge KF, Webster RG (2000) Characterization of the influenza A virus gene pool in avian species in southern China: was H6N1 a derivative or a precursor of H5N1? J Virol 74:6309–6315
145. Gillim-Ross L, Santos C, Chen Z, Aspelund A, Yang CF, Ye D, Jin H, Kemble G, Subbarao K (2008) Avian influenza H6 viruses productively infect and cause illness in mice and ferrets. J Virol 82:10854–10863
146. Chen Z, Santos C, Aspelund A, Gillim-Ross L, Jin H, Kemble G, Subbarao K (2009) Evaluation of live attenuated influenza a virus H6 vaccines in mice and ferrets. J Virol 83:65–72
147. Zitzow LA, Rowe T, Morken T, Shieh WJ, Zaki S, Katz JM (2002) Pathogenesis of avian influenza A (H5N1) viruses in ferrets. J Virol 76:4420–4429
148. Katz JM, Lu X, Frace AM, Morken T, Zaki SR, Tumpey TM (2000) Pathogenesis of and immunity to avian influenza A H5 viruses. Biomed Pharmacother 54:178–187
149. Govorkova EA, Rehg JE, Krauss S, Yen HL, Guan Y, Peiris M, Nguyen TD, Hanh TH, Puthavathana P, Long HT et al (2005) Lethality to ferrets of H5N1 influenza viruses isolated from humans and poultry in 2004. J Virol 79:2191–2198
150. Lipatov AS, Hoffmann E, Salomon R, Yen HL, Webster RG (2006) Cross-protectiveness and immunogenicity of influenza A/Duck/Singapore/3/97(H5) vaccines against infection with A/Vietnam/1203/04(H5N1) virus in ferrets. J Infect Dis 194:1040–1043
151. Webby RJ, Perez DR, Coleman JS, Guan Y, Knight JH, Govorkova EA, McCain-Moss LR, Peiris JS, Rehg JE, Tuomanen EI et al (2004) Responsiveness to a pandemic alert: use of reverse genetics for rapid development of influenza vaccines. Lancet 363:1099–1103
152. van Riel D, Munster VJ, de Wit E, Rimmelzwaan GF, Fouchier RAM, Osterhaus ADME, Kuiken T (2006) H5N1 virus attachment to lower respiratory tract. Science 312:399
153. Shinya K, Ebina M, Yamada S, Ono M, Kasai N, Kawaoka Y (2006) Influenza virus receptors in the human airway. Nature 440:435–436
154. Nicholls JM, Chan MCW, Chan WY, Wong HK, Cheung CY, Kwong DLW, Wong MP, Chui WH, Poon LLM, Tsao SW et al (2007) Tropism of avian influenza A (H5N1) in the upper and lower respiratory tract. Nat Med 13:147–149
155. Hinshaw VS, Webster RG, Easterday BC, Bean WJ (1981) Replication of avian influenza A viruses in mammals. Infect Immun 34:354–361
156. Tumpey TM, Maines TR, Van Hoeven N, Glaser L, Solorzano A, Pappas C, Cox NJ, Swayne DE, Palese P, Katz JM et al (2007) A two-amino acid change in the hemagglutinin of the 1918 Influenza virus abolishes transmission. Science 315:655–659
157. Paniker CKJ, Nair CMG (1970) Infection with A2 Hong Kong influenza virus in domestic cats. Bull Wld Hlth Org 43:859–862
158. Paniker CKJ, Nair CMG (1972) Experimental infection of animals with influenza-virus types A and B. Bull Wld Hlth Org 47:461–463
159. Rimmelzwaan GF, van Riel D, Baars M, Bestebroer TM, van Amerongen G, Fouchier RA, Osterhaus AD, Kuiken T (2006) Influenza A virus (H5N1) infection in cats causes systemic disease with potential novel routes of virus spread within and between hosts. Am J Pathol 168:176–183
160. Karaca K, Swayne DE, Grosenbaugh D, Bublot M, Robles A, Spackman E, Nordgren R (2005) Immunogenicity of fowlpox virus expressing the avian influenza virus H5 gene (TROVAC AIV-H5) in cats. Clin Diagn Lab Immunol 12:1340–1342
161. Saito T, Lim W, Suzuki T, Suzuki Y, Kida H, Nishimura SI, Tashiro M (2002) Characterization of a human H9N2 influenza virus isolated in Hong Kong. Vaccine 20:125–133

162. Rimmelzwaan GF, Kuiken T, van Amerongen G, Bestebroer TM, Fouchier RA, Osterhaus AD (2001) Pathogenesis of influenza A (H5N1) virus infection in a primate model. J Virol 75:6687–6691

163. Kuiken T, Rimmelzwaan GF, Van Amerongen G, Osterhaus AD (2003) Pathology of human influenza A (H5N1) virus infection in cynomolgus macaques (*Macaca fascicularis*). Vet Pathol 40:304–310

164. Chen H, Li Y, Li Z, Shi J, Shinya K, Deng G, Qi Q, Tian G, Fan S, Zhao H et al (2006) Properties and dissemination of H5N1 viruses isolated during an influenza outbreak in migratory waterfowl in western China. J Virol 80:5976–5983

165. Rudenko L (2008) Live attenuated vaccine in Russia: advantages, further research and development. In: Katz JM (ed) Options for the control of influenza VI. International Medical Press Ltd, London, Atlanta, pp 122–124

166. Rudenko L, Desheva J, Korovkin S, Mironov A, Rekstin A, Grigorieva E, Donina S, Gambaryan A, Katlinsky A (2008) Safety and immunogenicity of live attenuated influenza reassortant H5 vaccine (phase I-II clinical trials). Influenza Other Respi Viruses 2:203–209

167. Kobasa D, Jones SM, Shinya K, Kash JC, Copps J, Ebihara H, Hatta Y, Hyun Kim J, Halfmann P, Hatta M et al (2007) Aberrant innate immune response in lethal infection of macaques wtih the 1918 influenza virus. Nature 445:319–323

168. Karron RA, Talaat K, Luke C, Callahan K, Thumar B, Dilorenzo S, McAuliffe J, Schappell E, Suguitan A, Mills K et al (2009) Evaluation of two live attenuated cold-adapted H5N1 influenza virus vaccines in healthy adults. Vaccine 27:4953–4960

169. Talaat KR, Karron RA, Callahan KA, Luke CJ, DiLorenzo SC, Chen GL, Lamirande EW, Jin H, Coelingh KL, Murphy BR et al (2009) A live attenuated H7N3 influenza virus vaccine is well tolerated and immunogenic in a Phase I trial in healthy adults. Vaccine 27:3744–3753

170. Karron RA, Callahan K, Luke C, Thumar B, McAuliffe J, Schappell E, Joseph T, Coelingh K, Jin H, Kemble G et al (2009) A live attenuated H9N2 influenza vaccine is well tolerated and immunogenic in healthy adults. J Infect Dis 199:711–716

171. Itoh Y, Shinya K, Kiso M, Watanabe T, Sakoda Y, Hatta M, Muramoto Y, Tamura D, Sakai-Tagawa Y, Noda T et al (2009) In vitro and in vivo characterization of new swine-origin H1N1 influenza viruses. Nature 460:1021–1025

172. Maines TR, Jayaraman A, Belser JA, Wadford DA, Pappas C, Zeng H, Gustin KM, Pearce MB, Viswanathan K, Shriver ZH et al (2009) Transmission and pathogenesis of swine-origin 2009 A(H1N1) influenza viruses in ferrets and mice. Science 325:484–487

173. Dormitzer PR, Rappuoli R, Casini D, O'Hagan D, et al (2009) Adjuvant is necessary for a robust immune response to a single dose of H1N1 pandemic flu vaccine in mice. PLoS Currents Influenza. http://knol.google.com/k/philip-r-dormitzer/adjuvant-is-necessary-for-a-robust/uhahw99c63lg/1?collectionId=28qm4w0q65e4w.1&position=13#. Accessed 9 Oct 2009

174. Clark TW, Pareek M, Hoschler K, Dillon H, Nicholson KG, Groth N, Stephenson I (2009) Trial of influenza A (H1N1) 2009 monovalent MF59-adjuvanted vaccine – preliminary report. N Engl J Med 361:2424–2435

175. Greenberg ME, Lai MH, Hartel GF, Wichems CH, Gittleson C, Bennet J, Dawson G, Hu W, Leggio C, Washington D, et al (2009) Response to a monovalent 2009 influenza A vaccine. N Engl J Med 361: 2405–2413

176. Munster VJ, de Wit E, van den Brand JM, Herfst S, Schrauwen EJ, Bestebroer TM, van de Vijver D, Boucher CA, Koopmans M, Rimmelzwaan GF et al (2009) Pathogenesis and transmission of swine-origin 2009 A(H1N1) influenza virus in ferrets. Science 325:481–483

177. Dawood FS, Jain S, Finelli L, Shaw MW, Lindstrom S, Garten RJ, Gubareva LV, Xu X, Bridges CB, Uyeki TM (2009) Emergence of a novel swine-origin influenza A (H1N1) virus in humans. N Engl J Med 360:2605–2615

178. Center for Biologics Evaluation and Research (2007) Guidance for industry. Clinical data needed to support the licensure of pandemic influenza vaccines. Food and Drug Administration, Rockville, MD, USA

179. European Agency for the Evaluation of Medicinal Products (EMEA) (2004) Guideline ondossier structure and content for pandemic influenza vaccine marketing authorization application. European Agency for the Evaluation of Medicinal Products, London
180. Federal Register 21 CFR Parts 314 & 610 (2002) New drug and biological drug products; evidence needed to demonstrate effectiveness of new drugs when human efficacy studies are not ethical or feasible. USA
181. Loosli CG (1948) The pathogenesis and pathology of experimental air-borne influenza virus A infections in mice. J Infect Dis 84:153–168
182. O'Neill E, Krauss SL, Riberdy JM, Webster RG, Woodland DL (2000) Heterologous protection against lethal A/HongKong/156/97 (H5N1) influenza virus infection in C57BL/6 mice. J Gen Virol 81:2689–2696
183. Pushko P, Tumpey TM, Bu F, Knell J, Robinson R, Smith G (2005) Influenza virus-like particles comprised of the HA, NA, and M1 proteins of H9N2 influenza virus induce protective immune responses in BALB/c mice. Vaccine 23:5751–5759
184. Smith H, Sweet C (1988) Lessons from human influenza from pathogenicity studies with ferrets. Rev Infect Dis 10:56–75
185. Smith W, Stuart-Harris CH (1936) Influenza infection of man from the ferret. Lancet 228:121–123

Live Attenuated Influenza Vaccine

Harry Greenberg and George Kemble

Abstract The development of the live, attenuated influenza vaccine (LAIV), based on the cold-adapted (*ca*), attenuated *ca* A/Ann Arbor/6/60 and *ca* B/Ann Arbor/1/66 backbones, has spanned several decades. The vaccine contains three vaccine strains, two attenuated influenza A strains and one attenuated influenza B strain; these vaccine strains are genetic reassortants, each harboring two gene segments from the currently circulating wild type virus conferring the appropriate antigens (e.g., A/H3N2, A/H1N1 or B) and the remaining six gene segments of the live, attenuated influenza A or influenza B donor virus. Both donor viruses have complex genetic signatures that control the key biological traits of the resulting genetic reassortants, including temperature-sensitivity in vitro and attenuation in an animal model, and the overall attenuation of the vaccine. Studies in humans have demonstrated that the attenuated vaccine strains can elicit humoral antibodies as well as cellular immunity; both responses are generally more readily detectable in children than in adults. A number of different clinical studies in children and adults have shown that this vaccine can reduce the burden of influenza illness in vaccinated subjects, including seasons in which the circulating wild type strain has antigenically drifted from the antigens included in the vaccine. These attributes of the live vaccine, as well as others including the ability to produce substantially more vaccine doses per egg than inactivated influenza vaccine make it a potentially useful platform to generate an effective vaccine, to combat both annual seasonal influenza and future influenza pandemics.

H. Greenberg
Departments of Medicine and Microbiology and Immunology, Stanford University School of Medicine, Stanford, CA, USA
Veterans Affairs Palo Alto Health Care System, Palo Alto, CA, USA
G. Kemble (✉)
MedImmune, Mountain View, CA, USA
e-mail: kembleg@medimmune.com

G. Del Giudice and R. Rappuoli (eds.), *Influenza Vaccines for the Future*, 2nd edition,   273
Birkhäuser Advances in Infectious Diseases,
DOI 10.1007/978-3-0346-0279-2_11, © Springer Basel AG 2011

1   Introduction

One key principle in determining whether a vaccination strategy is likely to be an effective means of controlling disease is the observation that natural infection by a pathogen results in the protection of the individual from subsequent illness with the same or highly related pathogens. Natural infection with wild-type influenza virus elicits a highly protective and long-lasting immune response that protects the individual from suffering influenza illness following re-exposure to the same, or similar, strain of influenza. Historical analyses have shown that individuals infected with an A/H1N1 strain in the early 1950s were protected from illness when the same virus circulated nearly 25 years later [1]. Despite this long-lasting and highly effective immunity, adults are susceptible to influenza-like illness on a regular basis. This apparent paradox is not due to waning or ineffective immune responses, but rather to the fact that the influenza virus continually evolves in the human population by undergoing genetic changes in all its genes including those encoding the major antigens on the virion surface, which are targets of protective immunity. These ongoing changes in the two surface proteins, the hemagglutinin (HA) and neuraminidase (NA) glycoproteins, lead to antigenic drift. At some point, the newly evolved drifted influenza strain differs sufficiently from its progenitor so that the immunity built up to the progenitor in the human population is no longer capable of efficiently reacting with the new influenza strain, and the drifted variant is now poised to cause a new epidemic wave of disease. This evolving pattern of mutations resulting in diminished immunity in the population takes place continually and it is the rule that every year at least one of the circulating influenza strains has mutated sufficiently to require a change in vaccine content in order to preserve high levels of efficacy.

Immune responses to influenza can be measured in many different compartments including IgG and IgA in the serum, secretory IgA in the nasal secretions, and T, B, and NK cells in the periphery as well as various lymphoid tissues, especially those in the respiratory tree. Functional antibodies that neutralize the virus or prevent it from binding its cognate receptor, frequently designated as hemagglutination inhibiting (HAI) antibodies, can be found in the serum and occasionally in nasal secretions; the cellular immune responses and additional antibody responses target a variety of regions on the viral HA, NA, and other proteins encoded by the virus particularly M, NP, and NS. Several of the immune responses to the virus have been correlated with protection from disease, especially the quantity of HAI or neutralizing antibodies in the serum and potentially, the titer of serum antibody to NA as well. Despite the presence of these multiple measures of immunological memory and effector functions and availability of substantial information correlating some of these measures with protection, the fundamental role each has in preventing illness following re-exposure to influenza remains to be elucidated. Due to the complexity of the immune response to influenza infection and the lack of a detailed mechanistic understanding of the specific components of the response that provide protection, designing an optimally effective vaccine that targets only a limited subset of viral peptides or antigens has been difficult. A vaccine strategy that effectively mimics the immune response elicited by natural infection would be expected to provide an effective, cross-reactive, and long-lasting immunity.

2 Background

Inactivated influenza vaccines were first put into use over 50 years ago for the military. In the late 1960s, the process of cold-adaptation was applied to the influenza virus for the purpose of generating a live attenuated vaccine with the hope that such a vaccine would generate broader and higher levels of immunity. In 2003, following decades of research and development, a live attenuated influenza vaccine (LAIV) based on the *ca* A/Ann Arbor/6/60 and *ca* B/Ann Arbor/1/66 backbones was licensed in the United States. This vaccine is used for active immunization to prevent influenza-like illness in children and adults, 2 through 49 years of age and is currently manufactured in specific pathogen-free embryonated chicken eggs. The vaccine is a trivalent blend of three LAIV vaccine strains, A/H1N1, A/H3N2, and B, each recommended annually by the U.S. Public Health Service to antigenically match the strains expected to circulate in the upcoming influenza season. The material is blended such that the dose of each strain is approximately 7 $\log_{10}$ of infectious particles and is filled into sprayer devices that deliver the vaccine liquid into the nasal passages. The original licensed vaccine formulation was stored frozen until immediately prior to use, while the current vaccine is stored at refrigerator temperature, 2–8°C. The first 2 decades of LAIV clinical studies, many sponsored by the U.S. National Institutes of Health (NIH), were performed using monovalent and bivalent formulations of the vaccine delivered by nasal drop rather than spray and have been extensively reviewed previously [2]. This chapter describes the key studies during the development and characterization of the trivalent formulation of LAIV.

3 Development of Cold-Adapted Influenza Vaccine Strains

Live attenuated vaccines that are delivered by the same route of entry as the wild-type pathogen are expected to induce an immune response that is similar to the natural pathogen without eliciting the typical signs or symptoms associated with illness. This approach does not require a predetermined knowledge of the identity or structure of the crucial protective antigens nor a defined mechanistic understanding of the immune effector functions that mediate protection. In the 1960s, John Maassab at the University of Michigan set out to attenuate influenza virus for vaccine use through a process designated cold-adaptation. Forcing the virus to replicate efficiently at lower than normal temperatures was predicted to result in changes to its genetic makeup making it less fit to replicate at normal and elevated body temperatures, thereby attenuating the strain. The A/Ann Arbor/6/60 (H2N2) strain was isolated from a patient and serially passaged at reduced temperatures in both eggs and chicken cells along with biologically cloning the progeny at several intervals [3]. Biological characterization demonstrated that the resulting virus was cold adapted (*ca*), as defined by its ability to replicate to titers at 25°C that were within 2 $\log_{10}$ of titers obtained at 33°C, and temperature sensitive (*ts*), as defined by replication of the virus at 39°C that was debilitated by at least 2 $\log_{10}$ compared to its replication at 33°C [4]. These newly acquired properties of the cold-adapted progeny, designated *ca* A/Ann Arbor/6/60, distinguished it from its parent as well as most wild-type influenza strains. The spectrum of temperatures at which the *ca* virus replicated well was lower than the wild-type viruses that caused disease. Of note, further characterization of *ca* A/Ann Arbor/6/60 in the highly susceptible ferret model demonstrated that it was attenuated compared to wild-type influenza viruses. In contrast to the parental wild-type A/Ann Arbor/6/60 strain, the *ca* virus was unable to replicate in the lung tissues of ferrets or elicit signs of influenza-like illness [5]. Following the success of adaptation of influenza A, Massaab and his colleagues later isolated and cold-adapted an influenza B virus in a similar manner. This virus, designated *ca* B/Ann Arbor/1/66, had similar *ca*, *ts* and attenuated (*att*) properties as its influenza A counterpart [6]. This virus was even more restricted at higher temperatures than *ca* A/Ann Arbor/6/60, in that it was significantly restricted in replication at temperatures as low as 37°C. These two strains provide the genetic background of all LAIV strains, imparting their *ca*, *ts*, and *att* properties to the vaccine.

The influenza virus genome is comprised of eight different RNAs or gene segments. Individual monovalent LAIV strains are derived by combining the gene segments encoding the two surface glycoproteins, HA, and NA, of a contemporary field isolate of influenza with the remaining six internal gene segments of the appropriate *ca* master donor virus (MDV), either *ca* A/Ann Arbor/6/60 or *ca* B/Ann Arbor/1/66. The resulting 6:2 genetic reassortant combines the attenuation inherent to the MDVs with the antigens needed to elicit a neutralizing immune response that should prevent disease caused by currently circulating strains of influenza.

4 Basic Properties of the Vaccine

4.1 Genetic Basis of Biological Properties of the Vaccine

Sequence analysis and comparison of the genomes of *ca* A/Ann Arbor/6/60 and *ca* B/Ann Arbor/1/66 to their respective parental strains confirmed that a number of changes had accumulated during passage. Which of these changes were key to controlling the newly acquired biological properties was not immediately evident by merely examining the sequences. The first studies to determine the genetic basis of the *ca*, *ts*, and *att* phenotypes were performed by creating reassortant viruses *via* co-infection of a cell with two biologically distinct strains, the MDV, and, typically, a wild-type field isolate. The gene segments of these two viruses would reassort and the resulting progeny could be isolated from each other and independently characterized for retention, loss, or modification of the specific phenotype and their genetic composition. Because there was little control over which segments would reassort, distinct strains were used to facilitate selection and screening for the desired progeny. These studies helped identify the contribution of the PB1 gene segment to the *ts* phenotype of *ca* A/Ann Arbor/6/60. Sometimes, however, the results were misleading in that the loss of a biological property was not due to a specific mutation or set of mutations but rather to an incompatibility of gene segments from two diverse parental influenza strains, also known as the constellation effect. One of the best-documented constellation effects was observed with the MDV-A M gene segment. A recombinant wild-type virus harboring the M gene segment of MDV-A was attenuated in animals leading to the conclusion that specific mutations in this gene segment were responsible for the *att* phenotype [7]. However, later work showed that a similar recombinant harboring the M gene segment from the parental wild-type A/Ann Abor/6/60 strain was also attenuated [8]. The biological result was not due to the one nucleotide difference in the M gene segments between *ca* and wild-type pair; rather the phenotype was due to the inability of the MDV-A M gene segment, derived from A/Ann Arbor/6/60, to interact optimally with the other gene segments of the divergent field isolate.

The introduction of reverse genetics enabled biological traits to be associated with specific nucleotides without having to account for potential problems caused by constellation effects. Using reverse genetics recombinant viruses are derived directly from cDNA clones of the viral gene segments, allowing recombinant viruses with only one or a defined set of changes to be produced. No selection system or extensive screening procedure is required to obtain the desired recombinant virus; the genome of the recombinant virus accurately reflects the genetic content of the cDNAs used to produce it. To map and study the impact of the genetic changes between the wild-type and *ca* virus pairs two derivatives of A/Ann Arbor/6/60 and B/Ann Arbor/1/66 were produced, one contained the nucleotides present in the MDV, the other encoded either eight or nine amino acid changes in the internal gene segments, respectively, representing the wild-type progenitors. The changes resulting in the wild-type progenitor sequence were expected to result in non-*ts* and non-*att* properties; these properties were confirmed following biological characterization. Recombinant viruses were then derived by making individual or grouped changes in the wild-type or MDV version of the strain and evaluating the resulting phenotype. The culmination of these studies demonstrated that five nucleotide positions distributed between the PB1, PB2, and NP gene segments of A/Ann Arbor/6/60 controlled both the *ts* and *att* properties [9]. Studies with B/Ann Arbor/1/66 revealed that three positions (two in PA and one in NP) controlled the *ts* phenotype, an additional two nucleotides in M controlled the *att* phenotype, and another subset of three changes in PA and PB2 were responsible for the *ca* phenotype [10,11]. The robustness of these genetic signatures was demonstrated by placing only the minimal set of changes into divergent influenza strains and demonstrating the accompanying transfer of the biological traits. For example, the five changes responsible for controlling *ts* and *att* of MDV-A were introduced into A/PR8/34 (H1N1) and the resulting recombinant virus acquired both the *ts* and *att* phenotypes [12]. Similar studies with *ca* B/Ann Arbor/1/66 and B/Yamanashi/166/98 were conducted with similar results [10].

Further molecular and biochemical studies have revealed several different mechanistic differences between the MDVs and their respective parental wild type strains responsible for the poor replication at elevated temperatures. For MDV-A, the amount of mRNA and protein synthesized at the restrictive temperature is not significantly impacted; however, the amount of vRNA was reduced and a significant block in the export of vRNP from the nucleus was observed. Virions released from cells incubated at 39°C, were highly irregular in shape and the quantity of M1 protein was greatly reduced [13]. All of these defects combine to restrict the replication of MDV-A at higher temperatures. Studies of MDV-B revealed defects in the RNA polymerase functions of the PA and NP proteins at the restrictive temperature (37°C) resulting in poor protein synthesis and vRNA production at this temperature. The M1 protein of MDV-B was also contributing to restricted replication at higher temperatures and was packaged inefficiently into virions at higher temperatures [14]. The fundamental mechanisms restricting the replication of these vaccine strains at elevated temperatures are a result of the complex genetic signatures underlying them and work at multiple points of the replication cycle to provide a robust and stable set of attenuating changes to the vaccine.

4.2   Genetic Stability of the Vaccine in Manufacturing

Influenza virus, like other RNA viruses, has an RNA-dependent RNA polymerase that lacks a proofreading function. Picking and sequencing individual plaque isolates demonstrated an observed mutation frequency of one change per 10,000 nucleotides resulting in a rate of 1.5×10^{-5} mutations per replication cycle [15]. Because of this inherent capacity of influenza virus gene segments to change, the genetic stability of the vaccine was characterized both within the context of the manufacturing process as well as following intranasal administration. In general, manufacturing of the bulk vaccine only requires the seed material to be passaged once or twice in embryonated eggs. To evaluate the stability of the genetic elements during manufacture, the genomic sequences of bulk vaccine and its progenitor seed materials were analyzed, represented by over nine different strains distributed among nine independent seed materials and over 50 bulk vaccine stocks. Comparisons of the genomic sequences of these materials demonstrated that in all cases the bulk vaccine was identical to the seed material. These data demonstrated that the vaccine's genetic composition is stable and unchanged within the parameters used to manufacture the vaccine on a large scale [16].

4.3 Genetic Stability of the Virus In the Respiratory Tract and Transmission

Following intranasal administration, the vaccine virus infects and replicates in epithelial cells of the upper respiratory tract resulting in an immune response. Characterizing the genetic stability of the vaccine in humans is an important element for understanding the properties of the vaccine. Over the course of multiple decades studying monovalent, bivalent, and trivalent formulations of this vaccine in clinical studies, no revertants of the vaccine have been identified [2]. To evaluate the stability of the vaccine following replication in the upper respiratory tract of humans, a prospective shedding and transmissibility study was designed. Young children were selected due to the relatively longer duration and greater level of shedding of the vaccine following administration. Therefore, this population was expected to represent the most permissive setting for detecting revertants if they were to arise. In the study of genetic stability, 98 children aged 9–16 months were vaccinated with LAIV and nasal swab samples were taken at frequent and regular intervals. Of the children in the study, 86% shed at least one of the three strains in the vaccine with peak titers ranging from 1 to 8 days post vaccination and the last isolate shed 21 days post vaccination. The *ca* and *ts* phenotypes were preserved in all the shed viruses tested (135 of 250 isolates were tested) [17]. Of the isolates, 54 were chosen at random and their genomes sequenced in their entirety and compared to the sequences of the strains used to vaccinate the children. These analyses revealed that some genetic changes had occurred in a majority of shed isolates and in some cases the mutations were shared by multiple isolates [18]. These changes could have arisen during replication in the upper respiratory tract or could have preexisted in the vaccine material at a level not detected by sequence analysis of the bulk material. To address the latter hypothesis, samples of the vaccine material were obtained, amplified by RT-PCR and individual clones were sequenced. Interestingly, in most cases, the change(s) evident in the isolate shed from the child were representative of changes that preexisted in the bulk vaccine material. Despite the presence of these mutations, all isolates invariably retained their characteristic biological properties, confirming the exquisite genetic stability that had been previously described in observational studies.

A corollary concern associated with genetic stability and vaccine shedding is the potential for person-to-person transmission of the virus. Shedding of the vaccine from an individual is a necessary predecessor for transmission to an unimmunized contact; however, shedding is not necessarily sufficient for transmission to occur. The study of the genetic stability of the vaccine in children was also designed to assess the probability of transmission of the vaccine virus. Young children in a daycare setting were expected to increase the likelihood of detecting a transmission event should it occur due to the relatively high level of shed virus in respiratory secretions, relatively high level of susceptibility to vaccine take, and the general absence of hygienic practices among young children that generally inhibit transmission of viruses in adults. In addition to the 98 children vaccinated with LAIV,

97 children received placebo. These children were intermixed and placed in cohorts that played together in a daycare environment for at least 4 h every day for 3 or more days each week. Nasal swabs were obtained at regular and frequent intervals from each child and the presence of vaccine virus was assessed. Vaccine virus was recovered from 80% of the vaccinated children and in only one confirmed case from a placebo recipient [17]. The influenza B vaccine virus recovered from the placebo recipient was shed on only one day and matched the genetic signature of an influenza B vaccine virus shed by a vaccinated member of the same playgroup several days earlier. The transmitted vaccine isolate was shown to retrain its characteristic *ca* and *ts* properties and exhibited the attenuation phenotype in ferrets; additionally, the placebo child had no signs or symptoms distinguishable from other children in the study. These results were applied to the Reed-Frost model that indicated a 0.58% probability of vaccine transmission occurring from a single contact of a vaccinated young child with an unvaccinated young child [17]. The likelihood of transmission from a vaccinated adult would be expected to be substantially lower, since adults shed virus less frequently and in lower amounts than children. In a study designed to characterize the shedding of LAIV in adults, vaccine was recovered from nasal swabs of only 50% of individuals 3 days after vaccination and, by 10 days after vaccination only 5% of individuals had vaccine detectable in their nasal swab [19]. This low probability of transmission combined with the vaccine's genetic stability give additional confidence in the use of LAIV in children.

5 Basis of the Immune Response

Infection with wild-type influenza virus leaves the individual with a strong immunological memory that will prevent the same or antigenically similar variant from causing disease again in the same individual for decades. This immunological memory can be detected in many different compartments including local mucosal immunity, serum antibody, and T cells. The immune response to vaccination with LAIV has been studied in multiple different settings and the immune response is qualitatively similar but quantitatively less than that elicited by natural infection; immunity can be documented by mucosal IgA, serum HAI and neutralizing antibodies and cellular T and B cell responses. This observation is not surprising given that the vaccine stimulates immunity by replication in the upper respiratory tract similar to that of the wild-type virus. Despite finding evidence for vaccine-induced immunity at both local and systemic compartments, the specific functional role of any particular immune response and validated correlates of protection from influenza disease in vaccinated individuals remains unproven.

LAIV elicits a robust immune response in young children, particularly those that are seronegative for influenza at the time of vaccination [20–22]. Seroconversion rates, measured by the presence of hemagglutination inhibition antibody in the serum, can be as high as 80–90% or more in young children after two doses of vaccine. Seroconversion rates typically are lower for children or adults who have preexisting antibody at the time of vaccination [23]; however, at least in children, preexisting immunity from prior influenza infection or vaccination does not appear to result in reduced protection following LAIV [24]. The presence of antibody at the time of immunization may both limit the extent of replication of the vaccine in the upper airways, evidenced by lower rates of shedding, as well as mask the boosting of the immune response using relatively crude measures of immunogenicity such as HAI antibody in the serum. Other immune responses to LAIV have been documented in children including secretory IgA in nasal secretions and IgG and IgA antibody secreting cells (ASC) in the circulation 7–10 days after immunization [25,26]. T cellular immune responses have been evaluated in children. Children 6 to <36 months of age had measurable IFNγ secreting cells in their PBMCs by 13 days after LAIV; these responses were not evident in children vaccinated intranasally with heat inactivated LAIV or intramuscularly with inactivated vaccine [27]. Following immunization of children aged 5–9 years, blood was collected at 10 and 28 days post vaccination and stimulated with the A/H3N2 strain ex vivo. Both the CD4 and CD8 influenza-specific T cells were increased in these children compared to their pre-vaccination values; additionally, these increases were greater than those observed for TIV-immunized children in the same study and the CD8$^+$ T cells induced by vaccination underwent a number of specific phenotypic changes [28–30]. Of note, in this same population antibody-secreting B cells were also detected in the periphery within 7–10 days post vaccination.

Immunological markers in adults have been more difficult to measure. Virtually all adults have had multiple encounters with wild-type influenza and potentially influenza vaccination during their lifetime and have readily measurable levels of influenza antibody in their serum prior to vaccination. In contrast to studies in young seronegative children, vaccination of adults with LAIV infrequently produces a measurable increase in serum HAI antibody titers. Recent studies on T cell immunity following vaccination had similar results. One study found no demonstrable increases in CD4, CD8, or NK activity when cells were stimulated ex vivo either 10 or 28 days following vaccination, however virus specific CD8 cells were shown to have decreased CD27 expression following LAIV, consistent with the presence of effector CD8 cells [28,29]. A separate study described a modest twofold increase in IFNγ PBMCs by ELISPOT in both adult LAIV and TIV vaccines [31]. Flow cytometry of these cells demonstrated that the cells elicited by LAIV, but not TIV, had a CD4$^+$/CD69$^+$/CD18$^+$/MIP1α phenotype consistent with functional tissue-tropic lymphocytes. The levels of pre-vaccination influenza specific CD4 and CD8 cells increased with age of the subjects and that adults had significantly higher baseline quantities than children [29]. The level of pre-vaccination influenza specific CD4$^+$ T cells seems to be a critical determinant of whether or not vaccines experience a subsequent rise in either CD4 or CD8 T cells [30]. The relatively modest absolute increase in the cellular response may again be due to a higher level of influenza immunity prior to vaccination in adults; however, despite the lack of a large quantitative increase, the phenotypic properties of the virus specific T cells appear to be influenced by LAIV. In contrast to the T cell and

HAI responses, adults were generally shown to have increased influenza-specific antibody-secreting B cells in the blood 7–10 days post LAIV vaccination. While only 16% of the adults had a serological response measured by a fourfold or greater increase of HAI antibody following immunization, approximately 80% of the subjects had a measurable increase in the number of influenza-specific IgG-secreting antibody-secreting cells in the periphery and this was true for individuals who had been vaccinated in the prior year as well [25,26]. These data clearly demonstrated that LAIV elicited a readily detectable B cell response in most adults, which is consistent with the clinical experience that LAIV is highly efficacious in an adult population aged 18–49 [32,33]. The current immunological markers typically used to assess the function of influenza vaccines, such as HAI in the serum, may not be sufficiently sensitive nor monitor the appropriate compartment to detect a functional immune response to LAIV and better assays that can readily distinguish between preexisting and vaccine induced antibody are needed.

Vaccine studies often rely on correlate markers to demonstrate that the vaccine will perform as expected under the conditions being studied. A robust correlate of protection is an immunological marker that when present coincides with protection from disease upon subsequent exposure to the wild-type virus and the lack of which correlates with susceptibility to illness. Due to high rates of efficacy demonstrated for LAIV combined with the difficulty in using traditional serum-based influenza assays to measure an immune response in adults, these markers have been difficult to identify for LAIV. In children, particularly young children who are immunologically naive to influenza, vaccination with LAIV elicits a robust immune response that can be detected in multiple compartments. One study evaluated potential correlates of protection by inoculating children with either vaccine or placebo, followed by a relatively long interval at which point samples were taken to measure humoral immune responses, and then these children were given another dose of a monovalent vaccine strain. By correlating the level of immune response using a variety of immunological assay systems with shedding of the challenge vaccine virus, the utility of the various assays for predicting protective immunity was assessed [20]. First, the presence of serum HAI antibody strongly correlated with the absence of shed challenge virus. These data demonstrated that at least in children, a positive correlation existed for this marker. However, when the two groups of children who were seronegative at the time of challenge were compared, there was a significant increase in the number of children who shed vaccine in the placebo group compared to the LAIV-vaccinated group. These data led to the conclusion that the absence of serum HAI did not correlate completely with susceptibility. The same trends were observed when secretory nasal IgA was used as the marker. The observation that the presence of these markers correlates with protection helps identify potentially useful immunological measures; however, the observation that absence of these markers does not correlate with susceptibility argues that other important immune mechanisms may be overlooked when only serum HAI or secretory IgA are evaluated. Studies have demonstrated that LAIV elicits both humoral and cellular immune responses and those responses can be found at both mucosal and peripheral sites. A more recent study combining clinical efficacy with the IFNγ ELISPOT assay has suggested that a threshold number of IFNγ secreting T cells may eventually be identified that correlates with vaccine induced protection from disease, however, further studies will be needed to define the threshold among different strains, seasons and populations [27]. The functional immune mediators that govern protection from disease have not yet been elucidated, may be multi-factorial and may differ among populations as well. Further study will be needed to identify practical correlates for vaccine efficacy as well as detailed immunological profiling to understand the functional components of an effective immune response and how it controls disease.

6 Performance of the Vaccine in Clinical Studies

Vaccines derived from *ca* A/Ann Arbor/6/60 and *ca* B/Ann Arbor/1/66 have been extensively characterized in clinical studies. Prior to the mid 1990s, monovalent and bivalent forms of these vaccines were evaluated in over 15,000 subjects in a number of different clinical studies, many of which were sponsored by the NIH [2]. More recently, studies focused on both the frozen and refrigerator stable trivalent formulations of LAIV, have been conducted in a wide range of settings in individuals from 6 months to over 80 years of age.

The efficacy of the trivalent form of the vaccine has been evaluated in a number of settings in different age cohorts throughout the world. The vaccine has reproducibly been shown to prevent influenza-like illness (ILI) caused by all three influenza types including, A/H1N1, A/H3N2, and B. A meta analysis of placebo-controlled studies measured the mean efficacy of two doses in previously unvaccinated young children of 77% (95% CI: 72, 80), with efficacy of 85, 76, and 73% against A/H1N1, A/H3N2, and B, respectively; the mean efficacy of one dose in previously vaccinated children was 87% (95% CI: 81, 90) [27,34–42]. In addition, a single dose of vaccine, while not optimal, has been shown to provide a high degree of clinical efficacy among previously unvaccinated young children [37,43]. Interestingly, three studies were conducted in which LAIV was compared to TIV-vaccinated subjects. In the largest of these studies, which included over 8,000 children and approximately 491 isolates from children who had modified CDC-ILI, LAIV was shown to reduce the burden of illness by nearly 55% compared to TIV. Of note, all the A/H3N2 strains circulating in this study were antigenically mismatched to the two vaccines and the children vaccinated with LAIV had 79% fewer cases of modified CDC-ILI compared to the TIV group [35]. In two other studies, one conducted in children with recurrent respiratory illness and the other in older children with asthma, LAIV was also shown to be more efficacious than TIV [38,39].

Two placebo controlled field studies in adults have been reported using either effectiveness endpoints [44] or culture confirmed prevention of influenza-like illness in adults 60 years or older [45]. Additional comparative efficacy and effectiveness trials in adults have evaluated LAIV and TIV. In a study in which serosusceptible subjects were challenged with wild type virus, laboratory documented illness was observed in 7% of LAIV recipients, 13% of TIV recipients, and 45% of placebo subjects demonstrating that both vaccines were efficacious [33]. In a series of field studies in young adults TIV was shown to be more efficacious than LAIV, however, both groups had less illness than observed in the placebo group. In a study conducted in the 2007–2008 influenza season, 1952 subjects were enrolled and the inactivated vaccine was shown to have an efficacy of 72% (95% CI, 48–84) compared to the placebo and LAIV had an efficacy of 29% (95%CI, −14 to 55) compared to the placebo resulting in a relative efficacy of 60% (95% CI 33–77) for the inactivated vaccine [46–48]. These two vaccines have also been studied in military personnel, who can be exposed to high rates of morbidity due to influenza illness. In a retrospective cohort analysis, LAIV was more effective than TIV at preventing influenza illness in recruits and TIV was slightly more effective in nonrecruits [49]. In an analysis of more than a million nonrecruits, TIV was more effective at lowering health care encounters for pneumonia and influenza than LAIV, and the latter was shown effective in only one of the three seasons analyzed. However, LAIV was effective in the subset of vaccine naïve service members and similar to the effectiveness of TIV [50]. The varying results of these studies in adults compared to studies in children, where LAIV has appeared to be more efficacious than TIV, may reflect the interaction of LAIV with the immune system of the adult host.

In controlled studies, the most common adverse events in children were runny nose or nasal congestion, low-grade fever, decreased activity, and decreased appetite. In the youngest children, who received two doses of vaccine, no significant differences were observed following the second dose. In adults, the most common adverse events are runny nose/nasal congestion, cough, and sore throat, which were all short lived. The reactogenicity of the vaccine is consistent with replication of a live attenuated virus in the nasal epithelium of the subject. In a large safety database study performed in Northern California Kaiser Hospital system, a 3.5-fold increase in asthma events were noted within 42 days of vaccination in the pre-specified age stratum of 18–35 months [51]. The observation was further investigated in the large efficacy study of LAIV and TIV in young children. In this latter study in the age stratum less than 24 months of age (6–23 months) there were 3.2% of children in the LAIV group who had medically attended wheezing events within 42 days of vaccination compared to 2.0% in the TIV group. This difference was significant. There was no significant difference in rates after 42 days or in the children 24 months of age or older [36].

7 LAIV Results During Circulation of Antigenically Different Strains

Influenza virus continually evolves; changes in the HA molecule give rise to variants that are capable of escaping from the preexisting immunity in the population. Predicting which of these variant drifted strains will give rise to the next epidemic wave of seasonal influenza is an annual challenge addressed by the global public health authorities. In general, matching the vaccine antigen to the upcoming season's influenza strain should result in the best opportunity to produce effective influenza vaccines; however, because of the continuous nature of antigenic drift, it is not a rare instance when strains chosen for inclusion in the vaccine do not match well with the epidemic virus. Immunity elicited by LAIV may provide for a larger margin of error for antigenic mis-matching than occurs after inactivated vaccine administration. LAIV has been shown to provide protection against significantly antigenically drifted variants in several clinical settings. In 1997–1998, children were immunized with a trivalent blend of LAIV containing the A/Wuhan/359/95 (H3N2) strain. The virus that circulated in the community that year was designated A/Sydney/05/97 (H3N2) and was antigenically quite distinct from the H3N2 antigen contained in the vaccine. Despite this level of mismatch, the vaccine conferred efficacy greater than 85% against the A/Sydney/05/97 H3N2 virus [35]. That same season, LAIV was also shown to be effective in adults by monitoring febrile upper respiratory tract infections and other associated medical utilizations [44]. In the head to head study of LAIV and TIV in children, LAIV reduced modified CDC-ILI caused by an antigenically drifted A/H3N2 strain by 79% compared to TIV [36].

While the relationship between the importance of serum antibody responses and protection remains undefined, evaluation of the reactivity of serum antibodies to drifted strains have been performed [21,22]. Young children vaccinated with LAIV develop a robust immune response to vaccination that can be measured by a rise in serum HAI and neutralizing antibody titers. Children immunized with LAIV containing the A/Panama/2007/99 strain developed high levels of HAI and neutralizing antibody following one dose; in contrast only a minority of children receiving one dose of TIV with the same antigen responded to the vaccine. Notably, antibodies from children vaccinated with LAIV had significant reactivity to the drift variant that circulated through the community that year, the A/Fujian-like (H3N2), whereas the children receiving TIV had little to no reactivity to this strain [22].

8   Influenza Vaccination on Large Population

Influenza vaccination has been shown to have indirect benefits to others in the community who are not vaccinated. In Japan, the implementation of mandatory TIV vaccination of school-aged children with inactivated influenza vaccine resulted in a significant drop in the rate of pneumonia and influenza (P and I) mortality in the elderly. The rate of P and I mortality remained low for the duration of mandatory vaccination program and returned to higher baseline levels within 2 years after the program was abolished, demonstrating the powerful impact of reducing the burden of illness in young children on the community at large [52,53]. Two large field studies of the LAIV vaccine have been reported in which the impact of vaccination on both the vaccinated population and the non-vaccinated population were studied.

In a large open-label study in the Temple-Bolton area of Texas, LAIV was administered to several thousand children and the rates of medically attended acute respiratory illness were measured and compared to a similar control community. LAIV vaccination significantly reduced illness in the vaccinated individuals in the intervention community even in years in which a drift strain circulated through the community [54,55].

9 LAIV Technology and Pandemic Preparedness

Application of an effective and widely available vaccine is the best solution to prevent significant morbidity and mortality resulting from a severe influenza pandemic. A vaccine that has the ability to elicit an immune response in individuals who have had limited exposure to influenza, protect against disease caused by drifted strains, ease of administration, and the potential for rapid large scale manufacturing would be a good solution for a pandemic response. LAIV technology has the potential to match these characteristics. Extending the understanding of seasonal LAIV is a major component of pandemic planning during interpandemic periods. The emergence of a pandemic is the result of adaptation of the novel pandemic strain to humans, regardless of whether it originated in fowl, swine or other species. This adaptation is demonstrated by efficient human to human transmission combined with efficient replication in respiratory tissues. These properties of the pandemic strain are similar to those of annual epidemic strains in susceptible humans and predict that the pandemic LAIV strain will perform in a predictable manner in the manufacturing infrastructure and, importantly, be effective in preventing influenza illness, similar to seasonal LAIV strains.

A second element of pandemic planning was undertaken to construct and characterize several prototype LAIV pandemic strains to subtypes that were not fully adapted to humans. These studies were focused on determining whether constructing libraries of potential LAIV pandemic strains long before a pandemic occurred. Several LAIV vaccine candidates have been constructed that express HA and NA of avian influenza subtypes H9N2, H7N3, and H5N1. By using reverse genetics to construct these vaccine candidates the HA of highly pathogenic wild-type H5N1 strains, as well as highly pathogenic strains of other subtypes, could be modified by removing the multibasic amino acid proteolytic cleavage site between HA1 and HA2, one of several virulence determinants in the wild type strains, prior to constructing the vaccine candidate. All of these prototype pandemic LAIV strains were highly attenuated in chickens, mice, and ferrets, and yet produced immune responses that protect mice and ferrets from challenge with antigenically similar as well as antigencially drifted strains [56–58]. Furthermore, murine studies of the H5N1 LAIV candidates with and without the multibasic cleavage site, demonstrated that removal of the multibasic cleavage site contributed to the attenuation of the vaccine candidate and, as a result of lower levels of replication, reduced the immunogenicity in this model [59]. These same prototype pandemic LAIV strains were evaluated in small scale clinical studies in adults. The replication of each of these vaccine candidates was highly restricted in seronegative adults and the immune response varied by strain. Measuring serum antibodies by a combination of HAI, ELISA, and neutralization assays resulted in 100, 90, and 52% of the subjects exhibiting evidence of immune response following two doses of the H9N2, H7N3, and H5N1 LAIVs, respectively [60–62]. Each of these prototype pandemic LAIVs was highly attenuated, likely more than typical seasonal LAIV strains, possibly as a result of the presenting the human immune system with a vaccine expressing atypical, avian HA and NA antigens.

Several aspects of a pandemic are likely to be different than a typical seasonal epidemic and these unique features will alter the normal course of actions taken by public health authorities as well as vaccine manufacturers. First, the pandemic is likely to spread quickly and on a global scale. An effective vaccine will need to be administered to a large portion of the world's population. Currently, the annual worldwide distribution of influenza vaccines is only adequate for 300 million doses, far short of the six billion people who will need the vaccine. In addition, the nature of the pandemic antigen will be atypical; it will be comprised of an HA that has not circulated previously and vaccine seed strains will need to be quickly assembled. LAIV technology has the potential, capacity to prevent disease caused by antigenically drifted strains. A second essential feature of an effective pandemic strategy is rapid and large-scale production capacities. The dose of $7 \log_{10}$ infectious particles of LAIV is a small antigenic mass compared to TIV. One dose of LAIV represents less than approximately 1% of an inactivated (15 µg) vaccine dose of antigen. This efficiency translates into the potential to rapidly produce large quantities of bulk LAIV compared to inactivated vaccine. The capacity to produce large amounts of LAIV rapidly combined with its potential for cross-protection, ease of administration and high degree of efficacy in immunologically naive populations make this a promising candidate for pandemic preparedness.

10   Conclusion

The utilization of this novel vaccine technology continues to be refined and improved. Recent studies in children should enable greater use of this vaccine in this highly susceptible and vulnerable population. The current manufacturing methods used to make LAIV are based on production technologies that are over 50 years old; more modern production methods including manufacturing in cell culture substrates, are being developed. In addition, the generation of the 6:2 reassortant viruses used to initiate seed strain is being refined and integrated with the use of reverse genetics technology. Finally, the attributes that make this vaccine effective in young children is being further explored and developed to apply to pandemic solutions.

References

1. Wright PF, Neumann G, Kawaoka Y (2007) Orthomyxovirus. In: Knipe DM, Howley PM (eds) Fields virology, 5th edn. Lippincott, Philadelphia
2. Murphy BR, Coelingh K (2002) Principles underlying the development and use of live attenuated cold-adapted influenza A and B virus vaccines. Viral Immunol 15(2):295–323
3. Maassab HF (1967) Adaptation and growth characteristics of influenza virus at 25°C. Nature 213(5076):612–4
4. Maassab HF (1968) Plaque formation of influenza virus at 25°C. Nature 219(5154):645–6
5. Maassab HF, Francis T Jr, Davenport FM, Hennessy AV, Minuse E, Anderson G (1969) Laboratory and clinical characteristics of attenuated strains of influenza virus. Bull World Health Organ 41(3):589–94
6. Maassab HF (1970) Developments of variants of influenza virus. In: Barry RD, Mahy BWJ (eds) The biology of large RNA viruses. Academic, London, pp 542–66
7. Snyder MH, Betts RF, DeBorde D, Tierney EL, Clements ML, Herrington D et al (1988) Four viral genes independently contribute to attenuation of live influenza A/Ann Arbor/6/60 (H2N2) cold-adapted reassortant virus vaccines. J Virol 62(2):488–95
8. Subbarao EK, Perkins M, Treanor JJ, Murphy BR (1992) The attenuation phenotype conferred by the M gene of the influenza A/Ann Arbor/6/60 cold-adapted virus (H2N2) on the A/Korea/82 (H3N2) reassortant virus results from a gene constellation effect. Virus Res 25(1–2):37–50
9. Jin H, Lu B, Zhou H, Ma C, Zhao J, Yang CF et al (2003) Multiple amino acid residues confer temperature sensitivity to human influenza virus vaccine strains (FluMist) derived from cold-adapted A/Ann Arbor/6/60. Virology 306(1):18–24
10. Hoffmann E, Mahmood K, Chen Z, Yang CF, Spaete J, Greenberg HB et al (2005) Multiple gene segments control the temperature sensitivity and attenuation phenotypes of ca B/Ann Arbor/1/66. J Virol 79(17):11014–21
11. Chen Z, Aspelund A, Kemble G, Jin H (2006) Genetic mapping of the cold-adapted phenotype of B/Ann Arbor/1/66, the master donor virus for live attenuated influenza vaccines (FluMist). Virology 345(2):416–23
12. Jin H, Zhou H, Lu B, Kemble G (2004) Imparting temperature sensitivity and attenuation in ferrets to A/Puerto Rico/8/34 influenza virus by transferring the genetic signature for temperature sensitivity from cold-adapted A/Ann Arbor/6/60. J Virol 78(2):995–8
13. Chan W, Zhou H, Kemble G, Jin H (2008) The cold adapted and temperature sensitive influenza A/Ann Arbor/6/60 virus, the master donor virus for live attenuated influenza vaccines, has multiple defects in replication at the restrictive temperature. Virology 380 (2):304–11
14. Chen Z, Aspelund A, Kemble G, Jin H (2008) Molecular studies of temperature-sensitive replication of the cold-adapted B/Ann Arbor/1/66, the master donor virus for live attenuated influenza FluMist vaccines. Virology 380(2):354–62
15. Parvin JD, Moscona A, Pan WT, Leider JM, Palese P (1986) Measurement of the mutation rates of animal viruses: influenza A virus and poliovirus type 1. J Virol 59(2):377–83
16. Buonagurio DA, Bechert TM, Yang CF, Shutyak L, D'Arco GA, Kazachkov Y et al (2006) Genetic stability of live, cold-adapted influenza virus components of the FluMist/CAIV-T vaccine throughout the manufacturing process. Vaccine 24(12):2151–60
17. Vesikari T, Karvonen A, Korhonen T, Edelman K, Vainionpaa R, Salmi A et al (2006) A randomized, double-blind study of the safety, transmissibility and phenotypic and genotypic stability of cold-adapted influenza virus vaccine. Pediatr Infect Dis J 25(7):590–5
18. Buonagurio DA, O'Neill RE, Shutyak L, D'Arco GA, Bechert TM, Kazachkov Y et al (2006) Genetic and phenotypic stability of cold-adapted influenza viruses in a trivalent vaccine administered to children in a day care setting. Virology 347(2):296–306
19. Talbot TR, Crocker DD, Peters J, Doersam JK, Ikizler MR, Sannella E et al (2005) Duration of virus shedding after trivalent intranasal live attenuated influenza vaccination in adults. Infect Control Hosp Epidemiol 26(5):494–500

20. Belshe RB, Gruber WC, Mendelman PM, Mehta HB, Mahmood K, Reisinger K et al (2000) Correlates of immune protection induced by live, attenuated, cold-adapted, trivalent, intranasal influenza virus vaccine. J Infect Dis 181(3):1133–7

21. Lee MS, Mahmood K, Adhikary L, August MJ, Cordova J, Cho I et al (2004) Measuring antibody responses to a live attenuated influenza vaccine in children. Pediatr Infect Dis J 23(9):852–6

22. Mendelman PM, Rappaport R, Cho I, Block S, Gruber W, August M et al (2004) Live attenuated influenza vaccine induces cross-reactive antibody responses in children against an a/Fujian/411/2002-like H3N2 antigenic variant strain. Pediatr Infect Dis J 23(11):1053–5

23. Block SL, Reisinger KS, Hultquist M, Walker RE, CAIV-T Study Group (2007) Comparative immunogenicities of frozen and refrigerated formulations of live attenuated influenza vaccine in healthy subjects. Antimicrob Agents Chemother 51(11):4001–8

24. Belshe RB, Toback SL, Tingting Y, Ambrose CS (2010) Efficacy of live attenuated influenza vaccine by age in children 6 months to 17 years of age. Influenza Other Respi Viruses 4:141–145

25. Sasaki S, Jaimes MC, Holmes TH, Dekker CL, Mahmood K, Kemble GW et al (2007) Comparison of the influenza virus-specific effector and memory B-cell responses to immunization of children and adults with live attenuated or inactivated influenza virus vaccines. J Virol 81(1):215–28

26. Sasaki S, He XS, Holmes TH, Dekker CL, Kemble GW, Arvin AM et al (2008) Influence of prior influenza vaccination on antibody and B-cell responses. PLoS ONE 3(8):e2975

27. Forrest BD, Pride MW, Dunning AJ, Capeding MR, Chotpitayasunondh T, Tam JS et al (2008) Correlation of cellular immune responses with protection against culture-confirmed influenza virus in young children. Clin Vaccine Immunol 15(7):1042–53

28. He XS, Holmes TH, Mahmood K, Kemble GW, Dekker CL, Arvin AM et al (2008) Phenotypic changes in influenza-specific CD8+ T cells after immunization of children and adults with influenza vaccines. J Infect Dis 197(6):803–11

29. He XS, Holmes TH, Zhang C, Mahmood K, Kemble GW, Lewis DB et al (2006) Cellular immune responses in children and adults receiving inactivated or live attenuated influenza vaccines. J Virol 80(23):11756–66

30. He XS, Holmes TH, Sasaki S, Jaimes MC, Kemble GW, Dekker CL et al (2008) Baseline levels of influenza-specific CD4 memory T-cells affect T-cell responses to influenza vaccines. PLoS ONE 3(7):e2574

31. Hammitt LL, Bartlett JP, Li S, Rahkola J, Lang N, Janoff EN et al (2009) Kinetics of viral shedding and immune responses in adults following administration of cold-adapted influenza vaccine. Vaccine 27(52):7359–66

32. Nichol KL, Mendelman PM, Mallon KP, Jackson LA, Gorse GJ, Belshe RB et al (1999) Effectiveness of live, attenuated intranasal influenza virus vaccine in healthy, working adults: a randomized controlled trial. JAMA 282(2):137–44

33. Treanor JJ, Kotloff K, Betts RF, Belshe R, Newman F, Iacuzio D et al (1999) Evaluation of trivalent, live, cold-adapted (CAIV-T) and inactivated (TIV) influenza vaccines in prevention of virus infection and illness following challenge of adults with wild-type influenza A (H1N1), A (H3N2), and B viruses. Vaccine 18(9–10):899–906

34. Belshe RB, Mendelman PM, Treanor J, King J, Gruber WC, Piedra P et al (1998) The efficacy of live attenuated, cold-adapted, trivalent, intranasal influenzavirus vaccine in children. N Engl J Med 338(20):1405–12

35. Belshe RB, Gruber WC, Mendelman PM, Cho I, Reisinger K, Block SL et al (2000) Efficacy of vaccination with live attenuated, cold-adapted, trivalent, intranasal influenza virus vaccine against a variant (A/Sydney) not contained in the vaccine. J Pediatr 136(2):168–75

36. Belshe RB, Edwards KM, Vesikari T, Black SV, Walker RE, Hultquist M et al (2007) Live attenuated versus inactivated influenza vaccine in infants and young children. N Engl J Med 356(7):685–96

37. Bracco Neto H, Farhat CK, Tregnaghi MW, Madhi SA, Razmpour A, Palladino G et al (2009) Efficacy and safety of 1 and 2 doses of live attenuated influenza vaccine in vaccine-naive children. Pediatr Infect Dis J 28(5):365–71
38. Fleming DM, Crovari P, Wahn U, Klemola T, Schlesinger Y, Langussis A et al (2006) Comparison of the efficacy and safety of live attenuated cold-adapted influenza vaccine, trivalent, with trivalent inactivated influenza virus vaccine in children and adolescents with asthma. Pediatr Infect Dis J 25(10):860–9
39. Ashkenazi S, Vertruyen A, Aristegui J, Esposito S, McKeith DD, Klemola T et al (2006) Superior relative efficacy of live attenuated influenza vaccine compared with inactivated influenza vaccine in young children with recurrent respiratory tract infections. Pediatr Infect Dis J 25(10):870–9
40. Rhorer J, Ambrose CS, Dickinson S, Hamilton H, Oleka NA, Malinoski FJ et al (2009) Efficacy of live attenuated influenza vaccine in children: a meta-analysis of nine randomized clinical trials. Vaccine 27(7):1101–10
41. Tam JS, Capeding MR, Lum LC, Chotpitayasunondh T, Jiang Z, Huang LM et al (2007) Efficacy and safety of a live attenuated, cold-adapted influenza vaccine, trivalent against culture-confirmed influenza in young children in asia. Pediatr Infect Dis J 26(7):619–28
42. Vesikari T, Fleming DM, Aristegui JF, Vertruyen A, Ashkenazi S, Rappaport R et al (2006) Safety, efficacy, and effectiveness of cold-adapted influenza vaccine-trivalent against community-acquired, culture-confirmed influenza in young children attending day care. Pediatrics 118(6):2298–312
43. Block SL, Toback SL, Yi T, Ambrose CS (2009) Efficacy of a single dose of live attenuated influenza vaccine in previously unvaccinated children: A post hoc analysis of three studies of children aged 2 to 6 years. Clin Ther 31(10):2140–7
44. Nichol KL, Mendelman PM, Mallon KP, Jackson LA, Gorse GJ, Belshe RB et al (1999) Effectiveness of live, attenuated intranasal influenza virus vaccine in healthy, working adults: a randomized controlled trial. JAMA 282(2):137–44
45. De Villiers PJ, Steele AD, Hiemstra LA, Rappaport R, Dunning AJ, Gruber WC et al (2009) Efficacy and safety of a live attenuated influenza vaccine in adults 60 years of age and older. Vaccine 28(1):228–34
46. Ohmit SE, Victor JC, Teich ER, Truscon RK, Rotthoff JR, Newton DW et al (2008) Prevention of symptomatic seasonal influenza in 2005–2006 by inactivated and live attenuated vaccines. J Infect Dis 198(3):312–7
47. Monto AS, Ohmit SE, Petrie JG, Johnson E, Truscon R, Teich E et al (2009) Comparative efficacy of inactivated and live attenuated influenza vaccines. N Engl J Med 361(13):1260–7
48. Ohmit SE, Victor JC, Rotthoff JR, Teich ER, Truscon RK, Baum LL et al (2006) Prevention of antigenically drifted influenza by inactivated and live attenuated vaccines. N Engl J Med 355(24):2513–22
49. Eick AA, Wang Z, Hughes H, Ford SM, Tobler SK (2009) Comparison of the trivalent live attenuated vs inactivated influenza vaccines among US military service members. Vaccine 27(27):3568–75
50. Wang Z, Tobler S, Roayaei J, Eick A (2009) Live attenuated or inactivated influenza vaccines and medical encounters for respiratory illnesses among US military personnel. JAMA 301(9):945–53
51. Bergen R, Black S, Shinefield H, Lewis E, Ray P, Hansen J et al (2004) Safety of cold-adapted live attenuated influenza vaccine in a large cohort of children and adolescents. Pediatr Infect Dis J 23(2):138–44
52. Reichert TA (2002) The Japanese program of vaccination of schoolchildren against influenza: implications for control of the disease. Semin Pediatr Infect Dis 13(2):104–11
53. Sugaya N, Takeuchi Y (2005) Mass vaccination of schoolchildren against influenza and its impact on the influenza-associated mortality rate among children in Japan. Clin Infect Dis 41(7):939–47

54. Piedra PA, Gaglani MJ, Riggs M, Herschler G, Fewlass C, Watts M et al (2005) Live attenuated influenza vaccine, trivalent, is safe in healthy children 18 months to 4 years, 5 to 9 years, and 10 to 18 years of age in a community-based, nonrandomized, open-label trial. Pediatrics 116(3):e397–407

55. Halloran ME, Piedra PA, Longini IM Jr, Gaglani MJ, Schmotzer B, Fewlass C et al (2007) Efficacy of trivalent, cold-adapted, influenza virus vaccine against influenza A (fujian), a drift variant, during 2003–2004. Vaccine 25(20):4038–45

56. Suguitan AL Jr, McAuliffe J, Mills KL, Jin H, Duke G, Lu B et al (2006) Live, attenuated influenza A H5N1 candidate vaccines provide broad cross-protection in mice and ferrets. PLoS Med 3(9):e360

57. Joseph T, McAuliffe J, Lu B, Vogel L, Swayne D, Jin H et al (2008) A live attenuated cold-adapted influenza A H7N3 virus vaccine provides protection against homologous and heterologous H7 viruses in mice and ferrets. Virology 378(1):123–32

58. Chen H, Subbarao K, Swayne D, Chen Q, Lu X, Katz J et al (2003) Generation and evaluation of a high-growth reassortant H9N2 influenza A virus as a pandemic vaccine candidate. Vaccine 21(17–18):1974–9

59. Suguitan AL Jr, Marino MP, Desai PD, Chen LM, Matsuoka Y, Donis RO et al (2009) The influence of the multi-basic cleavage site of the H5 hemagglutinin on the attenuation, immunogenicity and efficacy of a live attenuated influenza A H5N1 cold-adapted vaccine virus. Virology 395(2):280–8

60. Talaat KR, Karron RA, Callahan KA, Luke CJ, DiLorenzo SC, Chen GL et al (2009) A live attenuated H7N3 influenza virus vaccine is well tolerated and immunogenic in a phase I trial in healthy adults. Vaccine 27(28):3744–53

61. Karron RA, Callahan K, Luke C, Thumar B, McAuliffe J, Schappell E et al (2009) A live attenuated H9N2 influenza vaccine is well tolerated and immunogenic in healthy adults. J Infect Dis 199(5):711–6

62. Karron RA, Talaat K, Luke C, Callahan K, Thumar B, Dilorenzo S et al (2009) Evaluation of two live attenuated cold-adapted H5N1 influenza virus vaccines in healthy adults. Vaccine 27(36):4953–60

Cell Culture-Derived Influenza Vaccines

Philip R. Dormitzer

Abstract Conventional egg-based vaccine manufacture has provided decades of safe and effective influenza vaccines using the technologies of the 1930–1960s. Concerns over the vulnerability of the egg supply in the case of a pandemic with a high pathogenicity avian influenza strain have spurred the development and licensure of mammalian cell culture-based influenza vaccines, the first major technological innovation in influenza vaccine since the mid-twentieth century. Mammalian cell culture provides a readily expansible, secure substrate for influenza vaccine manufacture, free from the need to suppress the bioburden associated with eggs. Most current cell culture-based vaccines still rely on seed viruses isolated in eggs. Conversion to a fully egg-free process is likely to increase the range of seed viruses available and improve the match between vaccine seed strains and circulating strains. The risk of adventitious agent introduction during manufacture in thoroughly characterized mammalian cell substrates is certainly low and probably significantly lower than the risks in egg-based manufacture. In clinical trials, cell-based influenza vaccines have proven safe and equivalent in immunogenicity to egg-based influenza vaccines. The higher containment that is possible with cell-based production proved valuable during the 2009 pandemic, when large-scale production of vaccine bulks could begin in cell culture manufacturing systems at biosafety level 3, while egg-based production was delayed, waiting for the biosafety level of the pandemic stain to be decreased. For cell-based production to replace egg-based production of influenza vaccine, the new technology will need to demonstrate its robustness over multiple strain changes and its economic competitiveness.

P.R. Dormitzer
Novartis Vaccines and Diagnostics, 350 Massachusetts Avenue, Cambridge, MA 02139, USA
e-mail: philip.dormitzer@novartis.com

G. Del Giudice and R. Rappuoli (eds.), *Influenza Vaccines for the Future*, 2nd edition,   293
Birkhäuser Advances in Infectious Diseases,
DOI 10.1007/978-3-0346-0279-2_12, © Springer Basel AG 2011

1   Introduction

Like the Apollo space program, the global system of influenza vaccine production is a testament to what could be accomplished with mid-twentieth century technology, applied with creativity and determination. Even in the early twenty-first century, the goal of making a vaccine that is updated with up to three new strains every year (indeed, twice a year due to the staggered seasonality of the hemispheres), combined with the occasional pandemic vaccine, remains the most impressive feat in vaccinology. We look at the first manned moon landing in 1969 and ask "How could this have been accomplished with a guidance system that had only 38K of memory?" Although biology has undergone a revolution since that time, most of the technology in conventional influenza vaccines dates from the Apollo era or earlier [1]. Indeed, some of the technologies are relics of an agrarian past. Conventional flu vaccine manufacture still relies on production in fertilized chicken eggs, release tests based on bleeding immunized sheep, and immunogenicity tests based on mixing human serum with red blood cells from turkeys, chickens, horses, or guinea pigs. The standard assays required for testing flu vaccine antigen content (single radial immunodiffusion – SRID) and immunogenicity (hemagglutination inhibition – HI) have simple visual read-outs that require no analytic equipment more sophisticated than a ruler. Their simplicity is a strength, but it also reflects their exclusive reliance on components that were readily available around the time of the Second World War. The assays carry with them the variability and idiosyncrasies of tests devised before well-defined, pure biological reagents and modern analytical tools were generally available.

Fertilized, farm-fresh eggs (used 9–12 days after laying) are an essential component of conventional influenza vaccine production. They are used to isolate viruses from clinical specimens, generate reassortant viruses, and produce the prodigious quantities of virions from which vaccine antigens are extracted. With a productivity of approximately one vaccine dose per egg, the hundreds of millions of doses administered each year require hundreds of millions of eggs delivered on schedule within a defined manufacturing season. Planning starts approximately a year in advance. Chickens must beget more chickens to have enough inseminated female chickens to lay all the fertilized eggs needed. Were a severe avian influenza outbreak to wipe out the flocks, influenza vaccine production would collapse. What if the same virus strain also caused severe disease in humans? Because high pathogenicity H5N1 avian influenza strains can infect humans with high lethality (but, fortunately, with low transmissibility between humans thus far), such a scenario is well within the realm of possibility. Concerns about the reliability of the egg supply have been a chief motivator for efforts to develop cell culture-based influenza vaccine production. The reasons for the continued use of eggs include regulatory barriers, decades of experience with egg-based manufacture, technical challenges with cell-based manufacture, and the large investments required for updating the dominant technology used for human influenza vaccine production.

The barriers are surmountable. Veterinary flu vaccines have been produced in cultured mammalian cells for years. One cell culture-based seasonal trivalent influenza vaccine (Optaflu® from Novartis Vaccines) is licensed in the EU, and two cell culture-based monovalent influenza vaccines (Celtura® from Novartis Vaccines and Celvapan® from Baxter) were deployed as part of the 2009 pandemic response. Attenuated seed virus generation by reverse genetics for high pathogenicity H5N1 strains [2] and the introduction of cell culture-based manufacturing are the first fundamental technological advances in influenza vaccines used for protection of human populations since the discovery of efficient growth of influenza viruses in embryonated chicken eggs in 1930s and early 1940s [3], the introduction of "splitting" to separate HA and NA from nucleocapsids in 1964 [4], the development of cold-adapted live-attenuated vaccines in 1965 [5], the purification of influenza viruses by continuous flow ultracentrifugation in 1967 [6], and the generation of high growth vaccine strains by reassortment in 1969 [7].

2 Isolation of Viruses for Vaccine Seeds

Avian influenza viruses are the primary repository of influenza genetic diversity, although pigs and horses are also potential sources of zoonotic strains [8]. All 16 hemagglutinin (HA) and 9 neuraminidase (NA) types of influenza A viruses circulate among waterfowl. In birds, influenza is primarily an enteric infection. Avian influenza viruses enter bird intestinal epithelial cells after attaching to α-2,3-linked sialoside moieties on cell surface glycoproteins or glycolipids [9]. The adaptation of avian viruses to infect humans efficiently requires mutation of the receptor binding region of HA so that it can mediate virus attachment to α-2,6-linked sialosides, the predominant sialoside linkage found in epithelium of the human upper respiratory tract [10]. A difference in overall contour facilitates the discrimination between these glycans by HA [11]. Sialosides with an α-2,3 linkage have a gently angled structure (resulting in a "cone-like" topology); those with an α-2,6 linkage have a more sharply angled structure (resulting in an "umbrella-like" topology) [12]. Nevertheless, a single amino acid substitution can be sufficient to alter the preferred sialoside linkage bound by HA [13, 14]. Because the dominant antibodies that neutralize influenza virus bind near the sialoside recognition site [15, 16], preventing virus attachment to cells [17], adaptation of influenza virus to replication in humans alters the antigenic region that is the main target of protective antibodies.

When human respiratory secretions are injected into eggs to isolate influenza viruses for vaccine seeds, the virus must re-adapt from its mammalian sialoside specificity to replicate efficiently in avian cells by attaching to α-2,3-sialosides. The allantoic cavity is the principal compartment of embryonated chicken eggs in which influenza viruses replicate [3]. The cells of the allantoic membrane primarily bear α-2,3-sialosides on their surface [14]. To facilitate adaptation to growth in the allantoic cavity, virus from clinical samples is selectively inoculated into the relatively small amniotic cavity [3]. The amniotic membrane bears both α-2,6- and α-2,3-linked sialosides [14]. Presumably, replication in the amniotic cavity expands the number and sequence diversity of viruses, increasing the chances that a variant will arise with α-2,3-sialoside specificity and become competent for replication in the allantoic cavity [18]. This switch in receptor specificity again selects for a mutation in the dominant antigenic region around the receptor binding site. Thus, egg-isolated virus strains and the HA in vaccines derived from them have an obligate sequence difference and, in many cases, a detectable antigenic difference from the strains that cause human disease [19, 20]. During the adaptation of B strains to growth in eggs, the loss of a glycosylation site can cause substantial antigenic changes [21].

MDCK cells bear both α-2,3- and α-2,6-sialosides on their surface [14]. Therefore, MDCK cells support the isolation of the influenza viruses shed from the mammalian cells that line the human respiratory tract with no obligate change in receptor specificity [22]. Accordingly, isolation rates are substantially higher on MDCK cells than in eggs and are higher still on an MDCK cell line (MDCK-SIAT1) engineered by the introduction of the cDNA of human 2,6-sialyltransferase to increase the proportion of α-2,6-linked sialosides on the cell surface [23, 24]. The difference in isolation rates is greatest for H3N2 strains. Since the introduction of this subtype into humans in the pandemic of 1968, H3N2 influenza viruses have adapted to human hosts and correspondingly become more difficult to isolate in eggs. In 2006 and 2007, of 264 H3N2 clinical specimens inoculated into both eggs and MDCK cells at the US Centers for Disease Control, only 4% produced viral isolates in eggs, but 65% produced viral isolates in MDCK cells [25]. The greater permissiveness of mammalian cells for the isolation of human influenza strains has led to their widespread use to isolate viruses for the purpose of epidemiologic surveillance. However, as of the writing of this chapter, these cell-isolated viruses may not be used for vaccine seeds. If a promising strain is identified by culture on mammalian cells, an attempt is made to re-isolate the strain from the original clinical specimen in eggs. This re-isolation is not always successful, restricting the range of seeds available for manufacture, including mammalian cell-based manufacture, to only those viruses that can be isolated in eggs [25]. Once the virus, shed from humans and isolated in mammalian cells during surveillance, is adapted to eggs and sent to cell-based vaccine manufacturers, it must then be re-adapted back to growth in mammalian cells. The re-adaptation typically requires two to three passages. This alternation between substrates wastes weeks during the tight annual vaccine production cycle, and there is no evidence that egg-adaptive mutations in the receptor binding region revert upon subsequent passage in mammalian cells.

Has the limited choice of influenza vaccine seeds imposed by continued reliance on egg isolation had a public health impact? During the 2003/2004 influenza season, no strain that was antigenically "like" the A/Fujian/411/2002 H3N2 strain could be isolated in eggs in time for vaccine preparation [26]. Therefore, the vaccine contained the antigenically mismatched A/Panama/2007/99 H3N2 strain. Retrospective analyses indicated that the vaccine had decreased effectiveness compared to vaccines produced during years in which the vaccine H3N2 strain was well matched to circulating H3N2 strains [27]. That influenza season was unusually severe and marked by an increase in pediatric mortality [28]. The degree to which a better matched vaccine would have ameliorated the severity of the influenza season is unknown.

As of the writing of this chapter, all seasonal influenza vaccines in commercial use are produced from influenza strains that have been passaged through chicken eggs, regardless of whether their final manufacture is in cultured mammalian cells or chicken eggs. Therefore, the strains have one or more amino acid sequence differences and possibly antigenic differences from the circulating strains that cause human disease. How significant are such differences for the efficacy of influenza virus vaccines? There are no human data to definitively answer this question, because no influenza virus that has been propagated exclusively in mammalian cells has been used in a clinical trial that could answer this question. The results of animal studies are not definitive. Infection of small numbers of ferrets with egg-isolated and MDCK-isolated viruses cross-protects the animals against influenza viruses grown on either substrate equivalently, but such experiments would only detect gross differences in efficacy [22]. Immunization of ferrets with formalin-inactivated MDCK-grown virus provided better protection against either MDCK-grown or egg-grown challenge virus than did formalin-inactivated, antigenically distinguishable egg-grown virus with a single amino acid difference in HA from the MDCK-grown virus [29]. Although this finding suggests that an all-cell-produced vaccine might have enhanced efficacy, this interpretation is tempered by the obscure mechanism of the difference in efficacy, which did not correspond to the degree of antigenic relatedness of the immunizing and challenge viruses.

Manufacturers and World Health Organization (WHO) Collaborating Centers have launched a multilateral effort to introduce mammalian cells to vaccine seed isolation [30]. Manufacturers of non-live influenza vaccines obtain strains from a common set of Collaborating Center laboratories, and it is impractical for these laboratories to isolate viruses for each manufacturer on a different cell line. Therefore, comparative studies will identify a common cell line that permits efficient isolation followed by ready adaptation to the various manufacturing processes. Isolation of viruses in mammalian cells could even facilitate the adaptation of additional strains to growth in eggs. The expanded pools following efficient isolation and expansion in mammalian cells will have greater sequence diversity than the small number of viable viruses present in the original clinical specimens. This higher viral titer and expanded diversity could increase the likelihood of successful adaptation of a strain to eggs.

3 Reassortment and Backbone Selection

To make vaccine seeds for inactivated vaccines from influenza type A strains, the HA and NA genome segments of egg-adapted clinical isolates are reassorted onto the "backbone" of A/Puerto Rico/34, an attenuated, egg-adapted strain [7]. To make the cold-adapted live-attenuated vaccine manufactured by MedImmune, the HA and NA of A strains are reassorted onto A/Ann Arbor/6/60 and the HA and NA of B strains are reassorted onto B/Ann Arbor/1/66 [31]. During reassortment, the individual genome segments of two or more viruses that coinfect a single cell are swapped in the progeny viruses – a mating process analogous to the assortment of chromosomes during sexual reproduction of eukaryotes. The backbone consists of the genome segments encoding the matrix protein (M), the nonstructural proteins (NS1 and NS2), the nucleoprotein (NP), and the polymerase complex (PA, PB1, and PB2). Selective pressure against viruses bearing the HA and NA of the backbone strain is provided by antisera specific for these determinants [7]. An ideal 6:2 reassortant would contain only the HA and NA genome segments of the clinical strain on a full set of other genome segments from the intended backbone donor. In practice, one or more backbone genome segments of the clinical isolate may also be incorporated into reassortants that are selected for manufacture.

The egg-adapted current backbones may not be optimal for the productivity of mammalian cell-based manufacture. Therefore, new cell-adapted backbones may increase the efficiency of manufacture in mammalian cells. Wild type strains that grow efficiently in mammalian cells can provide donors for cell-adapted backbones. Because HA and NA are major determinants of viral growth, selecting a backbone donor requires comparing consistent sets of HA and NA pairs on alternative backbones. With a donor selected, reverse genetics can allow the rational modification of backbone genome segments to increase productivity from cultured cells and consistency of downstream processing. Efficient polymerase complexes resulting in rapid replication in mammalian culture are found in some highly virulent strains, such as the 1918 pandemic H1N1 virus and a variant of PR8 with high virulence for mice [32, 33]. Therefore, safe vaccine manufacture with engineered highly productive strains may be facilitated by additional mutations, such as NS-1 deletions or truncations, that attenuate the viruses in mammals, including humans, while preserving replication efficiency in some mammalian cells [34–36]. In principle, reverse genetic engineering of backbones could produce strains that are produced more efficiently in eggs, too. The ability to manipulate both the cell and the virus in cell culture production systems allows greater application of modern viral and cellular genetics to cell optimize influenza virus vaccine manufacture.

4 Eliminating Adventitious Agents from Vaccine Seeds

The possibility that passage of influenza viruses from human secretions through eggs provides a "filter" that prevents the propagation of potential adventitious agents has been cited as a potential advantage of egg-based seed generation and manufacture [25]. Concerns have been also been raised that mammalian cell substrates themselves could be a source of adventitious agents [25]. From 1955 to 1963, simian virus 40 (SV40) from primary monkey kidney cells contaminated batches of polio virus vaccine [37], although there is no evidence of increased cancer risk in those who received polio virus vaccine during that period [38].

Cell lines, like intact organisms, can harbor endogenous retroviruses with the potential for re-activation [39]. For this reason, cell lines used for manufacture of vaccines undergo rigorous testing for absence of adventitious agents. Tests include PCR for known agents, enzymatic assays to detect viral polymerases, electron microscopy to detect viral particles, inoculation of cultured cells, and inoculation of animals [40]. In addition, influenza vaccine processing includes steps to elimi-nate or inactivate adventitious agents [41]. Chickens also carry viruses, such as Rous sarcoma virus, avian leukosis virus, and reticuloendotheliosis virus, which could potentially be introduced into egg-based manufacturing. Therefore, eggs used for manufacture of vaccines for the USA must be derived from flocks certified to be free of a list of specific pathogens, the production process must be shown to eliminate listed agents, or the absence of the agents from the vaccine must be demonstrated [40]. There is a key difference in the adventitious agent testing possible for cultured cell-based and egg-based production. Representative frozen aliquots of a banked continuous cell line are scrutinized for microbiological safety before a vaccine produced from equivalent frozen aliquots of that cell bank is licensed. It is not possible to apply the same level of rigor in biological control to the hundreds of millions of freshly laid eggs used in vaccine production each year.

A model for systematically assessing the risk of adventitious agent introduction by different influenza vaccine manufacturing schemes has been developed [41]. This model weighs the relative ability of viruses found in human nasopharyngeal secretions or in chickens to propagate in eggs or several mammalian cell types, to survive inactivation and other steps in the downstream manufacturing process, and to cause disease in a vaccine recipient. Analyzing available data using this model indicates that the use of eggs to isolate the virus seeds used for mammalian cell manufacture does not "filter" viruses found in human nasopharyngeal secretions more effectively than isolation on MDCK cells, but it does risk introducing avian viruses [41]. The physical hardiness of some prevalent avian viruses, particularly reoviruses, renders chicken eggs a chief potential source of agents that are relatively difficult to inactivate during the downstream processing of vaccines [41]. The risk that an adventitious agent will survive manufacturing is much lower for split and subunit influenza vaccines than it is for live-attenuated vaccines. The manufacture of non-live vaccines includes a virus inactivation step, typically using chemical agents such as β-propiolactone or formalin. These agents would destroy the infec-tivity of a live-attenuated vaccine. In addition, the "splitting" of most inactive influenza virus vaccines (except for whole virus vaccines) – that is, the use of detergents and sometimes solvents to separate HA and NA from the nucleoprotein core – adds an additional inactivation step that is particularly effective against enveloped viruses [41].

The use of eggs rather than cultured cells for isolation of seed viruses and for reassortment precludes the use of plaque purification, the chief technique employed in modern research settings to ensure that a viral isolate is clonal – a pure, genetically homogeneous (to the degree that an RNA viral quasi-species can be homogeneous) population derived from a single infectious virus. The direct visual-ization and harvesting of well-separated plaques on a cultured cell monolayer under a semi-solid overlay allow assurance that a single virus isolate is being obtained and allow the selective isolation of viruses with a large-plaque phenotype, which may correlate with efficient growth in culture [42]. Although decades of experience indicate that the blind technique of terminal dilution cloning in eggs is an adequate procedure to provide seeds for safe influenza vaccine manufacture, plaque purification in cultured cells can provide an additional margin of safety and selectivity.

The risk that adventitious agents from clinical samples could be introduced into the vaccine manufacturing process will be eliminated when the generation of influenza vaccine seeds by gene synthesis followed by reverse genetic rescue becomes the industry standard [2, 43, 44]. Current reverse genetics rescue starts by reverse transcribing DNA clones from viral RNA purified directly from an original clinical specimen or from a cultured virus isolated from such a specimen. The purification of RNA under biochemically harsh conditions can greatly reduce the risk of adventitious agent carry-over. Generating influenza-encoding DNA by chemical synthesis [44] could completely eliminate this source of adventitious agent risk. In a synthetic scheme, wet bench experiments with patient-derived specimens would generate sequence information. That information would be transmitted electronically to a DNA synthesizer to make the nucleic acids used for virus rescue, breaking any conceivable chain of adventitious agent transmission from an influenza patient's specimen to a vaccine lot. This advance in the hygiene of influenza vaccine manufacture will only be possible by using cultured cells to generate vaccine seeds.

5 The Cell Lines Used in Influenza Vaccine Manufacture

Several influenza vaccine manufactures are developing cell-based production processes. Details of these processes and their productivities are, in general, proprietary. Thus, a survey of the published literature would necessarily give an incomplete and dated impression of the state of the field. Therefore, this section does not attempt to provide a comprehensive review or a comparison of the relative merits of different manufacturers' processes, but rather focuses on key issues in the development of cell-based influenza vaccine manufacturing. Detailed reviews of information that can be gleaned from published literature on different manufacturers' cell-based processes are available [45, 46]. This chapter also does not review advances in making recombinant influenza virus vaccines containing antigens from sources other than cultured influenza viruses. Such candidates as purified protein, virus-like particle, vectored, and peptide vaccines are reviewed elsewhere [47].

As of the writing of this chapter, five cell culture-based influenza vaccines have been licensed for human use. In 2001, Influvac TC® from Solvay (seasonal trivalent, split, produced in MDCK cells) was licensed in the Netherlands but was never commercially distributed due to manufacturing delays [45, 48]. In 2002, Influject® from Baxter (seasonal trivalent, whole virion, produced in Vero cells) was licensed in the Netherlands, but subsequent phase II/III trials of this vaccine were suspended due to a higher than expected rate of fever and associated symptoms among vaccinees [45]. In 2007, Optaflu® from Novartis Vaccines (seasonal trivalent, subunit, produced in MDCK cells) was approved in the EU [48]. In 2009, Celvapan® from Baxter (H1N1 pandemic, monovalent, whole virion, produced in Vero cells) was authorized in the EU and sold commercially. In 2009, Celtura® from Novartis Vaccines (H1N1 pandemic, monovalent, MF59-adjuvanted subunit, produced in MDCK cells) was authorized in the EU and sold commercially. Sanofi Pasteur, MedImmune, GlaxoSmithKline, and Crucell all have or have had programs to develop cell culture-based influenza vaccines.

There are distinctions between the approaches taken by different manufacturers. For example, Celvapan® is produced by growing wild type virus in Vero cells that adhere to microcarrier beads. Of the cell lines used to produce influenza vaccines, Vero cell lines have the longest history in vaccine manufacture. They were first used to produce inactivated poliovirus and rabies virus vaccines in the 1980s [49]. Because Vero cells at limited passage number do not form tumors when injected into infant nude mice [50], the defined passage cells used in vaccine manufacture are considered nontumorigenic. Optaflu® and Celtura® are produced from conventional egg-isolated and (for type A strains) reassorted seeds obtained from WHO Collaborating Centers and adapted to replicate efficiently in a proprietary MDCK 33016 cell line that grows in suspension culture [48]. Infection of suspended, rather than adherent, cells simplifies the upstream manufacturing process. The Crucell vaccine candidate is also produced in a suspension cell line, PER.C6, a human fetal retinoblast cell line that was immortalized through a defined genetic manipulation – the introduction of the E1 minigene of adenovirus type 5 [51]. Vero and MDCK cells were isolated as continuous cell lines by empiric techniques [46, 48, 49].

Influenza viruses, with the exception of highly pathogenic H5N1 strains with furin-cleavable HA, require the addition of exogenous trypsin to cleave HA into the HA1 and HA2 fragments, activating the viruses for infection [16]. Trypsin can also loosen the attachment of adherent cells, a constraint during manufacture with adherent cell lines. The cell culture media used in influenza vaccine manufacture are always serum-free and generally animal-product-free, although trypsin and insulin of animal origin may be added. Animal-product-free preparations of insulin and trypsin-like serine proteases are available [52]. There is a movement toward chemically defined media, in which the chemical constituents are explicitly determined, for cell-based vaccine manufacture.

6   Egg-Based and Cell-Based Vaccine Production Processes

Production in cell culture allows greater control of infection parameters than egg-based production. Media can be changed or supplemented during the infection. Oxygen content, nutrient levels, agitation, and pH can be monitored and adjusted. Known multiplicities of infection can be optimized for each virus strain. A production process based on the infection of cultured cells is more complicated than the "fermentations" for most biopharmaceutical products [53]. In a standard fermentation, all cells in a culture produce a product continuously, possibly after induction, with steady accumulation of a secreted or retained product. Standard parameters of cell number, cell viability, pH, lactate production, oxygen tension, and product accumulation are monitored. All of these parameters are also relevant for an influenza virus infection. However, the proportion of cells infected, the production and release of virus, level of tryptic activity, and virus-mediated cell lysis must also be monitored for a well-controlled process. To limit the size of the viral seed pool, starting multiplicities of infection may be very low. Therefore, as the infection progresses, the initially infected cells undergo apoptosis and lysis, while other cells are just entering productive infection. Rates of virus production, HA production, and virus release from cells may change during the infection as may the morphology of released virus and even the viral genotype. The infection continues during the separation of virus and cells at harvest, which may take hours for thousands of liters of infected culture. The product, for a subunit influenza vaccine, is primarily HA. Yet, HA on budded spherical viruses, on budded filamentous viruses, or retained on cell membranes may behave quite differently in the downstream purification. Therefore, a number of virological parameters, in addition to the usual fermentation parameters, must be monitored to optimize infection. Successful optimization requires close collaboration between virologists and process engineers.

The complexity of infection optimization is particularly challenging due to the short interval between the announcement of a new vaccine strain and the start of the validation runs needed for a seasonal vaccine update or pandemic vaccine release. The constraints of a product license limit the permissible adjustments of the process. Therefore, a well-designed and efficient optimization strategy is an essential component of cell-based influenza vaccine manufacture. For egg-based processes, many years of experience provide a database of historically optimal parameters for each viral subtype. This experience provides a starting point for annual optimization of egg-based production. That experience is now being obtained, in a compressed time frame, for mammalian cell-based manufacturing processes.

The material harvested after infection in egg-based and cell-based manufacture is different. Egg allantoic fluid has a high content of non-influenza proteins. Harvested cell culture medium is relatively low in protein but contains the trypsin or trypsin-like proteases added to promote virus spread. The harvest of cell culture production is sterile (except for influenza virus); the harvest of egg-based production is not. Flocks of chickens used to produce the eggs used in vaccine manufacture may be specific pathogen-free or may meet the lower standards of a "clean" flock [82]. Neither specific pathogen-free nor clean chickens are germ-free. They have an abundant bacterial flora. Eggs pass through a hen's cloaca (Latin for "sewer"), the common orifice of the chicken digestive, urinary, and reproductive tracts. Eggs for vaccine manufacture are sanitized to reduce their bacterial load.

The infection in cell-based manufacture takes place in one or a few large (up to thousands of liters) tanks or bags. The cell culture medium is sterile, and sterility is maintained throughout the closed production process. In contrast, egg-based manufacture is an inherently open process. Each egg must be opened for virus inoculation, incubated after the shell is breached, and then accessed again to harvest the virus-containing allantoic fluid. In this process, there is opportunity for the introduction of agents from workers or the environment and for the leakage of influenza virus-containing fluid from the eggs. Although a variety of precautions are taken to minimize the microbiological risks of egg-based manufacture, problems with bioburden do occur. *Serratia marcescens* contamination of vaccine bulks at an egg-based manufacturing plant in Liverpool, UK, led to a plant shutdown and a severe influenza vaccine shortage in the USA in 2004–2005 [54]. If one of the tanks used in cell-based manufacture were contaminated, the loss of product would be large, but readily detected and isolated. Monitoring the sterility of the vast number of eggs used in traditional manufacture is a much more daunting challenge. To limit bioburden, traditional manufacture requires the addition of antibiotics to eggs at the time of virus inoculation and in some cases the addition of antimicrobial compounds to downstream process streams, until the final sterile filtration of the product. Cell-based manufacture can be antibiotic-free from start to finish.

After the virus is separated from other components of the harvests, most commonly by continuous flow ultracentrifugation [6], the downstream processing of egg-based and cell-based vaccines is similar, except for the bioburden control requirements for egg-based vaccines and one additional requirement for cell-based vaccines – control of host cell DNA size and content. Continuous cell lines, because they have been immortalized, have undergone one of the phenotypic changes associated with malignant transformation. Suspension cell lines are anchorage independent, another phenotype associated with malignancy. Some suspension cell lines, including MDCK 33016 cells, can establish tumors in highly immunodeficient infant nude mice [51, 55]. The tumors are not mouse tumors; they are foci of MDCK cells. Such dog kidney cell colonies would be rejected in a mouse (or human) with a functioning immune system. Downstream processing of vaccines produced in MDCK 33016 cells includes multiple filtration steps, detergent treatment steps, and an inactivation step. The redundancy of the cell removal is such that the risk of any individual ever receiving an MDCK cell from immunization, even if every individual who has ever lived or will live until the sun burns out were to receive a flu vaccine every year for 100 years, is estimated at less than 1 in 10^{12} [55].

There is a theoretical concern that the genes responsible for the immortalization and, in some cases, anchorage independence of production cell lines could be present in cell culture-based vaccines, be taken up by host cells, and bring about malignant transformation of a host cell [56]. To alleviate this concern, the host cell DNA content of cell culture-produced, injectable vaccines is limited to less than 10 ng per dose, and the size of DNA fragments is limited to less than 200 base pairs, precluding the presence of intact oncogenes [40, 49, 56]. DNA can be eliminated at multiple process steps, including cell separation, virus purification, splitting, and chromatographic polishing. To further limit DNA content, in-process material may be digested with benzonase. β-Propiolactone, commonly used to inactivate viruses in vaccines, also fragments DNA [57]. The use of a known immortalizing gene to generate "designer" cell lines, such as PER.C6 [58], creates the ability to assay the presence or absence of the specific transforming gene in a vaccine product.

Although the transmissibility of some tumors by cell-free extracts was first demonstrated by studying chicken sarcomas in 1911 [59] (the phenomenon is now known to be caused by an avian virus, Rous sarcoma virus, that transmits the src oncogene [60]), the tumorigenic potential of chicken cells and the oncogenic potential of residual egg DNA in vaccines have not been subjected to the same level of scrutiny as that applied to cultured cells and residual cultured cell DNA. Egg-based influenza vaccines are not subject to restrictions on host cell DNA content or size. This disparity reflects, in part, the more relaxed safety standards that prevailed when egg-based vaccines were first introduced. In fact, clinical experience has given no indication that human immunization with any vaccine, whether produced in eggs or cultured cells, predisposes to tumors.

HA content is a key release criterion for inactivated influenza vaccines. For subunit vaccines, this determination is based on SRID, a technique in which the vaccine antigen, often treated with Zwittergent, is placed in a well cut into in a layer of agarose that has been impregnated with a polyclonal, strain-specific sheep antiserum against a crude preparation of HA [61]. As the vaccine antigen diffuses into the gel, a zone of immunoprecipitation between the vaccine antigen and the sheep antiserum forms, visible by scattered light or protein staining. The diameter of the zone of precipitation is considered proportional to the antigen content of the vaccine. The sheep antisera and fixed virus antigen standards are provided by regulatory authorities. The antigens used as standards and as sheep immunogens are produced in eggs. Therefore, a question has been raised whether such egg-produced reagents are suitable for the assay of mammalian cell-produced influenza vaccine antigens [25]. The use of cell-produced reagents to assay cell-produced vaccines would normalize for any relevant differences in posttranslational modification between mammalian cell-produced and egg-produced HA. On the other hand, if the sheep are immunized with HA that is not completely pure, the elicited antiserum will contain a mixture of antibodies against HA and antibody against the impurities from the egg or mammalian cell substrate. Use of reagents derived from the same platform as the vaccine antigen may therefore result in an overestimation of HA content, because both HA and substrate-derived contaminants could form immunoprecipitates. Cross-platform assays (using antisera against egg-produced HA to assay cell-produced vaccines and vice versa) could prevent this potential error. Limited data on cross-platform potency assays have been published [62], and further studies comparing same platform and cross-platform immunoassays are needed to guide this regulatory decision.

7 Clinical Testing of Cell-Based Influenza Vaccines

Clinical trials have assessed the safety, immunogenicity, and protective efficacy of cell-based influenza vaccines [49, 63–74]. In published trials, the reactogenicity, safety, and immunogenicity of cell culture-based vaccines appear to be generally equivalent to those of comparable egg-based vaccines – whether MDCK-produced, Vero-produced, seasonal trivalent, H5N1 monovalent prepandemic, H1N1 monovalent pandemic, subunit, split, whole virus, or adjuvanted. There are modest exceptions. In trials of a Vero-produced H5N1 whole virion vaccine, there appeared to be less reactogenicity compared with historical egg-produced comparators [65, 74]. In one trial of a MDCK-produced seasonal subunit vaccine, there was a modest increase in mild-to-moderate local pain on injection relative to an egg-produced comparator [63], although this has not been a consistent finding with such vaccines. The cell culture-produced vaccines could differ from each other and from egg-produced vaccines in their posttranslational processing, particularly glycosylation. Although the more authentic glycosylation of mammalian cell-based vaccine antigens is a potential advantage, the clinical data to date do not provide evidence that these differences significantly affect immunogenicity, as measured by HI or single radial hemolysis (SRH). This is consistent with absence of functional anti-influenza antibodies known to bind glycans. Glycan masking of influenza epitopes has been well documented [16, 75], and HA produced in different substrates does vary in the bulkiness of its oligosaccharides [76], raising the possibility of antigenically relevant differences. It remains to be determined whether differences in immunogenicity between egg-produced and mammalian cell-produced vaccines will emerge when mammalian cell isolation is substituted for egg isolation of seed viruses for seasonal influenza vaccines produced in cell culture.

A chief contra-indication to immunization with egg-based vaccines is egg allergy, which has a prevalence estimated between 0.5% and 2.5% [77]. Egg allergens include ovalbumin and ovomucoid, contaminants found in egg-based vaccines [77]. This risk is eliminated by the use of mammalian cell-based vaccines. Questions have been raised whether immunization with influenza vaccines produced in MDCK cells, which were derived from a dog kidney, might elicit hypersensitivity responses in those with dog allergies. The chief dog allergens are found in dander and saliva [78] and have not been detected in MDCK cells [79]. Cultured mast cells, sensitized with IgE from dog-allergic human subjects, do not degranulate upon exposure to an MDCK-produced influenza vaccine [79]. The prevalence of dog allergy is high. Although data are incomplete for many groups, dog allergy is reported in greater than 4% of some pediatric populations in developed countries [78, 80]. Yet, no excess of hypersensitivity reactions has been observed in clinical comparisons of immunization with MDCK-produced and egg-produced influenza vaccines. Thus, available data do not indicate that the dog origin of MDCK cells increases the risk of hypersensitivity reactions to MDCK-produced vaccines. Does immunization with vaccines produced in mammalian cells increase the likelihood of autoimmunity? The clinical trial experience with cell-based influenza vaccines gives no indication of such an association. The more than 25 years of benign experience with simian Vero cell-produced vaccines, including inactivated rabies vaccines and inactivated and live polio vaccines, is particularly encouraging [49].

8   Role of Cell-Based Vaccine Manufacture in the Response to the Swine-Origin H1N1 Influenza Pandemic

The 2009–2010 swine-origin influenza pandemic provided a live test of cell-based manufacture for pandemic response. Much of the initial motivation for the development of cell-based manufacture was the need to ensure a rapid expansion of the influenza vaccine supply in the event of a pandemic. In fact, in 2009, egg supplies were not the main limiting factor for vaccine supply. This was, in part, the result of pandemic preparedness activities to increase the egg supply. However, the early days of the pandemic were marked by a scramble among influenza vaccine manufacturers to secure sufficient supplies of suitable eggs. The slower pace of identifying a suitable vaccine seed, producing sheep antisera for vaccine release, and staffing vaccine production facilities was more rate limiting, giving time to obtain the eggs needed. As these slower components of the pandemic response are accelerated in the future, egg supply could again become limiting, and the egg supply remains vulnerable to catastrophic depletion in an avian influenza pandemic.

Cell culture-based influenza vaccine manufacturing demonstrated a little antici-pated benefit during the 2009 H1N1 pandemic response. In the opening days of the pandemic, the pathogenicity of the pandemic strain was not known, and reassortants on a PR8 backbone were not available. Consequently, in the EU, the pandemic strain was handled at BSL3, precluding large-scale testing or production in open egg-based manufacturing. Vaccine bulks could, however, be produced at scale in BSL3 cell culture-based manufacturing facilities in Marburg, Germany (Novartis Vaccines), and Bohumil, Czech Republic (Baxter). The first GMP batch of a H1N1 vaccine candidate produced by a manufacturer that adheres to western quality standards was produced on June 12, 2009, at the Novartis flu cell culture facility. The time advantage provided by cell culture production enabled the start of clinical trials [66] with a reassortant-derived cell culture vaccine even before calibrated SRID reagents had been supplied by regulatory authorities. In contrast, large-scale pro-duction at Novartis egg-based manufacturing facilities in Siena and Rosia, Italy, could not start until the beginning of July, once reassortant strains were available and biosafety levels had been lowered. Had the Novartis MDCK-based manufacturing facility also been licensed for growth of genetically modified organisms at BSL3, a potential influenza vaccine seed rescued by reverse genetics on a PR8 backbone in Novartis research laboratories in mid-May could have been tested at scale weeks before reassortant seeds were received from a WHO collaborating center.

9   The Future of Cell-Based Influenza Vaccine Manufacture

The replacement of egg-based influenza vaccine manufacture by cell culture-based manufacture seems inevitable, but the pace remains uncertain. The current system of egg-based manufacture involves the isolation of viruses from the nasal and pharyngeal washings of people with respiratory illnesses, the propagation of these viruses in fertilized eggs tainted with the other cloacal output of chickens, and an open manufacturing process in which bacterial contamination must be suppressed with antimicrobial agents. Thus, egg-based conventional vaccines, although demonstrated to be acceptably safe through decades of experience, are the product of a process that can reasonably be described as "earthy." When reverse genetic seeds remove human secretions from the process (except as a source of sequence information); cell culture removes the possibility of contamination by chicken flora; and closed manufacturing processes greatly reduce the risk of contamination during manufacture, regulatory authorities and even the public may demand that their vaccine supply be produced by processes that conform to modern standards.

Economics is likely to drive the pace of the transition. Much of the development of flu cell culture has been funded through public–private partnerships to increase pandemic preparedness. The US Flu Cell Culture Facility, which recently opened in Holly Springs, North Carolina, was built through a partnership between Novartis and the United States Biological Advanced Development and Planning Authority [81]. To expand to exclusively privately funded manufacturing sites, flu cell culture vaccines will need to be profitable. The productivity of flu cell culture is a moving target, as the technology matures. The reliability of strain changes in the egg-based processes is built on decades of experience. Egg-based processes will only be abandoned when the new mammalian cell-based processes prove that they can deliver sufficient vaccine supplies as reliably as the egg-based processes.

Finally, the current hybrid egg-based seed generation followed by cell-based production process is proving to be as safe, immunogenic, and effective as the all egg-based processes. Collaborative efforts between manufacturers and public health agencies are underway to enable the production of entirely cell-based reassortant vaccines [25, 30]. In some years, in which no well-matched egg isolate is available, the cell-based vaccines could be dramatically more effective than egg-based vaccines. In non-mismatch years, the HA antigens of cell-based vaccines are expected to be more similar to the HA antigens of circulating strains by at least one amino acid in the receptor binding site. The clinical impact of this improved strain match remains to be determined. Finally, in the event of an avian influenza pandemic that causes high mortality among both humans and chickens, the advantages of cell-based manufacture could be of historic importance.

Acknowledgments I thank Giuseppe Del Giudice (Novartis Vaccines and Diagnostics) for his contribution to this chapter.

References

1. Oxford J, Lambkin-Williams W, Gilbert A (2008) Influenza vaccines have a short but illustrious history. In: Rappuoli R, Del Giudice G (eds) Influenza vaccines for the future. Birkhaeuser, Basel, pp 31–64
2. Nicolson C, Major D, Wood JM, Robertson JS (2005) Generation of influenza vaccine viruses on Vero cells by reverse genetics: an H5N1 candidate vaccine strain produced under a quality system. Vaccine 23:2943–2952

3. Burnet FM (1941) Growth of influenza virus in the allantoic cavity of the chick embryo. Aust J Exp Biol Med Sci 19:291–295

4. Davenport FM, Hennessy AV, Brandon FM, Webster RG, Barrett CD Jr, Lease GO (1964) Comparisons of serological and febrile responses in humans to vaccination with influenza viruses or their hemagglutinins. J Lab Clin Med 63:5–13

5. Alexandrova GI, Smorodintsev AA (1965) Obtaining of an additionally attenuated vaccinating cryophilic influenza strain. Roum Rev Inframicrobiol 2:179

6. Reimer CB, Baker RS, Van Frank RM, Newlin TE, Cline GB, Anderson NG (1967) Purification of large quantities of influenza virus by density gradient centrifugation. J Virol 1: 1207–1216

7. Kilbourne ED (1969) Future influenza vaccines and use of genetic recombinants. Bull WHO 41:643–645

8. Webster RG, Bean WJ, Gorman OT, Chambers TM, Kawaoka Y (1992) Evolution and ecology of influenza A viruses. Microbiol Rev 56:152–179

9. Connor RJ, Kawaoka Y, Webster RG, Paulson JC (1994) Receptor specificity in human, avian, and equine H2 and H3 influenza virus isolates. Virology 205:17–23

10. Shibya K, Ebina M, Yamada S, Ono M, Kasai N, Kawaoka Y (2006) Avian flu: influenza virus receptors in the human airways. Nature 440:435–436

11. Chandrasekaran A, Srinivasan A, Raman R, Viswanathan K, Raguram S, Tumpey TM, Sasisekharan V, Sasisekharan R (2008) Glycan topology determines human adaptation of avian H5N1 virus hemagglutinin. Nat Biotechnol 26:107–113

12. Eisen MB, Sabesan S, Skehel JJ, Wiley DC (1997) Binding of the influenza A virus to cell-surface receptors: sstructures of five hemagglutinin-sialyloligosaccharide complexes determined by X-ray crystallography. Virology 232:19–31

13. Rogers GN, Paulson JC, Daniels RS, Skehel JJ, Wilson IA, Wiley DC (1983) Single amino acid substitutions in influenza hemagglutinin change receptor binding specificity. Nature 304:76–78

14. Ito T, Sizuki Y, Takada A, Kawamoto A, Otsuki K, Masuda H, Yamada M, Suzuki T, Kida H, Kawaoka Y (1997) Differences in sialic acid-galactose linkages in the chicken egg amnion and allantois influence human influenza virus receptor specificity and variant selection. J Virol 71:3357–3363

15. Wiley DC, Wilson IA, Skehel JJ (1981) Structural identification of the antibody-binding sites of Hong Kong influenza hemagglutinin and their involvement in antigenic variation. Nature 289:373–378

16. Skehel JJ, Wiley DC (2000) Receptor binding and membrane fusion in virus entry: the influenza hemagglutinin. Annu Rev Biochem 69:531–569

17. Knossow M, Gaudier M, Douglas A, Barrère B, Bizebard T, Barbey C, Gigant B, Skehel JJ (2002) Mechanism of neutralization of influenza virus infectivity by antibodies. Virology 302:294–298

18. Robertson JS, Nicolson C, Major D, Robertson EW, Wood JM (1993) The role of amniotic passage in the egg-adaptation of human influenza virus is revealed by haemagglutinin sequence analyses. J Gen Virol 74:2047–2051

19. Schild GC, Oxford JS, de Jong JC (1983) Evidence of host-cell selection of influenza virus antigenic variants. Nature 303:706–709

20. Katz JM, Wang M, Webster RG (1990) Direct sequencing of the HA gene of influenza (H3N2) virus in original clinical samples reveals sequence identity with mammalian cell-grown virus. J Virol 64:1808–1811

21. Robertson JS, Naeve CW, Webster RG, Bootman JS, Newman R, Schild GC (1985) Alterations in the hemagglutinin associated with adaptation of influenza B virus to growth in eggs. Virology 143:166–174

22. Katz JM, Naeve CW, Webster RG (1987) Host cell-mediated variation in H3N2 influenza viruses. Virology 156:386–395

23. Mastrosovich M, Mastrosovich T, Carr J, Roberts NA, Klenk HD (2003) Overexpression of the alpha-2, 6-sialyltransferase in MDCK cells increases influenza virus sensitivity to neuraminidase. J Virol 77:8418–8425

24. Oh DY, Barr IG, Mosse JA, Laurie KL (2008) MDCK-SIAT1 cells show improved isolation rates for recent human influenza viruses compared to conventional MDCK cells. J Clin Microbiol 46:2189–2194

25. Minor PD, Engelhardt OG, Wood JM, Robertson JS, Blayer S, Colegate T, Fabry L, Heldens JG, Kino Y, Kistner O, Kompier R, Makizumi K, Medema J, Mimori S, Ryan D, Schwarz R, Smith JS, Sugawara K, Trusheim H, Tsai TF, Krause R (2009) Current challenges in implementing cell-derived influenza vaccines: implications for production and regulation, July 2007, NIBSC, Potters Bar, UK. Vaccine 27:2907–2913

26. Widjaja L, Ilyushina N, Webster RG, Webby RJ (2006) Molecular changes associated with adaptation of human influenza A virus in embryonated chicken eggs. Virology 350:137–145

27. CDC (2004) Preliminary assessment of the effectiveness of the 2003–2004 inactivated influenza vaccine – Colorado, December 2003. MMWR Morb Mortal Wkly Rep 53:8–11

28. CDC (2004) Update: influenza-associated deaths reported among children aged <18 years – United States, 2003–2004 influenza season. MMWR Morb Mortal Wkly Rep 52:1286–1288

29. Katz JM, Webster RG (1989) Efficacy of inactivated influenza A virus (H3N2) vaccines grown in mammalian cells or embryonated eggs. J Infect Dis 160:191–198

30. (1995) Cell culture as a substrate for the production of influenza vaccines: memorandum from a WHO meeting. Bull World Health Org 73: 431–435

31. Maassab HF (1969) Biological and immunologic characteristics of cold-adapted influenza virus. J Immunol 102:728–732

32. Watanabe T, Watanabe S, Shinya K, Kim JH, Hatta M, Kawaoka Y (2009) Viral RNA polymerase complex promotes optimal growth of 1918 virus in the lower respiratory tract of ferrets. Proc Natl Acad Sci USA 106:588–592

33. Grimm D, Staeheli P, Hufbauer M, Koemer I, Martinez-Sobrido L, Solorzano A, Garcia-Sastre A, Haller O, Kochs G (2007) Replication fitness determines high virulence of influenza A virus in mice carrying functional Mx1 resistance gene. Proc Natl Acad Sci USA 104:6806–6811

34. Talon J, Salvatore M, O'Neill RE, Nakaya Y, Zheng H, Muster T, Garcia-Sastre A, Palese P (2000) Influenza A and B viruses expressing altered NS1 proteins: a vaccine approach. Proc Natl Acad Sci USA 97:4309–4314

35. Wacheck V, Egorov A, Groiss F, Pfeiffer A, Fuereder T, Hoeffmayer D, Kundl M, Popow-Kraupp T, Redberger-Fritz M, Mueller CA, Cinatl J, Michaelis M, Geiler J, Bergmann M, Romanova J, Roethl E, Morokutti A, Wolschek M, Ferko B, Seipetl J, Dick-Gudenus R, Muster T (2010) A novel type of influenza vaccine: safety and immunogenicity of replication-deficient influenza virus created by deletion of the interferon antagonist NS1. J Infect Dis 201:354–362

36. Garcia-Sastre A, Egorov A, Matassov D, Brandt S, Levy DE, Durbin JE, Palese P, Muster T (1998) Influenza A virus lacking the NS1 gene replicates in interferon-deficient systems. Virology 252:324–330

37. Shah K, Nathanson N (1976) Human exposure to SV40: review and comment. Am J Epidemiol 103:1–12

38. Stratton K, Almario DA, McCormick M (eds) (2002) Institute of medicine report. Immunization safety review: SV40 contamination of poliovaccine and cancer. The National Academy of Sciences, Washington, DC

39. Miyazawa T (2010) Endogenous retroviruses as potential hazards for vaccines. Biologicals. doi:10.1016j

40. U.S. Department of Health and Human Services, Food and Drug Administration, Center for Biologics Evaluation and Research (2010) Guidance for industry. Characterization and qualification of cell substrates and other biological materials used in the production of viral vaccines for infectious disease indications. Office of Communication, Outreach, and Development, Rockville, MD

41. Gregersen JP (2008) A risk assessment model to rate the occurrence and relevance of adventitious agents in the production of influenza vaccines. Vaccine 26:3297–3304
42. Scholtissek C, Stech J, Krauss S, Webster RG (2002) Cooperation between the hemagglutinin of avian viruses and the matrix protein of human influenza A viruses. J Virol 76:1781–1786
43. Wood JM, Robertson JS (2004) From lethal to life-saving vaccine: developing inactivated vaccines for pandemic influenza. Nat Rev Microbiol 2:842–847
44. Smith HO, Hutchison CA, Pfannkoch C, Venter JC (2003) Generating a synthetic genome by whole genome assembly: phiX174 bacteriophage from synthetic oligonucleotides. Proc Natl Acad Sci USA 100:15440–15445
45. Patriarca P (2007) Use of cell lines for the production of influenza virus vaccines: an appraisal of technical, manufacturing, and regulatory considerations. IVR/WHO report, Geneva, Switzerland
46. Genzel Y, Reichel U (2009) Continuous cell lines as a production system for influenza vaccines. Expert Rev Vaccines 8:1681–1692
47. Tripp RA, Tompkins SM (2008) Recombinant vaccines for influenza virus. Curr Opin Investig Drugs 9:836–845
48. Doroshenko A, Halperin S (2009) Trivalent MDCK cell culture-derived influenza vaccine Optaflu (Novartis vaccines). Expert Rev Vaccines 8:679–688
49. Barrett PN, Mundt W, Kistner O, Howard MK (2009) Vero cell platform in vaccine production: moving towards cell culture-based viral vaccines. Expert Rev Vaccines 8:607–618
50. Levenbook IS, Petricciani JC, Elisberg BL (1984) Tumorigenicity of Vero cells. J Biol Stand 12:391–398
51. Shin SI, Freedman VH, Risser R, Pollack R (1975) Tumorigenicity of virus-transformed cells in nude mice is correlated specifically with anchorage independent growth in vitro. Proc Natl Acad Sci USA 72:4435–4439
52. Keenan J, Pearson D, Clynes M (2006) The role of recombinant proteins in the development of serum-free media. Cytotechnology 50:49–56
53. Schulze-Horsel J, Schulze M, Agalaridis G, Genzal U, Reichl U (2009) Infection dynamics and virus-induced apoptosis in cell culture-based influenza vaccine production – flow cytometry and mathematical modeling. Vaccine 27:2712–2722
54. Centers for Disease Control and Prevention (CDC) (2004) Updated interim influenza vaccination recommendations – 2004–05 influenza season. MMWR Morb Mortal Wkly Rep 53:1183–1184
55. Chiron (2005) Use of MDCK cells for the manufacture of inactivated influenza virus vaccines. VRBPAC 16 Nov 05 meeting. Available from http://www.fda.gov/ohrms/dockets/ac/05/slides/5-4188S1_5.pdf
56. Petricciani JC, Regan PJ (1987) Risk of neoplastic transformation from cellular DNA: calculations using the oncogene model. Dev Biol Stand 68:43–49
57. Morgeaux S, Tordo N, Gontier C, Perrin P (1993) Beta-propiolactone treatment impairs the biological activity of residual DNA from BHK-21 cells infected with rabies virus. Vaccine 11:82–90
58. Pau MG, Ophorst C, Koldijk MH, Schouten G, Mehtali M, Uytdehaag F (2001) The human cell line PER.C6 provides a new manufacturing system for the production of influenza vaccines. Vaccine 19:2716–2721
59. Rous P (1911) A sarcoma of the fowl transmissible by an agent separable from the tumor cell. J Exp Med 13:397–411
60. Stehelin D, Varmus HE, Bishop JM, Vogt PK (1976) DNA related to the transforming gene(s) of avian sarcoma viruses is present in normal avian DNA. Nature 260:170–173
61. Schild GC, Wood JM, Newman RW (1975) A single radial immunodiffusion technique for the assay of haemagglutinin antigen. WHO Bull 52:223–231
62. Wood JM, Dunleavy U, Newman RW, Riley AM, Robertson JS, Minor PD (1999) The influence of the host cell on standardization of influenza vaccine potency. Dev Biol Stand 98:183–188

63. Szymczakiewicz-Multanowska A, Groth N, Bugarini R, Lattanzi M, Casula D, Hilbert A, Tsai T, Podda A (2009) Safety and immunogenicity of a novel influenza subunit vaccine produced in mammalian cell culture. J Infect Dis 200:841–848

64. Keitel W, Groth N, Lattanzi M, Praus M, Hilbert AK, Borkowski A, Tsai TF (2010) Dose ranging of adjuvant and antigen in a cell culture H5N1 influenza vaccine: safety and immunogenicity of a phase 1/2 clinical trial. Vaccine 28:840–848

65. Ehrlich HJ, Muller M, Oh HM, Tambyah PA, Joukhadar C, Montomoli E, Fisher D, Berezuk G, Fritsch S, Low-Baselli A, Vartian N, Bobrovsky R, Pavlova BG, Pollabauer EM, Kistner O, Barrett PN, Baxter H5N1 Pandemic Influenza Vaccine Clinical Study Team (2008) A clinical trial of a whole-virus H5N1 vaccine derived from cell culture. New Engl J Med 358:2573–2584

66. Clark TW, Pareek M, Hoschler K, Dillon H, Nicholson KG, Groth N, Stephenson I (2009) Trial of 2009 influenza A (H1N1) monovalent MF59-adjuvanted vaccine. New Engl J Med 361:2424–2435

67. Reisinger KS, Block SL, Izu A, Groth N, Holmes SJ (2009) Subunit influenza vaccines produced from cell culture or in embryonated chicken eggs: comparison of safety reactogenicity, and immunogenicity. J Infect Dis 200:849–857

68. Kistner O, Barrett PN, Mundt W, Schober-Bendixen S, Dorner F (1998) Development of a mammalian cell (Vero) derived candidate influenza virus vaccine. Vaccine 16:960–968

69. Halperin SA, Smith B, Mabrouk T, Germain M, Trepanier P, Hassell T, Treanor J, Gauthier R, Mills EL (2002) Safety and immunogenicity of a trivalent, inactivated, mammalian cell culture-derived influenza vaccine in healthy adults, seniors, and children. Vaccine 20:1240–1247

70. Palache AM, Scheepers HSJ, de Regt V, van Ewijk P, Baljet M, Brands R, van Scharrenburg GJM (1999) Safety, reactogenicity, and immunogenicity of Madin Darby canine kidney cell-derived inactivated influenza subunit vaccine. A meta-analysis of clinical studies. Dev Biol Stand 98:115–125

71. Groth N, Montomoli E, Gentile C, Manini I, Bugarini R, Podda A (2009) Safety, tolerability and immunogenicity of a mammalian cell-culture-derived influenza vaccine: a sequential phase I and phase II clinical trial. Vaccine 27:786–791

72. Palache AM, Brands R, van Scharrenburg GJ (1997) Immunogenicity and reactogenicity of influenza subunit vaccines produced in MDCK cells or fertilized chicken eggs. J Infect Dis 176(Suppl 1):S20–S23

73. Halperin SA, Nestruck AC, Eastwood BJ (1998) Safety and immunogenicity of a new influenza vaccine grown in mammalian cell culture. Vaccine 16:1331–1335

74. Ehrlich HJ, Muller M, Fritsch S, Zeitlinger M, Berezuk G, Low-Baselli A, van der Velden MV, Pollbauer EM, Martisch F, Pavlova BG, Tambyah PA, Oh HM, Montomoli E, Kistner O, Noel Barrett P (2009) A cell culture (Vero)-derived H5N1 whole-virus vaccine induces cross-reactive memory responses. J Infect Dis 200:1113–1138

75. Wei CJ, Boyington JC, Dai K, Houser KV, Pearce MB, Kong WP, Yang ZY, Tumpey TM, Nabel GJ (2010) Cross-neutralization of 1918 and 2009 influenza viruses: role of glycans in viral evolution and vaccine design. Sci Transl Med 24:24ra21

76. Scharzer J, Rapp E, Hennig R, Genzel Y, Jordan I, Sandig V, Reichl U (2009) Glycan analysis in cell culture-based influenza vaccine production: influence of host cell line and virus strain on the glycosylation pattern of viral hemagglutinin. Vaccine 27:4325–4336

77. Tey D, Heine RG (2009) Egg allergy in childhood: an update. Curr Opin Allergy Clin Immunol 9:244–250

78. Tubiolo VC, Beall GN (1997) Dog allergy: understanding our 'best friend'? Clin Exp Allergy 27:354–357

79. Wanich N, Bencharitiwong R, Tsai T, Nowak-Wegrzyn AH (2009) In vitro assessment of the allergenicity of novel influenza vaccine produced in dog kidney cells in subjects with dog allergy. J Allergy Clin Immunol 123:S114

80. Ronmark E, Perzanowski M, Platts-Mills T, Lundback B (2003) Four-year incidence of allergic sensitization among schoolchildren in a community where allergy to cat and dog dominates sensitization: Report from the Obstructive Lung Disease in Northern Sweden Study Group. J Allergy Clin Immunol 112:747–754
81. Novartis Media Releases (2009) US Department of Health and Human Services awards Novartis USD 486 million contract to build manufacturing facility for pandemic flu vaccine. http://www.novartis.com/newsroom/media-releases/en/2009/1282432.shtml
82. Kock M, Seemann G (2008) Fertile eggs – a valuable product for vaccine production. Lohmann Inf 43:37–40

Conserved Proteins as Potential Universal Vaccines

Alan Shaw

Abstract In the current climate of an emerging pandemic (October 2009) and the need to vaccinate large populations in a short period of time, the traditional egg-based inactivated vaccine has been pushed, in terms of manufacturing capacity, to a remarkable degree. However, the enormity of the challenge to produce enough vaccine to cover areas beyond USA and Europe has led many investigators to look for a less cumbersome vaccine. Conserved antigens are clearly an attractive alternative as they offer the prospect of protection against a wider variety of influenza challenges.

1 Introduction

Influenza vaccines based on inactivated virus propagated in eggs have been available since the 1940s. The main antigenic component in these vaccines is viral hemagglutinin (HA). The low-fidelity replication apparatus of the influenza virus creates mutations, some of which alter the antigenic structure of HA, allowing the virus to escape the host's immune response. This leads to a need to update the composition of the vaccine to reflect the antigenic profile of the "new" HA. And the cycle starts over again.

With the advent of modern rDNA methods and a growing understanding of the immunology of influenza, there has been a movement to identify and develop influenza vaccine targets that could circumvent the need for annual revamping of a vaccine. Ideally, these new targets would be conserved antigens not subject to immunological drift. Several candidates fit this specification; the M2 ion channel, the HA cleavage site, a fusion intermediate of HA, and a couple of "internal"

A. Shaw
VaxInnate, 3 Cedar Brook Drive, Suite # 1, Cranbury, NJ 08512, USA
e-mail: Alan.Shaw@vaxinnate.com

G. Del Giudice and R. Rappuoli (eds.), *Influenza Vaccines for the Future*, 2nd edition, 313
Birkhäuser Advances in Infectious Diseases,
DOI 10.1007/978-3-0346-0279-2_13, © Springer Basel AG 2011

proteins that could be good T-cell targets. This review will concentrate on the approaches that have either been in clinical studies or are likely to be clinical candidates, but the most promising laboratory studies will be highlighted as well.

2 M2 Ion Channel

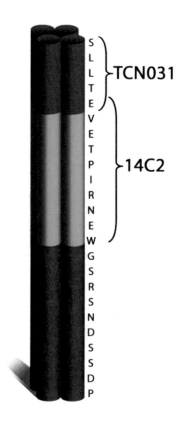

Fig. 1 M2e

The M2 Ion channel is a 97 amino acid transmembrane protein that serves as a conduit for protons to affect a change of pH inside the virion. M2 was discovered by Robert Lamb at The Rockefeller University in 1980 when nucleic acid sequencing became feasible. M2 was identified as a second, overlapping, reading frame on gene segment 7 of the influenza virus, which had been previously identified as the gene for the M (matrix) protein [1]. The existence of a corresponding protein was confirmed by making antiserum against a peptide segment of M2 coupled to KLH and finding an immunoreactive species in infected cell lysates [2]. M2 was shown to be an ion channel [3], the target of the original antivirals, amantidine, and rimantidine. Antibody against the M2 ectodomain, the 23 amino acid section of M2 protruding out from the surface of the virus and displayed on infected cells,

H1/H2/H3 consensus			SLLTEVETPIRNEWGSRSNDSSDP
A/New Caledonia/20/99	H1N1	human	SLLTEVETPIRNEWGCRCNDSSDP
A/Aichi/470/68	H3N1	human	SLLTEVETPIRNEWGCRCNDSSDP
A/Ann Arbor/6/60	H2N2	human	SLLTEVETPIRNEWGCRCNDSSDP
A/Berkeley/1/68	H2N2	human	SLLTEVETPIRNEWGCRCNDSSDP
A/Puerto Rico/8/34	H1N1	human	SLLTEVETPIRNEWGCRCNGSSDP
A/Wisconsin/3523/88	H1N1	human	SLLTEVETPIRNEWGCKCNDSSDP
A/Hebei/19/95	H3N2	human	SLLTEVETPIRNEWECRCNGSSDP
A/Swine/2009	H1N1	human	SLLTEVETPTRSEWECRCSDSSDP
H5 consensus			SLLTEVETPIRNEWESRSSDSSDP
A/Viet Nam/1203/2004	H5N1	human	SLLTEVETPIRNEWECRCSDSSDP
A/Chicken/Nakorn-Patom/Thailand	H5N1	avian	SLLTEVETPIRNEWECRCSDSSDP
A/Thailand/1(KAN-1)/04	H5N1	avian	SLLTEVETPIRNEWECRCSDSSDP
A/Duck/1525/81	H5N1	avian	SLLTEVETPIRNGWECKCSDSSDP
A/Hong Kong/156/97	H5N1	human	SLLTEVETLIRNGWGCRCSDSSDP
A/Chicken/New York/95	H7N2	avian	SLLTEVETPIRNGWECKCSDSSDP
A/Chicken/Hong Kong/G9/97	H9N2	avian	SLLTEVETPIRNGWGCRCSGSSDP
A/Hong Kong/1073/99	H9N2	human	SLLTEVETLIRNGWECKCRDSSDP

Table 1 Sequence alignment of M2e of influenza viruses

reduces the rate of spread of the virus in culture [4] and in vivo [5]. Most importantly for vaccine applications, this ectodomain sequence is highly conserved among the influenza viruses that have infected man. The sequence is also well conserved among the avian influenza viruses (Table 1). The ectodomain of M2 is a parallel homotetramer [6]. This suggests that M2e has some higher order structure that may be of interest to the immune system.

M2 would appear to be an attractive vaccine target based on these properties. There are, however, some drawbacks. The ectodomain is short, only 23 amino acids, and present in a very low copy number on the virus. Infected cells display a much higher density of M2. M2 does not elicit a robust immune response in humans during the course of natural infection [7]. Serum surveys of healthy individuals reveal a 3–5% rate of weak seropositivity for anti-M2e (VaxInnate, unpublished results). These serum surveys were used by Merck and by VaxInnate (see below) to establish background levels of anti-M2e in normal adults. The screening assays were based on recognition of the M2e as a peptide bound to a plastic microtiter plate.

So, here we have an interesting, conserved flu virus antigen, but it is not very immunogenic. Several solutions to this problem have been developed.

One of the earliest attempts at making an M2e vaccine began in the 1990s in Walter Fiers' group at the University of Ghent, Belgium. Prior experience with hepatitis B core antigen showed that HBcAg would self-assemble into a particulate array. Further studies showed that a variety of peptide antigens could be inserted into or appended to HBcAg as a means of making these peptides more immunogenic. Fiers' team produced a vaccine that carried three tandem copies of the 23aa M2 ectodomain at the C-terminus of each HBcAg monomer that should display 240 copies of M2e trimer per assembled particle. This vaccine provided good protection in murine challenge studies [8]. Acambis licensed this vaccine from the University of Ghent and carried out further development. Three versions of this vaccine were tested in a clinical trial in 2007. Volunteers received two doses of vaccine. The first

Fig. 2 Acambis HBcAg

vaccine contained the HBcAg-M2e particle alone. The second contained the particle adsorbed to alum, and the third contained the particle on alum with added QS21 adjuvant. Ninety percent seroconversion against M2e was seen in the third vaccine group (http://www.cidrap.umn.edu/cidrap/content/influenza/general/news/jan0408vaccine.html). In 2008, Acambis became part of Sanofi-Pasteur vaccines where this program may be advanced with a different adjuvant.

Walter Fiers' group has gone on to make a second M2e vaccine that is based on the leucine zipper structure GCN4 fused to the M2e sequence. This molecule forms a tetramer that mimics the tetrameric structure on M2e on infected cells [9]. This vaccine also protected mice from lethal challenge.

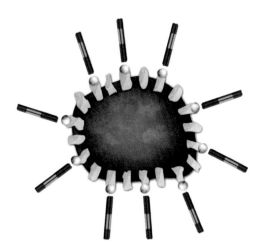

Fig. 3 OMPC-Merck

Merck Research Laboratories made an M2e vaccine candidate based on the 23aa M2e sequence, produced as a synthetic peptide, conjugated to the surface of the outer membrane protein complex (OMPC) of *Neisseria meningitidis*, the carrier moiety of Merck's *Haemophilus influenzae* group B conjugated polysaccharide vaccine [10]. This vaccine prevented severe disease and death in mice upon challenge with a lethal intranasal dose of live influenza virus. Follow-on work explored the effect of conjugating the M2e peptide to OMPC via the N- or the C-terminus. The C-terminal coupling was superior. For comparison with other known M2e vaccines, Merck also made a virus-like particle vaccine based on the hepatitis B core antigen with the M2e sequence inserted into the immunodominant epitope. This vaccine was immunogenic in mice but less so in monkeys, compared to the OMPC conjugates [11].

The M2e-OMPC conjugate vaccine was tested in a clinical study. Two dose levels of the conjugate were delivered IM on aluminum adjuvant in a three-dose series. Both dose levels were immunogenic, with the higher dose giving a higher antibody titer. A follow-on study of this vaccine incorporating a virosomal adjuvant including QS21 showed further improvement in immunogenicity.

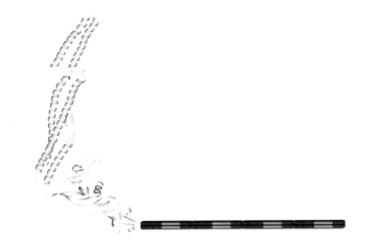

Fig. 4 Flagellin, VaxInnate

VaxInnate began their influenza vaccine program with a peptide-based M2e candidate, VAX101. The short length of the ectodomain, 23 amino acids, allows the production of the antigen by standard peptide synthesis methods. VaxInnate's strategy for making vaccines relies on covalent attachment of a Toll-Like Receptor ligand to the antigen of interest. The TLR2 ligand, tri-palmitoyl cysteine (Pam3-cys) was attached to the M2e peptide sequence N-terminus. Mice vaccinated with two doses of 3-300 μg of Pam3-cys were protected from a lethal challenge with mouse adapted H1N1 PR8/34 virus, although solid protection required a large dose.

Alanine scanning across the M2e sequence showed that the target of protective immunity was in the center of the M2e peptide and corresponded with the binding site of the14C2 monoclonal antibody described by Zebedee and Lamb [12]. Formulation of the lipopeptide was difficult, as one might imagine. Addition of a surfactant improved the immunological performance of the lipopeptide. This suggested that the N-terminal location of the Pam3-cys could be suboptimal; the natural presentation of M2e has the N-terminus free and the C-terminus at the virus or cell membrane. Attempts to make M2e peptide with Pam3-cys at the C-terminus were unsuccessful, largely due to the constraints of commercial peptide synthesis techniques.

Lagging slightly behind the peptide program, VaxInnate had a second vaccine platform based on fusing the genetic sequence for flagellin to the genetic sequence of a given protein antigen. Flagellin is the ligand for TLR5 and serves to target the chimeric protein to antigen-presenting cells where it is processed and presented to T-cells. This strategy has been applied to the M2e sequence. Four tandem copies of the M2e sequence are appended to the C-terminus of flagellin. This protein is produced very efficiently in *E. coli*. Mice vaccinated with 0.3–3.0 µg doses (a two-dose regimen) are protected from death and significant disease following a lethal intranasal challenge with live influenza virus [13]. This vaccine was taken into clinical trials in normal healthy adults. The first study established 1 µg of the fusion protein delivered intramuscularly as the optimal dose with 100%. Note that the M2e component of the fusion protein is one-sixth of the total mass; so in terms of antigen mass, this optimal dose is about 0.17 µg. Lower doses were explored by various injection methods (subcutaneous, intradermal, intramuscular), and greater than 50% seroconversion was achieved with as little as 30 ng given ID twice 28 days apart [14].

There are multiple views of how to use a vaccine based on M2e. Can it be a stand-alone vaccine administered once very few years? Should M2e be used as an adjunct to an HA-based vaccine? Either way, some protective value of M2e vaccination on its own will need to be demonstrated. As a first step to answer the latter question, VaxInnate carried out an adjunct study with Drs. Keipp Talbot and Kathy Edwards at Vanderbilt [15]. A licensed trivalent inactivated vaccine (TIV), Fluvirin, was co-administered with 1ug of the M2e vaccine, VAX102, at the same anatomical site. Antibody raised against the M2e component was similar in quantity to what was seen in the initial study of VAX102 alone, suggesting that TIV does not interfere with the immune response to M2e. The hemagglutination-Inhibiting antibody (HAI) directed to the HA component of TIV was somewhat elevated, about 50% greater (but not quite statistically significantly, due to the small sample size of 20 per group) for both the A/H1N1 and the A/H3N2 components in the presence of M2e. Interestingly, the HAI titers were unaffected by M2e. This suggests that an M2e vaccine could be combined with an HA based vaccine.

Fig. 5 Dynavax, CpG

Dynavax has a vaccine candidate based on the conserved nucleoprotein, a T-cell target, fused genetically to eight tandem copies of the M2e ectodomain. This chimeric protein is then conjugated chemically to a proprietary CpG-containing oligonucleotide, a TLR9 ligand. While there is no published information on this vaccine, it contains the elements of what should be a good candidate. This vaccine has just recently (July 2010) entered clinical studies.

Fig. 6 Bacteriophage QB, Cytos

Cytos AG has produced an M2e vaccine based on appending the M2e sequence to the immunodominant epitope of hepatitis B core antigen similar to but earlier than Merck's effort. While the resulting virus-like particle was immunogenic in mice, protection was deemed inferior to classical inactivated vaccine [16]. Cytos has recently initiated a second M2e vaccine effort based on expressing the M2e sequence on the surface of the dsRNA bacteriophage Qβ [17]. This vaccine, delivered intranasally to mice, afforded superior protection when compared to the core protein vaccine.

Gerhard at the Univeristy of Pennsylvania, and **Zhao** and colleagues at the State Key laboratory in Beijing have produced peptide arrays of the M2e sequence using multi-antennerary "lysine tree" strategies [18–20] and report heterosubtypic protection in animals.

Theraclone has discovered a pair of human monoclonal antibodies that recognize the N-terminus of the M2e homo tetramer [21]. These antibodies, TCN031 and TCN032 were derived from individuals seropositive against M2e expressed on HEK293 cells. Both antibodies bind to M2e on HEK293 cells, and they bind to M2e on the surface of influenza virus, something that the original 14C2 monoclonal antibody did not do well. Conversely, TCN031 and TCN032 do not bind to the monomeric M2e peptide bound to plastic, the usual format for measuring M2e antibody. Alanine scanning of the M2e peptide sequence shows that TCN031 and

032 bind to the first five amino acids, SLLTE. This suggests that the N-terminal tip of the M2e has a conformation that is not found in monomeric peptides. In mice, post challenge administration of these antibodies affords 60-80% protection from death, suggesting a potential for therapeutic use.

It is worth noting here that the Theraclone screen, based on M2e expressed on HEK293 cells identified only antibodies that recognized the tip of the M2e homo-tetramer; no clones recognizing the central 14C2 epitope were found. Conversely, the VaxInnate serological screen identified reactivity to monomeric peptide bound to plastic. The lesson to be learned here is that the design of your screen determines what you find!

3   Mechanism of Action

From the body of work described above, it is clear that the M2 ectodomain can be immunogenic when the right techniques are applied. Antibody raised against M2e can protect animals from death and severe disease.

What is known about M2e immunity that can be cobbled together into a mechanism of action?

First, we know from Zebedee and Lamb that a monoclonal antibody, 14C2, can reduce the rate of spread of influenza virus in a plaque assay. In the presence of 14C2, you get the same number of plaques as you would without the antibody, but the plaques are much smaller. This implies some sort of direct effect on the virus.

We know that there is very little M2 in or on an influenza virus; copy number has been estimated to be 10–20 per virion. On the other hand, there is a substantial amount of M2 displayed on infected cells. This suggests that an antibody-dependent cellular cytotoxicity activity may be involved. Antibody-mediated killing of infected cells could reduce replication and mitigate disease. The Merck group, however, looked at the disease in mice lacking NK cell function and saw no significant difference.

Perhaps, like many things in biology, the effect is due to a combination of effector mechanisms.

The other question is, "What do antibodies recognize when they react with M2e?" The monoclonal 14C2 has been shown to recognize the central portion of the M2e sequence. Passive immunization with 14C2 confers protection in mice. Monoclonal antibodies derived from mice at Merck [22] and from humans at TheraClone and at Kirin pharma [23] recognize the N-terminus of the M2e sequence in the form of a dimer. These latter monoclonals also bind to M2e on the surface of the virus while 14C2 does not. Whether this makes a difference in humans in vivo remains to be seen. The next step for the M2e vaccine is a double-blind, placebo-controlled field study to reveal efficacy against influenza disease.

4 Other Conserved Antibody Targets

There are two non-M2 targets for antibody-mediated immunity on the surface of influenza virus. The first is the "cleavage site" or "cleavage fragment," an approximately 25 amino acid sequence on the stalk of the hemagglutinin molecule. As part of the virus entry into the cell, a proteolytic cleavage must take place at a specific site on the stalk of HA [24, 25]. The cleavage site contains a number of basic amino acids on the N-terminal side and a highly conserved hydrophobic sequence on the C-terminal side as shown in Table. 2. Proteolytic cleavage at one of the basic residues results in the hydrophobic segment rearranging itself into a hydrophobic pocket and triggering a transition state of the structure of the stalk. This is an attractive target, but the hydrophobic nature of the sequence may present a challenge with respect to formulation.

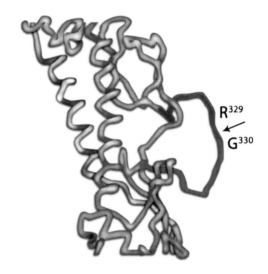

R^{329}

G^{330}

Fig. 7 HA cleavage site, Merck

Table 2 Highly conserved hydrophobic sequence in the C-terminal part of the HA cleavage site		
PAKLLKERGFFGAIAGFLEE	FLU B	
NIPSIQSRGLFGAIAGFIEE	FLU A	H1
NVPEKQTRGIFGAIAGFIEE	Flu A	H3
RERRRKKRGLFGAIAGFIEE	FLU A	H5
NVPQIESRGLFGAIAGFIEE	FLU A	H2

As part of their attempt to make a truly universal influenza vaccine, the Merck group developed a peptide that covered the cleavage site and conjugated it to OMPC. Since the M2 of the influenza B virus does not have an ectodomain, the cleavage site of the B virus was the only other available alternative. OMPC-cleavage site vaccine was quite effective at preventing disease and death in challenge studies [26].

The second conserved element of hemagglutinin is a newly identified transition state of the stalk. Two groups, one at Harvard [27] and the other at Crucell [28, 29], isolated, broadly neutralizing monoclonal antibodies from humans, which recognize and stabilize the transition state of the postcleavage stalk. These antibodies bind to a hydrophobic pocket just underneath the globular head of HA. Interestingly, these antibodies all use the VH1-69 germline gene. Given the breadth of influenza viruses that can be neutralized by these antibodies, the authors raise the possibility of using these antibodies as passive prophylaxis or passive therapy at doses in the range of 10–15 mg/kg. There could be some limitation to this approach. As a model, we can consider Synagis®, a neutralizing monoclonal antibody against respiratory syncytial virus developed by MedImmune. Synagis is dosed at 15 mg/kg in premature or otherwise fragile infants. Multiplying the price of a dose of Synagis to treat a 100 kg adult (the author) yields a cost of just over $20,000. Significant economies in manufacturing will be needed. Alternatively, if one could stabilize this antibody target without having the antibody in the way, a new active vaccination approach becomes available. An elegant engineering effort at Merck, the Indian Institute of Science, and the Nehru Center for Advanced Scientific Research [30] has yielded a subtype-specific vaccine candidate.

Given the overall conserved general structure of the HA molecule, it is tempting to think in terms of a "consensus" HA vaccine. One good example of this is a DNA vaccine with a sequence that encodes a consensus H5 hemagglutinin [31]. Plasmid DNA carrying the consensus H5 HA sequence provided broad protection against challenge with reassortant H5 virus and with some wild-type H5 viruses. If the dose of DNA required to achieve this type of response can be attained in man, this becomes an attractive idea.

5   T-Cell Targets

While the major focus of influenza immunology has been on antibody responses to surface-exposed antigens, there has been no shortage of effort directed to T-cell targets of "internal" influenza proteins. Matrix protein and nucleoprotein have received the most attention due to their conservation and relative abundance in the virion and in infected cells. Early attempts at vaccination against internal proteins, NP in particular, began with "naked DNA" as the immunogen [32]. Intramuscular injection of plasmid DNA encoding NP conferred protection in mice against nasal challenge. Work continues today to refine this attractive idea by including the matrix protein along with NP [33] and with formulations of DNA that enhance uptake. Presentation of the NP sequence via viral vectors of various types has also been explored with promising results [34].

Any discussion of influenza immunity eventually leads to a "T-cells versus humoral immunity" debate. Most of the time, the debate ends in recognition that antibody plus T-cells of the appropriate phenotype are both desirable for optimal protection.

6 Summary

In the current climate following the 2009 pandemic and the need to vaccinate large populations in a short period of time, the traditional egg-based inactivated vaccine has been pushed, in terms of manufacturing capacity, to a remarkable degree. However, the enormity of the challenge to produce enough vaccine to cover areas beyond the USA and Europe has led many investigators to look for a less cumbersome vaccine. Conserved antigens are clearly an attractive alternative [35] as they offer the prospect of protection against a wider variety of influenza challenges.

References

1. Lamb RA, Lai CJ, Choppin PW (1981) Sequences of mRNAs derived from genome RNA segment 7 of influenza virus: colinear and interrupted mRNAs code for overlapping proteins. Proc Natl Acad Sci USA 78:4170–4174
2. Lamb RA, Choppin PW (1981) Identification of a second protein (M2) encoded by RNA segment 7 of influenza virus. Virology 112:729–737
3. Pinto LH, Holsinger LJ, Lamb RA (1992) Influenza virus M2 protein has ion channel activity. Cell 69:517–528
4. Zebedee SL, Lamb RA (1988) Influenza A virus M2 protein: monoclonal antibody restriction of virus growth and detection of M2 in virions. J Virol 62:2762–2772
5. Treanor JJ, Tierney EL, Zebedee SL, Lamb RA, Murphy BR (1990) Passively transferred monoclonal antibody to the M2 protein inhibits influenza A virus replication in mice. J Virol 64:1375–1377
6. Holsinger LJ, Lamb RA (1991) Influenza virus M2 integral membrane protein is a homotetramer stabilized by formation of disulfide bonds. Virology 183: 32–43
7. Feng J, Zhang M, Mozdzanowska K, Zharikova D, Hoff H, Wunner W, Couch RB, Gerhard W (2006) Influenza A virus infection engenders a poor antibody response against the ectodomain of matrix protein 2. Virol J 3:102
8. Neirynck S, Deroo T, Saelens X, Vanlandschoot P, Jou WM, Fiers W (1999) A universal influenza A vaccine based on the extracellular domain of the M2 protein. Nat Med 5:1157–1163
9. De Filette M, Martens W, Roose K, Deroo T, Vervalle F, Bentahir M, Vandekerckhove J, Fiers W, Saelens X (2008) An influenza A vaccine based on tetrameric ectodomain of matrix protein 2. J Biol Chem 283:11382–11387
10. Fan J, Liang X, Horton MS, Perry HC, Citron MP, Heidecker GJ, Fu TM, Joyce J, Przysiecki CT, Keller PM et al (2004) Preclinical study of influenza virus A M2 peptide conjugate vaccines in mice, ferrets, and rhesus monkeys. Vaccine 22:2993–3003
11. Fu TM, Grimm KM, Citron MP, Freed DC, Fan J, Keller PM, Shiver JW, Liang X, Joyce JG (2009) Comparative immunogenicity evaluations of influenza A virus M2 peptide as recombinant virus like particle or conjugate vaccines in mice and monkeys. Vaccine 27:1440–1447
12. Zebedee SL, Lamb RA (1989) Growth restriction of influenza A virus by M2 protein antibody is genetically linked to the M1 protein. Proc Natl Acad Sci U S A 86:1061–1065
13. Huleatt JW, Nakaar V, Desai P, Huang Y, Hewitt D, Jacobs A, Tang J, McDonald W, Song L, Evans RK et al (2008) Potent immunogenicity and efficacy of a universal influenza vaccine candidate comprising a recombinant fusion protein linking influenza M2e to the TLR5 ligand flagellin. Vaccine 26:201–214
14. Turley C, Taylor DN, Tussey L, Kavita U, Johnson C, Rupp RE, Wolfson J, Stanberry L, Shaw AR (2010) Safety and Immunogenicity of a recombinant M2e-flagellin influenza vaccine (STF2.4XM2e) in healthy adults

15. Talbot HK, Rock MT, Johnson C, Tussey L, Kavita U, Shanker A, Shaw AR, Taylor DN (2010) Immunopotentiation of trivalent influenza vaccine when given with VAX102, a recombinant influenza M2e vaccine fused to the TLR5 ligand flagellin. PLoS

16. Jegerlehner A, Schmitz N, Storni T, Bachmann MF (2004) Influenza A vaccine based on the extracellular domain of M2: weak protection mediated via antibody-dependent NK cell activity. J Immunol 172:5598–5605

17. Bessa J, Schmitz N, Hinton HJ, Schwarz K, Jegerlehner A, Bachmann MF (2008) Efficient induction of mucosal and systemic immune responses by virus-like particles administered intranasally: implications for vaccine design. Eur J Immunol 38:114–126

18. Tam JP (1988) Synthetic peptide vaccine design: synthesis and properties of a high-density multiple antigenic peptide system. Proc Natl Acad Sci USA 85:5409–5413

19. Mozdzanowska K, Feng J, Eid M, Kragol G, Cudic M, Otvos L, Jr, Gerhard W (2003) Induction of influenza type A virus-specific resistance by immunization of mice with a synthetic multiple antigenic peptide vaccine that contains ectodomains of matrix protein 2. Vaccine 21:2616–2626

20. Zhao G, Sun S, Du L, Xiao W, Ru Z, Kou Z, Guo Y, Yu H, Jiang S, Lone Y et al (2009) An H5N1 M2e-based multiple antigenic peptide vaccine confers heterosubtypic protection from lethal infection with pandemic H1N1 virus. Virol J 7:151

21. Grandea AG, 3rd, Olsen OA, Cox TC, Renshaw M, Hammond PW, Chan-Hui PY, Mitcham JL, Cieplak W, Stewart SM, Grantham ML et al Human antibodies reveal a protective epitope that is highly conserved among human and nonhuman influenza A viruses. Proc Natl Acad Sci U S A

22. Fu TM, Freed DC, Horton MS, Fan J, Citron MP, Joyce JG, Garsky VM, Casimiro DR, Zhao Q, Shiver JW et al (2009) Characterizations of four monoclonal antibodies against M2 protein ectodomain of influenza A virus. Virology 385:218–226

23. Wang R, Song A, Levin J, Dennis D, Zhang NJ, Yoshida H, Koriazova L, Madura L, Shapiro L, Matsumoto A et al (2008) Therapeutic potential of a fully human monoclonal antibody against influenza A virus M2 protein. Antiviral Res 80:168–177

24. Garten W, Bosch FX, Linder D, Rott R, Klenk HD (1981) Proteolytic activation of the influenza virus hemagglutinin: The structure of the cleavage site and the enzymes involved in cleavage. Virology 115:361–374

25. Kawaoka Y, Webster RG (1988) Sequence requirements for cleavage activation of influenza virus hemagglutinin expressed in mammalian cells. Proc Natl Acad Sci USA 85:324–328

26. Bianchi E, Liang X, Ingallinella P, Finotto M, Chastain MA, Fan J, Fu TM, Song HC, Horton MS, Freed DC et al. (2005) Universal influenza B vaccine based on the maturational cleavage site of the hemagglutinin precursor. J Virol 79: 7380–7388

27. Sui J, Hwang WC, Perez S, Wei G, Aird D, Chen LM, Santelli E, Stec B, Cadwell G, Ali M et al (2009) Structural and functional bases for broad-spectrum neutralization of avian and human influenza A viruses. Nat Struct Mol Biol 16:265–273

28. Throsby M, van den Brink E, Jongeneelen M, Poon LL, Alard P, Cornelissen L, Bakker A, Cox F, van Deventer E, Guan Y et al (2008) Heterosubtypic neutralizing monoclonal antibodies cross-protective against H5N1 and H1N1 recovered from human IgM+ memory B cells. PLoS ONE 3:e3942

29. Ekiert DC, Bhabha G, Elsliger MA, Friesen RH, Jongeneelen M, Throsby M, Goudsmit J, Wilson IA (2009) Antibody recognition of a highly conserved influenza virus epitope. Science 324:246–251

30. Bommakanti G, Citron MP, Hepler RW, Callahan C, Heidecker GJ, Najar TA, Lu X, Joyce JG, Shiver JW, Casimiro DR et al Design of an HA2-based Escherichia coli expressed influenza immunogen that protects mice from pathogenic challenge. Proc Natl Acad Sci USA

31. Chen MW, Cheng TJ, Huang Y, Jan JT, Ma SH, Yu AL, Wong CH, Ho DD (2008) A consensus-hemagglutinin-based DNA vaccine that protects mice against divergent H5N1 influenza viruses. Proc Natl Acad Sci USA 105:13538–13543

32. Donnelly JJ, Friedman A, Martinez D, Montgomery DL, Shiver JW, Motzel SL, Ulmer JB, Liu MA (1995) Preclinical efficacy of a prototype DNA vaccine: enhanced protection against antigenic drift in influenza virus. Nat Med 1:583–587

33. Epstein SL, Tumpey TM, Misplon JA, Lo CY, Cooper LA, Subbarao K, Renshaw M, Sambhara S, Katz JM (2002) DNA vaccine expressing conserved influenza virus proteins protective against H5N1 challenge infection in mice. Emerg Infect Dis 8:796–801

34. Hoelscher MA, Singh N, Garg S, Jayashankar L, Veguilla V, Pandey A, Matsuoka Y, Katz JM, Donis R, Mittal SK et al (2008) A broadly protective vaccine against globally dispersed clade 1 and clade 2 H5N1 influenza viruses. J Infect Dis 197:1185–1188

35. Rappuoli R, Del Giudice G, Nabel GJ, Osterhaus AD, Robinson R, Salisbury D, Stohr K, Treanor JJ (2009) Public health. Rethinking influenza. Science 326:50

Emulsion-Based Adjuvants for Improved Influenza Vaccines

Derek T. O'Hagan, Theodore Tsai, and Steven Reed

Abstract Emulsions have a long history of use as potent and effective adjuvants in humans for a range of vaccines, particularly for influenza. Although older mineral oil- and water-in-oil-based emulsion adjuvants did not have an overall safety and tolerability profile to allow them to be acceptable for widespread use, a newer generation of oil-in-water adjuvants has been recently developed, based on the use of the biodegradable oil squalene. These adjuvants have shown particular value in the development of new generation vaccines to offer enhanced protection against both seasonal and pandemic strains of influenza virus. The first oil-in-water emulsion adjuvant included in an approved flu vaccine was MF59, which was originally licensed in Europe in 1997 as an improved influenza vaccine for the elderly. In the very recent past, MF59 and related adjuvants have shown their value by offering the possibility of significant antigen dose reductions and higher potency products in the face of the H1N1 pandemic emergency and other pandemic threats. The recent H1N1 global problem allowed the opportunity for widespread use of emulsion-based adjuvants in a range of population groups in a number of countries, in which strict monitoring of safety was the norm. Importantly, this widespread use allowed the safety profile of squalene-based emulsion adjuvants to be further substantiated in large and diverse populations of humans, including young children and pregnant women. It is our confident prediction that the coming years will see wider use and further licensures for oil-in-water emulsion adjuvants, particularly for improved flu vaccines.

D.T. O'Hagan (✉) and T. Tsai
Novartis Vaccines and Diagnostic, 350 Massachussetts Avenue, Cambridge, MA 02139, USA
e-mail: derek.ohagan@novartis.com
S. Reed
IDRI, 1124 Columbia Street, Seattle, WA 98104, USA

G. Del Giudice and R. Rappuoli (eds.), *Influenza Vaccines for the Future*, 2nd edition,   327
Birkhäuser Advances in Infectious Diseases,
DOI 10.1007/978-3-0346-0279-2_14, © Springer Basel AG 2011

1   Introduction

Early in the development of nonliving vaccines for widespread human use, modifications were made to enhance their potency. Martin Arrowsmith, the protagonist of the great American novel by the same name, is described as spending time in the laboratory preparing "lipovaccines" and expressing dismay at those promoting the superiority of vaccines suspended in "ordinary salt solutions" [1]. The first generation of lipovaccines, consisting of homogenized dried bacterial cells in lipids, reported in 1916 [2–4], was developed to overcome the relatively poor efficacy of killed bacterial vaccines, which required relatively high doses and multiple injections to induce protective immunity. This was in contrast to an alternative vaccine platform used at the time, live-attenuated organisms (e.g., smallpox) that provided suitable protection from a single dose.

Lipovaccine technology was used in human subjects, including the US military [5, 6], as an approach to increase vaccine potency, which enabled the use of decreasing doses of bacterial cells (dose sparing), as well as decreasing the number of injections required for protection (doseage sparing). These same challenges remain with us today, particularly in relation to influenza vaccines. Currently, a majority of influenza vaccines are produced in eggs, and since this technology is limited in capacity, global supplies are inadequate. Therefore, dose sparing through the use of adjuvants is a safe and practical means to increase vaccine supply. Also, because of the need to respond rapidly to new influenza outbreaks and to reduce the number of doses required to achieve the desired immune response, this represents another role for adjuvants in influenza vaccines. In addition, adjuvants are used to broaden the immune response to emerging influenza variants, as well as to increase responses in the elderly. This early history in lipovaccines has ultimately led to the development of safe emulsions as vaccine adjuvants.

2   Emulsion Technologies

Emulsions are defined as liquid dispersions of two immiscible phases, usually oil and water, either of which may comprise the dispersed phase or the continuous phase to provide water-in-oil (w/o) or oil-in-water (o/w) emulsions, respectively. Emulsions are generally unstable and need to be stabilized by surfactants, which lower interfacial tension and prevent coalescence of the dispersed droplets. Stable emulsions can be prepared through the use of surfactants that orientate at the interface between the two phases and reduce interfacial tensions, since surfactants comprise both hydrophobic and hydrophilic components. Although charged surfactants are excellent stabilizers, nonionic surfactants are widely used in pharmaceutical emulsions due to their lower toxicity and lower sensitivity to the destabilizing effects of formulation additives. Surfactants can be defined by their ratio of hydrophilic to hydrophobic components (hydrophile to lipophile balance, HLB), which gives information on their relative affinity for water and oil phases. At the high end of the scale, surfactants are predominantly hydrophilic and can be used to stabilize o/w emulsions. In contrast, oil-soluble surfactants are at the lower end of the scale and are used mainly to stabilize w/o emulsions. Polysorbates (Tweens) are commonly used surfactants with HLB values in the 9–16 range, while sorbitan esters (Spans) have an HLB in the range of 2–9. Extensive pharmaceutical experience has shown that a mixture of surfactants offers maximum emulsion stability, probably due to the formation of more rigid films at the interface. The physico-chemical characteristics of emulsions, including droplet size, viscosity, and so on, are controlled by a variety of factors, including the choice of surfactants, the ratio of continuous to dispersed phases, and the method of preparation. For an emulsion to be used for administration as an injection, stability and viscosity are important parameters, as too is sterility. In general, stability is enhanced by having smaller sized droplets, while viscosity is decreased by having a lower volume of the dispersed phase.

3   The History of Emulsions as Adjuvants

3.1   Water-in-Oil Emulsion Adjuvants

The desire to increase vaccine potency through association with lipid ultimately resulted in the development of emulsions as adjuvants. Following on from the lipovaccine technology, Freund et al. in 1937 [7] and for many years thereafter [8–10] developed and used w/o emulsions, in which antigen was suspended in an aqueous phase and then emulsified into oil. Early versions included paraffin oil, with Arlacel A as the emulsifier, and the inclusion of dried *Mycobacterium* cells. Such emulsions are still referred to as complete Freund's adjuvant (CFA or FCA). Typically, these emulsions contain >50% mineral oil. Freund also developed emulsions that did not contain bacteria (incomplete Freund's adjuvant, FIA or IFA) and applied them experimentally to a number of vaccine preparations [8–11]. However, CFA was found to be unacceptably reactogenic for use in human vaccines, and sensitization to the mycobacterial component compromised its ability to be used in booster immunizations. Painful local reactions with frequent granuloma formation were observed when administered subcutaneously, and extensive granuloma formations and nerve involvement could occur when administered intramuscularly [12].

Improvements in oil-rich emulsions were made through eliminating the myco-bacterial cells from the preparations and improving the quality of oils used (IFA) [13]. Unfortunately, preclinical studies suggested that oil-based adjuvants could be tumorigenic [14].

Nevertheless, IFA was developed and tested clinically for influenza vaccines, including landmark papers from Jonas Salk and others [15–17] that were proceeded by nonhuman primate studies for safety evaluation, followed by large-scale human trials [18–20]. Overall these studies demonstrated the acceptable safety and potency of the adjuvanted vaccines, including a dose sparing effect (up to 1,000-fold), the durability of antibody responses, and, in Salk's studies, a suggestion of increased breadth of response. Importantly, long-term follow-up studies of army recruits (approximately 18,000 having received IFA-adjuvanted influenza vaccine) indicated that there were no serious safety effects attributable to the vaccines [21, 22]. Hence, these studies helped to alleviate the concerns raised in the preclinical models that oil-based adjuvants could be tumorgenic. Nonetheless, human vaccines containing IFA did not manage to gain broad acceptance. During the period of 1964–1965, 900,000 persons in the UK received a licensed seasonal influenza vaccine containing IFA, and 40 individuals developed local nodular reactions, 9 of which required surgical treatment [23]. On the basis of these observations, which had similarities to the reactions observed in experimental animals, the influenza vaccine was withdrawn from the market [23, 24]. This withdrawal essentially killed the future use of mineral oil-based emulsion adjuvants for human vaccines. Nevertheless, subsequent long-term (35 years) analysis of the army recruits who had received the mineral oil emulsions has shown that not only were there no significant adverse events associated with the emulsion, but there was also a statistically significant reduction in certain forms of cancers in the recruits who had received the adjuvant [25].

It had been thought that much of the toxicity associated with the w/o emulsions was related to the presence of free fatty acids, either in the source materials or resulting from hydrolysis over time of the oil or surfactant [26]. Attempts to improve upon these emulsions included the development of adjuvant 65, an emulsion which consisted of approximately 50% peanut oil [27]. This formulation was tested in 182 volunteers with influenza vaccine and gave higher antibody titers of increased duration than those seen in individuals receiving aqueous vaccine, with minimal differences in reactogenicity between the two groups. Follow-up studies led to influenza vaccine formulated in adjuvant 65 being given to more than 16,000 individuals, with increased immune responses, broadening of immune responses, and greater persistence of antibody responses [28]. Local reactions were deemed to be minor with the adjuvant 65 vaccine, comparing favorably to unadjuvanted vaccine. Interestingly, a formulation combining adjuvant 65 with polyI:polyC, now known to activate innate immunity through toll-like receptor 3 (TLR3), significantly added to the potency of the adjuvant for influenza vaccine in monkeys [28]. However, the use of peanut oil emulsions did not advance significantly, partly due to potential safety concerns in individuals with peanut allergies.

More recent w/o emulsions have included the introduction of purified, metabolizable oils and emulsifiers. The most notable of these emulsions is the Montanide adjuvants (Seppic, Paris, France) which are based on purified squalene and squalane, emulsified with a highly purified mannide mono-oleate surfactant. These emulsions have been evaluated in clinical trials, notably for malaria, HIV, and cancer [29]. Several clinical studies with ISA51 and ISA720, two of the Montanide adjuvants, have reported on potent immune responses, although safety results seem to be somewhat questionable, with the incidence of adverse events and severe adverse events increasing with antigen dose and the number of administrations [30, 31].

Despite the ultimate failure of the oil-rich emulsions in terms of their adoption into licensed human vaccines, the use of these preparations demonstrated, over several decades, the value of adjuvanting influenza and other human vaccines. However, the mechanism through which w/o emulsions potentiate the immune response to vaccine antigens is unclear. It had been thought that these high oil content w/o emulsions functioned through a "depot effect," releasing antigen over time, but such a concept appears inconsistent with the observed kinetics of immune responses following administration of such adjuvanted vaccines. Moreover, studies in which immune response remained strong despite early excision of injections sites in experimental animals would argue against this mechanism of action [32]. The real mechanism of effect may be due to a combination of increased antigen uptake through association with lipid and antigen-presenting cell (APC) activation from the adjuvant components. However, overall safety concerns with high lipid content adjuvants led to emphasis on the development of standardized methods to produce formulations with lower lipid content that would neither form depots nor induce local granuloma and/or ulceration.

3.2   Oil-in-Water Emulsion Adjuvants: The Early Years

Extensive experience in animals and humans, first with lipovaccines, then with w/o emulsions, firmly established the value of formulations containing lipid for adjuvanting nonliving organisms to create more effective vaccines. Attempts to improve the safety of oil-containing adjuvant formulations while maintaining potent immune-stimulatory properties have led to the development of several emulsions with reduced oil content, typically <5% oil. The most advanced of these o/w emulsion adjuvants is MF59, but several others are in various stages of development. The underlying principles behind the development of these new emulsion formulations, in addition to using lower amounts of oil, were to use metabolizable oil (as opposed to mineral oil in the Freund's formulations), use nontoxic emulsifiers instead of Arlacel A, and retain efficacy while reducing toxicity.

Early work by Ribi et al. [33] described the use of o/w emulsions containing squalene, a metabolizable cholesterol precursor obtained from shark liver and Tween 80 surfactant, to which immunostimulants such as monophosphoryl lipid A (MPL), a glycolipid purified from Gram-negative bacteria and/or mycobacterial-derived components were added to create the Ribi adjuvant systems. Another early program was described by Syntex Corp. (reviewed by Allison [34]), which developed the Syntex adjuvant formulation (SAF), a 5% squalane, prepared by hydrogenation of squalene, o/w emulsion, that included polysorbate (Tween) 80 as an emulsifier. Formulations also included Pluronic 121, a block copolymer, and muramyl dipeptide (MDP), an adjuvant peptide based on a structure derived from mycobacterial cell walls. Thus, SAF and the Ribi adjuvant series (reviewed in [35]) represented a significant advance over the Freund formulations. Both Ribi adjuvants and SAF were tested in a variety of animal models, and the safety and efficacy profile was appropriate to allow clinical evaluation in cancer vaccine trials. Other early emulsions included the Hjorth formulations, which were also squalene based (reviewed in [34]).

The most advanced o/w emulsions currently in development include AS03 (reviewed below) and AS02 from GSK. AS03 is a squalene- and vitamin E-based emulsion, extensively developed for influenza vaccines. AS02 is an emulsion that contains MPL and QS-21, a purified molecule derived from saponin. The addition of MPL and QS-21 results in a formulation that has potent B- and T-cell-stimulating properties and has been extensively tested with malaria vaccine candidates, among others. AS02 and AS01, a liposomal formulation, are in advanced clinical development for malaria, tuberculosis, and other vaccine candidates [29].

3.3   The Development of Oil-in-Water Emulsions as Adjuvants for Flu Vaccines

There are several reasons why adding adjuvants to influenza vaccines is important. These include (1) to enhance protective immune responses in the elderly, the population in which the vast majority of influenza deaths occur, (2) to allow antigen dose sparing to increase the global vaccine supply, (3) to induce a rapid immune response in the case of the emergence of a pandemic, and (4) to induce a broader immune response to protect against serotypes not present in the administered vaccine. Today, the o/w adjuvant formulations represent the best approach to provide the necessary safety profile while fulfilling at least some of these performance criteria.

There are currently two o/w emulsions in licensed influenza vaccines and at least two others in development. By far the greatest experience is with MF59 (Novartis Vaccines and Diagnostics), which is a component of licensed vaccines for both seasonal and pandemic influenza and will be discussed in detail below. The other emulsion adjuvant that is a component of a licensed pandemic influenza vaccine is AS03 (GlaxoSmithKline Biologicals, GSK). Other emulsions in development for influenza vaccines include AF03 (Sanofi Pasteur) and SE (Infectious Disease Research Institute) (Table 1). All of these emulsions are squalene based, generally with a content of 2–4%, with different surfactants to stabilize the emulsions. In addition, AS03 contains α-D-tocopherol (vitamin E), which has been claimed to have adjuvant properties of its own. The mechanisms of action of o/w adjuvants will be discussed later with respect to MF59 but generally include activation of APCs leading to increased antigen uptake, increase of cytokine production, and influencing APC migration to draining lymph nodes through upregulation of chemokine receptors.

Table 1 The most advanced o/w emulsion adjuvants

Adjuvant emulsions of oil in water: content per adult dose

MF59 (Novartis)
  Squalene 9.75 mg, polysorbate 80 1.175 mg, sorbitan trioleate 1.175 mg
AS03 (GSK)
  Squalene 10.68 mg, DL-α-tocopherol 11.86 mg, polysorbate 80 4.85 mg
AF03 (Sanofi Pasteur)
  Squalene-containing emulsion (2.5% emulsion) no further details published
AS02 (GSK)
  Squalene 10.68 mg, DL-α-tocopherol 11.86 mg, polysorbate 80 4.85 mg
  3-D monophosphoryl lipid-A (10–50 μg depending on application)
  QS21 (10–25 μg depending on application)

AS03 (GSK) is being developed for both seasonal and pandemic influenza vaccines, including prepandemic vaccines to prime individuals against H5 to induce at least partial immunity against related influenza variants. Vaccines containing ASO3 have been evaluated in thousands of individuals. ASO3-H5N1 vaccine has been reported to be safe in both adults and children [36, 37]. The vaccine has been reported to be immunogenic, to be dose sparing, and to induce cross-clade immune responses [38, 39]. Pandemrix™, an AS03-H5N1 pandemic vaccine, and Prepandrix™, a prepandemic AS03-H5N1 vaccine, have been approved in Europe.

In addition to the pandemic influenza studies, AS03 has been developed for enhancing the efficacy of Fluarix™, GSK's seasonal influenza vaccine, in the elderly. A current trial is in progress to determine the effect of ASO3 in enhancing protection against disease. Early studies compared adjuvanted versus unadjuvanted Fluarix™ and indicated the possibility that including ASO3 could lead to increased T-cell responses and broadened serological responses in elderly subjects (reviewed in [24]). At an earlier stage of development is AF03, the Sanofi Pasteur squalene-based emulsion. This adjuvant has been evaluated as a H5N1 vaccine candidate in 251 healthy adults [40], was found to be adequately safe and immunogenic, and demonstrated both a dose sparing and an immune broadening effect [24]. These results further enforce the utility of o/w emulsions as a safe and effective approach to enhance vaccine potency. The use of such adjuvants comprises an important and necessary solution to develop vaccines to emerging threats, such as pandemic influenza, which cannot be adequately addressed with traditional, unadjuvanted vaccines.

Although emulsions are the most advanced novel adjuvants, many other attempts have been made to develop successful adjuvants based on a range of related technologies. In the 1980s, a number of groups worked on the development of new adjuvant formulations, including emulsions, ISCOMs, liposomes, and microparticles [41]. These approaches had the potential to be more potent and effective adjuvants than insoluble aluminum salts, which were the only adjuvants included in licensed human vaccines at that time. Unfortunately, alum has been shown to be a poor adjuvant for split and subunit influenza vaccines, which comprise the majority of the currently licensed products. Many of the novel adjuvant approaches contained immune potentiators of natural or synthetic origin, which were included to enhance the potency of the adjuvant. However, the inclusion of immune potentiators often raised concerns about the safety of the adjuvant technology. On the basis of the long history of emulsions as adjuvants, including FIA, several groups investigated the development of improved emulsion formulations as adjuvants. As discussed, Syntex developed an o/w emulsion adjuvant (SAF) using the biodegradable oil, squalane, to deliver a synthetic immune potentiator, called N-acetyl-muramyl-L-threonyl-D-isoglutamine (threonyl-MDP) [42]. The closely related immune potentiator, N-acetyl-L-alanyl-D-isoglutamine (MDP), had been originally identified in 1974 as the minimal structure isolated from the peptidoglycan of mycobacterial cell walls, which had adjuvant activity [43]. However, MDP was pyrogenic and induced uveitis in rabbits [44], making it unacceptable as an adjuvant for human vaccines. Therefore, various synthetic derivatives of MDP were produced, in an effort to identify an adjuvant molecule with an acceptable safety profile; threonyl-MDP was one of these synthetic compounds. More recently, it has been shown that MDP activates immune cells through interaction with the nucleotide-binding domain, which acts as an intracellular recognition system for bacterial components [45]. In addition to threonyl-MDP, SAF also contained a pluronic polymer surfactant (L121), which was included to help bind antigens to the surface of the emulsion droplets. Unfortunately, clinical evaluations of SAF as an adjuvant for an HIV vaccine showed it to have an unacceptable profile of reactogenicity [46]. As an alternative to SAF, Chiron vaccines used squalene, a similar biodegradable oil, to develop an o/w emulsion as a delivery system for an alternative synthetic MDP derivative, muramyl-tripeptide phosphatidylethanolamine (MTP-PE). MTP-PE was lipidated to allow it to be more easily incorporated into lipid-like formulations and to reduce toxicity [47]. Unfortunately, clinical testing also showed that emulsions of MTP-PE displayed an unacceptable level of reactogenicity, which made them unsuitable for routine clinical use [48, 49]. Although the emulsion formulation of MTP-PE enhanced antibody responses against influenza vaccine in humans, the level of adverse effects observed made this adjuvant unsuitable for widespread clinical use [48]. Nevertheless, additional clinical studies undertaken at the same time highlighted that the squalene-based emulsion alone (MF59), without any added immune potentiator, was well tolerated and had comparable immunogenicity to the formulation containing the MTP-PE [49, 50]. These observations resulted in the further development of the MF59 o/w emulsion vehicle alone as a vaccine adjuvant.

In preclinical studies with influenza vaccine, it was confirmed that the immune potentiator, MTP-PE, was not required for MF59 to be an effective adjuvant [51]. A key early study highlighted the ability of MF59 adjuvant to enhance protective immunity to flu virus challenge [52]. The use of MF59 adjuvant allowed a dose reduction of flu vaccine (50- to 200-fold lower doses) and improved protection against challenge for more than 6 months after vaccination [52]. MF59-induced enhanced antibody titers in comparison with flu vaccine alone, even at very low antigen dose. Moreover, the addition of MF59 to flu vaccine offered improved survival against challenge with influenza virus in mice and also reduced viral titers in the lungs of challenged mice. The enhanced protection afforded by the inclusion of MF59 in the vaccine was long lived and allowed a significant dose reduction in the amount of antigen needed to induce protection. Moving beyond the mouse model, MF59 was also shown to be an effective adjuvant for flu vaccine in a range of alternative preclinical animal models [51]. Importantly, in follow-up studies, it was shown that MF59 was able to enhance the immune responses to flu vaccines in both young and old animals [53]. Old mice (18 months old in these studies) typically have poor responses to flu vaccines, as do elderly humans, but the inclusion of MF59 in the vaccine restored the response of the old mice back up to the level of response achieved in young mice. Moreover, MF59 was also shown to induce a potent T-cell response to the flu vaccine, in both young and old mice. Pushing the mouse model further, MF59 was also shown to be an effective adjuvant in old mice, which had previously been infected with influenza, a situation more similar to that found in humans, who are often reinfected annually with circulating flu strains [53]. These preclinical studies highlighted the huge potential of MF59 to be used as an adjuvant for an improved flu vaccine, potentially allowing antigen dose reduction, while enhancing protective antibody and T-cell responses, for extended time periods. The ability of MF59 adjuvant to offer a significant reduction in the protective dose for flu vaccines has subsequently become very important in the pandemic flu vaccine setting.

The small droplet size of MF59 adjuvant emulsion, generated through the use of a microfluidizer in the preparation process, is crucial to the potency of the adjuvant, and also enhances emulsion stability and allows the formulation to be sterile filtered for clinical use. Overall, our early clinical experience with o/w emulsions served to highlight the need for careful selection of immune potentiators to be included in adjuvant formulations. The experience with MF59 showed that o/w emulsions can be highly effective adjuvants, with an acceptable safety profile, which may not need the addition of immune potentiators.

4 The Current Status of Emulsion Adjuvants for Flu Vaccines

MF59 is a safe and potent emulsion-based vaccine adjuvant that has been licensed in more than 20 countries, for more than 12 years, for use in an influenza vaccine focused on elderly subjects (Fluad®). The safety profile of MF59 is well established clinically through a large safety database (>26,000 subjects) and through pharmacovigilance evaluations of greater than 55 million doses that have been distributed. The MF59 adjuvant has a significant impact on the immunogenicity of flu vaccines in the elderly, who generally respond poorly to traditional influenza vaccines due to age-related impairment of their immune responses called immunosenescence.

Moving beyond the elderly population, the MF59 adjuvant has also been shown to have a significant impact on the immune response to flu vaccines in adults who are chronically ill with a range of diseases and, consequently, also respond poorly to traditional flu vaccines. Moreover, Fluad also shows enhanced immunogenicity in very young subjects, while displaying a similar reactogenicity profile to licensed vaccines in this population. Moving beyond seasonal flu vaccines, MF59 has also been shown to have a significant impact on the immunogenicity of potential pandemic flu vaccines and has enabled vaccines to achieve titers that might be expected to offer protection, with relatively low doses of vaccine. Moreover, the addition of MF59 to the vaccine allows for more broad cross-reactivity against viral strains not actually included in the vaccine. This is a key attribute, since it is difficult to predict exactly which strain might emerge and cause a pandemic. MF59 adjuvant recently received approval for licensure in Europe for all 27 member states for inclusion in a pandemic vaccine against H1N1 (Focetria®) for use in all subjects aged 6 months and older. This same vaccine adjuvant is also under consideration for approval for inclusion in a prepandemic vaccine (Aflunov®). Beyond its use in influenza vaccines, MF59 adjuvant has also been shown to be a potent adjuvant for a wide range of alternative vaccines, including those based on recombinant proteins, particulate antigens, and protein–polysaccharide conjugates. In most studies in which a comparison has been made, MF59 has been shown to be more potent for both antibody and T-cell responses than aluminum-based adjuvants. Moreover, clinical evaluations have established that the MF59 adjuvant is safe in a wide range of subjects from only a few days old to greater than 100 years of age. Hence, MF59 has broad potential to be used as a safe and effective vaccine adjuvant for a broad range of vaccines to be used in populations with a wide age range. The use of o/w adjuvants represents an important and necessary solution to develop vaccines to emerging threats, such as pandemic influenza, which cannot be adequately addressed with traditional, unadjuvanted vaccines.

4.1   The Composition of MF59

MF59 is a low oil content o/w emulsion. The oil used for MF59 is squalene, which is a naturally occurring substance found in plants and in the livers and skin of a range of species, including humans. Squalene is an intermediate in the human steroid hormone biosynthetic pathway and is a direct synthetic precursor to cholesterol. Therefore, squalene is biodegradable and biocompatible, since it is naturally occurring. Shark liver oil comprises 80% squalene and shark liver provides the natural source of the squalene, which is used to prepare MF59. MF59 also contains two nonionic surfactants, Tween 80 and Span 85, which are designed to optimally stabilize the emulsion droplets. Citrate buffer is also used in MF59 to stabilize pH. Although single-vial formulations can be developed with vaccine antigens dispersed directly in MF59, MF59 can also be added to antigens immediately before their administration. Although a less favorable option, combination before administration may be necessary to ensure optimal antigen stability for some antigens but not for flu.

4.2   Manufacturing of MF59

Details of the manufacturing process for MF59 at the 50-l scale have previously been described [54]. The process involves dispersing Span 85 in the squalene phase and Tween 80 in the aqueous phase before high-speed mixing to form a coarse emulsion. The coarse emulsion is then passed repeatedly through a microfluidizer to produce an emulsion of uniform small droplet size (165 nm), which can be sterile filtered and filled into vials. Methods have also been published to allow the preparation of MF59 on a small scale for use in research studies [55]. MF59 is extensively characterized by various physicochemical criteria after preparation.

4.3   The Mechanism of Action of MF59 Adjuvant

Early studies designed to determine the mechanism of action of MF59 focused on the possibility of the creation of a "depot" effect for coadministered antigen, since there had been suggestions that emulsions may retain antigen at the injection site. However, early work showed that an antigen depot was not established at the injection site and that the emulsion was cleared rapidly [56]. The lack of an antigen depot with MF59 was confirmed in later studies [57], which also established that MF59 and antigen were cleared independently. Subsequently, it was thought that perhaps the emulsion acted as a "delivery system" and was responsible for promoting the uptake of antigen into APCs. This theory was linked to earlier observations with SAF, which contained a pluronic surfactant that was thought to be capable of binding antigen to the emulsion droplets to promote antigen uptake [42]. However, studies with recombinant antigens showed that MF59 was an effective adjuvant, despite no evidence of binding of the antigens to the oil droplets [56]. A direct effect of MF59 on cytokine levels in vivo was also observed in separate studies, suggesting that the delivery method alone was too simplistic an explanation [58]. To gain a better understanding of the mechanism of action of MF59, we have studied the early steps of the immune response on human cells *in vitro* and in mouse muscle in vivo. We have shown that there are at least two human target cells for MF59, monocytes and granulocytes, and that MF59 has a range of effects, including increased antigen uptake, the release of chemoattractants, and the promotion of cell differentiation. The observation of increased antigen uptake is in line with previous findings in mice [59]. The most readily induced chemoattractant was the chemokine, CCL2, which is involved in cell recruitment. Previous work had shown a reduction of MF59-induced cell recruitment into the muscle in CCR2-deficient mice [60], which is consistent with our observations on human cells. Moreover, experiments on gene expression profiles at the injection site are also consistent with the key role of chemokines [61]. In addition, CCL2 was found in serum after injection of MF59 into mouse muscle, providing further consistency between *in vitro* and *in vivo* observations. MF59 also induces phenotypic changes on human monocytes that are consistent with a maturation process toward immature dendritic cells (DCs). There is an impressive consistency between data obtained in vitro from human cells and data obtained in vivo from mouse. These observations suggest that MF59 induces a local proinflammatory environment within the muscle, which promotes the induction of potent immune responses to coadministered vaccines. Figure 1 summarizes the mechanism of action of MF59.

Hence, we conclude that during vaccination, adjuvants like MF59 augment the immune response at a range of intervention points. Through induction of chemokines, they increase recruitment of immune cells to the injection site, they augment antigen uptake by monocytes at the injection site, and they enhance differentiation of monocytes into DCs, which represent the gold-standard cell type for priming naive T cells. A particularly important feature of MF59 is that it strongly induces the homing receptor CCR7 on maturing DCs, thus facilitating their migration into draining lymph nodes where they can trigger the adaptive immune response specific

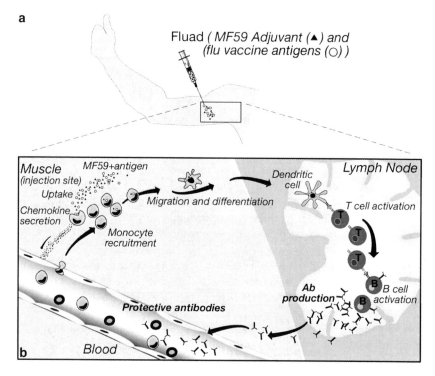

Fig. 1 A model for the mechanism of action of MF59 following immunization with the licensed seasonal influenza vaccine containing MF59 (Fluad). Adapted from [62]

to the vaccine antigen. Nevertheless, further studies are necessary to better define the precise mechanism of action of MF59 and these studies are ongoing.

4.4 Preclinical Experience with MF59

Preclinical experience with MF59 is extensive and has been reviewed on several occasions previously [63–65]. MF59 has been shown to be a potent adjuvant in a diverse range of species, in combination with a broad range of vaccine antigens, to include recombinant protein antigens, isolated viral membrane antigens, bacterial toxoids, protein–polysaccharide conjugates, peptides, and virus-like particles. MF59 is particularly effective for inducing high levels of antibodies, including functional titers (neutralizing, bactericidal, and opsonophagocytic titers) and is generally more potent than alum.

In one study, we directly compared MF59 and alum for several different vaccines and confirmed that MF59 was generally more potent, although alum performed well for bacterial toxoids, particularly diphtheria toxoid [66]. MF59 has also shown enhanced potency over alum when directly compared in nonhuman primates with protein–polysaccharide conjugate vaccines [67] and with a recombinant viral antigen [55]. In preclinical studies, MF59 is the most potent adjuvant for flu vaccines in comparison with various readily available alternatives (Fig. 2). In one study, we compared a number of adjuvants for flu vaccine in mice and showed that MF59 significantly outperforms alternatives, including alum, for both antibody and T-cell responses [68]. Moreover, we have recently shown that MF59 offers enhanced protection against challenge with pandemic flu strains in mice [69], which is consistent with our earlier work on interpandemic strains [52]. Moreover, heterologous protection is achieved against challenge strains in ferrets [70]. In addition to immunogenicity studies, extensive preclinical toxicology studies have been undertaken with MF59 in combination with a range of different antigens in a

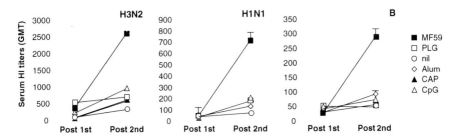

Fig. 2 Serum hemagglutination inhibition titers in mice against the three strains of influenza virus included in seasonal vaccines (H3N2, H1N1, and B) in combination with adjuvants. The adjuvants evaluated included MF59 o/w emulsion, aluminum (Alum), calcium phosphate (CAP), poly-lactide co-glycolide microparticles (PLG), CpG oligonucleotide (CpG), and the vaccine alone (nil). MF59 was the most potent adjuvant for all three strains

number of species. In these studies, it has been shown that MF59 is neither mutagenic nor teratogenic and did not induce sensitization in an established guinea pig model to assess contact hypersensitivity. The favorable toxicological profile established for MF59 allowed extensive clinical testing for MF59 with a number of different vaccine candidates and the approval of a flu vaccine containing MF59 in Europe in 1997.

4.5 Clinical Experience with MF59 Adjuvant: Fluad Seasonal Influenza Vaccine

Fluad, an MF59-adjuvanted seasonal influenza vaccine, was licensed in Italy in 1997 and is now registered in 29 countries worldwide. Fluad was approved on the basis of a clinical development program in more the 20,000 subjects that showed the MF59-adjuvanted vaccine was well tolerated and more immunogenic than conventional nonadjuvanted seasonal trivalent inactivated vaccines (TIV). The adjuvanted vaccine was associated with a low incidence of transient local adverse reactions that were mostly mild or moderate in severity and that did not increase in incidence following subsequent immunizations over 3 years [71, 72] (Fig. 3). Compared with unadjuvanted inactivated vaccine comparators, only local pain, erythema, induration, and myalgia occurred significantly more often in the adjuvanted vaccine recipients, while other systemic adverse events including fever and malaise occurred at similar frequencies in both groups [73]. In most countries where it is registered, Fluad is indicated only for adults over 60 or 65 years of age.

Fluad was initially developed for vaccination of senior adults to fill the medical need for an improved influenza vaccine for this age group in whom conventional

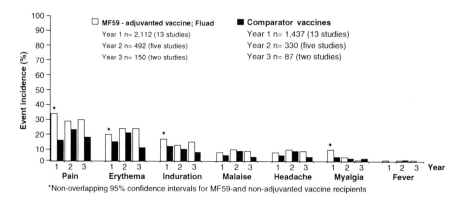

Fig. 3 MF59-adjuvanted vaccine, Fluad, was well tolerated in the elderly after three consecutive annual vaccinations. A meta-analysis of 20 prospective, randomized, observer-blinded clinical studies in elderly subjects (>65 years); subjects received up to three doses 1 year apart of MF59-adjuvanted subunit influenza vaccine or nonadjuvanted subunit or split vaccines. Adapted from [71]

influenza vaccines are less immunogenic and less efficacious compared with younger adults [74, 75]. For this reason, most of the early clinical trials with Fluad were performed in subjects over 65 years old. In this population, the adjuvanted vaccine has induced higher geometric mean titers (GMT), seroconversion rates, and seroprotection rates compared with unadjuvanted vaccine comparators, depending on the vaccine strain composition. GMT HI antibody responses to Fluad typically have been 1.5- to 2.0-fold higher than to unadjuvanted comparators (Fig. 4, right panel). Importantly, higher antibody responses have been seen in the subset of even more frail older adults over 75 years of age [78]. In addition, Fluad has been evaluated in a number of small trials in other patient populations in whom immune response to TIV frequently is lower. Fluad has provided higher HI antibody responses, to varying degrees, in patients with renal transplantation, patients on chronic glucocorticoid therapy, HIV-infected patients on therapy, and senior adults with chronic diseases [79–81].

Fluad is also more immunogenic in adults 18–60 years old with chronic diseases than nonadjuvanted vaccines. In one published study, geometric mean ratios were higher for the MF59-adjuvanted group (Fluad) for all three vaccine strains [82] (Fig. 5).

In addition to augmenting the antibody response in senior adults, MF59 also induces antibody responses that are more broadly cross-reactive [82, 83]. Broader reactivity in a seasonal influenza vaccine is a genuine advantage, as influenza viruses regularly undergo antigenic drift, resulting in periodic mismatches between strains contained in the vaccine and those prevailing in the community. In studies of responses to the H3N2 component of Fluad versus an unadjuvanted vaccine that was otherwise identical (AGRIPPAL), Ansaldi showed that Fluad induced HI titers

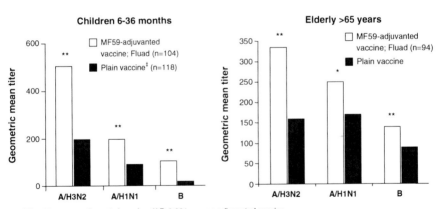

*$P<0.01$ vs. non-adjuvanted vaccine; **$P<0.001$ vs. non-adjuvanted vaccine

Fig. 4 MF59-adjuvanted vaccine induces strong immune responses in children and the elderly. *Left panel* shows an observer-blinded, randomized study in healthy children (6 to <36 months; $N = 222$) involving two doses 4 weeks apart of Fluad ($n = 104$) or a nonadjuvanted plain split vaccine ($n = 118$). *Right panel* shows a randomized, observer-blinded study in elderly (>65 years; $N = 192$) involving a single dose of Fluad ($n = 94$) or a nonadjuvanted plain subunit vaccine ($n = 98$). Adapted from [76] and [77]

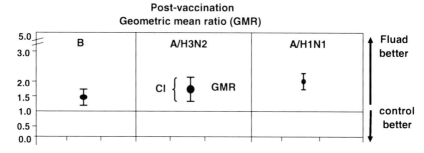

Fig. 5 Fluad is more immunogenic in adults 18–60 years old with chronic diseases than non-adjuvanted vaccines. Geometric mean ratio of hemagglutination inhibition response in adults (18–60 years of age) with chronic diseases who were immunized with an MF59-adjuvanted subunit influenza vaccine versus a similar group immunized with a vaccine without MF59. Geometric mean ratios were higher for the MF59-adjuvanted group (Fluad) for all three vaccine strains. Adapted from [82]

that were significantly higher, not only to the H3N2 strain contained in the vaccine, but also to the predominant circulating H3N2 strains that circulated in each of the following 3 years, each of which had drifted yet further from the previous year's strain [84]. Moreover, the Fluad-induced HI responses to the H3N2 strains that circulated 1 and 2 years later would have met CHMP criteria for the annual update for those strains, whereas responses to the unadjuvanted vaccine would not have [85]. The implication of these observations is that the adjuvanted vaccine might mitigate against the poorer antibody responses expected from periodic mismatches of the recommended vaccine composition with strains circulating in the community. To put this in another way, the adjuvanted vaccine induced cross-reactive antibody responses to strains that emerged 1–3 years in the future and that might not yet have circulated in nature at the time the vaccine was manufactured. The fact that these results pertained to H3N2 strains is significant, as that subtype contributes disproportionately more to seasonal influenza morbidity and mortality than the other subtypes.

The higher HI antibody titers induced by Fluad could be expected to lead to increased vaccine efficacy over unadjuvanted TIV. A large-scale observational study of 150,000 senior adults, comparing the effectiveness of Fluad with AGRIPPAL in reducing influenza-related hospitalizations, showed a 23% lower rate of pneumonia and influenza hospitalizations in recipients of Fluad compared to unadjuvanated vaccine [86]. Puig-Barbera et al. have described the effectiveness of Fluad (compared with no influenza vaccination) in preventing emergency admissions for pneumonia, cardiovascular events, and cerebrovascular events in senior adults [87, 88]. The two case–control studies over two seasons showed significantly reduced hospitalization rates for pneumonia, cardiovascular disease, and cerebrovascular disease in adults over 65 years of age with adjusted odds ratios of 0.31, 0.13, and

0.07, respectively. Considerable efforts were made to account for confounding host factors and to focus the analysis on the period when influenza virus circulated in the community, addressing many of the criticisms of observational studies of influenza vaccination in the elderly [74].

At the other end of the age spectrum, young children also have a reduced immune response to TIV, necessitating two doses for primary immunization, and compared with healthy young adults, vaccine efficacy also is lower in this age group [76]. Two trials have evaluated the safety and immunogenicity of Fluad compared with a licensed inactivated split vaccine comparator in young children (6–36 months of age and 6–59 months of age who had never received influenza vaccine). In the first trial, the MF59-adjuvanted vaccine induced significantly higher HI antibody titers at every time point that was studied – 3 weeks after dose one, 3 weeks after dose two, and 6 months after dose two – for all three subtypes (Fig. 4, left panel). In the Fluad group, the proportion of subjects achieving an HI titer of 40 or higher exceeded 70% for all three subtypes, meeting the CHMP criterion for seroprotection in young adults (NB: the CHMP has not established annual update criteria for children). Moreover, for the H3N2 strain, 91% of the Fluad recipients reached an HI titer of ≥ 40 after just one dose, suggesting the possibility that, for some antigens, the addition of MF59 could sufficiently augment the immune response in immunologically naïve hosts to obviate the need for a two dose primary schedule. The sustained higher HI antibody response, for at least 6 months following the second dose, could be important, as the seasonal influenza vaccine is routinely being delivered in August (in the northern hemisphere), 6 months before the usual peak month of transmission in February of the following calendar year, and 8 months before the usual end of seasonal transmission in April. This is of particular importance for pediatric vaccination as influenza B often is transmitted in the spring, and children under 14 years of age are affected disproportionately by that subtype.

In the same pediatric study, cross-reactive responses to Fluad and the split vaccine comparator also were evaluated in HI tests against strains that were antigenically mismatched to those in the vaccine [76]. As was seen in senior adults, Fluad recipients mounted significantly higher HI antibody titers to mismatched strains of all three subtypes and, for the H1N1 and H3N2 subtypes, CHMP criteria would have been met for geometric mean ratio response (Fig. 6). For the B subtype, however, the antigenic variant that was chosen for testing was not a heterovariant but was a representative of the B/Yamagata lineage while the vaccine contained a B/Victoria lineage strain. The GMT HI titer elicited to the Victoria lineage-virus was just 11, showing that the antigenic distance between the two B lineages is too great for the adjuvant effect of MF59 to bridge.

A randomized controlled efficacy field trial comparing Fluad versus active comparators in More than 3,000 6-<72 month old children showed that Fluad was 86% efficacious against laboratory confirmed influenza while unadjuvanted influenza vaccine was 43% efficacious, resulting in a relative efficacy of Fluad over unadjuvanted vaccine of 75% (Novartis data on file).

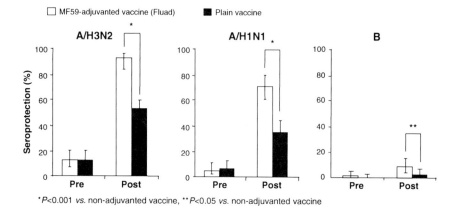

□ MF59-adjuvanted vaccine (Fluad)   ■ Plain vaccine

*P<0.001 vs. non-adjuvanted vaccine, **P<0.05 vs. non-adjuvanted vaccine

Fig. 6 Fluad induced higher levels of cross-reactive antibodies against heterovariant strains in children. Observer-blinded study in children (6–36 months, $n = 222$); involving two 0.25-ml doses of Fluad ($n = 104$) or a nonadjuvanted plain split influenza vaccine ($n = 118$) administered 4 weeks apart and immune responses were measured against strains not included in the vaccine. Adapted from [76]

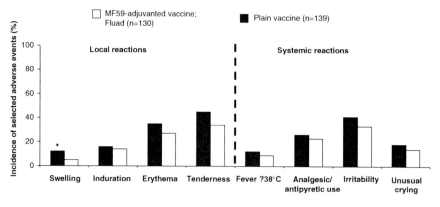

□ MF59-adjuvanted vaccine;   ■ Plain vaccine (n=139)
  Fluad (n=130)

Fig. 7 The MF59-adjuvanted influenza vaccine, Fluad, was well tolerated in children. Observer-blinded study in children (6–36 months, $n = 269$); two doses of Fluad ($n = 130$) or a nonadjuvanted plain split influenza vaccine ($n = 139$), 4 weeks apart. Adapted from [76]

Fluad was shown to be well tolerated among infants and children in these trials (Fig. 7). Although local adverse events occurred more often in the Fluad recipients, only induration at the injection site occurred at a significantly higher frequency [76]. The safety of novel adjuvants in this age group has elicited concern, particularly with respect to the potential for exacerbation of or induction of autoimmune phenomena. All of the pediatric Fluad trials and trials of adjuvanted pandemic influenza vaccines (see below) have been under the oversight of independent data monitoring boards; thus far, none of the trials have been interrupted for safety reasons and no significant safety signals have emerged. Nevertheless, larger scale safety evaluations are needed.

4.6 Safety Evaluations of MF59

More than 45 million doses of Fluad have been distributed commercially, and an analysis of pharmacovigilance reports for the product was undertaken for an interval that covered the distribution of approximately 27 million doses [73]. Reports of all adverse events, SAEs, and certain adverse events of specific interest, including allergic events, acute disseminated encephalomyelitis, encephalitis, Guillain–Barré syndrome, other neurologic events, and blood–vascular disorders occurred at a low rate, well below those reported in the literature. Because of the passive nature and low sensitivity of reporting to pharmacovigilance systems, proportional reporting with similar vaccines is a more appropriate indicator of safety signals than comparisons with rates from epidemiological studies. Compared with AGRIPPAL, the unadjuvanted influenza vaccine counterpart to Fluad, reporting rates of the above adverse events were similar, indicating no detectable increase in risk for these adverse advents associated with MF59 (unpublished data, Novartis Vaccines).

A more systematic analysis of the safety of MF59 has been undertaken, by compiling data from 64 clinical trials in which MF59-adjuvanted and unadjuvanted influenza antigens were studied, providing an opportunity to evaluate the safety of MF59 in isolation. The database, comprising approximately 27,998 subjects, included mainly older adults (65%) in the MF59-adjuvanted group [89]. The median duration of follow-up was approximately 6 months. The analysis focused on SAEs, including hospitalizations and deaths, and specific events, such as the new onset of chronic disease, autoimmune disorders, and cardiovascular events. When randomized trials were examined, reports for these outcomes were no higher in the MF59-adjuvanted group compared with the unadjuvanted vaccine recipients.

Additional safety data on events leading to hospitalization will be forthcoming from the observational study in senior adults mentioned above.

Despite the absence of any data indicating the induction of antibodies against squalene contained in vaccines, some members of the public have associated administration of vaccines and of squalene with Gulf War syndrome. Several reviews of the available data and an epidemiological study among Navy Seabees found no association between squalene antibodies and symptoms of Gulf War syndrome [89]. In addition, it is not generally appreciated that naturally occurring antibodies to squalene are present among healthy individuals. A study comparing anti-squalene antibodies in recipients of Fluad and AGRIPPAL found no difference in antibody rises in the two groups, indicating that MF59 adjuvant neither raises the levels of preexisting antibodies nor induces new antibody responses against squalene [90].

4.7 Pandemic Influenza Vaccines Containing MF59: Avian Influenza Viruses

Pandemic viruses, by definition, are of a subtype that is not currently circulating and usually are novel antigens to which the majority of the population is immunologically naïve. Stimulating a protective immune response to such novel antigens has been shown to require two or more primary vaccination doses, and at least for the H5N1 subtype, requires formulations containing a larger quantity of antigen per dose than is present in the 15 µg/strain contained in the seasonal vaccine. The logistical challenges of vaccinating entire populations with two pandemic vaccine doses and the capacity of individual countries and the world to produce sufficient quantities of viral antigen are challenges to public health systems and governments. Moreover, as with seasonal strains, the H5N1 virus continues to evolve into numerous genetically and antigenically distinguishable clades thus far – with further antigenic variation among viruses within the clades.

These difficulties potentially can be ameliorated or even overcome by the use of emulsion-adjuvanted formulations. Ninety micrograms of unadjuvanted H5N1 split HA in two doses provides an HI antibody titer ≥ 40 in % of healthy young adult subjects, and administering more doses with even higher antigen content is marginally successful in inducing high antibody titers in a suitable proportion of subjects [91]. Substantially less antigen can be used – as little as 1.9 or 3.75 µg of HA is equally or more immunogenic when adjuvanted with an emulsion adjuvant [40, 92–94]. A head-to-head study comparing a subunit H5N1 that was administered unadjuvanted or adjuvanted with alum or with MF59 found that 15 µg of MF59-adjuvanted antigen was significantly more immunogenic than either 45 µg of unadjuvanted or 30 µg of alum-adjuvanted antigen, indicating the potential for MF59 to provide dose sparing, and furthermore, the ineffectiveness of alum as an adjuvant in this circumstance [95].

The antigen dose sparing potential of MF59 was seen even more dramatically in a study of an H9N2 avian influenza virus vaccine in which formulations containing 3.75–30 µg of antigen were studied [96]. All of the adjuvanted formulations provided significantly higher neutralizing antibody titers compared with their unadjuvanted counterparts. Of interest, one dose of the adjuvanted 15-µg formulation was more immunogenic than two doses of the unadjuvanted 15-µg vaccine, suggesting again the potential for MF59 to boost primary antibody responses sufficiently, at least for some antigens, that just one dose could be clinically useful. The ability of 3.75 µg of H5N1 antigen to induce potent immune response in humans in the presence of MF59 adjuvant was recently shown using a flu cell culture-derived influenza vaccine (Fig. 8).

One licensed MF59-adjuvanted H5N1 vaccine has been registered (Focetria®), under the European "mock-up" procedure, as a pandemic vaccine to be used upon a pandemic declaration. Its registration was based on a series of clinical trials using a 7.5-µg HA formulation containing MF59 in the same quantity present in the licensed adjuvanted seasonal vaccine. Two doses in adults 18–64 years old, adolescents 9–17 years old, children 3–8 years old, and infants 6 months to 2 years old

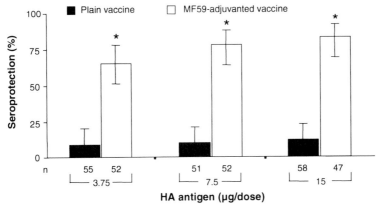

Fig. 8 Hemagglutinin inhibition (HI) assay in healthy young adults 3 weeks post-second vaccination with a cell culture-derived H5N1 vaccine. Observer-blinded, randomized study in adults (18–40 years of age; $n = 695$), involving two doses 21 days apart of MF59-adjuvanted vaccine or nonadjuvanted vaccine with 3.75, 7.5, or 15 μg cell culture-grown influenza A/H5N1 HA. Adapted from [97]

provided HI antibody titers meeting CHMP criteria ([58], Novartis data on file). The approved formulation used a reverse genetics-derived clade 1 A/Vietnam/1204/2007 (H5N1) strain; however, formulations using clade 2.2 and clade 2.3.4 also have been manufactured.

Recipients of the clade 1 A/Vietnam/1204 vaccine mentioned above not only made antibodies at putatively protective levels to the vaccine antigen after primary two-dose vaccination, but in young adults, also elicited cross-reactive antibodies to that degree to a clade 2.2 antigen. The responses in senior adults nearly reached those levels. In a pseudotype neutralization assay, primary responses in young adults also were shown to be broadly reactive to clade 2.1 and 2.3.4 antigens [98].

The cross-reactivity of antibody responses of an H5N1 vaccine to viruses in other clades is of considerable importance, as it would be desirable if individuals primed against an antigen in one clade could respond with an anamnestic response to protective antibody levels after a single booster dose of the vaccine produced against the H5N1 virus that actually emerged in a pandemic. Because the emergence of such a pandemic cannot be predicted, induction of persistent immune memory, lasting years, would be desirable.

Data in support of such priming are available from a small cohort of young adults who were immunized with MF59-adjuvanted or unadjuvanted H5N3 vaccine, an antigen that is antigenically related to H5N1 clade 0 viruses [99–101]. The subjects were reconvened 7 years later and boosted with two doses of adjuvanted H5N1 vaccine [102]. Those who previously had received unadjuvanted H5N3 vaccine needed both doses to produce significant neutralizing antibodies to the H5N1 clade 1 antigen. On the other hand, those who previously had been primed with the adjuvanted H5N3 vaccine responded within 7 days of the first adjuvanted H5N1 dose with high levels of neutralizing antibodies not only to the homologous clade 1 antigen but also to the H5N3 (clade 0 equivalent) antigen and to clade 2.1, 2.2, and 2.3.4 antigens, representing the clades responsible for nearly all the reported cases of H5N1 disease. Before booster vaccination, H5N1 viral-specific memory B cells were present more abundantly among the subjects who had been primed years earlier. Memory B cells were higher in number and peaked earlier among the adjuvant-primed subjects at day 21 after vaccination – correlating with the levels of neutralizing antibodies.

These observations are consistent with preclinical data in ferrets, mentioned below, of the priming effect of adjuvanted seasonal vaccine on responses to A/CA/ 07/2009 (H1N1) antigen. Together, the data suggest that MF59 broadens the primary and memory immune responses to coadministered antigens. This attribute of the adjuvanted response could be of practical importance in the context of prepandemic preparation, as persons at high risk or critical infrastructure workers who were primed potentially could be protected with a single dose of pandemic vaccine as soon as it became available, even if the respective antigens were at some antigenic distance.

Operationally, it is of interest that the priming schedule for the MF59-adjuvanted H5N1 vaccine could be separated between dose one and dose two by as long as 1 year (Novartis data on file). Importantly, when the second dose was derived from a different clade (dose one was a clade 1 antigen, and dose two was a clade 2.2 antigen), responses to the second dose were highly cross-reactive to both antigens, meeting CHMP criteria for viruses in both clades (Novartis data on file). This suggests that annual revaccinations with updated formulations representing newly emerging clades could lead to protection with a single dose. These annual updates could be administered in conjunction with annual seasonal influenza vaccination, as coadministration of the adjuvanted H5N1 vaccine and seasonal inactivated vaccine did not interfere with responses to either seasonal or avian influenza antigens (Novartis data on file).

The still emerging H1N1 pandemic has focused attention on the inadequacy of the global influenza vaccine supply, as the WHO has estimated that only 4.9 billion monovalent doses of vaccine can be produced by all manufacturers within a year. But that assessment is based on the formulations proposed by manufacturers which includes approximately two billion doses that are adjuvanted [103]. The antigen dose sparing potential of MF59 and other emulsion adjuvants on responses to H5N1 virus has been contrasted with alum which has shown variable results. While emulsion adjuvants such as MF59 have provided high immune responses independent of antigen dose above approximately 6 μg, antibody responses to unadjuvanted and alum-adjuvanted vaccines correlate with antigen dose, and to achieve putatively protective levels, amounts of subunit and split antigens greater than 15 μg are needed [104].

As adjuvants themselves must be manufactured with some lead time, minimizing the quantity of adjuvant in a pandemic formulation to its smallest effective dose would be desirable. In fact, when combined with 3.75 or 7.5 μg of subunit H5N1 antigen, half of the usual quantity of MF59 contained in Fluad was as immunogenic as 15 μg of antigen with a full dose of MF59 [97] (Fig. 8).

4.8 Pandemic H1N1 Virus

The distant antigenic and genetic relationship of seasonal H1N1 virus to the novel pandemic H1N1 virus suggested that adjuvanted vaccines would be needed to stimulate protective immunity. The early emerging clinical trial data, however, have indicated that a single 15-μg dose of unadjuvanted hemagglutinin is sufficient to stimulate putatively protective antibody levels in both young and senior adults but that two doses were needed in children [105].

As has been seen with H5N1 virus, the addition of MF59 to subunit H1N1 antigen allowed for considerable antigen sparing [106]. A 3.75-μg antigen dose adjuvanted with a full antigen of MF59 was highly immunogenic in adults after either one or two doses. A combination of 7.5 μg of antigen with half of the usual complement of MF59 was as immunogenic as 15 μg of unadjuvanted antigen and unlike the unadjuvanted vaccine, a single adjuvanted dose was highly immunogenic in children. The former formulation used a cell culture-derived antigen which could prove to be the future of influenza antigen production.

The pandemic H1N1 virus has not yet been observed to drift antigenically from the viruses that were isolated from Mexico and California early in the outbreak. However, as the virus becomes resurgent in the northern hemisphere in the fall of 2009, increased immune pressure from persons who were infected in the spring potentially could lead to emergence of heterovariantsthat might not be neutralized by the current A/CA07/2009 formulated vaccine. When such heterovariants emerge, as they surely will, it will be of considerable interest to determine if the MF59-adjuvanted vaccine provides cross-neutralization.

The public health deployment of these MF59-adjuvanted vaccines in millions of doses will provide additional safety data as well as effectiveness data that should greatly aid further evaluations of the adjuvant in clinical practice. Thus far, no safety signals have emerged with the distribution of more than 100 million doses of MF59-adjuvanted pandemic H1N1 vaccine, which has included tens of thousands of pregnant women and children.

4.9 MF59 with Other Antigens

MF59 has been used in early phase clinical trials with a number of antigens, including herpes simplex, HIV, hepatitis B and C, and cytomegalovirus (CMV) candidate vaccines [107–110]. Of note, MF59 has been used as an adjuvant for pediatric vaccines with CMV and HIV viral antigens [110–113]. Seronegative toddlers immunized with CMV glycoprotein B showed antibody titers that were higher than those found in adults naturally infected with CMV. Moreover, the MF59-adjuvanted vaccine was well tolerated in this age group. A recent phase II study of the MF59-adjuvanted CMV glycoprotein B vaccine in naïve post-postpartum women showed promise, demonstrating 50% efficacy (95% confidence interval,

7–73) in preventing infection, expressed as person-years over a 42-month follow-up interval [114].

4.10   Additional Oil-in-Water Adjuvant Formulations in Development

One of the advantages of o/w emulsion formulations is that they can be used with other immune-stimulating molecules to further improve vaccines to address critical issues for which current vaccines may not be optimal. Examples include combining o/w emulsions with toll-like receptor (TLR) agonists. In development are AS03-MPL (GSK) and MF59 combinations (Novartis), as well as MPL-SE (Infectious Disease Research Institute), which has been clinically evaluated in therapeutic vaccines against leishmaniasis. Such combination adjuvants may help address issues critical to future vaccine development, including rapid response, induction of protective immunological memory, broadening of the immune response, and, of increasing importance, developing more effective vaccines for the elderly. In preclinical studies, combination adjuvant formulations based on MF59 have been shown to induce enhanced immune responses, particularly enhanced T-cell responses with a more Th1 profile [115].

5   The Future

Although MF59 has been used in a licensed seasonal influenza vaccine in Europe for more than a decade, and ASO3 has recently obtained licensure as a pandemic and a prepandemic vaccine, neither of these adjuvants are yet licensed in the USA. The adoption of emulsion-adjuvanted vaccines in the USA will require further demonstrations of safety, particularly if a pediatric indication is sought or if repeated annual exposure is anticipated, as with seasonal influenza vaccine. However, the willingness of regulatory authorities and public health officials to accept the licensure of a vaccine containing the novel MPL adjuvant in a HPV vaccine is encouraging, as it provides a precedent for a licensure path and for clinical acceptance of other novel adjuvants, including emulsions.

References

1. Lewis S (1924) Arrowsmith. Signet Classics Edition, Chapter 27, p 305
2. Pinoy LMa (1916) Les vaccins en emulsion dans les corps gras ou 'lipo-vaccins'. Soc Biol Fil seace du 4 mars, t. LXXIX, 79:201–203
3. Pinoy LMa (1916) Application a l'homme des vaccines en emulsion dans les corps gras (lipo-vaccins). Soc Biol Fil seance du 6 mars, t. LXXIX, 79:352–354

4. Achard CaF C (1916) Sur l'emploi des corps gras comme vehicules des vaccines microbiens. Soc Biol Fil seace du 4 mars , t. LXXIX, 79:209–211
5. Whitmore ER (1919) Lipovaccines, with special reference to public health work. Am J Public Health 9:504–507
6. Lewis PA, Dodge FW (1920) The sterilization of lipovaccines. J Exp Med 31:169–175
7. Freund J, Casals J, Hosmer EP (1937) Sensitization and antibody formation after injection of turbecle bacili and paraffin oil. Proc Soc Exp Biol Med 37:509–513
8. Freund J, McDermott K (1942) Sensitization to horse serum by means of adjuvants. Proc Soc Exp Biol Med 49:548–553
9. Freund J, Walter A (1944) Saprophytic acidfast bacilli and paraffin oil as adjuvants in immunization. Proc Soc Exp Biol Med 56:47–50
10. Freund J, Bonanto M (1944) The effect of paraffin oil, lanolin-like substances and killed tubercle bacilli on immunization with diphtheric toxoid and bact. Typhosum. J Immunol 48:325–334
11. Freund J, Bonanto M (1946) The duration of antibody-formation after injection of killed typhoid bacilli in water-in-oil emulsion. J Immunol 52:231–234
12. Hilleman MR (1966) Critical appraisal of emulsified oil adjuvants applied to viral vaccines. Prog Med Virol 8:131–182
13. Jansen T, Hofmans MP, Theelen MJ, Schijns VE (2005) Structure-activity relations of water-in-oil vaccine formulations and induced antigen-specific antibody responses. Vaccine 23:1053–1060
14. Murray R, Cohen P, Hardegree MC (1972) Mineral oil adjuvants: biological and chemical studies. Ann Allergy 30:146–151
15. Friedewald WF (1944) Enhancement of the immunizing capacity of influenza virus vaccines with adjuvants. Science 99:453–454
16. Salk JE, Bailey ML, Laurent AM (1952) The use of adjuvants in studies on influenza immunization. II. Increased antibody formation in human subjects inoculated with influenza virus vaccine in a water in-oil emulsion. Am J Hyg 55:439–456
17. Salk JE, Laurent AM, Bailey ML (1951) Direction of research on vaccination against influenza; new studies with immunologic adjuvants. Am J Public Health 41:669–677
18. Henle W, Henle G (1945) Experiments on vaccination of human beings against epidemic influenza. Proc Soc Exp Biol Med 59:181
19. Stuart-Harris CH, Andrews CH, Andrews BE et al (1955) Antibody responses and clinical reactions with saline and oil adjuvant influenza virus vaccines. Br Med J 2(4950):1229–1232
20. Salk JE (1953) Use of adjuvants in studies on influenza immunization. 3. Degree of persistence of antibody in human subjects two years aftr vaccination. JAMA 151:169–1175
21. Beebe GW, Simon AH, Vivona S (1964) Follow-up study on army personnel who received adjuvant influenza virus vaccine 1951–1953. Am J Med Sci 247:385–405
22. Beebe GW, Simon AH, Vivona S (1972) Long-term mortality follow-up of Army recruits who received adjuvant influenza virus vaccine in 1951–1953. Am J Epidemiol 95:337–346
23. Stuart-Harris CH (1969) Adjuvant influenza vaccines. Bull World Health Organ 41:617–621
24. Vogel FR, Caillet C, Kusters IC, Haensler J (2009) Emulsion-based adjuvants for influenza vaccines. Expert Rev Vaccines 8:483–492
25. Page W (1993) Long-term followup of army recruits immunized with Freund's incomplete adjuvanted vaccine. Vaccine Res 2:141–149
26. Page M, Vella C, Corcoran T, Dilger P, Ling C, Heath A, Thorpe R (1992) Restriction of serum antibody reactivity to the V3 neutralizing domain of HIV gp120 with progression to AIDS. AIDS 6:441–446
27. Smith JW, Fletcher WB, Peters M, Westwood M, Perkins FJ (1975) Response to influenza vaccine in adjuvant 65-4. J Hyg (Lond) 74:251–259
28. Hilleman MR (1969) The roles of early alert and of adjuvant in the control of Hong Kong influenza by vaccines. Bull World Health Organ 41:623–628

29. Coler RN, Carter D, Friede M, Reed SG (2009) Adjuvants for malaria vaccines. Parasite Immunol 31:520–528
30. Lawrence GW, Saul A, Giddy AJ, Kemp R, Pye D, Ulanova M, Tarkowski A, Hahn-Zoric M, Hanson LA, Moingeon P (1997) Phase I trial in humans of an oil-based adjuvant Seppic Montanide ISA 720. Vaccine 15:176–178
31. Saul A, Lawrence G, Smillie A, Rzepczyk CM, Reed C, Taylor D, Anderson K, Stowers A, Kemp R, Allworth A et al (1999) Human phase I vaccine trials of 3 recombinant asexual stage malaria antigens with Montanide ISA720 adjuvant. Vaccine 17:3145–3159
32. Freund J (1951) The effect of paraffin oil and mycobacteria on antibody formation and sensitization: a review. Am J Clin Pathol 21:645–656
33. Ribi E, Meyer TJ, Azuma I, Parker R, Brehmer W (1975) Biologically active components from mycobacterial cell walls. IV. Protection of mice against aerosol infection with virulent mycobacterium tuberculosis. Cell Immunol 16:1–10
34. Allison AC (1999) Squalene and squalane emulsions as adjuvants. Methods 19:87–93
35. Stills HF Jr (2005) Adjuvants and antibody production: dispelling the myths associated with Freund's complete and other adjuvants. ILAR J 46(3):280–293
36. Rumke HC, Bayas JM, de Juanes JR, Caso C, Richardus JH, Campins M, Rombo L, Duval X, Romanenko V, Schwarz TF et al (2008) Safety and reactogenicity profile of an adjuvanted H5N1 pandemic candidate vaccine in adults within a phase III safety trial. Vaccine 26:2378–2388
37. Ballester A, Garces-Sanchez M, Planelles Cantarino MV et al (2008) Pediatric safety evaluation of an AS-adjuvanted H5N1 vaccine in children aged 6–9 years: a phase II study. Presented at 26th annual meeting of the European Society of Infectious Disease, Graz, Austria, 13–17 May, 2008
38. Leroux-Roels I, Borkowski A, Vanwolleghem T, Drame M, Clement F, Hons E, Devaster JM, Leroux-Roels G (2007) Antigen sparing and cross-reactive immunity with an adjuvanted rH5N1 prototype pandemic influenza vaccine: a randomised controlled trial. Lancet 370:580–589
39. Leroux-Roels I, Bernhard R, Gerard P, Drame M, Hanon E, Leroux-Roels G (2008) Broad clade immmunity induced by an adjuvanted clade 1 rH5N1 pandemic influenza vaccine. PLoS ONE 3(2):e1665
40. Levie K, Leroux-Roels I, Hoppenbrouwers K, Kervyn AD, Vandermeulen C, Forgus S, Leroux-Roels G, Pichon S, Kusters I (2008) An adjuvanted, low-dose, pandemic influenza A (H5N1) vaccine candidate is safe, immunogenic, and induces cross-reactive immune responses in healthy adults. J Infect Dis 198:642–649
41. Vogel FR, Powell MF (1995) A compendium of vaccine adjuvants and excipients. In: Powell MF, Newman MJ (eds) Vaccine design: the subunit and adjuvant approach. Plenum, New York, pp 141–228
42. Allison AC, Byars NE (1986) An adjuvant formulation that selectively elicits the formation of antibodies of protective isotypes and of cell-mediated immunity. J Immunol Methods 95:157–168
43. Ellouz F, Adam A, Ciorbaru R, Lederer E (1974) Minimal structural requirements for adjuvant activity of bacterial peptidoglycan derivatives. Biochem Biophys Res Commun 59:1317–1325
44. Waters RV, Terrell TG, Jones GH (1986) Uveitis induction in the rabbit by muramyl dipeptides. Infect Immun 51:816–825
45. Fritz JH, Ferrero RL, Philpott DJ, Girardin SE (2006) Nod-like proteins in immunity, inflammation and disease. Nat Immunol 7:1250–1257
46. Kenney RT, Edelman R (2004) New generation vaccines. Marcel Dekker, New York
47. Wintsch J, Chaignat CL, Braun DG, Jeannet M, Stalder H, Abrignani S, Montagna D, Clavijo F, Moret P, Dayer JM et al (1991) Safety and immunogenicity of a genetically engineered human immunodeficiency virus vaccine. J Infect Dis 163:219–225

48. Keitel W, Couch R, Bond N, Adair S, Van Nest G, Dekker C (1993) Pilot evaluation of influenza virus vaccine (IVV) combined with adjuvant. Vaccine 11:909–913

49. Keefer MC, Graham BS, McElrath MJ, Matthews TJ, Stablein DM, Corey L, Wright PF, Lawrence D, Fast PE, Weinhold K et al (1996) Safety and immunogenicity of Env 2-3, a human immunodeficiency virus type 1 candidate vaccine, in combination with a novel adjuvant, MTP-PE/MF59. NIAID AIDS Vaccine Evaluation Group. AIDS Res Hum Retroviruses 12:683–693

50. Kahn JO, Sinangil F, Baenziger J, Murcar N, Wynne D, Coleman RL, Steimer KS, Dekker CL, Chernoff D (1994) Clinical and immunologic responses to human immunodeficiency virus (HIV) type 1SF2 gp120 subunit vaccine combined with MF59 adjuvant with or without muramyl tripeptide dipalmitoyl phosphatidylethanolamine in non-HIV-infected human volunteers. J Infect Dis 170:1288–1291

51. Ott G, Barchfeld GL, Van Nest G (1995) Enhancement of humoral response against human influenza vaccine with the simple submicron oil/water emulsion adjuvant MF59. Vaccine 13:1557–1562

52. Cataldo DM, Van Nest G (1997) The adjuvant MF59 increases the immunogenicity and protective efficacy of subunit influenza vaccine in mice. Vaccine 15:1710–1715

53. Higgins DA, Carlson JR, Van Nest G (1996) MF59 adjuvant enhances the immunogenicity of influenza vaccine in both young and old mice. Vaccine 14:478–484

54. Ott G (2000) The adjuvant MF59: a ten year perspective. In: O'Hagan D (ed) Vaccine adjuvants: preparation methods and research protocols. Humana, Totowa, pp 211–228

55. Traquina P, Morandi M, Contorni M, Van Nest G (1996) MF59 adjuvant enhances the antibody response to recombinant hepatitis B surface antigen vaccine in primates. J Infect Dis 174:1168–1175

56. Ott G, Barchfeld GL, Chernoff D, Radhakrishnan R, van Hoogevest P, Van Nest G (1995) MF59: design and evaluation of a safe and potent adjuvant for human vaccines. In: Powell MF, Newman MJ (eds) Vaccine design: the subunit and adjuvant approach. Plenum, New York, pp 277–296

57. Dupuis M, McDonald DM, Ott G (1999) Distribution of adjuvant MF59 and antigen gD2 after intramuscular injection in mice. Vaccine 18:434–439

58. Valensi JP, Carlson JR, Van Nest GA (1994) Systemic cytokine profiles in BALB/c mice immunized with trivalent influenza vaccine containing MF59 oil emulsion and other advanced adjuvants. J Immunol 153:4029–4039

59. Dupuis M, Murphy TJ, Higgins D, Ugozzoli M, Van Nest G, Ott G, McDonald DM (1998) Dendritic cells internalize vaccine adjuvant after intramuscular injection. Cell Immunol 186:18–27

60. Dupuis M, Denis-Mize K, LaBarbara A, Peters W, Charo IF, McDonald DM, Ott G (2001) Immunization with the adjuvant MF59 induces macrophage trafficking and apoptosis. Eur J Immunol 31:2910–2918

61. Mosca F, Tritto E, Muzzi A, Monaci E, Bagnoli F, Iavarone C, O'Hagan D, Rappuoli R, De Gregorio E (2008) Molecular and cellular signatures of human vaccine adjuvants. Proc Natl Acad Sci USA 23:23

62. Seubert A, Monaci E, Pizza M, O'Hagan DT, Wack A (2008) The adjuvants aluminum hydroxide and MF59 induce monocyte and granulocyte chemoattractants and enhance monocyte differentiation toward dendritic cells. J Immunol 180:5402–5412

63. Ott G (2000) Vaccine adjuvants: preparation methods and research protocols. In: O'Hagan D (ed) Vaccine adjuvants: preparation methods and research protocols. Humana, Totowa

64. Podda A, Del Giudice G (2003) MF59-adjuvanted vaccines: increased immunogenicity with an optimal safety profile. Expert Rev Vaccines 2:197–203

65. Podda A, Del Giudice G, O'Hagan DT (2005) A safe and potent adjuvant for human use. In: Schijns V, O'Hagan DT (eds) Immunopotentiators in modern vaccines. Elsevier, Amsterdam, p 149

66. Singh M, Ugozzoli M, Kazzaz J, Chesko J, Soenawan E, Mannucci D, Titta F, Contorni M, Volpini G, Del Guidice G et al (2006) A preliminary evaluation of alternative adjuvants to alum using a range of established and new generation vaccine antigens. Vaccine 24: 1680–1686

67. Granoff DM, McHugh YE, Raff HV, Mokatrin AS, Van Nest GA (1997) MF59 adjuvant enhances antibody responses of infant baboons immunized with *Haemophilus influenzae* type b and *Neisseria meningitidis* group C oligosaccharide-CRM197 conjugate vaccine. Infect Immun 65:1710–1715

68. Wack A, Baudner BC, Hilbert AK, Scheffczik H, Ugozzoli M, Singh M, Kazzaz J, Del Giudice G, Rappuoli R, O'Hagan DT (2008) Combination adjuvants for the induction of potent, long-lasting antibody and T cell rsponses to influenza vaccine. Vaccine 26:552–561

69. Subbarao K, McAuliffe J, Vogel L, Fahle G, Fischer S, Tatti K, Packard M, Shieh WJ, Zaki S, Murphy B (2004) Prior infection and passive transfer of neutralizing antibody prevent replication of severe acute respiratory syndrome coronavirus in the respiratory tract of mice. J Virol 78:3572–3577

70. Forrest HL, Khalenkov AM, Govorkova EA, Kim JK, Del Giudice G, Webster RG (2009) Single- and multiple-clade influenza A H5N1 vaccines induce cross protection in ferrets. Vaccine 27:4187–4195

71. Podda A (2001) The adjuvanted influenza vaccines with novel adjuvants: experience with the MF59-adjuvanted vaccine. Vaccine 19:2673–2680

72. Minutello M, Senatore F, Cecchinelli G, Bianchi M, Andreani T, Podda A, Crovari P (1999) Safety and immunogenicity of an inactivated subunit influenza virus vaccine combined with MF59 adjuvant emulsion in elderly subjects, immunized for three consecutive influenza seasons. Vaccine 17:99–104

73. Schultze V, D'Agosto V, Wack A, Novicki D, Zorn J, Hennig R (2008) Safety of MF59 adjuvant. Vaccine 26:3209–3222

74. Goodwin K, Viboud C, Simonsen L (2006) Antibody response to influenza vaccination in the elderly: a quantitative review. Vaccine 24:1159–1169

75. Simonsen L, Taylor RJ, Viboud C, Miller MA, Jackson LA (2007) Mortality benefits of influenza vaccination in elderly people: an ongoing controversy. Lancet Infect Dis 7: 658–666

76. Vesikari T, Pellegrini M, Karvonen A, Groth N, Borkowski A, O'Hagan DT, Podda A (2009) Enhanced immunogenicity of seasonal influenza vaccines in young children using MF59 adjuvant. Pediatr Infect Dis J 28:563–571

77. De Donato S, Granoff D, Minutello M, Lecchi G, Faccini M, Agnello M, Senatore F, Verweij P, Fritzell B, Podda A (1999) Safety and immunogenicity of MF59-adjuvanted influenza vaccine in the elderly. Vaccine 17:3094–3101

78. Squarcione S, Sgricia S, Biasio LR, Perinetti E (2003) Comparison of the reactogenicity and immunogenicity of a split and a subunit-adjuvanted influenza vaccine in elderly subjects. Vaccine 21:1268–1274

79. Banzhoff A, Nacci P, Podda A (2003) A new MF59-adjuvanted influenza vaccine enhances the immune response in the elderly with chronic diseases: results from an immunogenicity meta-analysis. Gerontology 49:177–184

80. Iorio AM, Francisci D, Camilloni B, Stagni G, De Martino M, Toneatto D, Bugarini R, Neri M, Podda A (2003) Antibody responses and HIV-1 viral load in HIV-1-seropositive subjects immunised with either the MF59-adjuvanted influenza vaccine or a conventional non-adjuvanted subunit vaccine during highly active antiretroviral therapy. Vaccine 21: 3629–3637

81. Baldo V, Baldovin T, Floreani A, Carraro AM, Trivello R (2007) MF59-adjuvanted influenza vaccine confers superior immunogenicity in adult subjects (18–60 years of age) with chronic diseases who are at risk of post-influenza complications. Vaccine 25:3955–3961

82. Baldo V, Baldovin T, Floreani A, Fragapane E, Trivello R (2007) Response of influenza vaccines against heterovariant influenza virus strains in adults with chronic diseases. J Clin Immunol 27:542–547

83. Del Giudice G, Hilbert AK, Bugarini R, Minutello A, Popova O, Toneatto D, Schoendorf I, Borkowski A, Rappuoli R, Podda A (2006) An MF59-adjuvanted inactivated influenza vaccine containing A/Panama/1999 (H3N2) induced broader serological protection against heterovariant influenza virus strain A/Fujian/2002 than a subunit and a split influenza vaccine. Vaccine 24:3063–3065

84. Ansaldi F, Canepa P, Parodi V, Bacilieri S, Orsi A, Compagnino F, Icardi G, Durando P (2009) Adjuvanted seasonal influenza vaccines and perpetual viral metamorphosis: the importance of cross-protection. Vaccine 27:3345–3348

85. Ansaldi F, Bacilieri S, Durando P, Sticchi L, Valle L, Montomoli E, Icardi G, Gasparini R, Crovari P (2008) Cross-protection by MF59-adjuvanted influenza vaccine: neutralizing and haemagglutination-inhibiting antibody activity against A(H3N2) drifted influenza viruses. Vaccine 26:1525–1529

86. Mannino S, Villa M, Weiss N, Apolone G, Rothman K (2010) Effectiveness of influenza vaccination with FLUAD versus a Sub-unit Influenza Vaccine Society for Epidemiologic Research (SER) Annual Meeting: Seattle, Washington – June 23–26, Ref 00002742

87. Puig-Barbera J, Diez-Domingo J, Perez Hoyos S, Belenguer Varea A, Gonzalez Vidal D (2004) Effectiveness of the MF59-adjuvanted influenza vaccine in preventing emergency admissions for pneumonia in the elderly over 64 years of age. Vaccine 23:283–289

88. Puig-Barbera J, Diez-Domingo J, Varea AB, Chavarri GS, Rodrigo JA, Hoyos SP, Vidal DG (2007) Effectiveness of MF59-adjuvanted subunit influenza vaccine in preventing hospitalisations for cardiovascular disease, cerebrovascular disease and pneumonia in the elderly. Vaccine 25:7313–7321

89. Pellegrini M, Nicolay U, Lindert K, Groth N, Della Cioppa G (2009) MF59-adjuvanted versus non-adjuvanted influenza vaccines: integrated analysis from a large safety database. Vaccine 27(49):6959–6965

90. Phillips CJ, Matyas GR, Hansen CJ, Alving CR, Smith TC, Ryan MA (2009) Antibodies to squalene in US Navy Persian Gulf War veterans with chronic multisymptom illness. Vaccine 27:3921–3926

91. Del Giudice G, Fragapane E, Bugarini R, Hora M, Henriksson T, Palla E, O'Hagan D, Donnelly J, Rappuoli R, Podda A (2006) Vaccines with the MF59 adjuvant do not stimulate antibody responses against squalene. Clin Vaccine Immunol 13:1010–1013

92. Beigel JH, Voell J, Huang CY, Burbelo PD, Lane HC (2009) Safety and immunogenicity of multiple and higher doses of an inactivated influenza A/H5N1 vaccine. J Infect Dis 200:501–509

93. Leroux-Roels G (2009) Prepandemic H5N1 influenza vaccine adjuvanted with AS03: a review of the pre-clinical and clinical data. Expert Opin Biol Ther 9:1057–1071

94. Banzhoff A, Gasparini R, Laghi-Pasini F, Staniscia T, Durando P, Montomoli E, Capecchi PL, di Giovanni P, Sticchi L, Gentile C et al (2009) MF59-adjuvanted H5N1 vaccine induces immunologic memory and heterotypic antibody responses in non-elderly and elderly adults. PLoS ONE 4:e4384

95. Bernstein DI, Edwards KM, Dekker CL, Belshe R, Talbot HK, Graham IL, Noah DL, He F, Hill H (2008) Effects of adjuvants on the safety and immunogenicity of an avian influenza H5N1 vaccine in adults. J Infect Dis 197:667–675

96. Atmar RL, Keitel WA, Patel SM, Katz JM, She D, El Sahly H, Pompey J, Cate TR, Couch RB (2006) Safety and immunogenicity of nonadjuvanted and MF59-adjuvanted influenza A/H9N2 vaccine preparations. Clin Infect Dis 43:1135–1142

97. Keitel W, Groth N, Lattanzi M, Praus M, Hilbert AK, Tsai TF (2009) Dose ranging of adjuvant and antigen in a cell culture H5N1 influenza vaccine: safety and immunogenicity of a phase 1/2 clinical trial. Vaccine 28:840–848

98. Alberini I, Del Tordello E, Fasolo A, Temperton NJ, Galli G, Gentile C, Montomoli E, Hilbert AK, Banzhoff A, Del Giudice G et al (2009) Pseudoparticle neutralization is a reliable assay to measure immunity and cross-reactivity to H5N1 influenza viruses. Vaccine 27:5998–6003

99. Nicholson KG, Colegate AE, Podda A, Stephenson I, Wood J, Ypma E, Zambon MC, Windon RG, Chaplin PJ, McWaters P et al (2001) Safety and antigenicity of non-adjuvanted and MF59-adjuvanted influenza A/Duck/Singapore/97 (H5N3) vaccine: a randomised trial of two potential vaccines against H5N1 influenza. Lancet 357:1937–1943

100. Stephenson I, Bugarini R, Nicholson KG, Podda A, Wood JM, Zambon MC, Katz JM (2005) Cross-reactivity to highly pathogenic avian influenza H5N1 viruses after vaccination with nonadjuvanted and MF59-adjuvanted influenza A/Duck/Singapore/97 (H5N3) vaccine: a potential priming strategy. J Infect Dis 191:1210–1215

101. Stephenson I, Nicholson KG, Colegate A, Podda A, Wood J, Ypma E, Zambon M (2003) Boosting immunity to influenza H5N1 with MF59-adjuvanted H5N3 A/Duck/Singapore/97 vaccine in a primed human population. Vaccine 21:1687–1693

102. Galli G, Hancock K, Hoschler K, DeVos J, Praus M, Bardelli M, Malzone C, Castellino F, Gentile C, McNally T et al (2009) Fast rise of broadly cross-reactive antibodies after boosting long-lived human memory B cells primed by an MF59 adjuvanted prepandemic vaccine. Proc Natl Acad Sci USA 106:7962–7967

103. Collin N, de Radigues X, Kieny MP (2009) Vaccine production capacity for seasonal and pandemic (H1N1) 2009 influenza. Vaccine 27:5184–5186

104. Manzoli L, Salanti G, De Vito C, Boccia A, Ioannidis JP, Villari P (2009) Immunogenicity and adverse events of avian influenza A H5N1 vaccine in healthy adults: multiple-treatments meta-analysis. Lancet Infect Dis 9:482–492

105. Greenberg ME, Lai MH, Hartel GF, Wichems CH, Gittleson C, Bennet J, Dawson G, Hu W, Leggio C, Washington D et al (2009) Response to a monovalent 2009 influenza A (H1N1) vaccine. N Engl J 361:2405–2413

106. Clark TW, Pareek M, Hoschler K, Dillon H, Nicholson KG, Groth N, Stephenson I (2009) Trial of 2009 influenza A (H1N1) monovalent MF59-adjuvanted vaccine. N Engl J Med 361:2424–2435

107. Heineman TC, Clements-Mann ML, Poland GA, Jacobson RM, Izu AE, Sakamoto D, Eiden J, Van Nest GA, Hsu HH (1999) A randomized, controlled study in adults of the immunogenicity of a novel hepatitis B vaccine containing MF59 adjuvant. Vaccine 17:2769–2778

108. Langenberg AG, Burke RL, Adair SF, Sekulovich R, Tigges M, Dekker CL, Corey L (1995) A recombinant glycoprotein vaccine for herpes simplex virus type 2: safety and immunogenicity [corrected] [published erratum appears in Ann Intern Med 1995 Sep 1; 123(5):395]. Ann Intern Med 122:889–898

109. Corey L, Langenberg AG, Ashley R, Sekulovich RE, Izu AE, Douglas JM Jr, Handsfield HH, Warren T, Marr L, Tyring S et al (1999) Recombinant glycoprotein vaccine for the prevention of genital HSV-2 infection: two randomized controlled trials. Chiron HSV Vaccine Study Group [see comments]. JAMA 282:331–340

110. Mitchell DK, Holmes SJ, Burke RL, Duliege AM, Adler SP (2002) Immunogenicity of a recombinant human cytomegalovirus gB vaccine in seronegative toddlers. Pediatr Infect Dis J 21:133–138

111. McFarland EJ, Borkowsky W, Fenton T, Wara D, McNamara J, Samson P, Kang M, Mofenson L, Cunningham C, Duliege AM et al (2001) Human immunodeficiency virus type 1 (HIV-1) gp120-specific antibodies in neonates receiving an HIV-1 recombinant gp120 vaccine. J Infect Dis 184:1331–1335

112. Borkowsky W, Wara D, Fenton T, McNamara J, Kang M, Mofenson L, McFarland E, Cunningham C, Duliege AM, Francis D et al (2000) Lymphoproliferative responses to recombinant HIV-1 envelope antigens in neonates and infants receiving gp120 vaccines. AIDS Clinical Trial Group 230 Collaborators. J Infect Dis 181:890–896

113. Cunningham CK, Wara DW, Kang M, Fenton T, Hawkins E, McNamara J, Mofenson L, Duliege AM, Francis D, McFarland EJ et al (2001) Safety of 2 recombinant human immunodeficiency virus type 1 (hiv-1) envelope vaccines in neonates born to hiv-1-infected women. Clin Infect Dis 32:801–807
114. Pass RF, Zhang C, Evans A, Simpson T, Andrews W, Huang ML, Corey L, Hill J, Davis E, Flanigan C et al (2009) Vaccine prevention of maternal cytomegalovirus infection. N Engl J Med 360:1191–1199
115. Baudner BC, Ronconi V, Casini D, Tortoli M, Kazzaz J, Singh M, Hawkins LD, Wack A, O'Hagan DT (2009) MF59 emulsion is an effective delivery system for a synthetic TLR4 agonist (E6020). Pharm Res 26:1477–1485

Lessons Learned from Clinical Trials in 1976 and 1977 of Vaccines for the Newly Emerged Swine and Russian Influenza A/H1N1 Viruses

Robert B. Couch

Abstract An explosive local outbreak of respiratory disease in the US in 1976 with a swine influenza A/H1N1-like virus [A/New Jersey/76 (H1N1)] and the appearance of another A/H1N1 virus [A/USSR/77 (H1N1)] in the subsequent year led to extensive clinical trials of vaccines as preparations for public use. Two whole virus (WV) and two subunit (SV) vaccines of each virus were evaluated in all age groups for safety and immunogenicity. A/NJ WV vaccines were more reactogenic and more immunogenic than SV vaccines, particularly in children. Increase in the dosage led to increase in reactogenicity, which was significant for one WV vaccine that contained more antigen, but reactogenicity was best associated with the presence of WV particles of varied morphology. Although this result led to the concept that WV vaccines were not acceptable for children, WV A/USSR vaccines were not reactogenic in adults as were the A/NJ vaccines.

Patterns of antibody response by age, dosage, and number of doses were similar for both A/NJ and A/USSR vaccines. Increasing dosage increased the frequency and magnitude of responses and µg of HA related better to this finding than CCA units used initially for dosage of A/NJ vaccines. "Primed" persons (exposed to A/H1N1 viruses circulating before 1957) responded to single doses of vaccine. One dose of WV vaccine induced acceptable antibody responses among most unprimed persons but SV vaccines required two doses. For WV vaccine, one dose of high dosage vaccine (60–118 µg HA) was as immunogenic as two doses of lower dosage among adults; two doses as low as 2.5 µg HA were immunogenic in children. For "primed" persons, doses of 15–20 µg HA induced adequate responses. These principles of dosage, morphology and priming as major determinants of reactogenicity and immune responses have been replicated since 1976–1977 and

R.B. Couch
Department of Molecular Virology and Microbiology, Baylor College of Medicine, One Baylor Plaza MS, BCM280, Houston, TX 77030, USA
e-mail: ikirk@bcm.edu

G. Del Giudice and R. Rappuoli (eds.), *Influenza Vaccines for the Future*, 2nd edition, Birkhäuser Advances in Infectious Diseases,
DOI 10.1007/978-3-0346-0279-2_15, © Springer Basel AG 2011

seem likely to be replicated again in the current vaccine trials with the 2009 swine-like A/H1N1 virus.

1 Introduction

An explosive outbreak of febrile respiratory disease in military recruits occurred at Fort Dix, New Jersey, USA in January 1976. The cause of the outbreak was identified as a swine influenza A (H1N1)-like virus; it was designated as A/NJ/76 (H1N1). The outbreak was about 4 weeks in duration and primarily involved basic combat training units as very little spread to other units occurred and none to the surrounding community. Influenza authorities, government officials, and industry representatives convened and concluded that an influenza pandemic caused by A/NJ-like (A/NJ) viruses was possible; authorities proceeded to organize vaccine production and clinical trials in preparation for immunization of the public during the fall of 1976. Those clinical trials were an extensive effort coordinated by the US. Federal Drug Administration (FDA), National Institutes of Allergy and Infectious Diseases (NIAID), and the Centers for Disease Control (CDC). Results of the trials are contained in a supplement to the *Journal of Infectious Diseases* [1].

Approximately 1 year later, a new strain of influenza A/H1N1 virus was identified in the Soviet Union as a cause of influenza outbreaks, particularly among younger people. The strain was designated A/USSR (H1N1) virus and was shown to be similar to A/H1N1 viruses detected almost 30 years earlier in humans and distinct from A/NJ (H1N1) virus [2]. Vaccine clinical trials with this virus were organized by NIAID to provide guidance for vaccines for public use; results are contained in *Reviews of Infectious Diseases* [3].

These two experiences with type A/H1N1 virus vaccines preceding a potential influenza A/H1N1 pandemic are the only large, organized vaccine clinical trials with the A/H1N1 influenza virus performed prior to the current trials with the A/H1N1 virus that emerged in Mexico in the spring of 2009. Both the A/NJ/76 and the newly emerged A/H1N1/2009 virus are swine-like viruses.

The author of this report participated in both the A/NJ/76 and A/USSR/77 vaccine trials which are summarized in this chapter.

2 The Vaccines and Evaluations

The transition from quantitation of vaccine antigen using chick cell agglutinating (CCA) units to quantitation of the HA of each strain in influenza vaccines occurred with the 1976–1977 vaccine trials. The A/NJ vaccines were formulated in CCA units and later tested for HA content; the A/USSR vaccines were formulated in μg of HA and later tested for CCA units.

The variables evaluated in the trials were vaccine manufacturer (four manufacturers), vaccine type [whole virus (WV) or split virus (SV)], dosage, schedule (one or two doses), and age (children, adults, or elderly people). Two manufacturers supplied whole virus vaccines [Merck Sharp and Dohme (MSD) and Merrell National[1] (MN) for A/NJ that became Connaught[1] (C) for A/USSR] and two supplied subunit vaccines [Wyeth and Parke-Davis (PD)]. In addition to CCA and HA content, evaluations of the vaccines included protein quantity, endotoxin content, and viral mass of each WV A/NJ vaccine [4, 5]. Clinical evaluations were of reactogenicity and serum anti-HA antibody responses.

3 Reactogenicity

A tendency for increasing reactogenicity with increasing dosage was noted in the A/NJ trials but the most striking difference was between the WV and SV vaccine reactogenicity in children [6]. The WV vaccines clearly induced greater systemic reactions in children than did the SV vaccines. Reactogenicity of 1,567 children in the initial single-dose trial is summarized in Table 1.

Systemic reactogenicity after the SV vaccines was considered similar to those of placebo recipients (although more mild local reactions occurred with SV). However, both WV vaccines induced significantly greater systemic reactogenicity among both the 3–5 and 6–10 year age groups, including more instances of fever; the MSD vaccine was more reactogenic than the MN vaccine. A later study with smaller numbers of children evaluated dosages considered acceptable for reactions (low dosages for WV vaccines); a second dose was given one month later, and an

Table 1 Comparison of reactogenicity among children after whole and split virus A/New Jersey (H1N1) inactivated influenza vaccines

Vaccine	Dosage[a]		Reaction Index[b]	
	CCA	HA	3–5 Years (N)	6–10 Years (N)
Wyeth	43		0.13 (30)	
	87	4	0.24 (38)	0.36 (50)
PD	54		0.04 (26)	
	108	–	0.07 (43)	0.20 (50)
MN	32		0.55 (64)	
	64	6	0.52 (50)	0.68 (95)
MSD	46		0.99 (71)	
	93	14	0.47 (17)	1.28 (89)
Placebo	–		0.25 (104)	0.28 (93)

PD Parke-Davis, *MN* Merrell National; *MSD* Merck Sharp and Dohme
[a]Approximate CCA and HA/0.5 ml; PD not available as was A/Swine/31 virus
[b]Mean score for fever, headache, malaise, abdominal symptoms; >0.6 considered significant

---

[1]Now Sanofi Pasteur.

11–18-year-old group was added. Dosages were not identical but the higher reactogenicity after WV vaccine was confirmed and included the 11–18 years group. The reaction index was generally lower after the second vaccination; this was most notable for the WV vaccines.

Reactogenicity among adults with A/NJ vaccines was varied among dosages and groups. Only the MSD vaccine exhibited a significant increase in systemic reactogenicity (Table 2); 12.8% of those given the 800 CCA vaccine developed fever, more than double that of any other group [7]. Local reactogenicity was increased with increasing dosages but severe reactions were rare.

A number of laboratory studies of the A/NJ vaccines were conducted to better understand the differences in reactogenicity and immunogenicity between vaccines. There was no correlation between systemic reactivity and endotoxin concentrations, rabbit pyrogenicity, protein concentration, or neuraminidase content [4]. Vaccine mass (for WV vaccines) was determined by chromatographic separations; an increase in mass correlated with an increase in reactivity of the MSD vaccines. In further comparisons of the two WV vaccines, electron micrographs of each WV vaccine were obtained; the MSD vaccine contained more intact virus particles with a greater variety of shapes than did the MN vaccine. This difference was associated with a greater HA concentration, a higher viral mass and greater systemic reactivity for the MSD vaccine than for similar CCA levels of the other WV vaccine and both SV vaccines.

A comparison of reactogenicity among adults for the A/NJ and A/USSR vaccines prepared by the four manufacturers is shown in Table 2. The Reaction Index (RI) for all A/USSR vaccines was low. There is no consistent increase in reactogenicity with increasing dosage, as seen with the A/NJ vaccines. An increase in the RI with increasing dosage was noted among children but all were clinically acceptable [8]. Protein and endotoxin content of the A/USSR vaccines varied; viral mass was not reported.

Table 2 Comparison of reactogenicity among adults after A/New Jersey (H1N1) and A/USSR (H1N1) inactivated influenza vaccines

Vaccine	Strain	CCA/HA[a]	RI[b]	CCA/HA[a]	RI[b]	CCA/HA[b]	RI[b]
Wyeth	NJ	174/8	0.47	335/23	0.30	661/65	0.28
	USSR	107/16	0.26	451/61	0.39		
PD	NJ	217/–	0.34	431/–	0.31	739/–	0.52
	USSR	102/10	0.37	296/43	0.00		
MN/CL	NJ	128/12	0.44	301/26	0.27	697/51	0.58
	USSR	186/19	0.34			805/59	0.31
MSD	NJ	185/28	0.49	356/60	0.90	736/118	2.15
	USSR	81/12	0.21	391/45	0.34		

PD Parke-Davis, *MN/CL* Merrell-National which became Connaught Laboratories and is currently Sanofi Pasteur, *MSD* Merck Sharpe and Dohme

[a]CCA/HA CCA = Actual chick cell agglutinating units/μg HA per 0.5 ml; PD for A/NJ not available as was A/Swine/31 virus

[b]Reaction Index = Mean score for fever, headache, malaise, nausea; A/USSR RI includes children

Table 3 Summary of reactogenicity of inactivated influenza A/H1N1 vaccines in the 1976–1977 clinical trials

Local reactions after all A/NJ vaccines and dosages in all age groups were clinically acceptable and were frequently within the range for placebo

A trend for increasing systemic reactions with increasing dosage of A/NJ vaccine was noted, particularly among children. The trend related better to HA than CCA content

Whole virus A/NJ vaccines induced more systemic reactions than split-product vaccines, particularly among children. These reactions occurred 6–24 h after vaccination

One whole virus A/NJ vaccine was the most reactogenic vaccine and related best to high viral mass, characterized as intact viral particles of varied morphology that included filamentous forms

Reactions after A/USSR vaccines were clinically acceptable and about the same for whole and split virus among both adults and children although children were only given SV vaccines

Systemic reaction scores were lower after a second dose than the first dose for both A/NJ and A/USSR vaccines

Reactions among adults after a dose of 200 CCA of the A/NJ vaccine and 20 µg of HA of the A/USSR vaccine were about the same and were low for all vaccines

This experience with influenza A/H1N1 inactivated vaccines provided an understanding of reactogenicity that influenced many vaccine-related decisions that are still in place today. A summary of the major findings regarding reactogenicity is in Table 3. WV vaccines are considered more reactogenic than SV vaccines, particularly among children. While this is generally true, the comparisons of reactions for the various vaccines and dosages in Table 2 indicate that this is not a uniform finding for WV vaccines, at least not among adults. Reports from trials of A/Hong Kong/68 (H3N2) vaccines indicate that high reactogenicity of WV vaccines was not seen among children [9]. Thus, rejecting WV vaccines as an option for inactivated vaccines because of a potential for increased reactogenicity seems inappropriate, particularly since available data suggest WV vaccines are generally more immunogenic than SV vaccines (see below). Using the HA content as a basis for vaccine dosage provided a better correlate for reactogenicity than did the CCA content.

4 Immunogenicity

Both one-dose and two-dose immunogenicity studies with the A/NJ inactivated vaccines were conducted in children [6]. The one-dose trial exhibited a high frequency and magnitude of responses only in the children given high dosages of WV vaccine, but the level of reactogenicity was considered unacceptable. For this reason, a second trial was done in children that used dosages of the vaccines considered acceptable for reactogenicity. Results of that two-dose trial are shown in Table 4. The lower dosages used were lower for both CCA and µg HA in each age group for the two WV vaccines than for the two SV vaccines. Despite the lower dosages, responses to the WV vaccines were similar to, and usually better than, those for higher dosages of SV vaccine. In general, responses to one dose were deficient for all vaccines at the dosages tested but were satisfactory after two doses.

Table 4 Comparison of serum antibody responses after A/New Jersey (H1N1) inactivated influenza vaccines among children

Vaccine	CCA/HA^a	AB 6–36 Mo^b		CCA/HA^a	Ab 3–5 Yr^b		CCA/HA^a	Ab 6–10 Yr^b		CCA/HA^a	Ab 11–18 Yr^b	
		GMT	% ≥40		GMT	% ≥40		GMT	% ≥40		GMT	% ≥40
Wyeth												
1 dose^c	97/6.5	5.4	0	194/13	8.6	7	194/13	8.7	43	388/26	14.1	25
2 doses^c		15.9	25		65.9	83		65.6	79		90.6	82
PD												
1 dose^c	95/10	9.3	10	190/20	17.1	18	190/20	23.3	38	380/40	27.9	41
2 doses^c		63.1	89		65.9	78		102.3	93		86.7	88
MN												
1 dose^c	26/2.5	18	21	51/5	13.2	15	102/10	14	13	204/20	17.8	52
2 doses^c		86.3	89		51.3	84		41.2	77		53.7	74
MSD												
1 dose^c	14/2.5	18.5	29	28/5	26.7	47	55/10	25.9	44	110/20	41.5	54
2 doses^c		59.8	92		82.4	91		54.7	84		71.6	96

PD Parke-Davis, *MN* Merrell-National which became Connaught Laboratories and is now Sanofi Pasteur, *MSD* Merck Sharpe and Dohme

[a] CCA/HA Actual chick cell agglutinating units/μ HA per 0.5 ml

[b] Ab responses by age group for GMT = geometric mean hemagglutination-inhibition titer and % with titer ≥1:40

[c] 1 dose no. = 10–63, 2 doses no. = 9–31

Despite an estimated dosage of only 2.5 μg of HA, the good responses to two doses of either of the two WV vaccines among 6–36-month-old children were notable. A 10 μg HA dosage of the PD vaccine was comparable for this age group but the responses to the 6.5 μg HA dosages of Wyeth vaccine were lower.

One- and two-dose trials were also performed in subjects aged 17–24 years (Table 5). A trend for increasing antibody responses with increasing dosages was seen and, as was seen in children, satisfactory antibody responses to one dose were seen only in the group given the higher dosages of the reactogenic MSD vaccine [7]. The HA dosages for the two higher CCA dosages of MSD vaccine were 60 and 118 μg. Responses to two doses of the lower dosages (20–40 μg HA) were satisfactory for the PD, MN, and MSD vaccines but not for the SV Wyeth vaccine that was of lower dosage (13 μg HA; GMT 39, 52% ≥1:40).

Older persons (≥25 years) were given a single dose only of each vaccine as these older persons exhibited good antibody responses to one dose [7]. Prior to vaccination, HAI titers ≥1:10 were seen in 15.4% of those aged 25–34 years, in 28.1% of those aged 35–51 years, and in 94.9% of those ≥52 years [7]. A comparison of responses of a group aged 22–43 years and one ≥52 years old at one clinical trial site is shown in Table 6 [10, 11]. Satisfactory responses to A/Swine/31 (H1N1) were seen among those given one and two doses of the inappropriate A/Swine/31 (H1N1) vaccine provided by Parke-Davis. Among those aged 22–43 years, vaccines of higher dosage (>20 μg HA) induced satisfactory responses and, again, were highest for the reactogenic MSD vaccines. One dose (12–60 μg HA) induced very good responses among all subjects ≥52 years of age.

Table 5 Comparison of serum antibody responses after A/New Jersey (H1N1) inactivated influenza vaccines among subjects 17–24 years of age

Vaccine	CCA/HA[a]	Antibody[b]		CCA/HA[a]	Antibody[b]		CCA/HA[a]	Antibody[b]	
		GMT	% ≥40		GMT	% ≥40		GMT	% ≥40
Wyeth									
1 dose[c]	174/8	10	21	335/23	19	31	661/65	24	34
2 doses[c]	194/13	39	52						
PD									
1 dose[c]	217/ND[d]	13	34	431/ND[d]	8	20	739/ND[d]	17	45
2 doses[c]	190/20	94	82						
MN									
1 dose[c]	128/12	22	44	301/26	34	51	697/51	31	46
2 doses[c]	204/20	72	83						
MSD									
1 dose[c]	185/28	42	56	356/60	66	84	736/118	82	91
2 doses[c]	221/40	125	94						

PD Parke-Davis, *MN* Merrell-National (which became Connaught Laboratories and is now Sanofi Pasteur), *MSD* Merck Sharpe and Dohme
[a]*CCA/HA* Actual chick cell agglutinating units/μ HA per 0.5 ml
[b]*GMT* Geometric mean hemagglutination-inhibiting antibody titer; ≥40 = % with ≥1:40 titer
[c]Separate clinical trials
[d]HA not determined as HA was A/Swine/31 not A/New Jersey/176

Table 6 Comparison of serum antibody responses after A/New Jersey (H1N1) inactivated influenza vaccines among subjects 22–43 and ≥52 years of age

Vaccine[a] Age Gp (yrs)[b]	CCA/HA[c]	Antibody[d]		CCA/HA[c]	Antibody[d]		CCA/HA[c]	Antibody[d]	
		GMT	% ≥40		GMT	% ≥40		GMT	% ≥40
Wyeth	174/8			335/23			661/65		
22–43		26	39		48	75		94	76
PD	217/ND[e]			431/ND[e]			739/ND[e]		
22–43		89	95		181	94		166	85
≥52		265	98		286	100			
MN	128/12			301/26			697/51		
22–43		40	63		26	57		84	75
≥52		164	98		232	100			
MSD	185/28			356/60			736/118		
22–43		105	78		174	100		136	88
≥52		191	98		246	98			
Placebo	0/0								
22–43		<10	0						
≥52		52	72						

[a]*PD* Parke-Davis, *MN* Merreill-National (which became Connaught Laboratories and is now Sanofi Pasteur), *MSD* Merck Sharpe and Dohme. 22–43 years given A/NJ vaccine; ≥52 years given A/NJ/76 – A/Victoria/75 (H3N2) vaccine. Wyeth vaccine not provided
[b]22–43 years, all 224 were <1:10 HAI in prevaccination sera; 6/405 ≥52 years were <52 years with a high risk condition
[c]*CCA/HA* Actual chick cell agglutinating units/μ HA per 0.5 ml
[d]*GMT* Geometric mean hemagglutination-inhibiting antibody titer; ≥40 = % with ≥1:40 titer
[e]HA not determined as HA was A/Swine/31 not A/New Jersey/76

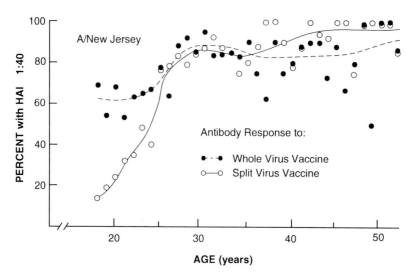

Fig. 1 Relation of age to serum hemagglutinin-inhibiting (HAI) antibody responses. Percent ≥ 1:40 after vaccination with one dose of vaccine. Results pooled for WV vaccines (Merrell-National and Merck Sharp and Dohme) and SV vaccines (Parke-Davis and Wyeth). Reprinted from Parkman et al. [7]

The effect of age on antibody responses is shown in Fig. 1 from Parkman et al. [7]. Responses after one dose are variable but generally good for both SV and WV vaccines among those ≥25 years of age but were significantly better after one dose of WV vaccine than for SV vaccines among those ≤25 years.

Antibody responses to A/USSR (H1N1) vaccines are summarized in Table 7. Dosages were proposed to be 2.3, 7, 20, and 60 μg of HA; results for the three lower projected dosages are shown in the table. Because of the reactogenicity seen with WV A/NJ vaccines, SV only was given to children; results for those aged 3–6 and 7–12 years are shown [8]. None of the single-dose groups developed a satisfactory response (1.3–16 μg HA), while all of the groups exhibited relatively good responses after two doses. Responses to one dose among those 20–25 years old of either SV or WV were generally good but not to the lower dosages (3–6.3 μg HA); responses to two doses were very good including to the lower dosages [12]. Responses to one dose of vaccine containing 10–19 μg HA in those aged 55–88 years old were very good and a second dose of the SV and WV vaccines added very little to the response. Patterns of responses to one and two doses of vaccine for all subjects in the A/USSR vaccine trials are shown in Fig. 2 from La Montagne et al. [5]. Increasing dosages induced increasing frequencies of antibody response for those ≤25 years after one dose. For those aged 13–25 years, an optimal response frequency, which was equivalent to the response after two doses of the middle-dose (10–19 μg HA) vaccine, was seen after one dose of a high-dosage (43–61 μg HA) vaccine; no benefit ensued in this age group from a second dose of a high-dosage vaccine. The greatest value for a second dose was seen in those ≤12 years of age.

A summary of the immunogenicity findings for the A/NJ and A/USSR vaccines is presented in Table 8. The patterns of serum anti-hemagglutinin antibody

Table 7 Comparison of serum antibody responses after A/USSR (H1N1) inactivated influenza vaccines

Vaccine[a] Age Gp (Yr)	CCA/ HA[b]	Antibody[c]		CCA/ HA[b]	Antibody[c]		CCA/ HA[b]	Antibody[c]	
		GMT	% ≥40		GMT	% ≥40		GMT	% ≥40
Split virus[d]	W 12/2			W 37/6			W 107/16		
3–6 yr[e]	PD 12/1.3			P 36/3.4			P 102/10		
1 dose		10	10		13	22		15	22
2 doses		40	61		54	71		57	67
7–12 yrs[e]									
1 dose		13	16		12	21		23	39
2 doses		37	74		36	58		57	77
Split virus[d]				W 37/6			W 107/16		
20–25 yr[e]				P 36/3.4			P 102/10		
1 dose					14	18		29	60
2 doses					38	82		48	87
55–88[e]									
1 dose								189	100
2 doses								215	100
Whole virus[d]				C 61/6.3			C 186/19		
20–25[e]				M 20/3			M 81/12		
1 dose					21	25		32	77
2 doses					48	83		47	92
55–88[e]									
1 dose								56	83
2 doses								61	87

[a]Vaccines were PD Parke-Davis, W Wyeth, Connaught Laboratories formerly Merrell-National, now Sanofi Pasteur, MSD Merck Sharpe and Dohme
[b]CCA/HA Actual chick cell agglutinating units/estimated μ HA per 0.5 ml
[c]Antibody responses; GMT geometric mean titer; % ≥40 = % with titer ≥1:40
[d]Vaccines were SV = Wyeth and Parke-Davis; responses to each vaccine pooled; WV = Connaught (C) and MSD (M); responses to each vaccine pooled
[e]No. – 3–6 = 18–31, 7–12 = 28–32 (86%.<1:10 pre); 20–25 = 12–17 (all <1:10); 55–88 = 21–23 (all >1:10 pre)

responses were similar for both the A/NJ and A/USSR vaccines. Increasing the dosage of the vaccine increased the frequency and magnitude of responses and one dose of a higher dosage vaccine frequently elicited a response similar to two doses of a lower dosage vaccine. For the most unprimed age groups (younger children), two doses were almost always needed. Among the healthy "primed," as defined by age, a single moderate dosage (15–20 μg HA) of vaccines was generally adequate. While WV vaccines appeared more immunogenic (and reactogenic) for A/NJ virus, they were not for A/USSR virus vaccines. The greater immunogenicity of A/NJ WV vaccines was at least partly due to a higher HA content.

5 Comment

These clinical trials in humans with two sets of influenza A/H1N1 vaccines constituted the most extensive experience with inactivated whole virus and split-product influenza virus vaccines ever undertaken to that time. Altogether, almost 10,000

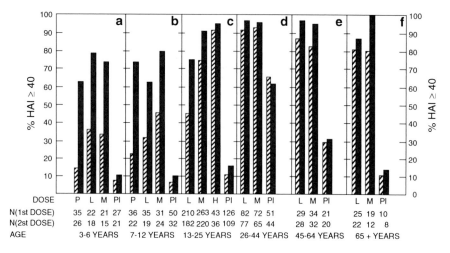

Fig. 2 Cumulative serum hemagglutinin-inhibiting (HAI) antibody responses to the A/USSR/77 (H1N1) vaccines for all subjects in the trial. Each panel illustrates the proportion of individuals who attained serum titers of HAI antibody of ≥1:40 after the first injection (hatched bar) and after the second injection (solid bar). The results are shown by ages of the subjects. The HA dosage of vaccine administered [P pediatric (SV, 1.3–2 μg), L low (SV, 3.4–6 μg; WV, 3–6.3 μg), M medium (SV, 10–16 μg; WV, 12–19 μg), or H high (SV, 43–61 μg; WV, 45–59 μg)], ages of the subjects, and number of individuals who received the first and second vaccinations are shown along the abscissa. Reprinted from La Montagne et al. [5]

Table 8 Summary of immunogenicity of inactivated influenza A/H1N1 vaccines

The patterns of antibody responses by age, dosage, and number of doses were similar for both sets of vaccines – A/New Jersey (H1N1) and A/USSR (H1N1) virus vaccines

Increasing the dosage of the vaccine increased the frequency and magnitude of antibody responses

Expressing vaccine dosage as μg of HA related better to responses than CCA units

A higher dosage of vaccine frequently induced antibody responses similar to two doses of lower dosage vaccine

Two doses of vaccine were almost always needed for satisfactory antibody responses among children

A single dose of vaccine of moderate dosage (15–20 μg HA) induced a satisfactory antibody response among those "primed" because of past exposure to related HA antigens. Age was adequate for this determination

Whole virus vaccines appeared more immunogenic than split-product vaccines for A/New Jersey virus but not for A/USSR viruses although A/USSR WV vaccines were not tested in children

Long-term persistence of A/H1N1 serum antibody was demonstrated (≥25 years)

volunteers of ages 0.5–100 years participated in the trials. The trials confirmed many immunological concepts noted in earlier vaccine trials and expanded our understanding of vaccine responses. Reinforced principles included that increasing dosage increases antibody responses and two doses of a low dosage (<10 μg HA) will induce a greater response than one dose. These concepts were particularly noticeable among persons unprimed to H1 antigens. Priming was shown to be a major determinant of enhanced responses and was age-related as determined by a likely exposure to H1 antigens of influenza viruses circulating in earlier years; the years of A/H1N1 circulation were 1918–1956.

Responses to WV vaccines were not consistent in that reactogenicity and immunogenicity were both greater for A/NJ vaccines than for SV vaccines while for A/USSR they were similar. The increased reactogenicity and immunogenicity for the A/NJ WV vaccines could at least partly be caused by the increased HA content. However, for the MSD vaccine, the increased reactogenicity and apparent immunogenicity appeared to be also partly due to the structure of virus particles in the vaccine; the MSD vaccine contained more intact particles, including both spherical and filamentous particles, than did the MN whole virus vaccine. This adjuvant-like effect for WV vaccines was not seen for the A/USSR WV vaccines but virus particle structure of the two WV vaccines was not provided. Thus, WV vaccines may not be uniformly more reactogenic nor immunogenic than SV vaccines.

A number of additional variables were examined with the A/NJ vaccines. An Intradermal (ID) immunization study with 0.1 ml vaccine of WV vaccine (MN vaccine) was inferior to 0.5 ml IM among unprimed persons after one dosage and an ID–IM sequence was no better than an IM–IM sequence [13]. Responses to ID were similar to IM among those with some prior antibody. Local reactions were greater among those given ID vaccine but systemic reactions were greater and also among those given vaccine IM. A number of studies were conducted among persons of "high risk" [1]. The increased reactogenicity of WV vaccines was noted in these studies, but the vaccines were otherwise considered safe and immunogenic in subjects with multiple sclerosis, asthma, pulmonary disease, heart disease, cancer, and some other disorders.

The 1976–1977 vaccine trials did not explore some other vaccine variables of interest; these include timing for dose two. A 3-week interval between doses is commonly used by European investigators but a 4-week interval was used in the two dose studies of 1976–1977. A 2-week interval between vaccinations was explored in the past and responses were suboptimal ([14, 15] and own unpublished data). The value of adjuvants was not explored in the 1976–1977 trials but incomplete Freund's adjuvant was used in some A/H2N2 vaccines in 1957 and an adjuvant effect was demonstrated [15]. It was stated that using IFA with a 100 CCA dose induced an antibody response similar to a 400 CCA dose without IFA.

Safety and immunogenicity trials are being performed currently with the newly emerged influenza A/H1N1 virus. All vaccines in the USA 2009 trials are split-product or subunit vaccines; it seems likely that these trials will largely confirm the conclusions from the earlier A/H1N1 vaccine trials. On the basis of the 1976–1977 experience, it is predicted that persons ≥ 85 years of age have a high likelihood of possessing antibody and resistance to the 2009 A/H1N1 virus and that most ≥ 55 years old will possess antibody and a high level of "priming"; a single vaccination of moderate dosage should induce substantial responses in both of these populations. Because other A/H1N1 viruses have been causing human infections since

1977 and have been included in seasonal vaccines, a considerable portion of the population should be primed and exhibit satisfactory antibody responses to a single moderate dosage (10–20 µg of HA) of vaccine although a higher dosage would probably induce a greater response. A precise age for priming cannot be set as in the 1976–1977 trials but lower frequencies of priming will be encountered with reducing age so that two doses of vaccine will be required for an adequate response in younger persons. If these predictions are correct, the concepts and principles developed in the 1976–1977 trials with inactivated influenza virus vaccines will be reinforced.

Acknowledgments Financial support: Research performed by the authors and summarized in this report was supported by Public Health Service Contract NO1-AI-30039 from the National Institute of Allergy and Infectious Diseases.
The content of this publication does not necessarily reflect the views or policies of the Department of Health and Human Services, nor does mention of trade names, commercial products, or organizations imply endorsement by the US Government.

References

1. (1977) Clinical studies of influenza vaccines – 1976. J Infect Dis 136(Supplement)
2. Kendal AP, Nobel GR, Skehel JJ, Dowdle WR (1978) Antigenic similarity of influenza A (H1N1) viruses from epidemics in 1977–1978 to "Scandinavian" strains isolated in epidemics of 1950–1951. Virology 89:632–636
3. (1983) Clinical studies of influenza vaccines – 1978. Rev Infect Dis 5: 721–764
4. Ennis FA, Mayner RE, Barry DW, Manischewitz JE, Dunlap RC, Verbonitz MW, Bozeman FM, Schild GC (1977) Correlation of laboratory studies with clinical responses to A/New Jersey influenza vaccines. J Infect Dis 136(Suppl):S397–S406
5. La Montagne JR, Noble GR, Quinnan GV, Curlin GT, Blackwelder WC, Smith JI, Ennis FA, Bozeman FM (1983) Summary of clinical trials of inactivated influenza vaccine – 1978. Rev Infect Dis 5:723–736
6. Wright PF, Thompson J, Vaughn WK, Folland DS, Sell SHW, Karzon DT (1977) Trials of influenza A/New Jersey/76 virus vaccine in normal children: an overview of age-related antigenicity and reactogenicity. J Infect Dis 136:S731–S741
7. Parkman PD, Hopps HE, Rastogi SC, Meyer HM Jr (1977) Session V. Summary of clinical studies. Summary of clinical trials of influenza virus vaccines in adults. J Infect Dis 136: S722–S730
8. Wright PF, Cherry JD, Foy HM, Glezen WP, Hall CB, McIntosh K, Monto AS, Parrott RH, Portnoy B, Taber LH (1983) Antigenicity and reactogenicity of influenza A/USSR/77 virus vaccine in children – a multicentered evaluation of dosage and safety. Rev Infect Dis 5:758–764
9. Glezen WP, Loda FA, Denny FW (1969) A field evaluation of inactivated, zonal-centrifuged influenza vaccines in children in Chapel Hill, North Carolina, 1968–69. Bull WHO 41: 566–569
10. Cate TR, Couch RB, Kasel JA, Six HR (1977) Clinical trials of monovalent influenza A/New Jersey/76 virus vaccine in adults: reactogenicity, antibody response, and antibody persistence. J Infect Dis 136:S450–S455
11. Cate TR, Kasel JA, Couch RB, Six HR, Knight V (1977) Clinical trials of bivalent influenza A/New Jersey/76-A/Victoria/75 vaccines in the elderly. J Infect Dis 136:S518–S525

12. Cate TR, Couch RB, Parker D, Baxter B (1983) Reactogenicity, immunogenicity, and antibody persistence in adults given inactivated influenza virus vaccines – 1978. Rev Infect Dis 5:737–747
13. Brown H, Kasel JA, Freeman DM, Moise LD, Grose NP, Couch RB (1977) The immunizing effect of influenza A/New Jersey/76 (Hsw1N1) virus vaccine administered intradermally and intramuscularly to adults. J Infect Dis 136:S466–S471
14. Bayne GM, Liu OC, Boger WP (1958) Asian influenza vaccine: effect of age and schedule of vaccination upon antigenic response. Am J Med Sci 236:290–299
15. Meiklejohn GN (1961) Asian influenza vaccination: dosage, routes, schedules of inoculation, and reactions. The Amer Rev of Resp Dis 83:175–177

Occurrences of the Guillain–Barré Syndrome (GBS) After Vaccinations with the 1976 Swine A/H1N1 Vaccine, and Evolution of the Concern for an Influenza Vaccine-GBS Association

Robert B. Couch

Abstract Implementation of a public vaccination program with A/New Jersey (A/NJ) vaccines in 1976 led to the recognition of an increased risk among vaccinated persons of developing the neurological disorder known as Guillain–Barré Syndrome (GBS). The attributable risk was 8.8 among adults in the 6-week period after vaccination or about 1 case per 100,000 vaccinations. Skepticism of the statistically significant association was resolved with a subsequent careful assessment in two states in the USA that confirmed the association. Subsequent efforts to confirm an association between other influenza vaccines and occurrence of GBS have mostly failed to identify an association but a suggestion of about one case of GBS per one million vaccinations has been reported. GBS has been associated with various other infections, illnesses, vaccinations, and other disorders. *Campylobacter jejuni* infections are accepted as inducing a risk for GBS and evidence suggests antiganglioside immune responses that react with the nerve myelin sheath as the mechanism. To assess this possibility, A/NJ and some other influenza vaccines were all shown to induce antiganglioside antibodies in mice; however, a relation of this finding in mice to GBS in humans has not been provided.

More recently, reports have indicated a risk of GBS after clinical influenza that is greater than the risk after influenza vaccination, suggesting that influenza vaccination may actually protect against GBS. Influenza, influenza vaccinations, and their role in the occurrences of GBS are evolving subjects. At present, however, occurrence of GBS cannot be considered an inherent risk of influenza vaccination.

R.B. Couch
Department of Molecular Virology & Microbiology, Baylor College of Medicine, One Baylor Plaza, MS: BCM280, Houston, TX 77030, USA
e-mail: ikirk@bcm.edu

G. Del Giudice and R. Rappuoli (eds.), *Influenza Vaccines for the Future*, 2nd edition,   373
Birkhäuser Advances in Infectious Diseases,
DOI 10.1007/978-3-0346-0279-2_16, © Springer Basel AG 2011

1 Introduction

In the fall of 1976, the USA implemented a public health program of vaccination of the population with influenza A/New Jersey/76 (A/H1N1) vaccines so as to prevent a pandemic from the swine-like (H1N1) virus that had emerged at Fort Dix, New Jersey, USA. That program led to recognition of a risk for occurrence after vaccination of the neurological disorder known as the Guillain–Barré Syndrome (GBS). That experience and the current status of understanding of an influenza vaccine-induced risk for GBS are summarized in this chapter.

2 The A/New Jersey Vaccine Experience

The nationwide campaign for immunization of all citizens in the USA in 1976 with influenza A/New Jersey (A/NJ) vaccine encountered numerous difficulties during the decision, organization, and implementation [1]. The final event that led to discontinuation of the public health program was the apparent increase in GBS among vaccine recipients in the setting of no apparent spread of the swine A/H1N1 virus in the population.

A subsequent organized GBS surveillance effort in the USA identified 1098 cases during the vaccination period, 532 of which had occurred after vaccination with the A/NJ vaccine [2]. Calculations of the risk of GBS after A/NJ vaccination indicated a significant attributable risk of 8.8 among adults in the 6-week period after vaccination. This corresponds to about one case per 100,000 vaccinations; almost 50 million vaccinations were performed. Additional convincing data for this association was the pattern of an increase in the GBS rate after vaccination with a peak in weeks two and three and a subsequent return to the baseline rate for unvaccinated persons (Fig. 1). Risk appeared about the same for all four vaccines distributed; a number of potential confounding variables were excluded.

Subsequent to this landmark report, substantive questions were raised about the quality of the data leading to the "conclusive" epidemiologic association [4, 5]. Additionally, other reports appeared that confirmed and did not confirm the association [3, 6, 7]. In an attempt to resolve the concerns, a new study was mounted to review all cases of GBS in the states of Michigan and Minnesota during the vaccination period using neurologists for formulating diagnostic criteria and reviewing all cases [8]. Some cases used in the earlier association report were discarded but the relative risk for GBS after A/NJ vaccination remained about as proposed in the initial report. Overall, these various reports provide very strong evidence that vaccination with A/NJ vaccine in 1976 increased the risk for developing the GBS in a 6–8 week period after vaccination; the reason(s) for this epidemiological association are unknown.

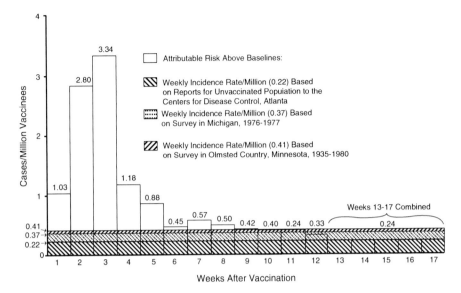

Fig. 1 Guillain–Barré syndrome, relative risks for population 18 years of age and over, by week of onset, after A/New Jersey/76 influenza vaccination, United States. Attributable risk and expected incidence evaluated from Olmsted County, Minnesota, Michigan, and USA data. Reprinted from Beghi et al. [3] with permission

3   The Guillain–Barré Syndrome

GBS is a neurologic disorder consisting of a constellation of neurological signs and symptoms of unknown cause. There are no objective tests available for establishing the diagnosis in a suspect case. Because of lack of uniformity in reported cases, the National Institute of Neurological and Communicative Disorders in the USA and the Brighton Collaboration provided definitions of the syndrome that have, for the most part, been followed [9–11]. The syndrome is an acute illness characterized by progressive motor weakness of more than one limb that tends to be symmetrical and is accompanied by areflexia. Mild sensory findings and cranial nerve involvement may be present and recovery generally occurs. Strong support for a diagnosis of GBS is provided by finding an albumin-cytologic dissociation (elevated protein and reduced cell concentration) in cerebrospinal fluid and typical electrophysiologic abnormalities.

   Descriptions of GBS over many years have specified that most cases (about two-thirds) seemed to follow an episode of illness [12, 13]. In a review of GBS, Lineman identified 1,100 cases with about 60% having followed an infection, most commonly a nonspecific acute viral or bacterial infection of the respiratory tract [14]. Subsequently, reports have emphasized enteric infections and noted association with cytomegalovirus infections, infectious mononucleosis and numerous other infections and maladies. Until recently (see later), it was thought that influenza infections did not convey a risk for GBS despite the fact that a history of an acute viral-like respiratory illness was known to be the most common preceding event [15–17]. Associations with immune disorders and events have included GBS after vaccinations; in earlier reports this association was seen more commonly after rabies vaccinations or use of tetanus antitoxin. Many vaccinations have now been followed by GBS and caused some concern; these have included various live and inactivated vaccines, viral and bacterial vaccines, protein and carbohydrate vaccines, and adjuvanted and nonadjuvanted vaccines (references available upon request).

Incidence rates of GBS have varied but are generally about one case per 100,000 persons per year [17, 18]. This low rate has compounded the problem of identifying and proving a specific cause for the GBS. Clusters of cases have been noted, sometimes associated with a discrete preceding outbreak such as acute gastroenteritis [19]. The GBS rate increases with increasing age is somewhat more common in males and, possibly, in whites.

4 A/New Jersey Vaccine and Pathogenesis of GBS

The GBS is primarily a polyradiculopathy of spinal nerve roots. The major histopathology is edema and disorganization of the myelin sheath that progresses to a mild inflammatory reaction and to Swann cell proliferation in the late stages [20]. Although the specific pathogenesis of the GBS is unknown, it is thought by most to be induced by an aberrant immune response that leads to reaction with the nerve myelin sheath or other axonal sites and results in impaired transmission of impulses through peripheral nerves [13, 21–23]. This possibility is most completely explored for GBS following *Campylobacter*-associated enteric infections [24, 25].

Campylobacter jejuni apparently has surface polysaccharides that are similar to gangliosides of nerve sheaths. Molecular mimicry is proposed as the mechanism for induction of ganglioside antibodies that react with nerve tissue and induce the pathology that impairs nerve impulse transmissions [24, 25]. Clusters of GBS associated with *Campylobacter* infections and gastroenteritis outbreaks that could be from *Campylobacter* infections are compatible with this specific infection being a precipitating event for GBS [24–26].

Whatever the exact relationship between infection, antiganglioside antibody, and GBS, it is clear that ganglioside antibody is not useful as a diagnostic test for GBS. Whether this antibody, in conjunction with cofactors or coevents, describes a pathogenetic circumstance for all cases of GBS is uncertain.

In an effort to assess ganglioside antibody induction and GBS after A/NJ vaccinations in 1976, residual A/NJ vaccines were used for immunizing mice and testing for ganglioside antibodies [27]. All of the A/NJ/76 vaccines induced anti-GM1 antibodies, the antibody of interest. If this antibody mediates GBS, it would be reasonable to suggest that a high frequency of responses occurs but that antibody mediates clinical disease in only a low percentage of cases. On the other hand, this same report immunized mice with two different seasonal influenza vaccines and two A/H5N1 vaccines and all induced ganglioside antibodies. Although the number of H5 vaccinations is too small for detecting a GBS relationship, no relationship was noted for the two seasonal vaccines evaluated. Unfortunately, this provocative report has apparently not been followed by the indicated studies that would further clarify the significance of the finding. Thus, at present, it seems most reasonable to regard the influenza vaccination–ganglioside antibody–GBS relationship as no more than an interesting, suggestive observation that could use resolution for significance.

5   GBS and Influenza Vaccines Other than the A/New Jersey Vaccine

In order to verify whether the A/NJ–GBS association was an inherent risk of influenza vaccination, the American Center for Disease Control conducted surveillance for an association in 3 years subsequent to the A/NJ vaccination period. For three separate vaccination years, 1978–1979, 1979–1980, 1980–1981, no association of GBS and prior influenza vaccination was detected [28–30]. However, a borderline significant increased risk for GBS was reported after an investigation of influenza vaccinations in 1992–1994 that was precipitated by reports to the vaccine adverse event reporting system (VAERS) in the USA; the validity of this association has been questioned [31, 32]. Some other reports of an association between influenza vaccination and GBS have appeared as well as reports of no association [32–38]. Reports utilizing VAERS data to support an association between influenza vaccinations and GBS have been refuted by the CDC with the strong statement "the rate of an adverse event cannot be approximated from VAERS data" [39]. Thus, although it is possible that a small increase in risk for the GBS may follow vaccination with seasonal influenza vaccines, present knowledge suggests it is equally plausible to believe that there is no increased risk.

6   Recent Contributions

The most significant recent data on the influenza vaccine–GBS relationship are data supporting a risk for GBS after influenza infections [32, 40, 41]. Using the self-controlled case series method, an approach thought superior to cohort and case–control studies, Stowe et al. [32] found the relative incidence of GBS after influenza vaccination in the UK General Practice Database to be 0.76 within 90 days while GBS after an influenza-like illness was 7.35 within 90 days and 16.67 within 30 days. A similar relationship for GBS and clinical influenza-like illness was reported by Sivaden-Tardy et al. (Fig. 2), which was further supported by

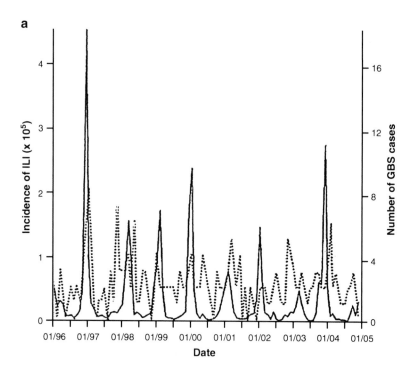

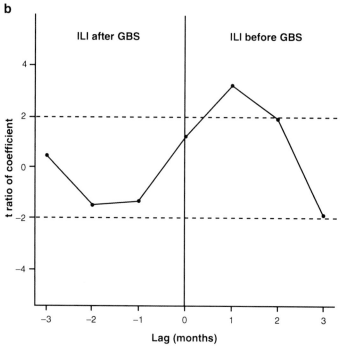

Fig. 2 (**a**) Monthly incidence of influenza-like illness (ILI; *solid line*) and Guillain–Barré syndrome (GBS; *dashed line*) caused by an unidentified agent. (**b**) *t* Ratios of lagged-regression

objective evidence of actual influenza virus infection in the GBS cases [41]. Also of interest is the fact that none of the GBS cases after influenza infection developed ganglioside antibodies. These recent reports offer the intriguing possibility that influenza vaccinations may protect against any increased risk for development of GBS after clinical influenza that might exist.

7 Comment

A summary of the evolution of the influenza vaccine-GBS association is provided in Table 1. It seems clear that vaccination with the A/NJ influenza vaccine in 1976 increased the risk for developing GBS; however, numerous efforts since 1976 have failed to confirm a risk for GBS attributable to influenza vaccination. Thus, it seems reasonable to conclude that GBS does not constitute an inherent risk from influenza vaccinations. It remains possible that influenza vaccinations may, on rare occasions, serve as a contributing factor to occurrences of GBS and account for the reports that have suggested a minor increase in risk; for some unknown reason, a contribution of significance occurred in the fall of 1976 when a nationwide vaccination campaign was conducted to prevent a pandemic with swine influenza

Table 1 Summary of the Influenza vaccine Guillain–Barré Syndrome (GBS) Association

- There was a significant increased risk for developing GBS in the 6–8 week period after vaccination with the swine influenza A/H1N1 vaccine in 1976. The reason(s) for this increase are unknown.
- The GBS is a paralytic syndrome of unknown cause that is presumed to be an immunopathologic disorder resulting from an immune reaction with nerve tissue.
- Ganglioside antibodies are proposed as the mediator for the immunopathological reaction leading to GBS and data are available to support this hypothesis, particularly for GBS after *Campylobacter* infections. Available data are insufficient to support this as a mechanism for an influenza vaccine-induced GBS.
- The many studies reported on influenza vaccinations and GBS since 1976 have either failed to identify a risk for GBS or have reported a very low frequency risk (~one per million vaccinations).
- Recent studies of influenzal illnesses and GBS have reported a significant risk for GBS after clinical influenza. GBS after a nonspecific acute viral-like respiratory illness has been an accepted association for decades.
- Influenza, influenza vaccinations, and their role in occurrences of GBS are evolving subjects. At present, occurrence of GBS cannot be considered an inherent risk of influenza vaccination.

←————————————————————————————————

Fig. 2 (continued) coefficients between residual cases of GBS caused by an unidentified agent at month t and ILI at month t – lag, for lags of −3 to 3. Horizontal dashed lines indicate 5% significance level of a two-sided test of association. Reprinted from Sivadon-Tardy et al. [41] with permission

A/H1N1 viruses. The similarities between the swine influenza A/H1N1 experience in 1976 and that ongoing at the present time (2009), with swine influenza A/H1N1 vaccinations are obvious. The magnitude of the current vaccination effort should clarify any relationship between influenza vaccination and occurrence of the GBS.

Acknowledgments Financial support: Research performed by the authors and summarized in this report was supported by Public Health Service Contract NO1-AI-30039 from the National Institute of Allergy and Infectious Diseases.

The content of this publication does not necessarily reflect the views or policies of the Department of Health and Human Services, nor does mention of trade names, commercial products, or organizations imply endorsement by the U.S. Government.

References

1. Sencer DJ, Millar JD (2006) Reflections on the 1976 swine flu vaccination program. Emerg Infect Dis 12:29–33
2. Schonberger LB, Bregman DJ, Sullivan-Bolyal JZ, Keenlyside RA, Ziegler DW, Retailliau HF, Eddins DL, Bryan JA (1979) Guillain–Barre syndrome following vaccination in the National Influenza Immunization Program. United States, 1976-1977. Am J Epidemiol 110:105–122
3. Johnson DE (1982) Guillain-Barre syndrome in the US Army. Arch Neurol 39:21–24
4. Beghi E, Kurland LT, Mulder DW, Wiederholt WC (1985) Guillain-Barre syndrome: clinicoepidemiologic features and effect of influenza vaccine. Arch Neurol 42:1053–1057
5. Kurland LT, Wiederholt WC, Kirkpatrick JW, Potter HG, Armstrong P (1985) Swine influenza vaccine and Guillain-Barre syndrome. Epidemic or Artifact? Arch Neurol 42:1089–1092
6. Marks JS, Halpin TJ (1980) Guillain-Barre syndrome in recipients of A/New Jersey influenza vaccine. JAMA 243:2490–2494
7. Breman JG, Hayner NS (1984) Guillain-Barre syndrome and its relationship to swine influenza vaccination in Michigan, 1976-1977. Am J Epidemiol 119:880–889
8. Safranek TJ, Lawrence DN, Kurland LT, Culver DH, Wiederholt WC, Hayner NS, Osterholm MT, O'Brien P, Hughes JM, Expert Neurology Group (1991) Reassessment of the association between Guillain-Barre syndrome and receipt of swine influenza vaccine in 1976-1977: Results of a two-state study. Am J Epidemiol 133:940–951
9. Asbury AK, Arnason BG, Karp HR, McFarlin DE (1978) Criteria for diagnosis of Guillain-Barre syndrome. Ann Neurol 3:565–566
10. Asbury AK, Cornblath DR (2004) Assessment of current diagnostic criteria for Guillain-Barre syndrome. Ann Neurol 27:S21–S24
11. Sejvar J, Cornblath D, Hughes R, The Brighton Collaboration Guillain-Barre Syndrome Working Group. (2008) The Brighton Collaboration case definition for Guillain-Barre syndrome as an adverse event following immunization. Inflammatory Neuropathy Consortium Meeting, Paris, France, July 2008
12. Kennedy RH, Danielson MA, Mulder DW, Kurland LT (1978) Guillain-Barre Syndrome. A 42-year epidemiologic and clinical study. Mayo Clin Proc 53:93–99
13. van Doorn PA, Ruts L, Jacobs BC (2008) Clinical features, pathogenesis, and treatment of Guillain-Barre syndrome. Lancet Neurol 7:939–950
14. Leneman F (1966) The Guillain-Barre syndrome. Arch Intern Med 118:139–144
15. Melnick SC, Flewett TH (1964) Role of infection in the Guillain-Barre syndrome. J Neurol Neurosurg Psychiatry 27:385–407
16. Langmuir AD (1979) Guillain-Barre syndrome: the swine influenza virus vaccine incident in the United States of America, 1976-77: preliminary communication. J R Soc Med 72:660–669

17. Schonberger LB, Hurwitz ES, Katona P, Holman RC, Bregman DJ (1981) Guillain-Barre syndrome: its epidemiology and associations with influenza vaccination. Ann Neurol 9:31–38
18. Black S, Eskola J, Siegrist C-A, Halsey N, MacDonald N, Law B, Miller E, Andrews N, Stowe J, Salmon D et al (2009) Importance of background rates of disease in assessment of vaccine safety during mass immunization with pandemic H1N1 influenza vaccines. Lancet 374 (9707):2115–2122
19. Sliman NA (1978) Outbreak of Guillain-Barre syndrome associated with water pollution. Br Med J 1:751–752
20. Haymaker W, Kernohan JW (1949) The Landry-Guillain-Barre syndrome: a clinical pathologic report of 50 fatal cases and a review of the literature. Medicine 28:59–141
21. Zhu J, Mix E, Link H (1998) Cytokine production and the pathogenesis of experimental autoimmune neuritis and Guillain-Barre syndrome. J Neuroimmunol 84:40–52
22. Ariga T, Miyatake T, Yu RK (2001) Recent studies on the roles of antiglycosphingolipids in the pathogenesis of neurological disorders. J Neurosci Res 65:363–370
23. McCarthy N, Giesecke J (2001) Incidence of Guillain-Barre syndrome following infection with *Campylobacter jejuni*. Am J Epidemiol 153:610–614
24. Yuki N (2001) Infectious origins of, and molecular mimicry in, Guillain-Barre and Fisher syndromes. Lancet Infect Dis 1:29–37
25. Rees JH, Soudain SE, Gregson NA, Hughes RAC (1995) *Campylobacter jejuni* infection and Guillain-Barre syndrome. BMJ 333:1374–1379
26. Bereswill S, Kist M (2003) Recent developments in *Campylobacter* pathogenesis. Curr Opin Infect Dis 16:487–491
27. Nachamkin I, Shadomy SV, Moran AP, Cox N, Fitzgerald C, Ung H, Corcoran AT, Iskander JK, Schonberger LB, Chen RT (2008) Anti-ganglioside antibody induction by swine (A/NJ/ 19976/H1N1) and other influenza vaccines: Insights into vaccine-associated Guillain-Barre syndrome. J Infect Dis 198:226–233
28. Hurwitz ES, Schonberger LB, Nelson DB, Holman RC (1981) Guillain-Barre syndrome and the 1978-1979 influenza vaccine. New Engl J Med 304:1557–1561
29. Kaplan JE, Katona P, Hurwitz ES, Schonberger LB (1982) Guillain-Barre syndrome in the United States, 1979-1980 and 1980-1981. JAMA 248:698–700
30. Kaplan JE, Schonberger LB, Hurwitz ES, Katona P (1983) Guillain-Barre syndrome in the United States, 1978-1981: Additional observations from the national surveillance system. Neurology 33:633–637
31. Lasky T, Terracciano GJ, Magder I, Koski CL, Ballesteros M, Nash D, Clark S, Haber P, Stolley PD, Schonberger LB et al (1998) The Guillain-Barre syndrome and the 1992-1993 and 1993-1994 influenza vaccines. New Engl J Med 339:1797–1802
32. Stowe J, Andrews N, Wise L, Miller E (2009) Investigation of the temporal association of Guillain-Barre syndrome with influenza vaccine and influenzalike illness using the United Kingdom general practice research database. Am J Epidemiol 169:382–388
33. Roscelli JD, Bass JW, Pang L (1991) Guillain-Barre syndrome and influenza vaccination in the US Army, 1980-1988. Am J Epidemiol 133:952–955
34. Centers for Disease Control and Prevention (2003) *Surveillance Summaries*, January 24, 2003. Surveillance for Safety after immunization: vaccine adverse event reporting system (VAERS) – United States, 1991–2001. *MMWR* 52 (No. SS-1)
35. Geier MR, Geier DA, Zahalsky AC (2003) Influenza vaccination and Guillain Barre syndrome. Clin Immunol 107:116–121
36. Haber P, DeStefano F, Angulo FJ, Iskander J, Shadomy SV, Weintraub E, Chen RT (2004) Guillain-Barre syndrome following influenza vaccination. JAMA 292:2478–2481
37. Souayah N, Nasar A, Fareed M, Suri K, Qureshi AI (2007) Guillain-Barre syndrome after vaccination in United States. A report from the CDC/FDA vaccine adverse event reporting system. Vaccine 25:5253–5255

38. Juurlink DN, Stukel TA, Kwong J, Kopp A, McGeer A, Upshur RE, Manuel DG, Moineddin R, Wilson K (2006) Guillain-Barre syndrome after influenza vaccination in adults. Arch Intern Med 166:2217–2221
39. Haber P, Slade B, Iskander J (2007) Letter to the editor. Vaccine 25:8101
40. Tam CC, O'Brien SJ, Petersen I, Islam A, Hayward A, Rodrigues LC (2007) Guillain-Barre syndrome and preceding infection with *Campylobacter*, influenza and Epstein-Barr virus in the general practice research database. PLoS ONE 2(4):e344. doi:10.1371/journal.pone.0000344
41. Sivadon-Tardy V, Orlikowski D, Porcher R, Sharshar T, Durand M-C, Enouf V, Rozenberg F, Caudie C, Annane D, van der Wert S et al (2009) Guillain-Barre syndrome and influenza virus infection. Clin Infect Dis 48:48–56

Human Monoclonal Antibodies for Prophylaxis and Treatment of Influenza

Wouter Koudstaal, Fons G. UytdeHaag, Robert H. Friesen, and Jaap Goudsmit

Abstract Influenza viruses continue to be challenging targets for vaccination because of their rapid evolution and tolerance for changes in their antigenic structure. Moreover, the efficacy of influenza vaccines is suboptimal in the elderly, the group at highest risk of infection-related complications, as a consequence of immunosenescence. In order to protect this vulnerable group, measures not reliant on the immune system are thus required. Passive immunotherapy using monoclonal antibodies (mAbs) represents such an approach, but its development has long been hampered by the lack of mAbs with potent heterosubtypic neutralizing activity. Recently, a novel class of antibodies has been discovered that, instead of blocking viral attachment to the host cell by binding to the globular head of the HA, recognize a site in the membrane-proximal HA stem and neutralize the virus by inhibiting the conformational changes required for membrane fusion and uncoating of the virus. In accordance with its functional importance, this epitope is conserved among all group 1 influenza viruses. Several such antibodies have already been shown to be protective in mice when given before and after lethal H5N1 or H1N1 challenge. These promising results justify clinical evaluation of broadly neutralizing anti-influenza mAbs as novel therapeutic agents for the prevention and treatment of influenza infections, which, given the increasing resistance to the leading antiviral drug oseltamivir, are urgently needed. The challenge that lies ahead now is the identification of an equally conserved epitope in group 2 influenza viruses, in particular, in those of the H3 subtype.

W. Koudstaal, F.G. UytdeHaag, R.H. Friesen, and J. Goudsmit (✉)
Crucell Holland BV, Archimedesweg 4-6, 2333 CN Leiden, The Netherlands
e-mail: jaap.goudsmit@crucell.com

G. Del Giudice and R. Rappuoli (eds.), *Influenza Vaccines for the Future*, 2nd edition,   383
Birkhäuser Advances in Infectious Diseases,
DOI 10.1007/978-3-0346-0279-2_17, © Springer Basel AG 2011

1 Introduction

Influenza viruses belong to the orthomyxovirus family of enveloped viruses with a segmented negative-stranded RNA genome [1]. Of the three influenza genera, influenza A, B, and C, only the first two are associated with significant human disease and cause annual epidemics during autumn and winter in temperate regions and circulate throughout the year in some tropical countries, with one or two peaks during rainy seasons. Such "seasonal influenza" is characterized by sudden onset of fever, cough, headache, muscle and joint pain, sore throat, and runny nose. Although most people recover within a week without requiring medical attention, influenza results in about three to five million cases of severe illness and up to 500,000 deaths worldwide every year, particularly among the very young, elderly, and chronically ill [2]. Furthermore, influenza A viruses occasionally cause major pandemics involving multitudes of these numbers of severe illness and death.

2 Immune Response to Influenza

Invading influenza A viruses are detected by "pattern recognition receptors" (PRRs) which recognize products of influenza virus replication and initiate a signaling cascade that culminates in an antiviral response. A pulmonary infiltrate of innate immune cells comprising natural killer cells, neutrophils, and macrophages mediates protection through both direct cytotoxicity of virus-infected cells and release of a torrent of innate immune molecules that limit infection. Furthermore, low-affinity "natural" antibodies form a first line of antibody-mediated defense against influenza by restricting virus dissemination and promote the recruitment of viral antigen to the secondary lymphoid organs [3, 4]. Here, an adaptive antibody response is mounted in germinal centers formed by B cells that, helped by T cells, clonally expand and differentiate. This differentiation is characterized by isotype switching (e.g., from IgM to IgG) and the introduction of random mutations in the Ig genes of the B cells (somatic hypermutation). Such somatic mutations result in B cell receptors with varying affinities for viral antigen. B cells able to bind the antigen with high affinity and specificity exit the germinal center as terminally differentiated antibody-producing plasma cells or memory B cells. Next to antibodies, cytotoxic CD8+ T cells have been shown to contribute to controlling influenza virus infection, but are outside the scope of this chapter (for review, see [5]).

3 Vaccination Against Influenza

The most effective way to prevent influenza or severe outcomes from the illness is vaccination and influenza vaccines have been available and used for more than 60 years [6]. Influenza viruses are challenging targets for vaccination as their segmented RNA genomes facilitate rapid evolution. Firstly, because RNA polymerases inherently lack proofreading capability, point mutations readily accumulate in the viral genome. When such mutations lead to amino acid substitutions in the major surface glycoproteins, hemagglutinin (HA) and neuraminidase (NA), the virus may escape from antibodies induced by vaccination against, or previous infection with, a closely related strain. This "antigenic drift" causes the regular occurrence of influenza epidemics by allowing viral escape from herd immunity. Furthermore, when two different influenza A viruses coinfect the same cell, progeny virions with new combinations of the RNA segments, so-called "reassortants", can be formed. When such a reassortant contains HA and NA segments derived from a bird or swine influenza virus against which humans have no immunity, an event called "antigenic shift", it may cause a pandemic, if it is easily transmitted from human to human. As vaccination is effective only when the vaccine viruses are well-matched with circulating viruses, the WHO biannually (once for the Northern, and once for the Southern Hemisphere) recommends a vaccine composition that targets the three most representative strains of each subtype (H1N1, H3N2, B) in circulation. Despite these recommendations, antigenic mismatches between the vaccine virus strain and the circulating strain may occur that negatively influence vaccine effectiveness [7]. Between 1997 and 2007, there were five occurrences of mismatch, and 11 occurrences of partial mismatch across the three vaccines strains in Europe and the USA [8]. Although influenza illness affects people of all ages, adults over 65 years of age account for approximately 90% of all influenza-related mortality [9]. Vaccination programs currently recommend that older adults should be vaccinated against influenza, as well as people who live with, or care for older adults [10, 11]. The benefit of vaccinating the elderly against influenza is subject of much controversy. Current influenza vaccines may be less effective among older adults than among younger adults [11–15] and prevent laboratory-confirmed influenza in only 30–40% of people over 65 years of age [12]. This is caused by changes that occur in the immune system with advancing age resulting in a reduced immune response and reduced capacity to produce antibodies [16, 17].

4 Treatment

For the treatment and/or prophylaxis of influenza infections, two classes of drugs are currently available: the adamantanes or M2 inhibitors, and the neuraminidase (NA) inhibitors. However, the adamantanes (amantadine and rimantadine) are associated with several toxicities, particularly of the central nervous system, rapid emergence of drug-resistant strains, and are not active against influenza B viruses [18, 19]. Compared to the adamantanes, the two licensed neuraminidase (NA) inhibitors zanamivir (Relenza) and oseltamivir (Tamiflu) are associated with little toxicity and are less prone to selecting for resistant influenza viruses [20, 21]. Nevertheless, emergence of resistance after oseltamivir treatment has been reported for both seasonal and avian influenza strains [22–26]. Moreover, during the 2007–2008 influenza season, oseltamivir resistance among H1N1 viruses increased significantly worldwide, apparently unrelated to oseltamivir use [27] and oseltamivir-resistant H1N1 viruses are now circulating on all major continents [28, 29]. Although, to date, these viruses are usually susceptible to zanamivir, the increasing use of zanamivir monotherapy because of the increasing resistance to oseltamivir may well lead to the development of zanamivir resistance [30]. In addition, the use of zanamivir is limited to patients who can actively use an inhaled drug, which excludes young children, impaired older adults, or patients with underlying airway disease [31]. Influenza thus continues to cause significant morbidity and mortality each year. The shortcomings of our current defenses against influenza are furthermore illustrated by the current pandemic. The causative subtype H1N1 virus from swine emerged from Mexico and the United States in March and April 2009, which was too late for inclusion in the vaccine for the Northern Hemisphere 2009/2010 influenza season. Although considerable effort has been expended and vaccines to combat this new H1N1 virus are now being produced, the availability is limited. Given the speed with which oseltamivir resistance became widespread among seasonal H1N1 viruses [31, 32], it is worrisome that the first oseltamivir resistant pandemic H1N1 influenza viruses have been isolated [33, 34]. Clearly, there is a dire need for more universal countermeasures to fight influenza.

5 Passive Immunotherapy

Pioneered by von Behring and Kitasato, in 1890, the administration of hyperimmune sera from immunized animals was shown to provide options for treatment of infectious diseases [35]. Serum therapy has been widely used for the treatment of bacterial and viral infections (reviewed in [36]). Unfortunately, therapy with animal-derived serum products often had significant side effects because of the immune response against the animal-derived antibodies, the most severe being serum sickness. Human convalescent-phase serum products were and are being used with greater success than animal-derived serum products for the prevention of viral infections such as rabies, hepatitis A and B, varicella zoster, and pneumonia caused by respiratory syncytial virus (RSV) [37]. The fact that neutralizing antibodies are a major component of immune protection against influenza and are the established immune correlate of protection for influenza vaccines implies that passive immunotherapy may also be a viable option for the treatment of influenza. This notion is supported by the observations that treatment with convalescent blood product improved the clinical outcome in severely ill influenza patients during the 1918 Spanish Flu pandemic as well as in influenza infections with the highly pathogenic avian H5N1 virus [38–40]. A major disadvantage of serum-derived polyclonal antibody preparations is that only a small proportion of the total IgG content will be directed against influenza and only a fraction hereof will be able to neutralize the virus. As a consequence, several doses are often required for effectiveness [41]. Other disadvantages include batch-to-batch variation and risk of pathogen transmission. Such limitations can be overcome by using monoclonal antibodies (mAbs). MAbs are attractive biologic drugs because of their exquisite specificity, low toxicity, and well-understood mechanisms of action [42]. Initially, therapeutic mAbs were generated in mice and used directly in humans. However, mouse antibodies often elicited human anti-mouse responses that impaired their efficacy. To attenuate such adverse reactions, mouse mAbs were engineered, somewhat laboriously, to look more like human antibodies. More recent improvements in mAb isolation, screening, and production technologies have provided access to fully human mAbs which has led to remarkable results in the treatment of cancer and inflammatory diseases [43]. However, only one mAb, palivizumab, is currently licensed for an infectious disease (RSV).

6 Neutralizing Influenza

Three viral proteins have domains that are exposed to the outside environment: HA, NA, and matrix protein 2 (M2). Whereas antibodies against NA and M2 have been shown to reduce viral titers, morbidity, and viral shedding [44–55], it is only antibodies directed against HA that can effectively neutralize the viral infection. Three HA monomers form a trimeric HA spike protruding from the viral membrane. Each monomer is synthesized as a precursor protein that is cleaved into two subunits (HA1 and HA2) by host cell proteases. The major part of the HA1 subunit forms a "globular head" region that contains the receptor-binding sites required for binding of the virus to target cells, while the HA2 subunit forms a membrane-anchored "stem region" which contains a fusion peptide and is responsible for viral entry by mediating fusion of the viral envelope with the endosomal membrane [56]. Antibodies against HA may neutralize the virus through blocking viral attachment to the sialyl receptors on host cells or through interfering with HA conformational changes at low pH within the endosome, thereby preventing fusion and uncoating of the virus [57–60]. No less than sixteen antigenically distinct subtypes of HA have been detected [61] which are principally defined as serotypes between which polyclonal antisera to the respective HA subtypes show little cross-neutralizing activity. Moreover, since the structures of HA antigenic sites do not only vary between, but also within subtypes, cross-neutralizing mAbs have been rarely described in the literature [62]. Interestingly, Yoshida et al. recently described a mouse mAb, recognizing an epitope on the globular head of HA that is shared between H1, H2, H3, H5, H9, and H13 subtypes [63]. Although passive immunization of mice with this mAb conferred heterosubtypic protection, escape mutants resistant to this antibody could be generated in a single round of *in vitro* selection [63]. To minimize such immunological escape, functionally important regions that are highly conserved among different HA subtypes should be targeted.

7   Human mAbs Against Influenza

MAbs against influenza viruses have been studied for decades, but their potential – and thus development – as "passive" immunotherapy for influenza has not progressed much because of the failure to produce fully human mAbs with broad heterosubtypic neutralizing activity. However, improved technologies have increased access to fully human mAbs [43, 64], and considerable effort has been directed to the generation of human mAbs against influenza viruses in recent years. A number of mAbs with potential for use in passive immunization have been described. Simmons et al. [65] used an improved method for EBV immortalization of memory B cells to isolate mAbs from human donors who recovered from H5N1 infection and described four mAbs with neutralizing activity *in vitro* and both prophylactic and therapeutic efficacy in mice challenged with highly pathogenic H5N1 viruses. Interestingly, two of the mAbs were able to neutralize both clade 1 and clade 2 H5N1 viruses *in vitro*, and three were therapeutically active *in vivo* against a virus from each of these clades [65]. Also, using EBV, but with a different protocol, Yu et al. [66] immortalized memory B cells from humans who had naturally been exposed to the 1918 pandemic virus and isolated five mAbs with potent neutralizing activity. These antibodies cross-reacted with the antigenically similar HA of a 1930 swine influenza strain, but did not cross-react with HAs of more recent human influenza viruses [66] An alternative approach was used by Sun et al. [67] who used a combinatorial Fab antibody phage library from a patient recovered from H5N1 infection and isolated one mAb able to neutralize only clade 2 H5N1 viruses and one able to cross-neutralize most of the clade 0, clade 1, and clade 2 viruses tested. Passive immunization of mice with either mAb resulted in protection from lethal infection with a clade 2 virus. Although some of these mAbs show some degree of cross-neutralizing activity between different viral isolates, they have in common that they bind to epitopes on the globular head of HA1 which, given the remarkable antigenic diversity of this region, limits their therapeutic potential.

Recently, however, Throsby et al. [68] described a panel of 12 mAbs that showed broad heterosubtypic neutralizing activity against antigenically diverse H1, H2, H5, H6, H8, and H9 influenza subtypes. Epitope mapping studies indicated that instead of binding to the globular head of the HA, these antibodies recognize a highly conserved hydrophobic pocket in the membrane-proximal stem of the HA. In a subsequent study, the epitope and mechanism of neutralization of one of these antibodies, CR6261, were characterized in detail [69]. Cocrystals of CR6261 Fab, with the HA ectodomains of human 1918 H1N1 pandemic virus (A/South Carolina/ 1/1918; SC1918/H1) and a highly pathogenic avian H5N1 virus (A/Vietnam/1203/ 2004; Viet04/H5), confirmed that CR6261 binds in the HA stem, distant from strain-specific antibodies (Fig. 1).

As mentioned, during viral maturation the HA monomers are proteolytically cleaved into two disulphide-linked chains, HA1 and HA2. Although HA1 consists primarily of the membrane-distal receptor domain, its N- and C- terminal regions extend towards the viral membrane and are intertwined with the exterior surface of HA2. HA2 constitutes the core fusion machinery in the stalk region and is

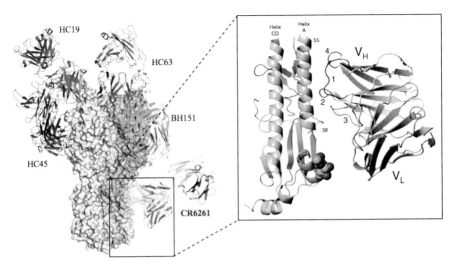

Fig. 1 Broadly neutralizing CR6261 binds in the HA stem distant from other strain-specific antibodies using only its heavy chain. *Left panel*: Comparison of the binding sites of Fab CR6261 (heavy chain in *yellow*, light chain in *orange*) that recognizes the HA stem and strain-specific antibodies that bind to the HA1 globular head domain. The strain-specific antibodies are depicted in *green* (BH151), *copper* (HC63), *dark red* (HC45), and *blue* (HC19). The HA trimer is shown as a surface representation, with the HA1 and HA2 from one protomer colored in *purple* and *cyan*, respectively. *Inset*: Close-up view of the interaction of CR6261 with the A-Helix. HA1 and HA2 are colored in *purple* and *cyan*, respectively. The HCDRs 1, 2, and 3 of the CR6261 VH domain are highlighted in *red*, *blue*, and *green*, respectively. CDR1 runs along the side of the A-helix, interacting with five consecutive helical turns. In contrast, the light chain makes no contacts with the HA. Adapted from Ekiert et al. [69]

dominated by the long central CD-helix (residues 75–126) that forms a trimeric coiled-coil and the shorter A-helix (residues 38–58). MAb CR6261 interacts primarily with the HA2 A-helix, but also contacts HA1 residues in the stem region. Unexpectedly, the interaction is mediated exclusively by the heavy chain (Fig. 1, inset). The HCDR1 forms hydrogen bonds with five consecutive turns of the A-helix and nonpolar interactions with a hydrophobic patch at the junction between the A-helix and HA1, whereas the conserved hydrophobic tip of the HCDR2, and a residue from HCDR3 interact with a second hydrophobic patch closer to the membrane– proximal end of the A-helix. Furthermore, residues of Framework Region 3 were shown to interact with the first hydrophobic patch.

The location of this epitope suggested that instead of blocking viral attachment to the host cell by binding to the globular head of the HA, CR6261 might mediate its neutralizing activity by inhibiting membrane fusion. Therefore, the ability of this mAb to prevent conversion of SC1918/H1 and Viet04/H5 HAs to the postfusion state upon exposure to low pH was assessed. Hereto, the fact that the HA prefusion state is highly protease resistant, while the postfusion state is much more susceptible to protease degradation [70] was exploited. Both HAs are converted to their protease-susceptible postfusion form at pH 4.9 or 5.3, but not at pH 8.0 (Fig. 2a, b, lanes 7–9).

Importantly, this conversion is prevented in the presence of CR6261 (Fig. 2a, b, lanes 10–12). In accordance, the cocrystal structure of SC1918/H1 HA with CR6261 was in the prefusion state, despite the fact that the crystals were grown at pH 5.3, below the pH of membrane fusion (Fig. 2c). The conformational changes

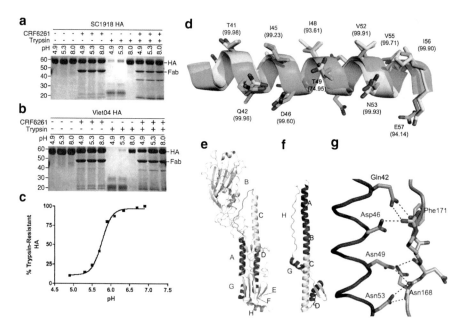

Fig. 2 CR6261 recognizes a functionally conserved epitope in the stalk region and inhibits the pH-induced conformational changes in the SC1918/H1 and Viet04/H5 HAs. CR6261 protects SC1918/H1 (**a**) and Viet04/H5 (**b**) HAs from the pH-induced protease sensitivity associated with membrane fusion. Exposure to low pH renders the SC1918/H1 and Viet04/H5 HAs sensitive to trypsin digestion (*lanes 7 and 8 versus 9*), but CR6261 prevents conversion to the protease susceptible conformation (*lanes 10–12*). The CR6261-SC1918-H1 crystals were grown at pH 5.3, which also indicates that CR6261 blocks the extensive pH-induced conformational changes. (**c**) Titration of SC1918-H1 trypsin resistance versus varying pH treatments (followed by neutralization to pH 8.0). The pH resulting in 50% conversion to protease sensitive conformation is 5.76 (95% confidence interval, 5.70–5.82). (**d**) Superposition of the A helix from SC1918/H1 (*green*) and an H3 HA (*yellow*) reveals that the CR6261-interacting surface is highly conserved among all subtypes. Helix positions are labeled according to the SC1918/H1 sequence, with the percent similarity across all subtypes (H1 to H16) from an analysis of 5,261 sequences) indicated in *parentheses*. (**e and f**) HA2 undergoes a dramatic and irreversible conformational change between the pre- and postfusion states. This results in the translocation of the A helix (*red*) from its initial position near the viral envelope (**e**) toward the target membrane at the opposite end of the HA trimer (**f**). The zipping-up of coil H along the outside of the A and B helices is thought to drive the fusion reaction. The orientation of helix C (*yellow*) is roughly identical in (**e**) and (**f**). (**g**) Polar residues on the CR6261-interacting surface of the A helix form a network of interactions with coil H in the postfusion state. These residues are well positioned to play a critical role in the late stages of membrane fusion, explaining the exceptional conservation the CR6261 epitope on the A helix

that are induced by low pH and lead to fusion of the viral envelope with the membrane of the endosomal vesicle are quite dramatic. Low pH exposure converts the connecting segment between the A- and CD-helices to an additional α-helical segment, extending the central HA2 trimeric coil towards the target endosomal membrane and dragging the A-helix and N-terminal fusion peptide along with it (Fig. 2e, f). Subsequent rearrangements in HA2 are thought to bring the viral and target membranes into close proximity for fusion [71, 72]. MAb CR6261 thus appears indeed to neutralize the virus by stabilizing the prefusion state and preventing the pH-dependent fusion of viral and cellular membranes. The heterosubtypic neutralizing activity of CR6261 is in line with the functional importance of its epitope, which seems to segregate with a previously characterized division of HAs in group 1 that includes the H1, H2, H5, H6, H8, H9, H11, H12, H13, and H16 subtypes, and group 2 that includes the H3, H4, H7, H10, H14, and H15 subtypes [61, 73–75]. In fact, the CR6261-interacting surface with the A-helix is highly conserved among all subtypes (Fig. 2d, g), possibly because polar residues in this area play a critical role in the late stages of membrane fusion. Parts of the epitope other than the A-helix dictate the apparent restriction of CR6261 to group 1 viruses.

Similar mAbs with a similar mode of action and heterosubtypic neutralizing activity against group 1 viruses have subsequently been described by Sui et al. [76]. Furthermore, Kashyap et al. [77] isolated three mAbs that neutralized both H1 and H5 viral strains from phage-display libraries generated from B cell populations of patients who survived H5N1 infection, although the mechanism of neutralization of the mAbs described in this study remains to be determined.

Strikingly, out of ~50 different human germline genes, all the heterosubtypic neutralizing mAbs are derived from the same germline gene, VH1-69 [68, 76, 77]. The unusual shape, and the unique presence of two hydrophobic residues at the tip, of the HCDR2 loop encoded by this germline gene contributes to the unusual ability of these mAbs to bind to conserved hydrophobic pockets. Preferential use of the VH1-69 germline is also reported for antibodies against HCV [78–80] and an HCV E2-specific antibody derived from the VH1-69 germline was shown to inhibit fusion [81]. Similarly, a broadly cross-neutralizing antibody directed against HIV protein gp41 blocks a conformational change that is necessary for fusion by insertion of the hydrophobic tip of its VH1-69 HCDR2 loop into a conserved hydrophobic pocket [82, 83]. Binding to hydrophobic pockets may thus represent a general mechanism by which mAbs can lock viral envelope proteins into a nonfusogenic conformation and neutralize viruses.

Why are broadly neutralizing mAbs not generally generated and expanded during successive rounds of influenza infection and repeated vaccination? The explanation may lie in the fact that the humoral immune response against influenza is highly restricted [84, 85], and focused on subtype and strain-specific epitopes [86, 87]. Put simply, the immunodominant antibody response directed against highly exposed epitopes (on the globular head) may overwhelm the antibody response to more conserved (and less exposed) epitopes.

The approaches that have recently led to the successful isolation of broadly neutralizing mAbs are in line with this notion as the single-chain variable region fragment (scFv) libraries screened for broadly neutralizing H5N1 mAbs were built from B cells of unimmunized donors [76], and donors recently vaccinated with the seasonal influenza vaccine [68], respectively. The fact that Kashyap et al. likely have found similar antibodies from a convalescent phage display library indicates that the restriction of the immune response to subtype and strain-specific epitopes is not complete. However, of the more than 300 unique antibodies reactive to H5N1 viral antigens recovered in this study were directed, only three were able to neutralize both H1 and H5 subtype viruses.

To access a more diverse immune repertoire, Throsby et al. constructed a combinatorial library built on a IgM$^+$/CD27$^+$ subset of B cells (Fig. 3).

This approach is based on the hypothesis that this subset contains a diverse repertoire of antibodies against conserved epitopes on pathogens. Despite their CD27 expression and mutated V genes, characteristics tightly linked to the memory B cell phenotype, the origin and role of this subset of B cells is controversial. It has been proposed that circulating B cells with this phenotype are linked to marginal zone B cells and have a primary role in T-independent immunity [89, 90], while others argue that they are formed as part of an intermediate differentiation step in

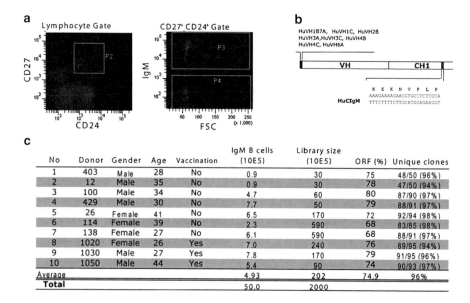

Fig. 3 Construction of IgM$^+$ memory B cell libraries. (a) Donor lymphocytes were isolated by Ficoll-plaque from heparinized blood and stained for the phenotypic markers CD27, CD24, and IgM. CD24$^+$ CD27$^+$ cells were gated and the IgM$^+$ cells within this gate sorted directly into Trizol for RNA extraction. (b) RT-PCR was performed using a pool of 5′ oligonucleotide primers that cover all VH gene families and a 3′ oligonucleotide primer that anneals in a region of the CH1 domain of Cμ distinct from other immunoglobulin isotypes. (c) Using cDNA generated in this way, 10 individual scFv libraries were constructed as described previously [88]. Donors 1020, 1030, and 1050 had been vaccinated with the Dutch 2005 seasonal influenza vaccine 7 days prior to collection of blood. All libraries demonstrated a high percentage of correct scFv ORF's and diversity based on unique HCDR3 sequence. Adapted from Throsby et al. [68]

normal T-dependent germinal center immune responses [91]. Anyway, several reports have highlighted a role for IgM in the early stages of protection from experimental influenza virus challenge, including subtypes to which mice are immunologically naïve [92–94]. The success in identifying mAbs with hetero-subtypic neutralizing activity that, given the limited number of somatic mutations, represent an immediate germline response [95], validates the strategy of using the IgM$^+$/CD27$^+$ subset of B cells. Although Sui et al. isolated mAbs with similar mode of action, they used a much larger library (in total 2.7×10^{10} vs. 2×10^8 members).

8 Perspectives

Development of clinical interventions against influenza is challenged by the large variety of viral subtypes, the antigenic variability within subtypes and the potential for antiviral resistance to emerge. The recently identified class of antibodies that neutralize influenza by inhibiting membrane fusion are particularly attractive candidates for mAb-based immunotherapy due to the fact that they are fully human, demonstrate potent neutralizing activity against a wide spectrum of viral subtypes, and, in accordance with the high conservation of the epitope, escape mutants resistant to these antibodies are not readily generated *in vitro* [68, 76]. Several such antibodies have already been shown to be protective in mice when given before and after lethal H5N1 or H1N1 challenge [68, 76] and clinical evaluation is currently being initiated. Since the broadly neutralizing human mAbs described here do not neutralize influenza viruses that belong to group 2, identification of a highly conserved epitope in these viruses would further enhance the applicability of human mAb-based passive immunotherapy. Such therapy would be particularly beneficial for the groups at the highest risk of severe disease due to seasonal influenza – the elderly and immunocompromised – but may also be indispensable for the general public during a pandemic.

Furthermore, highly conserved epitopes may provide leads for the design of antivirals or, given the progress in the field of structure-based antigen design [96], can potentially be exploited for the generation of a universal influenza vaccine.

References

1. Palese P, Shaw M (2007) Orthomyxoviridae: the viruses and their replication. In: D Knipe, P Howley (eds) Fields virology. Lippincott Williams & Wilkins, Philadelphia, pp 1647–1689
2. WHO (2009) Fact sheet 211: Influenza. World Health Organization, Geneva
3. Ochsenbein AF, Pinschewer DD, Odermatt B, Ciurea A, Hengartner H, Zinkernagel RM (2000) Correlation of T cell independence of antibody responses with antigen dose reaching secondary lymphoid organs: implications for splenectomized patients and vaccine design. J Immunol 164:6296–6302

4. Ochsenbein AF, Zinkernagel RM (2000) Natural antibodies and complement link innate and acquired immunity. Immunol Today 21:624–630

5. Mintern JD, Guillonneau C, Turner SJ, Doherty PC (2008) The immune response to Infleunza A viruses. In: Rappuoli R, Del Giudice G (eds) Influenza vaccines for the future. Birkhauser, Basel

6. Webby RJ, Sandbulte MR (2008) Influenza vaccines. Front Biosci 13:4912–4924

7. Carrat F, Flahault A (2007) Influenza vaccine: the challenge of antigenic drift. Vaccine 25:6852–6862

8. Ansaldi F, Bacilieri S, Durando P, Sticchi L, Valle L, Montomoli E, Icardi G, Gasparini R, Crovari P (2008) Cross-protection by MF59-adjuvanted influenza vaccine: neutralizing and haemagglutination-inhibiting antibody activity against A(H3N2) drifted influenza viruses. Vaccine 26:1525–1529

9. Thompson WW, Shay DK, Weintraub E, Brammer L, Bridges CB, Cox NJ, Fukuda K (2004) Influenza-associated hospitalizations in the United States. JAMA 292:1333–1340

10. van Essen GA, Palache AM, Forleo E, Fedson DS (2003) Influenza vaccination in 2000: recommendations and vaccine use in 50 developed and rapidly developing countries. Vaccine 21:1780–1785

11. Fiore AE, Shay DK, Broder K, Iskander JK, Uyeki TM, Mootrey G, Bresee JS, Cox NS (2008) Prevention and control of influenza: recommendations of the advisory committee on immunization practices (ACIP), 2008. MMWR Recomm Rep 57:1–60

12. Hannoun C, Megas F, Piercy J (2004) Immunogenicity and protective efficacy of influenza vaccination. Virus Res 103:133–138

13. Jefferson T, Smith S, Demicheli V, Harnden A, Rivetti A, Di Pietrantonj C (2005) Assessment of the efficacy and effectiveness of influenza vaccines in healthy children: systematic review. Lancet 365:773–780

14. Nichol KL, Nordin J, Mullooly J, Lask R, Fillbrandt K, Iwane M (2003) Influenza vaccination and reduction in hospitalizations for cardiac disease and stroke among the elderly. N Engl J Med 348:1322–1332

15. Jefferson T, Rivetti D, Rivetti A, Rudin M, Di Pietrantonj C, Demicheli V (2005) Efficacy and effectiveness of influenza vaccines in elderly people: a systematic review. Lancet 366: 1165–1174

16. Saurwein-Teissl M, Lung TL, Marx F, Gschosser C, Asch E, Blasko I, Parson W, Bock G, Schonitzer D, Trannoy E et al (2002) Lack of antibody production following immunization in old age: association with CD8(+)CD28(−) T cell clonal expansions and an imbalance in the production of Th1 and Th2 cytokines. J Immunol 168:5893–5899

17. Lazuardi L, Jenewein B, Wolf AM, Pfister G, Tzankov A, Grubeck-Loebenstein B (2005) Age-related loss of naive T cells and dysregulation of T-cell/B-cell interactions in human lymph nodes. Immunology 114:37–43

18. Moscona A (2008) Medical management of influenza infection. Annu Rev Med 59:397–413

19. Fleming DM (2001) Managing influenza: amantadine, rimantadine and beyond. Int J Clin Pract 55:189–195

20. Bright RA, Medina MJ, Xu X, Perez-Oronoz G, Wallis TR, Davis XM, Povinelli L, Cox NJ, Klimov AI (2005) Incidence of adamantane resistance among influenza A (H3N2) viruses isolated worldwide from 1994 to 2005: a cause for concern. Lancet 366:1175–1181

21. Moscona A (2005) Neuraminidase inhibitors for influenza. N Engl J Med 353:1363–1373

22. de Jong MD, Tran TT, Truong HK, Vo MH, Smith GJ, Nguyen VC, Bach VC, Phan TQ, Do QH, Guan Y et al (2005) Oseltamivir resistance during treatment of influenza A (H5N1) infection. N Engl J Med 353:2667–2672

23. Gubareva LV, Kaiser L, Matrosovich MN, Soo-Hoo Y, Hayden FG (2001) Selection of influenza virus mutants in experimentally infected volunteers treated with oseltamivir. J Infect Dis 183:523–531

24. Nicholson KG, Aoki FY, Osterhaus AD, Trottier S, Carewicz O, Mercier CH, Rode A, Kinnersley N, Ward P (2000) Efficacy and safety of oseltamivir in treatment of acute

influenza: a randomised controlled trial. Neuraminidase Inhibitor Flu Treatment Investigator Group. Lancet 355:1845–1850

25. Kiso M, Mitamura K, Sakai-Tagawa Y, Shiraishi K, Kawakami C, Kimura K, Hayden FG, Sugaya N, Kawaoka Y (2004) Resistant influenza A viruses in children treated with oseltamivir: descriptive study. Lancet 364:759–765

26. Stephenson I, Democratis J, Lackenby A, McNally T, Smith J, Pareek M, Ellis J, Bermingham A, Nicholson K, Zambon M (2009) Neuraminidase inhibitor resistance after Oseltamivir treatment of Acute Influenza A and B in Children. Clin Infect Dis 48(4):389–396

27. Dharan NJ, Gubareva LV, Meyer JJ, Okomo-Adhiambo M, McClinton RC, Marshall SA, St George K, Epperson S, Brammer L, Klimov AI et al (2009) Infections with oseltamivir-resistant influenza A(H1N1) virus in the United States. JAMA 301:1034–1041

28. CDC (2009) Update: influenza activity–United States, September 28, 2008–April 4, 2009, and composition of the 2009–10 influenza vaccine. MMWR Morb Mortal Wkly Rep 58:369–374

29. WHO (2009) Influenza A(H1N1) virus resistance to oseltamivir – 2008/2009 influenza season, northern hemisphere. WHO, Geneva

30. Poland GA, Jacobson RM, Ovsyannikova IG (2009) Influenza virus resistance to antiviral agents: a plea for rational use. Clin Infect Dis 48:1254–1256

31. Moscona A (2009) Global transmission of oseltamivir-resistant influenza. N Engl J Med 360:953–956

32. Enserink M (2009) Drug resistance. A 'wimpy' flu strain mysteriously turns scary. Science 323:1162–1163

33. CDC (2009) Oseltamivir-resistant novel influenza A (H1N1) virus infection in two immunosuppressed patients – Seattle, Washington, 2009. MMWR Morb Mortal Wkly Rep 58:893–896

34. Leung TW, Tai AL, Cheng PK, Kong MS, Lim W (2009) Detection of an oseltamivir-resistant pandemic influenza A/H1N1 virus in Hong Kong. J Clin Virol 46(3):298–299

35. von Behring E, Kitasato S (1991) The mechanism of diphtheria immunity and tetanus immunity in animals. 1890. Mol Immunol 28(1317):1319–1320

36. Casadevall A, Scharff MD (1995) Return to the past: the case for antibody-based therapies in infectious diseases. Clin Infect Dis 21:150–161

37. Casadevall A, Dadachova E, Pirofski LA (2004) Passive antibody therapy for infectious diseases. Nat Rev Microbiol 2:695–703

38. Luke TC, Kilbane EM, Jackson JL, Hoffman SL (2006) Meta-analysis: convalescent blood products for Spanish influenza pneumonia: a future H5N1 treatment? Ann Intern Med 145:599–609

39. Zhou B, Zhong N, Guan Y (2007) Treatment with convalescent plasma for influenza A (H5N1) infection. N Engl J Med 357:1450–1451

40. Kong LK, Zhou BP (2006) Successful treatment of avian influenza with convalescent plasma. Hong Kong Med J 12:489

41. Hemming VG (2001) Use of intravenous immunoglobulins for prophylaxis or treatment of infectious diseases. Clin Diagn Lab Immunol 8:859–863

42. ter Meulen J (2007) Monoclonal antibodies for prophylaxis and therapy of infectious diseases. Expert Opin Emerg Drugs 12:525–540

43. Lanzavecchia A, Corti D, Sallusto F (2007) Human monoclonal antibodies by immortalization of memory B cells. Curr Opin Biotechnol 18:523–528

44. Couch RB, Kasel JA, Gerin JL, Schulman JL, Kilbourne ED (1974) Induction of partial immunity to influenza by a neuraminidase-specific vaccine. J Infect Dis 129:411–420

45. Murphy BR, Kasel JA, Chanock RM (1972) Association of serum anti-neuraminidase antibody with resistance to influenza in man. N Engl J Med 286:1329–1332

46. Schulman JL, Khakpour M, Kilbourne ED (1968) Protective effects of specific immunity to viral neuraminidase on influenza virus infection of mice. J Virol 2:778–786

47. Webster RG, Reay PA, Laver WG (1988) Protection against lethal influenza with neuraminidase. Virology 164:230–237

48. Wang R, Song A, Levin J, Dennis D, Zhang NJ, Yoshida H, Koriazova L, Madura L, Shapiro L, Matsumoto A et al (2008) Therapeutic potential of a fully human monoclonal antibody against influenza A virus M2 protein. Antiviral Res 80:168–177

49. Treanor JJ, Tierney EL, Zebedee SL, Lamb RA, Murphy BR (1990) Passively transferred monoclonal antibody to the M2 protein inhibits influenza A virus replication in mice. J Virol 64:1375–1377

50. Frace AM, Klimov AI, Rowe T, Black RA, Katz JM (1999) Modified M2 proteins produce heterotypic immunity against influenza A virus. Vaccine 17:2237–2244

51. Mozdzanowska K, Feng J, Eid M, Kragol G, Cudic M, Otvos L Jr, Gerhard W (2003) Induction of influenza type A virus-specific resistance by immunization of mice with a synthetic multiple antigenic peptide vaccine that contains ectodomains of matrix protein 2. Vaccine 21:2616–2626

52. Tompkins SM, Zhao ZS, Lo CY, Misplon JA, Liu T, Ye Z, Hogan RJ, Wu Z, Benton KA, Tumpey TM et al (2007) Matrix protein 2 vaccination and protection against influenza viruses, including subtype H5N1. Emerg Infect Dis 13:426–435

53. Ernst WA, Kim HJ, Tumpey TM, Jansen AD, Tai W, Cramer DV, Adler-Moore JP, Fujii G (2006) Protection against H1, H5, H6 and H9 influenza A infection with liposomal matrix 2 epitope vaccines. Vaccine 24:5158–5168

54. Fan J, Liang X, Horton MS, Perry HC, Citron MP, Heidecker GJ, Fu TM, Joyce J, Przysiecki CT, Keller PM et al (2004) Preclinical study of influenza virus A M2 peptide conjugate vaccines in mice, ferrets, and rhesus monkeys. Vaccine 22:2993–3003

55. Neirynck S, Deroo T, Saelens X, Vanlandschoot P, Jou WM, Fiers W (1999) A universal influenza A vaccine based on the extracellular domain of the M2 protein. Nat Med 5:1157–1163

56. Skehel JJ, Wiley DC (2000) Receptor binding and membrane fusion in virus entry: the influenza hemagglutinin. Annu Rev Biochem 69:531–569

57. Barbey-Martin C, Gigant B, Bizebard T, Calder LJ, Wharton SA, Skehel JJ, Knossow M (2002) An antibody that prevents the hemagglutinin low pH fusogenic transition. Virology 294:70–74

58. Yoden S, Kida H, Kuwabara M, Yanagawa R, Webster RG (1986) Spin-labeling of influenza virus hemagglutinin permits analysis of the conformational change at low pH and its inhibition by antibody. Virus Res 4:251–261

59. Knossow M, Gaudier M, Douglas A, Barrere B, Bizebard T, Barbey C, Gigant B, Skehel JJ (2002) Mechanism of neutralization of influenza virus infectivity by antibodies. Virology 302:294–298

60. Kida H, Webster RG, Yanagawa R (1983) Inhibition of virus-induced hemolysis with monoclonal antibodies to different antigenic areas on the hemagglutinin molecule of A/seal/Massachusetts/1/80 (H7N7) influenza virus. Arch Virol 76:91–99

61. Fouchier RA, Munster V, Wallensten A, Bestebroer TM, Herfst S, Smith D, Rimmelzwaan GF, Olsen B, Osterhaus AD (2005) Characterization of a novel influenza A virus hemagglutinin subtype (H16) obtained from black-headed gulls. J Virol 79:2814–2822

62. Okuno Y, Isegawa Y, Sasao F, Ueda S (1993) A common neutralizing epitope conserved between the hemagglutinins of influenza A virus H1 and H2 strains. J Virol 67:2552–2558

63. Yoshida R, Igarashi M, Ozaki H, Kishida N, Tomabechi D, Kida H, Ito K, Takada A (2009) Cross-protective potential of a novel monoclonal antibody directed against antigenic site B of the hemagglutinin of influenza A viruses. PLoS Pathog 5:e1000350

64. Marasco WA, Sui J (2007) The growth and potential of human antiviral monoclonal antibody therapeutics. Nat Biotechnol 25:1421–1434

65. Simmons CP, Bernasconi NL, Suguitan AL, Mills K, Ward JM, Chau NV, Hien TT, Sallusto F, Ha do Q, Farrar J et al (2007) Prophylactic and therapeutic efficacy of human monoclonal antibodies against H5N1 influenza. PLoS Med 4:e178

66. Yu X, Tsibane T, McGraw PA, House FS, Keefer CJ, Hicar MD, Tumpey TM, Pappas C, Perrone LA, Martinez O et al (2008) Neutralizing antibodies derived from the B cells of 1918 influenza pandemic survivors. Nature 455:532–536

67. Sun L, Lu X, Li C, Wang M, Liu Q, Li Z, Hu X, Li J, Liu F, Li Q et al (2009) Generation, characterization and epitope mapping of two neutralizing and protective human recombinant antibodies against influenza A H5N1 viruses. PLoS One 4:e5476

68. Throsby M, van den Brink E, Jongeneelen M, Poon LL, Alard P, Cornelissen L, Bakker A, Cox F, van Deventer E, Guan Y et al (2008) Heterosubtypic neutralizing monoclonal antibodies cross-protective against H5N1 and H1N1 recovered from human IgM+ memory B cells. PLoS One 3:e3942

69. Ekiert DC, Bhabha G, Elsliger MA, Friesen RH, Jongeneelen M, Throsby M, Goudsmit J, Wilson IA (2009) Antibody recognition of a highly conserved Influenza virus epitope. Science 324(5924):246–251

70. Skehel JJ, Bayley PM, Brown EB, Martin SR, Waterfield MD, White JM, Wilson IA, Wiley DC (1982) Changes in the conformation of influenza virus hemagglutinin at the pH optimum of virus-mediated membrane fusion. Proc Natl Acad Sci USA 79:968–972

71. Bullough PA, Hughson FM, Skehel JJ, Wiley DC (1994) Structure of influenza haemagglutinin at the pH of membrane fusion. Nature 371:37–43

72. Chen J, Skehel JJ, Wiley DC (1999) N- and C-terminal residues combine in the fusion-pH influenza hemagglutinin HA(2) subunit to form an N cap that terminates the triple-stranded coiled coil. Proc Natl Acad Sci USA 96:8967–8972

73. Air GM (1981) Sequence relationships among the hemagglutinin genes of 12 subtypes of influenza A virus. Proc Natl Acad Sci USA 78:7639–7643

74. Russell RJ, Gamblin SJ, Haire LF, Stevens DJ, Xiao B, Ha Y, Skehel JJ (2004) H1 and H7 influenza haemagglutinin structures extend a structural classification of haemagglutinin subtypes. Virology 325:287–296

75. Nobusawa E, Aoyama T, Kato H, Suzuki Y, Tateno Y, Nakajima K (1991) Comparison of complete amino acid sequences and receptor-binding properties among 13 serotypes of hemagglutinins of influenza A viruses. Virology 182:475–485

76. Sui J, Hwang WC, Perez S, Wei G, Aird D, Chen LM, Santelli E, Stec B, Cadwell G, Ali M et al (2009) Structural and functional bases for broad-spectrum neutralization of avian and human influenza A viruses. Nat Struct Mol Biol 16:265–273

77. Kashyap AK, Steel J, Oner AF, Dillon MA, Swale RE, Wall KM, Perry KJ, Faynboym A, Ilhan M, Horowitz M et al (2008) Combinatorial antibody libraries from survivors of the Turkish H5N1 avian influenza outbreak reveal virus neutralization strategies. Proc Natl Acad Sci USA 105:5986–5991

78. Chan CH, Hadlock KG, Foung SK, Levy S (2001) V(H)1-69 gene is preferentially used by hepatitis C virus-associated B cell lymphomas and by normal B cells responding to the E2 viral antigen. Blood 97:1023–1026

79. Marasca R, Vaccari P, Luppi M, Zucchini P, Castelli I, Barozzi P, Cuoghi A, Torelli G (2001) Immunoglobulin gene mutations and frequent use of VH1-69 and VH4-34 segments in hepatitis C virus-positive and hepatitis C virus-negative nodal marginal zone B-cell lymphoma. Am J Pathol 159:253–261

80. Carbonari M, Caprini E, Tedesco T, Mazzetta F, Tocco V, Casato M, Russo G, Fiorilli M (2005) Hepatitis C virus drives the unconstrained monoclonal expansion of VH1-69-expressing memory B cells in type II cryoglobulinemia: a model of infection-driven lymphomagenesis. J Immunol 174:6532–6539

81. Haid S, Pietschmann T, Pecheur EI (2009) Low pH-dependent hepatitis C virus membrane fusion depends on E2 integrity, target lipid composition, and density of virus particles. J Biol Chem 284:17657–17667

82. Miller MD, Geleziunas R, Bianchi E, Lennard S, Hrin R, Zhang H, Lu M, An Z, Ingallinella P, Finotto M et al (2005) A human monoclonal antibody neutralizes diverse HIV-1 isolates by binding a critical gp41 epitope. Proc Natl Acad Sci USA 102:14759–14764

83. Luftig MA, Mattu M, Di Giovine P, Geleziunas R, Hrin R, Barbato G, Bianchi E, Miller MD, Pessi A, Carfi A (2006) Structural basis for HIV-1 neutralization by a gp41 fusion intermediate-directed antibody. Nat Struct Mol Biol 13:740–747

84. Wang ML, Skehel JJ, Wiley DC (1986) Comparative analyses of the specificities of anti-influenza hemagglutinin antibodies in human sera. J Virol 57:124–128
85. Wrammert J, Smith K, Miller J, Langley WA, Kokko K, Larsen C, Zheng NY, Mays I, Garman L, Helms C et al (2008) Rapid cloning of high-affinity human monoclonal antibodies against influenza virus. Nature 453:667–671
86. Caton AJ, Brownlee GG, Yewdell JW, Gerhard W (1982) The antigenic structure of the influenza virus A/PR/8/34 hemagglutinin (H1 subtype). Cell 31:417–427
87. Gerhard W, Yewdell J, Frankel ME, Webster R (1981) Antigenic structure of influenza virus haemagglutinin defined by hybridoma antibodies. Nature 290:713–717
88. Kramer RA, Marissen WE, Goudsmit J, Visser TJ, Clijsters-Van der Horst M, Bakker AQ, de Jong M, Jongeneelen M, Thijsse S, Backus HH et al (2005) The human antibody repertoire specific for rabies virus glycoprotein as selected from immune libraries. Eur J Immunol 35:2131–2145
89. Kruetzmann S, Rosado MM, Weber H, Germing U, Tournilhac O, Peter HH, Berner R, Peters A, Boehm T, Plebani A et al (2003) Human immunoglobulin M memory B cells controlling Streptococcus pneumoniae infections are generated in the spleen. J Exp Med 197:939–945
90. Weller S, Braun MC, Tan BK, Rosenwald A, Cordier C, Conley ME, Plebani A, Kumararatne DS, Bonnet D, Tournilhac O et al (2004) Human blood IgM "memory" B cells are circulating splenic marginal zone B cells harboring a prediversified immunoglobulin repertoire. Blood 104:3647–3654
91. Tangye SG, Good KL (2007) Human IgM+CD27+ B cells: memory B cells or "memory" B cells? J Immunol 179:13–19
92. Harada Y, Muramatsu M, Shibata T, Honjo T, Kuroda K (2003) Unmutated immunoglobulin M can protect mice from death by influenza virus infection. J Exp Med 197:1779–1785
93. Jayasekera JP, Moseman EA, Carroll MC (2007) Natural antibody and complement mediate neutralization of influenza virus in the absence of prior immunity. J Virol 81:3487–3494
94. Baumgarth N, Herman OC, Jager GC, Brown LE, Herzenberg LA, Chen J (2000) B-1 and B-2 cell-derived immunoglobulin M antibodies are nonredundant components of the protective response to influenza virus infection. J Exp Med 192:271–280
95. Kwong PD, Wilson IA (2009) HIV-1 and influenza antibodies: seeing antigens in new ways. Nat Immunol 10:573–578
96. Dormitzer PR, Ulmer JB, Rappuoli R (2008) Structure-based antigen design: a strategy for next generation vaccines. Trends Biotechnol 26:659–667

Part III
Economic and Social Implications

Learning from the First Pandemic of the Twenty-First Century

Giuseppe Del Giudice and Rino Rappuoli

Abstract The response to the first influenza pandemic of the twenty-first century was facilitated by years of preparation for a possible pandemic caused by the avian influenza H5N1. The threat of an H5N1 pandemic had led to an increase in manufacturing capacity, to the development of influenza vaccines made in cell culture instead of eggs, to the development of innovative adjuvants and to the establishment of clear rules to license pandemic vaccines. Most of these tools have been used and validated by the H1N1 pandemic. The main lesson learned is that oil-in-water adjuvants can be safely used in large scale and in all ages and conditions, including pregnant women. Adjuvants increase the titer of the antibody responses and broaden the epitopes recognized by antibodies so that they can neutralize also drifted viruses. In addition, they induce long lasting B- and T-memory cells. A further advantage of the use of adjuvants is the ability to use lower doses of vaccine, thus multiplying the manufacturing capacity up to fourfold. Cell-based vaccines have been established as a new technology to produce influenza vaccines.

Both adjuvants and cell cultures are expected to change not only the way we will address future pandemics but also the way we approach seasonal influenza, changing a field that has been stagnant for too many decades.

1 Of Birds and Humans: The Lessons Learned

Since 1580 at least ten influenza pandemics have occurred, with an average of one pandemic every 42 years. Analysis of the most recent and more accurate data predicts one pandemic every 30 years. The last pandemic was in 1968, 40 years ago. Therefore, common sense and mathematical models predicted that we had to be prepared for a new pandemic.

G. Del Giudice (✉) and R. Rappuoli
Research Center, Novartis Vaccines and Diagnostics, Via Fiorentina 1, 53100 Siena, Italy
e-mail: giuseppe.del_giudice@novartis.com, rino.rappuoli@novartis.com

G. Del Giudice and R. Rappuoli (eds.), *Influenza Vaccines for the Future*, 2nd edition,   401
Birkhäuser Advances in Infectious Diseases,
DOI 10.1007/978-3-0346-0279-2_18, © Springer Basel AG 2011

During the past 12 years, all the events that were expected to happen before a pandemic did happen, and all these events indicated avian influenza viruses as the most likely cause of the next pandemic. First, a new virus carrying a hemagglutinin (HA) with a new antigenic specificity (H5) that had never been isolated in humans before jumped from chicken into men in Hong Kong and killed six people in 1997. This early outbreak was contained by culling chickens in the Hong Kong area. The virus momentarily disappeared, but it was not dead at all: it was successfully breeding, multiplying, and expanding in birds in South East Asia [1], until it suddenly blew up again in humans in 2003 and 2004 in Vietnam, Thailand, Indonesia, China, etc. The virus had clearly escaped any control and was so widespread that culling hundreds of millions of chickens in the areas of outbreak had provided only a temporary relief and not been able to limit the spread of the virus. The virus in fact was spreading to the rest of Asia and outside, to Turkey, Egypt, and Africa through migratory birds as vectors. Concomitantly with this geographic spreading, the H5N1 virus, like all influenza viruses, underwent antigenic drift, and today many genetically and antigenically distinct clades and subclades of the virus have been identified. As of today, more than 440 cases have been reported since 2003, with an overall mortality rate higher than 60%. The appearance and the worldwide spreading of the swine-origin H1N1 virus have not stopped the transmission of the H5N1 virus from birds to humans. As recently as 2009, 47 new cases have been reported in Egypt (36 cases, 4 deaths), in China (seven cases, four deaths), and in Vietnam (four cases, all fatal) [2]. All human cases of H5N1 infection derived from close contacts with poultry. Although in a few cases close contacts among people may have caused the infection, till now the H5N1 virus has not adapted itself to humans and does not seem to represent an immediate threat, even less now with the appearance of the H1N1 virus of swine origin. One cannot, however, underscore the risk of the potential recombination of the two viruses as they both circulate in the same areas.

At the beginning of 2009, the threat of pandemic due to avian influenza viruses appeared very high, based both on the number of cases occurring in several countries and the very high fatality rate. Although the H5N1 virus was expected to be the most likely cause of the pandemic, other viruses such as the H9N2 and the H7N7 were also under strict observation because, despite the fewer cases they caused, they had the potential to kill the human host as in the case of the H7N7 virus [3, 4]. Because only H1, H2, and H3 viruses had been thus far reported from humans, it was reasonably supposed that human beings were immunologically naïve toward virus strains bearing novel HAs, there being the consequent intrinsic risk of high mortality in case of adaptation to humans. All these (and other) considerations prompted many academic laboratories, biotech companies and vaccine manufacturers to develop a plethora of vaccines potentially active against avian influenza viruses. More commonly, these vaccines are prepared in eggs, using the same technology used to manufacture seasonal influenza vaccines. The vaccines consist of whole inactivated virus, detergent-split virus, or purified HA and neuraminidase (NA) subunits. All these vaccines have been widely tested both in animal models and in extensive clinical studies. Some of them have also been approved in

Europe for a prepandemic use or for a pandemic use through a mock-up application which turned out instrumental for fast approval of vaccines against the pandemic H1N1 virus. Other approaches to the development of vaccines against avian influenza have been represented by live-attenuated vaccines, use of virus-like particles produced in baculovirus, vaccines based on conserved proteins such as the external domain of the M2 protein, and others, which, however, have been less extensively investigated in both the preclinical and clinical settings.

The results of many of the preclinical and clinical studies with H5N1-based vaccines have been discussed in details in recent chapters and reviews [5–7]. The question which now needs to be asked is: which lessons did we learn from the experience with vaccines against avian influenza? A corollary logical question is then: can we apply at least part of this learning to the development and use of the vaccine against the pandemic H1N1 influenza? And against the seasonal influenza in general? In the sections below, we attempt to analyze the new knowledge acquired and discuss how this knowledge could (and should) be exploited for the development of better vaccines against influenza and toward a better use of existing or novel influenza vaccines.

1.1 The Need of Adjuvants

A critical lesson learned from the trials of avian influenza vaccines is the importance of adjuvants, particularly of oil-in-water adjuvants in enhancing the quantity and in shaping the quality of protective antibody responses.

Adjuvants represent the best known way to enhance the immunogenicity of vaccines. Most vaccines which are licensed worldwide contain adjuvants. The influenza vaccine is one of the very few vaccines which are given without adjuvants. This is very likely because the individuals are already immunologically experienced with influenza antigens, thanks to previous annual vaccinations and/ or thanks to previous contacts (clinically overt or not) with the influenza virus. In such a context, the vaccination acts through the expansion of an already existing pool of memory cells without any need for further "help" from an adjuvant.

Aluminum salts (including alum) are the most utilized vaccine adjuvants worldwide and, until 2009, the only adjuvants admitted for human use in the USA. However, the use of these adjuvants to enhance the immunogenicity of influenza vaccine has consistently failed. Adsorption of influenza virus HA onto aluminum phosphate had been shown to increase the immunogenicity of the vaccine in mice [8]. However, when this aluminum phosphate-adsorbed influenza vaccine was tested in healthy military recruits, it did not enhance the antibody response over a nonadjuvanted vaccine [9]. Despite the failure demonstrated by these studies, during the 1960s and the 1970s, many of the influenza vaccines (whole-virion, split, or subunit) commercially available both in Europe and in the USA were still prepared together with aluminum salts. We had to wait until the 1980s to see the removal of these adjuvants based on the overwhelming evidence that the adjuvant did not increase the immunogenicity of the vaccine, while it increased its reactigenicity [10–12]. The potential use of aluminum salts has been recently reconsidered for the development of vaccines against the influenza virus A H5N1. Some controversial results have been reported. Indeed, if some enhancement was observed, it was lower than that provided by the oil-in-water adjuvants in dose sparing and in increasing the responsiveness to the vaccine at all ages, including elderly individuals [5–7].

In the 1950s, it had already been shown that the immunogenicity of influenza vaccines could be significantly enhanced by the use of mineral oil adjuvants. These adjuvants allowed significant dose sparing [13], enhancement of the antigen-specific antibody response [14], and persistence of these antibodies, which were still detectable 2–9 years later [15–17]. However, this adjuvant, which was nonmetabolizable and nonexcretable, caused serious adverse events, such as sterile abscesses in almost 3% of the vaccinees, and raised concerns about possible long-term effects. An almost 20-year follow-up of these subjects did not show any increased mortality attributable to the mineral oil adjuvant, not even in those subjects who had had sterile abscesses [18]. Nevertheless, the unacceptably high frequency of local side effects prevented for several years the development of novel, potent oil-based adjuvants. We had to wait until the mid-1990s to see the development of the first oil-in-water adjuvant, referred to as MF59 [19], which was finally licensed for use together with an inactivated subunit influenza vaccine in >65-year-old subjects [20]. The successful approach to the development of a strong and safe adjuvant such as MF59 was to reduce the amount of the oil in the emulsion from 50% to 4–5% and to replace the nonmetabolizable oil with a fully metabolizable, such as squalene, which is a physiological component of the human body, as precursor of cholesterol and adrenal hormones [20].

The very first demonstration of the need for adjuvants for the induction of an optimal response with avian influenza vaccines came in 2001 with the publication in the *Lancet* of the results of a clinical study using the nonpathogenic H5N3 as a source of antigens (at that time the reverse genetics was not available yet and the pathogenic H5N1 strains were lethal for embryonated eggs) and MF59 as the adjuvant [21]. This pioneer study provided most of the useful information available today and paved the way for further vaccine development. In essence, this study showed that the conventional, nonadjuvanted vaccine did not elicit a significant protective antibody response, as compared to the MF59-adjuvanted vaccine. These data were confirmed later by many other groups showing that even increasing the vaccine dosage to 90 µg or more did not induce neutralizing antibodies in the majority of the vaccinees [22]. Instead, the MF59-adjuvanted vaccine allowed three essential features for a pandemic vaccine: (1) dose-sparing not only for H5-based vaccines [21], but also for H9N2 vaccines [23]; (2) broadening of the neutralizing antibody response [24], and (3) induction of strong immunological memory [25–27]. All these features were typical of the oil-in-water adjuvant MF59, and not of adjuvants in general, since alum consistently failed to enhance the neutralizing antibody response to H5N1 at levels comparable to those achieved with MF59

[28, 29]. The very promising data obtained with MF59 prompted other groups to develop oil-in-water adjuvants also based on squalene [30]. One of these, referred to as AS03, is being actively utilized for the development of a vaccine against H5N1 [31] and is part of a vaccine used in various countries against the pandemic H1N1. Another squalene-based emulsion, referred to as AF03, is still at early stages of development [30] and is mainly addressed at the development of H5N1 influenza vaccines [32]. Many of the features exhibited by MF59-adjuvanted avian influenza vaccines were then reproduced also by these other oil-in-water adjuvants (see later).

1.2 Dose-Sparing and Increased Dose Availability

As mentioned earlier, the first consequence of using oil-in-water adjuvants in the formulation of avian influenza vaccine was the possibility to use amounts of antigens lower than 15 μg, the conventional dose of HA used in the seasonal vaccines. This was first demonstrated in adult volunteers immunized with 30, 15, or 7.5 μg of subunit H5N3 vaccine with or without MF59. Indeed, the highest antibody response was observed in the subjects who had received the lowest dose of the vaccine, 7.5 μg [21]. The possibility of dose sparing was then demonstrated also with the H9N2 vaccine. In this study, the levels of specific antibody obtained with 3.75 μg of MF59-adjuvanted vaccine after one single dose were similar to those reached after two doses of 30 μg of nonadjuvanted vaccine. A second dose of MF59-adjuvanted H9N2 vaccine significantly enhanced the level of neutralizing antibodies [23].

Similar data were then obtained by other groups using split H5N1 vaccines formulated with other oil-in-water adjuvants, such as AS03 and AF03. Indeed, doses as low as 3.75 μg or even 1.9 μg of H5N1 vaccine induced significant neutralizing antibody titers in vaccinated volunteers [32, 33].

Taken together, all these findings speak in favor of the possibility of significantly increasing the potential coverage of the human population vaccinated against a pandemic due to the significant reduction of the dose necessary to reach protective levels of circulating antibodies. In the absence of specific knowledge of the evolution of the threat of avian influenza, with the spread of the new H1N1 pandemic virus, and the need to still cover against the seasonal influenza, all this has seen in parallel an increased investment of vaccine manufacturers to enhance the capacity of production of both monovalent pandemic vaccines and trivalent seasonal vaccines. From the surveys conducted during summer 2009, it has been estimated that the total capacity worldwide for seasonal trivalent vaccines has increased from 400 to more than 900 million doses per year, with the potential to produce more than four billion doses in case of reduced output of seasonal trivalent vaccines [34]. This would translate into an increased availability of pandemic vaccine even for developing countries although, in this case, other issues (such as cost, storage, and distribution) will have to be solved by the cooperation between the World Health Organization (WHO), the governments, and other nongovernmental organizations [35].

1.3   Broadening of the Antibody Response to Drifted Influenza Virus Strains

An ideal vaccine against influenza should contain conserved internal viral proteins (e.g., NP, M1, and M2e) to induce protective immune responses against all the possible drifted (and possibly shifted) variants of the virus. This would avoid the continuous (almost yearly) change in the vaccine composition necessary to adapt the strains used for the vaccine to those which are expected to circulate that year. In addition, the use of conserved internal viral proteins would also induce cell-mediated immune effector mechanisms to complement the antibody response elicited against HA and NA present in the currently used vaccines. Despite various efforts toward this end (see the chapter "Conserved Internal Proteins as Potential Universal Vaccines" by A. Shaw), these vaccines have not turned the corner. However, the data available so far clearly show that inactivated influenza vaccines can confer significant seroprotection against drifted influenza virus strains when they are prepared together with oil-in-water adjuvants.

This was first shown with subunit H5N3 vaccines adjuvanted with MF59 [24]. While the subjects who had been vaccinated with the nonadjuvanted vaccine had no or very poor detectable neutralizing antibody responses against drifted H5N1 virus strains, the majority of the subjects with the MF59-adjuvanted H5N3 (clade 0-like, isolated in 1997) vaccine had protective levels of antibodies against the heterologous clade 1 virus isolated in Vietnam and in Thailand, isolated in 2003–2004. In summary, the data had shown that MF59-adjuvanted vaccines could induce protective immunity against viruses not fully matching the vaccine strain and could cover the antigenic drifts of the virus occurring over 6–7 years at least. This data opened the way to the concept of the prepandemic vaccination, in other terms the possibility to vaccinate before the formal declaration of the pandemic since the data proved that priming with a mismatched virus induced cross-protective immunity. For a rigorous foundation of the prepandemic vaccination approach, it remained to be shown that the immunological memory induced by this priming could also be boosted by a vaccine containing a drifted H5N1 virus. This was formally proven in subsequent studies and is discussed in the next section on adjuvant-driven immunological memory.

Subsequent studies with H5N1 vaccines have amply confirmed these pioneer findings originally obtained with H5N3-based vaccines. We know now that immunization with H5N1 (clade 1) vaccines containing MF59 or other oil-in-water adjuvants such as AS03 or AF03 induces antibodies against a wide panel of drifted strains, for example, those belonging to the subclades 2.1, 2.2, and 2.3 [32, 36–39]. The induction of antibodies against heterovariant virus strains in humans was paralleled by a stronger efficacy of the adjuvanted vaccine in ferrets challenged with various heterovariant H5N1 virus strains [40, 41]. Cross-clade antibody responses have also been reported in subjects vaccinated with whole-virion H5N1 vaccine unadjuvanted [42] or adjuvanted with aluminum hydroxide [43]. As none of these vaccines were tested in the same clinical study, it is difficult to make a direct comparison of the height and the extent of this antibody response.

Antigenic mismatch between the seasonal vaccine virus strains and the circulating virus strains is not a rare event, and it can affect influenza vaccine efficacy and effectiveness [44]. Mismatch is caused by the accumulation of point mutations at antigenic sites on the HA and NA proteins (antigenic drift), which occur between the time that WHO makes its recommendation for vaccine composition and the period of subsequent exposure to the circulating strain. This leads to the appearance of new antigenic determinants. Although occurring in both type A and type B viruses, the antigenic drift occurs more frequently in the influenza A (H3N2) viral subtype [44]. It has been shown that the antigenic drift causes a decrease in vaccine-induced immunogenicity in elderly people [45]. In older subjects with a high ($\geq$80%) postvaccination seroprotection rate against the homologous vaccine strain, the rate of sero-protection against the drifted circulating strains dropped to 4–75%, based on the circulating and on the vaccine strains, and on the age groups [46–48]. In addition, antigenic mismatch can have a strong impact on vaccine effectiveness, as demonstrated by a study for the period 1995–2005, when the vaccine effectiveness among older adults ($\geq$65 years of age) dropped during the seasons with a drifted strain (1997–1998 and 2002–2003) to values below 30% [49].

The finding that MF59-adjuvanted H5N3 vaccine induced neutralizing antibodies also against drifted H5N1 virus strains suggested that the same could take place with the seasonal influenza vaccines. Indeed, this was the case. Several clinical studies have now shown that the seasonal MF59-adjuvanted influenza vaccine induces strong antibody responses against heterovariant strains [46, 48, 50]. Thus, MF59-adjuvanted influenza vaccine provides greater seroprotection in the case of antigenic drift than nonadjuvanted vaccines. For example, significantly ($P < 0.0001$) more older adults receiving MF59-adjuvanted influenza vaccine containing A/Panama/2007/99 (H3N2) were seroprotected against the drifted variant A/Wyoming/3/2003 (H3N2) than those receiving nonadjuvanted split-virus vaccine or nonadjuvanted subunit vaccine (98%, 80% and 76%, respectively) [46]. The enhanced seroprotection against a large panel of drifted H3N2 [48] and B virus strains [50] has been confirmed in other studies in elderly people and more recently also in 6–36 month-old children [51], showing that this wide breadth of cross-protection is a general phenomenon induced by MF59.

1.4   Induction and Persistence of Immunological Memory

Vaccination with inactivated influenza vaccine without adjuvant works, thanks to an immunological memory, which is acquired with age through clinically overt or asymptomatic infections and is maintained via yearly vaccinations and/or subsequent contacts with the influenza viruses. It is difficult to discriminate which part of the memory is due to the infection and which one is provided by the vaccination. The development of vaccines against the avian H5N1 virus allowed to dissect the priming of the immune response, the induction of the immunological memory, and its persistence over time. In this context, it was possible to discover the critical role played by adjuvants and, in particular, by the oil-in-water adjuvant MF59, in the induction and persistence of immunological memory against influenza viruses.

Most of the clinical studies carried out so far with H5N1 vaccines have clearly shown their relatively poor immunogenicity, not only because of the need for strong oil-in-water adjuvants, but also because of the necessity of two doses of vaccines to induce protective titers of neutralizing antibodies in the majority of the vaccinees. The question was then to evaluate whether an immunological "signal" could be measured after one single dose of the H5N1 vaccine to formally show that successful priming had taken place, even if antibody titers were generally poor in most of the people. Indeed, after one single immunization with MF59-adjuvanted H5N1 subunit vaccine (clade 1), there was a significant increase in the frequency of HA-specific (using a panel of overlapping peptides spanning the entire length of the HA) central memory CD4+ T cells committed to produce IL-2 (with or without TNF-α), but not IFN-γ. The frequency of these cells did not increase after the second dose of the vaccine 3 weeks later and persisted at frequencies higher than baseline for 6 months, when it increased after a booster dose and was maintained at high levels later on [52]. It is interesting that these CD4+ T cells induced by the MF59-adjuvanted vaccines were mostly directed against epitopes which were conserved among the HA of the various H5N1 clades. Nevertheless, these cells also recognized epitope-containing sequences that varied in the HA of clade 2.1 and of H5N3 virus strains [52]. It is remarkable that a threefold increase in the frequency of H5-specific memory CD4+ T cells after a single dose of MF59-adjuvanted vaccine was predictive of a rise in neutralizing antibody titers above 1:80 after the booster dose 6 months after the first dose and also their persistence over time after the booster dose [52].

The persistence of the immunological memory can be demonstrated clinically by boosting individuals previously immunized with the same or a slightly different (heterovariant) vaccine. This, however, needs to wait for a sufficient long period of time between priming and boosting. The first example of this approach was shown in those subjects who had received the H5N3 vaccine adjuvanted with MF59 or otherwise [21]. Only the subjects previously primed with the adjuvanted vaccine exhibited a fast and strong rise in the titers of anti-H5N3 antibodies when boosted 16 months later with the same vaccine, whereas those who had received the nonadjuvanted vaccine mounted a detectable, but still much lower, response even after being boosted with the adjuvanted vaccine [25].

In order to understand how long the immunological memory at the B-cell level persisted over time and to evaluate the breadth of this memory in terms of cross-reactivity with H5N1 virus strains appeared from 2003 to 2007, these same subjects, and other subjects who had received the MF59-adjuvanted H5N3 vaccine twice [53], were boosted 6–8 years later with an MF59-adjuvanted subunit vaccine based on a clade 1 virus strain. Previously unprimed subjects immunized for the first time with the MF59-adjuvanted vaccine served as a control. One single injection with this vaccine induced a poor rise in the frequency of H5N1-specific memory B cells in previously unprimed subjects and also in the subjects who had been primed 8 years earlier with the nonadjuvanted H5N3 vaccine. The frequency of these cells increased (doubled) after a second booster dose. On the contrary, the frequency of memory B cells sharply and rapidly increased (up to 12% of all circulating IgG-producing memory B cells) after one single dose of the adjuvanted H5N1 vaccine in the subjects who had been previously primed with the MF59-adjuvanted H5N3 vaccine [27]. This significant and rapid increase in memory B cells was paralleled by a massive production of anti-H5N1 antibodies as detected by hemagglutination inhibition (HI), microneutralization (MN), and single radial hemolysis (SRH). Indeed, only 7 days after the booster dose with MF59-adjuvanted H5N1 clade 1 vaccine, all subjects primed 6–8 years earlier with the adjuvanted H5N3 vaccine had antibody titers that significantly exceeded the "protective" threshold of 1:40, not only against the homologous clade 1 virus strain, but also against other clade 1 strains, and against various strains belonging to the subclades 2.1, 2.2, and 2.3. A second dose of the vaccine did not increase the serum antibody response. This broad neutralizing antibody response persisted at high, protective levels for at least 6 months [26, 27]. Individuals who had been previously primed with the nonadjuvanted H5N3 vaccine mounted an anti-H5N1 antibody response post-boost but with slower kinetics and reaching levels much lower than the subjects primed 6–8 years earlier with the adjuvanted vaccine. Remarkably, all these subjects had post-booster antibody titers to the original priming H5N3 virus, similar to or lower than those detectable against the boosting H5N1 virus [26, 27], strongly suggesting that at least in these conditions — using the MF59 adjuvant in both the priming and boosting vaccine — no original antigenic sin was observed.

These data proved that strong immunological memory is induced upon vaccination with adjuvanted H5N1 influenza vaccines, that it persists for not less than 8 years, and that it can be strongly and rapidly boosted by a heterovariant adjuvanted vaccine. The induction of immunological memory and the possibility of boosting with heterovariant strains have now been shown with other vaccine combinations, using nonadjuvanted split vaccines [54], adjuvanted split vaccines [55], or nonadjuvanted whole-virion vaccines [56]. It is clear from all these data that the best responses are observed when both the priming and the boosting are performed with adjuvanted vaccines. However, it has been reported that the anti-H5N1 response after boosting with an AS03-adjuvanted split H5N1 vaccine can be negatively affected in subjects who had been previously primed with a non-adjuvanted heterovariant vaccine [55]. It is not clear whether this is due to an original antigenic sin. Should this be the case and considering that this was not observed with inactivated subunit vaccine adjuvanted with MF59 [26, 27], one can speculate that differences in the vaccine preparation (i.e., split versus purified subunit) and/or in the adjuvant preparation (i.e., AS03 versus MF59, which, although oil-in-water and squalene-based, contain substantial differences in their formulations – see chapter "Adjuvants for Influenza Vaccines: the Role of Oil-in-Water Adjuvants" by D.T. O'Hagan et al.) affect the antibody response to the influenza vaccine.

2 Of Pigs and Humans: How to Apply This Learning?

When the scientific community, the public health authorities, the regulatory agencies, and all national and international bodies were actively working on the preparedness plans to counteract the risks of an influenza pandemic caused by avian viruses, and when the discussions on the opportunity and feasibility of prepandemic vaccination with vaccines based on avian virus strains were at their peak, suddenly the alert of cases of influenza infections caused by an A/H1N1 influenza virus of swine origin in Northern America was given in April 2009. Soon the virus started to spread and in a couple of months affected all continents, until the WHO declared the pandemic. In several ways, the event of a pandemic caused by an A/H1N1 virus was unexpected. We were expecting a pandemic due to a non-H1/non-H3 virus. Most of the people were actively working on the development and stockpiling of H5N1-based vaccines because of the very high number of cases in wild and domestic birds in Asia, Europe, and Africa, and in the humans in Asia and Africa, which is an exceptionally high lethality rate. Some people were still pledging to prepare vaccines against other avian viruses, such as H9N2 and H7N7. Second, most of the preparedness focused on virus strains of avian origin, and none at all on strains from pigs. Finally, the entire community was watching at the appearance and evolution of novel influenza strains from Far-East Asia with a westward propagation, while the pandemic originated from the West and exhibited an eastward propagation.

As the virus strain causing the pandemic was an A/H1N1, which has coexisted with humans since the Spanish flu pandemic of 1918, the question was immediately asked as to whether there were similarities between this novel virus that had popped out from North American pigs and the A/H1N1 virus that composes the trivalent seasonal vaccines and whether the seasonal vaccine would have been able to induce antibodies to cross-react with the novel virus. The genetic analysis of the new, pandemic A/H1N1 virus and the prediction of the structure of its HA, inferred by the amino acid sequence, are clearly against this possibility (see the chapter "The Origin and Evolution of H1N1 Pandemic Influenza Viruses" by R.G. Webster et al.). In addition, the very first serological studies confirmed later by comprehensive studies using serum samples from subjects of all ages immunized with seasonal influenza vaccines showed that neutralizing antibodies induced by the seasonal inactivated influenza vaccine poorly recognized the novel A/H1N1 virus, suggesting that novel B-cell epitopes were expressed by this virus. More specifically, such cross-reactive antibodies were undetectable in children below the age of 9, while they were detectable in 12–22% of adults between 18 and 64 years of age and in 5% of older adults. Interestingly, a proportion of older adults had cross-reactive antibodies which preexisted the vaccination with the seasonal vaccines [57]. These findings are in agreement with the epidemiological observation that people older than 65 years are less susceptible to the novel A/H1N1 virus than younger people [58].

All these data suggest that a vaccine against the novel A/H1N1 virus was necessary since most of the people were clearly immunologically naïve (at least based on their neutralizing antibodies). On the other side, the data in the older subjects suggested that some immunological memory could exist between the novel and the seasonal A/H1N1 viruses. The questions then arose as to whether the lessons learned toward preparedness for a pandemic due to avian influenza viruses, such as H5N1 viruses, could be applied to the development and the use of vaccines against the novel A/H1N1 virus. The need for adjuvants and the induction and persistence of immunological memory will be discussed in the next sections.

2.1 Adjuvants and A/H1N1 Vaccines?

When the genetic data of the novel A/H1N1 virus became available and when the first data on the poor cross-reactivity of antibodies between seasonal and pandemic viruses were reported, it was immediately considered that the vaccine against this new virus had to share some key characteristics of the vaccines already developed against the avian H5N1 viruses. For example, because the vaccine was expected to be given to immunologically naïve individuals who had never seen this virus earlier, the vaccine had to contain a strong adjuvant, an oil-in-water adjuvants such as MF59 or AS03, and had to be given twice to reach sustained protective levels of antibodies that met the criteria fixed by regulatory agencies such as the FDA and the EMEA. These expectations, mainly the one related to the double doses required for effective priming, influenced the decision of national authorities on the number of doses required to cover the population included in the national plans of immunization.

It was, therefore, surprising to see the first results of the clinical trials when they were published. Indeed, in contrast to all expectations, the vaccine against the pandemic A/H1N1 virus was immunogenic (i.e., met the regulatory criteria for licensure) even in the absence of adjuvants when the dosage of antigen in the formulation was increased. In addition, and strikingly, one single dose was immunogenic enough to meet these criteria.

In a study carried out in Australia with a split-virion A/H1N1 influenza vaccine (A/California/7/2009) from CSL, 240 subjects aged between 18 and 64 years received twice either 15 or 30 µg of vaccine, 21 days apart. Three weeks after the first dose, 95% and 89% of subjects who had received 15 or 30 µg of vaccine, respectively, had HI antibody titers above 1:40. These percentages became 98% and 96%, respectively, after the second dose. The second dose of vaccine only slightly increased the geometric antibody titers already achieved after the first immunization [59]. It is interesting to note that the initiation of this study (last week of July 2009) coincided with the first pandemic wave in Australia, and one volunteer tested positive for the novel A/H1N1 infection during the 21 days after the first vaccination. In addition, the authors of this clinical study report that 45% of the subjects had received the 2009 seasonal influenza vaccine before being enrolled in the pandemic vaccine study [59]. It would then be important to understand the potential contribution of natural (subclinical) infection with the pandemic A/H1N1 virus and/or of the prior seasonal influenza infections and/or vaccination with the seasonal influenza vaccine in the priming of an immunological memory that would have been then boosted by the pandemic vaccination. Indeed, almost 27% of the subjects participating in this study had HI antibodies above the level of 1:40 at baseline [59].

The question still remains open even after a second study with the same split-virion nonadjuvanted vaccine from CSL. This study was carried out in Australia with the same dosages and the same dose regimen in 370 healthy infants and children 6 months to less than 8 years of age [60]. Again, after one single dose of nonadjuvanted split vaccine, 92.2% and 97.7% of children receiving 15 or 30 µg of vaccine, respectively, had HI antibody titers exceeding 1:40. The geometric antibody titers post-first dose ranged between 113 in those below 3 years with the lowest dose and 268 in those above 3 years with the highest dose. Unlike the previous study in adults [59], in the study in children the second dose significantly increased the levels of serum HI antibody titers [60]. This study was carried out (August 2009) in areas in Australia where the notification of the novel A/H1N1 influenza infection had started to decline. In addition, 40% of the infants and children enrolled had been previously vaccinated with the 2009 seasonal influenza vaccine. Finally, even before vaccination with the pandemic vaccine, a high proportion (9.2% to 33.3%) of the infants and children had levels of HI antibodies to the A/H1N1 virus in the ratio of 1:40.

The data from these two studies suggest that this H1N1 vaccine is particularly immunogenic at all ages, including in young children, and more immunogenic than the avian H5N1 vaccines, the seasonal H1N1 vaccines, and the swine H1N1 vaccines developed during the 1970s and used only in the USA. For the H5N1 vaccine, two doses were required to obtain a sustained "protective" antibody response at all ages. For the seasonal H1N1 vaccine, two doses are necessary to induce good priming in young children. For the swine H1N1 influenza vaccine of the 1970s, one dose was enough for adults, but two doses were required for children below the age of 9 [61]. The difference between these three vaccines and the pandemic one tested in Australia is that the H5N1 virus never circulated in areas where the vaccines were tested and there is no H5N1 vaccination ongoing. Similarly, the swine H1N1 virus of the 1970s did not circulate outside New Jersey. Furthermore, the H1N1 virus that appeared in 1918 disappeared in 1957. This means that the <24-year-old subjects who required two doses of vaccine had never been exposed to the H1N1 virus, which would reappear in 1997, after the study with the A/New Jersey H1N1 vaccine [61]. In addition, during the 1970s, seasonal influenza vaccination was not recommended in children and was poorly implemented even in adults. The seasonal H1N1 virus tends to circulate less than the H3N2 virus, depending on the seasons. Instead, the novel A/H1N1 virus was amply circulating during the period of study. To conclude, one cannot rule out that subclinical infection with the virus had happened and that this may have contributed to specific immunological priming that would have then been boosted by the vaccination.

This hypothesis has now been substantiated by a cross-sectional study carried out in the UK, which shows a high prevalence of anti-novel H1N1 antibodies during the first wave of infection [62] Unfortunately, in these clinical studies with the vaccine against the novel A/H1N1 virus, serum samples were not taken earlier than 21 days post-first dose to evaluate the kinetics of the antibody response, as performed, for example, in studies aimed at investigating the immunological memory induced by vaccinations with H5-based vaccines several years earlier [26, 27].

In a study carried out from July to August 2009 in China, 2,200 subjects received 7.5, 15, or 30 μg of a split-virion A/California/7/2009 H1N1 vaccine produced by Hualan Biological Bacterin Company and formulated with or without alum as an adjuvant [63]. Again, a single 15-μg administration without adjuvant was sufficient to induce HI antibody titers above 1:40 in 74.5% of subjects between 3 and 11 years of age, in 97% of those between 12 and 60 years of age, and in 79% of those 61 years of age or older. The GMTs were lower in the youngest group (3–11 years) compared with the older groups. As expected by the previous experience with H5N1 vaccines, the addition of alum did not influence the antibody response. It is interesting that, like in the Australian studies, a second dose of vaccine did not affect the HI antibody titers in the subjects 12 years of age or older, while it significantly enhanced the response in the younger group (3–11 years). It should be noted that the frequency of subjects with antibody titers higher than 1:40 before immunization was much lower than that found in the Australian trial, ranging between 1% and 6% [63]. Similar results were obtained in a much larger (>12,000 subjects) multicenter, double-blind, randomized, placebo-controlled study carried out from August to September 2009 in China, using the same formulations with or without alum, plus two whole-virion formulations containing 5 or 10 μg of HA plus alum [64]. Essentially, this larger trial reported immunogenicity data very similar to those of the first, smaller study in terms of seroprotection rates at baseline by age groups, seroprotection rates after the first and the second dose, and as GMT in the younger compared with the older groups after the first and the second dose. Interestingly, in this study, the addition of alum to the vaccine formulations clearly suppressed the antibody response in comparison with the same nonadjuvanted formulation. There is no evidence of an ongoing wave of pandemic at the time when these two studies were carried out. However, there was no information on the status of previous immunizations with seasonal vaccines or on the status of previous influenza infections, mainly in consideration of the high rates of asymptomatic infections.

Results not different from these reported from China were also obtained with a single dose of 6 μg of HA of a split-virion vaccine produced in Hungary and adjuvanted with aluminum phosphate, and given in August 2009 to 203 adults and 152 elderly individuals. The immunogenicity of this vaccine was not affected when it was given at the same time with a trivalent inactivated seasonal vaccine [65].

The effect of previous vaccination on the immune response to the pandemic vaccine before vaccination has been very well demonstrated in >18-year-old subjects who received a single dose of a split-virion vaccine from Sanofi-Pasteur in two randomized, placebo-controlled studies carried out in the USA during the first half of August 2009 (>800 adults/elderly). This effect was less, or not at all, evident in children below the age of 9 (>400 children) [66]. In these studies, seroprotection rates (HI titers above 1:40) were consistently higher than 90% in adults and elderly individuals who received 7.5, 15, or 30 μg of HA. However, these frequencies went down to 69% and 75% in children between 3 and 9 years and to 45% and 50% in 6–35-month-old children immunized with 7.5 or 15 μg of HA, respectively [66]. These studies strongly suggest that previous priming with seasonal vaccine may improve the immune responsiveness to subsequent vaccination with the pandemic A/H1N1 vaccine. Indeed, the antibody response was much lower in young children expressed both as seroprotection/seroconversion and as GMT. It is very likely that a second dose of vaccine would have significantly increased the immune response to vaccination. Unfortunately, the results of the second immunization were not reported.

Using a cell culture-derived subunit A/H1N1 vaccine from Novartis, it was possible to show in a study carried out in adults in the UK at the end of July that one dose of vaccine was sufficient to induce seroprotection in 72% and 52% by HI and in 76% and 67% by MN in subjects receiving 3.75 or 7.5 μg of HA without adjuvant, respectively. These percentages increased to >90% by HI and to 100% by MN in the subjects immunized with the same dosages of vaccine in the presence of the oil-in-water adjuvant MF59 [67]. An important finding of this study was that these antibody titers and seroprotection rates were reached just 2 weeks after the vaccination. As expected, a second dose of the vaccine increased the immunogenicity parameters. In a randomized study carried out in Costa Rica in 3–17-year-old children, both unadjuvanted (15 and 30 μg) and MF59-adjuvanted (7.5 μg) egg-derived A/H1N1 vaccines from Novartis met the criteria for immunogenicity. The vaccine with low antigen and adjuvant was clearly more immunogenic after one single dose than the higher dosages without adjuvant, in the younger age group (3–8 years of age) [68]. Seroprotection rates by HI higher than 98% were also reported after one single dose in 18–60-year-old adults vaccinated in Germany with a split-virion vaccine from GSK given without adjuvant or adjuvanted with the AS03, squalene-based adjuvant [69].

2.2   Immunological Memory: Priming by Previous Influenza Infection/Vaccination

As mentioned above, the results of these trials are surprising. On the basis of the poor antigenic similarities between seasonal and pandemic A/H1N1 viruses and the poor cross-reactivity between the two viruses, it was expected that more than one priming dose would have to be administered, mainly in young children, and that strong adjuvants, such as MF59 or other oil-in-water emulsions, would be needed.

One hypothesis that could explain these findings is that a certain level of cross-priming takes place through natural infections (clinically overt or asymptomatic) or through vaccination with trivalent seasonal vaccines that contain the A/H1N1 virus component. These hypotheses are clearly motivated not only by the results of the clinical studies in the USA with the Sanofi-Pasteur vaccine [66] but also by the Australian trials carried out during the eve of the A/H1N1 pandemic in a population that had largely received the seasonal influenza vaccines [59, 60].

This hypothesis has been now formally proven in ferrets. Animals immunized with two doses, 1 month apart, of seasonal trivalent inactivated vaccine with or without MF59 did not mount any detectable antibody response against the novel A/H1N1 virus, either by HI or by MN. HI and MN antibodies became detectable in the ferrets that had received the seasonal influenza vaccine first followed by the nonadjuvanted A/H1N1 vaccine 1 month later. Intermediate antibody titers were achieved with one single dose of MF59-adjuvanted A/H1N1 vaccine. However, the strongest HI and MN antibody response was detected in those ferrets first primed with the seasonal vaccines (better if adjuvanted with MF59) followed by the A/H1N1 vaccine adjuvanted with MF59 [70]. A striking finding of this study was that this antibody response was mirrored by the decrease in the A/H1N1 viral load in the upper and lower respiratory tract. Indeed, if two doses of the seasonal vaccine were totally unable to affect the viral load in the lungs and in the throats of the ferrets, previous priming with seasonal vaccine followed by the nonadjuvanted A/H1N1 or a single immunization with the adjuvanted vaccine in unprimed animals significantly reduced the viral load in the lungs. However, previous priming with the MF59-adjuvanted seasonal vaccine followed by vaccination with the MF59-adjuvanted A/H1N1 vaccine totally prevented the viral colonization not only in the lower, but also in the upper respiratory tracts [70].

A few conclusions can be drawn from this study. First, a previous priming via vaccination (or very likely via previous clinically overt or asymptomatic influenza infection) significantly enhances the immunogenicity and the efficacy of the A/H1N1 vaccine. Second, this priming is not necessarily evident through the detection of cross-reacting antibodies. It is likely that this priming takes place through cross-reactive CD4+ T cells primed by the seasonal vaccination that provide help to B cells to produce antibodies to the A/H1N1 virus after boosting with this vaccine. It is known that seasonal and novel A/H1N1 viruses share several CD8+ T cell epitopes [71]. The same can easily be the case for CD4+ epitopes. Another, not mutually exclusive, hypothesis is that low-affinity, cross-protective memory B cells or high-affinity, but rare, memory B cells primed by seasonal vaccination are further expanded by the adjuvanted 2009 A/H1N1 vaccine. The known effect of MF59 in inducing CD4+ T cells and memory B cells can be in favor of these hypotheses [27, 52]. These hypotheses, however, are difficult to address in ferrets but could be approached in well-designed clinical trials, ideally in populations who are immunologically naïve to influenza, such as young children. Finally, the best immunogenicity and efficacy of the A/H1N1 vaccine (prevention of viral infection both in the lung and in the upper respiratory tract) is observed when all vaccines are given in the presence of MF59. This finding suggests that if nonadjuvanted H1N1 vaccines are immunogenic enough to meet all the criteria required for licensure of the vaccines, the use of adjuvants, and of MF59, in particular, can dramatically affect the quality of the immune response, thereby improving the efficacy of the vaccine.

2.3 Shaping of the Repertoire of the Influenza B-Cell Epitopes by Adjuvants

The progress in the understanding of the mechanisms of action of certain families of adjuvants has tremendously boosted the research in a field which, until very recently, has remained very empirical and mostly confined to the mere observation of in vitro and in vivo effects. The discovery that several adjuvant families exert their action through binding to toll-like receptors and that the most utilized adjuvants, the aluminum salts, work via the inflammasome using pathways involving IL-1β [72] has paved the way for further development of novel adjuvants, with the ultimate goal of evoking the most appropriate immune response depending on the targeted vaccine.

As a matter of fact, not all adjuvants exert their action in the same manner. For example, there are adjuvants that neither interact with toll-like receptors nor follow the inflammasome pathway. One of these adjuvants is the oil-in-water MF59, which exerts its immunopotentiating effects at local (muscle) level and then at the level of draining lymph nodes, without interacting with toll-like receptors or with inflammasome [73]. We have mentioned several times earlier the effects of the adjuvants on the enhancement of the immune response to seasonal and pandemic (avian and swine) influenza vaccines. One question that still remains unanswered is through which mechanisms MF59 broadens the immune response when it favors the production of antibodies that are able to neutralize not only the homologous virus strain present in the vaccine, but also a large panel of virus strains that underwent antigenic drift in their HA, sometimes over a large period of time [26, 27, 48]. One simplistic hypothesis would be that this is due to the larger amount of antibodies induced by MF59, which would now be able to cross-neutralize drifted virus strains. Another hypothesis is that MF59 affects the quality of the immune response by inducing antibodies against epitopes in the vaccine antigens that otherwise would have not been recognized if the vaccine was without adjuvants or with other adjuvants.

To answer this question and to elucidate if and how MF59 affected the antibody repertoire against influenza antigens, serum samples from subjects vaccinated with plain, with aluminum hydroxide-adjuvanted, or with MF59-adjuvanted H5N1 vaccines were analyzed by whole-genome fragment phage display libraries (GFPDL) followed by surface plasmon resonance technologies. The results obtained were striking [74]. While sera from subjects vaccinated with nonadjuvanted or with aluminum-adjuvanted vaccines mostly recognized fragments of the HA2 region, the oil-in-water adjuvant MF59 induced epitope-spreading from HA2 to HA1 and allowed the appearance of antibodies to neuraminidase. Moreover, a nearly 20-fold increase in the frequency of HA1/HA2 specific phage clones was observed in sera after MF59-adjuvanted vaccine administration when compared with responses after the administration of unadjuvanted or alum-adjuvanted H5N1 vaccines. Additionally, MF59-adjuvanted vaccines induced a two- to threefold increase in the frequency of antibodies reactive with properly folded HA1 (28-319), a fragment that absorbed most neutralizing activity in immune sera [74]. It is important to note that this fragment was recognized by cross-reacting neutralizing monoclonal antibodies and by sera from immune subjects who had recovered from a natural infection with the H5N1 virus [75]. The adjuvant-dependent increased binding to conformational HA1 epitopes correlated with broadening of cross clade neutralization and predicted improved in vivo protection. Finally, antibodies against potentially protective epitopes in the C-terminal of neuraminidase, close to the sialic acid binding enzymatic site, were also induced primarily following vaccination with MF59-adjuvanted vaccine, but not with plain nor with alum-adjuvanted vaccines [74].

These data clearly show that MF59 profoundly shapes the repertoire of the B-cell epitopes recognized by protective antibodies that are not only directed against HA but also against the NA. Remarkably, this is not an effect merely linked to the quantity of antibodies induced and would not be detected by the conventional serological assays used to evaluate the immunogenicity of influenza vaccine and, ultimately, to license them. As a direct consequence of these findings, it is very likely that the same principle applies to all influenza vaccines, including the vaccine against the novel pandemic A/H1N1 virus. These analyses are now in progress with a special focus on the priming of the B-cell repertoire (for example in young children) as compared with the boosting of this repertoire (for example at older ages).

3   Conclusions: Rethinking Influenza

The threat of avian influenza and the reality of the influenza pandemic due to a virus of swine origin have had a tremendous impact on the field of influenza in general. It has boosted a striking technological progress. The reverse genetics has been developed which has allowed the preparation of virus seeds suitable for the preparation of vaccines [76]. Vaccines have been produced and licensed using in vitro cell cultures instead of the conventional embryonated eggs [77]. The use of pseudoparticles has permitted a rapid and safe evaluation of neutralizing and cross-neutralizing antibodies against wide panels of virus strains [38]. The role of adjuvants in the preparation of stronger influenza vaccines is being better understood and has pushed various vaccine manufacturers to develop their own adjuvants for influenza vaccines after the original introduction of MF59 in the influenza vaccine arena in 1997 (this chapter and chapter "Adjuvants for Influenza Vaccines: the Role of Oil-in-Water Adjuvants" by D.T. O'Hagan et al.).

This influenza pandemic is teaching us a lot on the gaps and the needs that still remain in the field of influenza vaccine development and in the field of influenza vaccination. These needs will force us to completely rethink influenza as a whole, from the understanding of the virus biology and evolution (could we predict the appearance of an A/H1N1 pandemic virus? From pigs? From North America?) to the vaccine preparation (more attention to novel delivery systems, to internal conserved proteins, etc.), from the methodologies to appropriately analyze the protective immune response evoked by the different vaccines in different age groups (more emphasis on cell-mediated immunity, on the priming of the immune response in younger ages, on the persistence of memory, and on counteracting the waning of the immune responsiveness in the elderly) to a more precise understanding of the epidemiology in developing countries (in the tropics, influenza does not exhibit the seasonal peaks of transmission as in temperate climates), from the present use of the influenza vaccines, which is oriented toward the elderly, to a broader, universal use of these vaccines [78].

A last word should be added concerning the safety of influenza vaccines and of adjuvanted influenza vaccines in particular. Thanks to the need to implement the pandemic vaccination in a large proportion of the world, important clinical research has been undertaken, with the intrinsic risk of observing a high rate of coincidental side effects, to quantify the baseline risk of acquiring a large panel of diseases (chronic, neurological, autoimmune, etc.) in various populations [79–81]. The information available so far on the use of the pandemic A/H1N1 vaccines in several million individuals worldwide strongly speaks in favor of the safety of these vaccines. This very good safety also applies to vaccines adjuvanted with MF59 or with AS03, which represent the vaccines mostly utilized in Europe. This information is particularly important due to the particular risks caused by the pandemic A/H1N1 infection in some populations such as children [82] and pregnant women [83]. The safety of these adjuvants, for example, MF59, has been shown in these groups of people as well [84, 85]. More data are being reported from the experience in some countries such as the UK [86], and further data will become available in the next months. It is hoped that through this experience and learning, vaccine adjuvants will become more and more useful in the development of other novel vaccines.

References

1. Deng G, Li Z, Tian G, Li Y, Jiao P, Zhang L, Liu Z, Webster RG, Yu K (2004) The evolution of H5N1 influenza viruses in ducks in southern China. Proc Natl Acad Sci USA 101:10452–10457
2. World Health Organization (2010) http://www.who.int/csr/disease/avian_influenza/ Timeline_10_01_04.pdf. Accessed 12 Jan 2010
3. Gillim-Ross L, Subbarao K (2006) Emerging respiratory viruses: challenges and vaccine strategies. Clin Microbiol Rev 19:614–636
4. Washington D, Basser RL (2009) Response to a monovalent 2009 influenza A (H1N1) vaccine. N Engl J Med 361:2405–2413
5. Rappuoli R, Del Giudice G (2008) Waiting for a pandemic. In: Rappouli R, Del Giudice G (eds) Influenza vaccines for the future. Birkhaeuser, Basel, pp 261–279
6. Leroux-Roels I, Leroux-Roels G (2009) Current status and progress of prepandemic and pandemic influenza vaccine development. Expert Rev Vaccines 8:401–415
7. Keitel WA, Atmar RL (2009) Vaccines for pandemic influenza: summary of recent clinical trials. Curr Top Microbiol Immunol 333:431–451
8. Davenport FM (1968) Antigenic enhancement of ether-extracted influenza virus vaccines by AlPO$_4$. Proc Soc Exp Biol Med 127:587–590
9. Davenport DM, Hennessy AV, Askin FB (1968) Lack of adjuvant effect of AlPO$_4$ on purified influenza virus hemagglutining in man. J Immunol 100:1139–1140

10. Werner J, Kuwert EK, Stegmaier R, Simbock H (1980) Local and systemic antibody response after vaccination with 3 different types of vaccines against influenza. II: Neuraminidase inhibiting antibodies. Zentralbl Bakteriol A 246:1–9

11. D'Errico MM, Grasso GM, Romano F, Montanaro D (1988) Comparison of anti-influenza vaccines: whole adsorbed trivalent, trivalent subunit and tetravalent subunit. Boll Ist Sieroter Milan 67:283–289

12. Ionita E, Lupulescu E, Alexandrescu V, Matepiuc M, Constantinescu C, Cretescu L, Velea L (1989) Comparative study of the immunogenicity of aqueous versus aluminium phosphate adsorbed split influenza vaccine C.I. Arch Roum Pathol Exp Microbiol 48:265–273

13. Hennessy AV, Davenport FM (1961) Relative merits of aqueous and adjuvant influenza vaccines when used in a two-dose schedule. Public Health Rep 76:411–419

14. Salk JE, Bailey ML, Laurel AM (1952) The use of adjuvants in studies on influenza immunization. II. Increased antibody formation in human subjects inoculated with influenza virus vaccine in a water-in-oil emulsion. Am J Hyg 55:439–456

15. Salk JE (1953) Use of adjuvants in studies on influenza immunization. III. Degree of persistence of antibody in subjects two years after vaccination. JAMA 151:1169–1175

16. Davis DJ, Philip RN, Bell JA, Voegel JE, Jensen DV (1961) Epidemiological studies on influenza in familiar and general population groups 1951–1956. III. Laboratory observations. Am J Hyg 73:138–147

17. Davenport FM, Hennessy AV, Bell JA (1962) Immunologic advantages of emulsified influenza virus vaccines. Mil Med 127:95–100

18. Beebe GW, Simon AH, Vivona S (1972) Long-term mortality follow-up of Army recruits who received adjuvant influenza virus vaccine in 1951–1953. Am J Epidemiol 95:337–346

19. Van Nest GA, Steimer KS, Haigwood NL, Burke RL, Ott G (1992) Advanced adjuvant formulations for use with recombinant subunit vaccines. In: Brown F, Chanock RM, Greenberg HS, Lerner RA (eds) Vaccines 92. Cold Spring Harbor Laboratory Press, Cold Spring Habor, pp 57–62

20. Podda A, Del Giudice G, O'Hagan DT (2005) MF59: a safe and potent adjuvant for human use. In: Schijns V, O'Hagan DT (eds) Immunopotentiators in modern medicines, chapter 9. Elsevier Press, Amsterdam, p 149

21. Nicholson KG, Colegate AE, Podda A, Stephenson I, Wood J, Ypma E, Zambon MC (2001) Safety and antigenicity of non-adjuvated and MF59-adjuvanted influenza A/Duck/Singapore/97 (H5N3) vaccine: a randomized trial of two potential vaccines against H5N1 influenza. Lancet 357:1937–1943

22. Treanor JJ, Campbell JD, Zangwill KM, Rowe T, Wolff M (2006) Safety and immunogenicity of an inactivated subvirion influenza A (H5N1) vaccine. N Engl J Med 354:1343–1351

23. Atmar RL, Keitel WA, Patel SM, Katz JM, She D, El Sahly H, Pompey J, Cate TR, Couch RB (2006) Safety and immunogenicity of nonadjuvanted and MF59-adjuvanted influenza A/H9N2 vaccine preparations. Clin Infect Dis 43:1135–1142

24. Stephenson I, Bugarini R, Nicholson KG, Podda A, Wood JM, Zambon MG, Katz JM (2005) Cross-reactivity to highly pathogenic avian influenza H5N1 viruses after vaccination with nonadjuvanted and MF59-adjuvanted influenza A/Duck/Singapore/97 (H5N3) vaccine: a potential priming strategy. J Infect Dis 191:1210–1215

25. Stephenson I, Nicholson KG, Colegate A, Podda A, Wood J, Ypma E, Zambon M (2003) Boosting immunity to influenza H5N1 with MF59-adjuvanted H5N3 A/Duck/Singapore/97 vaccine in a primed human population. Vaccine 21:1687–1693

26. Stephenson I, Nicholson KG, Hoschler K, Zambon MC, Hancock K, DeVos J, Katz JM, Praus M, Banzhoff A (2008) Antigenically distinct MF59-adjuvanted vaccine to boost immunity to H5N1. N Engl J Med 359:1631–1633

27. Galli G, Hancock K, Hoschler K, DeVos J, Praus M, Bardelli M, Malzone C, Castellino F, Gentile C, McNally T, Del Giudice G, Banzhoff A, Brauer V, Montomoli E, Zambon M, Katz J, Nicholson K, Stephenson I (2009) Fast rise of broadly cross-reactive antibodies after

boosting long-lived human memory B cells primed by an MF59 adjuvanted prepandemic vaccine. Proc Natl Acad Sci USA 106:7962–7967

28. Bernstein DI, Edwards KM, Dekker CL, Belshe R, Talbot HK, Graham IL, Noah DL, He F, Hill H (2008) Effects of adjuvants on the safety and immunogenicity of an avian influenza H5N1 vaccine in adults. J Infect Dis 197:667–675

29. Bresson JL, Perronne C, Launay O, Gerdil C, Saville M, Wood J, Noeschler K, Zambon MC (2006) Safety and immunogenicity of an inactivated split-virion influenza A/Vietnam/1194/2004 (H5N1) vaccine: phase I randomized trial. Lancet 367:1657–1664

30. Vogel FR, Caillet C, Kusters IC, Haensler J (2009) Emulsion-based adjuvants for influenza vaccines. Expert Rev Vaccines 8:483–492

31. Leroux-Roels G (2009) Prepandemic H5N1 influenza vaccine adjuvanted with AS03: a review of the pre-clinical and clinical data. Expert Opin Biol Ther 9:1057–1071

32. Levie K, Leroux-Roels I, Hoppenbrouwers K, Kervyn AD, Vandermeulen C, Forgus S, Leroux-Roels G, Pichon S, Kusters I (2008) An adjuvanted, low-dose, pandemic influenza A (H5N1) vaccine candidate is safe, immunogenic, and induces cross-reactive immune responses in healthy adults. J Infect Dis 198:642–649

33. Leroux-Roels I, Borkowski A, Vanwolleghem T, Dramé M, Clement F, Hons E, Devaster JM, Leroux-Roels G (2007) Antigen sparing and cross-reactive immunity with an adjuvanted rH5N1 prototype pandemic influenza vaccine: a randomized controlled trial. Lancet 370:580–589

34. Collin N, de Radiguès X, World Health Organization H1N1 Vaccine Task Force (2009) Vaccine production capacity for seasonal and pandemic (H1N1) 2009 influenza. Vaccine 27:5184–5186

35. World Health Organization (2010) Pandemic (H1N1)2009 vaccine deployment update – 23 December 2009. http://www.who.int/csr/disease/swineflu/vaccines/ h1n1_vaccination_deployment_update_20091223.pdf. Accessed 13 Jan 2010

36. Leroux-Roels I, Bernhard R, Gérard P, Dramé M, Hanon E, Leroux-Roels G (2008) Broad clade 2 cross-reactive immunity induced by an adjuvanted clade 1 rH5N1 pandemic influenza vaccine. PLoS ONE 3:e1665

37. Banzhoff A, Gasparini R, Laghi-Pasini F, Staniscia T, Durando P, Montomoli E, Capecchi P, Di Giovanni P, Sticchi L, Gentile C, Hilbert A, Brauer V, Tilman S, Podda A (2009) MF59®-adjuvanted H5N1 vaccine induces immunologica memory and heterotypic antibody responses in non-elderly and elderly adults. PLoS ONE 6:e4364

38. Alberini I, Del Tordello E, Fasolo A, Temperton NJ, Galli G, Gentile C, Montomoli E, Hilbert AK, Banzhoff A, Del Giudice G, Donnelly JJ, Rappuoli R, Capecchi B (2009) Pseudoparticle neutralization is a reliable assay to measure immunity and cross-reactivity to H5N1 influenza viruses. Vaccine 27:5998–6003

39. Chu DW, Hwang SJ, Lim FS, Oh HM, Thongcharoen P, Yang PC, Bock HL, Dramé M, Gillard P, Hutagalung Y, Tang H, Teoh YL, Ballou RW, H5N1 Flu study group for Hong Kong, Singapore, Taiwan and Thailand (2009) Immunogenicity and tolerability of an AS03-adjuvanted prepandemic influenza vaccine: a phase III studying a large population of Asian adults. Vaccine 27:7428–7435

40. Baras B, Stittelaar KJ, Simon JH, Thoolen RJ, Mossman SP, Pistoor FH, van Amerongen G, Wettendorff MA, Hanon E, Osterhaus AD (2008) Cross-protection against lethal H5N1 challenge in ferrets with an adjuvanted pandemic influenza vaccine. PLoS ONE 3:e1401

41. Forrest HL, Khalenkov AM, Govorkova EA, Kim JK, Del Giudice G, Webster RG (2009) Single- and multiple-clade influenza A H5N1 vaccines induce cross-protection in ferrets. Vaccine 27:4187–4195

42. Ehrlich HJ, Mueller M, Oh HM, Tambyah PA, Joukhadar C, Montomoli E, Fisher D, Berezuk G, Fritsch S, Loew-Baselli A, Vartian N, Bobrovsky R, Pavlova BG, Poellabauer EM, Kistner O, Barrett PM, Baxter H5N1 pandemic influenza vaccine clinical study team (2008) A clinical trial of a whole-virus H5N1 vaccine derived from cell culture. N Engl J Med 358:2573–2584

43. Wu J, Fang HH, Chen JT, Zhou JC, Feng ZJ, Li CG, Qiu YZ, Liu Y, Lu M, Liu LY, Dong SS, Gao Q, Zhang XM, Wang N, WD Y, Dong XP (2009) Immunogenicity, safety, and cross-

reactivity of an inactivated, adjuvanted, prototype pandemic influenza (H5N1) vaccine: a phase II, double-blind, randomized trial. Clin Infect Dis 48:1087–1095

44. Carrat F, Flahault A (2007) Influenza vaccine: the challenge of antigenic drift. Vaccine 25:6852–6862

45. de Jong JC, Beyer WE, Palache AM, Rimmelzwaan GF, Osterhaus AD (2000) Mismatch between the 1997/1998 influenza vaccine and the major epidemic A(H3N2) virus strain as the cause of an inadequate vaccine-induced antibody response to this strain in the elderly. J Med Virol 61:94–99

46. Del Giudice G, Hilbert AK, Bugarini R, Minutello A, Popova O, Toneatto D, Schoendorf I, Borkowski A, Rappuoli R, Podda A (2006) An MF59-adjuvanted inactivated influenza vaccine containing A/Panama/1999 (H3N2) induced broader serological protection against heterovariant influenza virus strain A/Fujian/2002 than a subunit and a split influenza vaccine. Vaccine 24:3063–3065

47. Kojimahara N, Maeda A, Kase T, Yamaguchi N (2006) Cross-reactivity of influenza A (H3N2) hemagglutination-inhibition antibodies induced by an inactivated influenza vaccine. Vaccine 24:5966–5969

48. Ansaldi F, Bacilieri S, Durando P, Sticchi L, Valle L, Montomoli E, Icardi G, Gasparini R, Crovari P (2008) Cross-protection by MF59-adjuvanted influenza vaccine: neutralizing and hemagglutination-inhibiting antibody activity against A(H3N2) drifted influenza viruses. Vaccine 26:1525–1529

49. Legrand J, Vergu E, Flahault A (2006) Real-time monitoring of the influenza vaccine field effectiveness. Vaccine 24:6605–6611

50. Camilloni B, Neri M, Lepri E, Iorio AM (2009) Cross-reactive antibodies in middle-aged and elderly volunteers after MF59-adjuvanted subunit trivalent influenza vaccine against B viruses of the B/Victoria or B/Yamagata lineages. Vaccine 27:4099–4103

51. Vesikari T, Pellegrini M, Karvonen A, Groth N, Borkowski A, O'Hagan DT, Podda A (2009) Enhanced immunogenicity of seasonal influenza vaccines in young children using MF59 adjuvant. Pediatr Infect Dis J 28:563–571

52. Galli G, Medini D, Borgogni E, Zedda L, Bardelli M, Malzone C, Nuti S, Tavarini S, Sammicheli C, Hilbert AK, Brauer V, Banzhoff A, Rappuoli R, Del Giudice G, Castellino F (2009) Adjuvanted H5N1 vaccine induces early CD4+ T cell response that predicts long-term persistence of protective antibody levels. Proc Natl Acad Sci USA 106:3877–3882

53. Stephenson I, Zambon MC, Rudin A, Colegate A, Podda A, Bugarini R, Del Giudice G, Minutello A, Bonnington S, Holmgren J, Mills KH, Nicholson KG (2006) Phase I evaluation of intranasal trivalent inactivated influenza vaccine with nontoxigenic *Escherichia coli* enterotoxin and novel biovector a mucosal adjuvants, using adult volunteers. J Virol 80:4962–4970

54. Zangwill KM, Treanor JJ, Campbell JD, Noah DL, Ryea J (2008) Evaluation of the safety and immunogenicity of a booster (third) dose of inactivated subvirion H5N1 influenza vaccine in humans. J Infect Dis 197:580–583

55. Leroux-Roels I, Roman F, Forgus S, Maes C, De Boever F, Dramé M, Gillard P, van der Most R, Van Mechelen M, Hanon E, Leroux-Roels G (2010) Priming with AS03-adjuvanted H5N1 influenza vaccine improves the kinetics, magnitude and durability of the immune response after a heterologous booster vaccination: an open non-randomised extension of a double-blind randomized primary study. Vaccine 28:849–857

56. Ehrlich HJ, Mueller M, Fritsch S, Zeitlinger M, Berezuk G, Loew-Baselli A, van der Velden MV, Poellabauer EM, Maritsch F, Pavlova BG, Tambyah PA, Oh HM, Montomoli E, Kistner O, Noel Barrett P (2009) A cell culture (Vero)-derived H5N1 whole-virus vaccine induces cross-reactive memory responses. J Infect Dis 200:1113–1118

57. Hancock K, Veguilla V, Lu X, Zhong W, Butler EN, Sun H, Liu F, Dong L, DeVos J, Gargiullo PM, Brammer TL, Cox NJ, Tumpey TM, Katz JM (2009) Cross-reactive antibody responses to the 2009 pandemic H1N1 influenza virus. N Engl J Med 361:1945–1952

58. Health Protection Agency (2009) Weekly international summary. http://www.hpa.org.uk/web/HPAwebFile/HPAweb_C/1252326272372.Accessed 3 Sept 2009
59. Greenberg ME, Lai MH, Hartel GF, Wichems CH, Gittleson C, Bennet J, Dawson G, Hu W, Leggio C, Washington D, Basser RL (2009) Response to a monovalent 2009 influenza A (H1N1) vaccine. N Engl J Med 361:2405–2413
60. Nolan T, McVernon J, Skeljo M, Richmond P, Wadia U, Lambert S, Nissen M, Marshall H, Booy R, Heron L, Hartel G, Lai M, Basser R, Gittleson G, Greenberg M (2009) Immunogenicity of a monovalent 2009 influenza A(H1N1) vaccine in infants and children. JAMA 303 (1):37–46
61. Pandemic Working Group of the MRC (UK) Committee on Influenza and Other Respiratory Virus Vaccines (1977) Antibody response and reactogenicity of graded doses of inactivated influenza A/New Hersey/76 whole-virus vaccine in humans. J Infect Dis 136:S475–S483
62. Miller E, Hoschler K, Hardelid P, Stanford E, Andrews N, Zambon M (2010) Incidence of 2009 pandemic influenza A H1N1 infection in England: a cross-sectional serological study. Lancet 375:1100–1108. doi:10.1016/S0140-6736(09)62126-7
63. Zhu FC, Wang H, Fang HH, Yang JG, Lin XJ, Liang XF, Zhang XF, Pan HX, Meng FY, Hu YM, Liu WD, Li CG, Li W, Zhang X, Hu JM, Peng WB, Yang BP, Xi P, Wang HQ, Zheng JS (2009) A novel influenza A (H1N1) vaccine in various age groups. N Engl J Med 361:2414–2423
64. Liang XF, Wang HQ, Wang JZ, Fang HH, Wu J, Zhu FC, Li RC, Xia SL, Zhao YL, Li FJ, Yan SH, Yin WD, An K, Feng DJ, Cui XL, Qi FC, Ju CJ, Zhang YH, Guo ZJ, Chen PY, Chen Z, Yan KM, Wang Y (2009) Safety and immunogenicity of 2009 pandemic influenza A H1N1 vaccines in China: a multicentre, double-blind, randomized, placebo-controlled trial. Lancet 375:56–66
65. Vajo Z, Tamas F, Sinka L, Jankovics I (2010) Safety and immunogenicity of a 2009 pandemic influenza A H1N1 vaccine when administered alone or simulataneously with the seasonal influenza vaccine for the 2009–2010 influenza season: a multicentre, randomized controlled trial. Lancet 375:49–55
66. Plennevaux E, Sheldon E, Blatter M, Reeves-Hoché MK, Denis M (2009) Immune response after a single vaccination against 2009 influenza A H1N1 in USA: a preliminary report of two randomized controlled phase 2 trials. Lancet 375:41–48
67. Clark TW, Pareek M, Hoschler K, Dillon H, Nicholson KG, Groth N, Stephenson I (2009) Trial of 2009 influenza A (H1N1) monovalent MF59-adjuvanted vaccine. N Engl J Med 361:2424–2435
68. Arguedas A, Soley C, Lindert K (2009) Responses to 2009 H1N1 vaccine in children 3 to 17 years of age. N Engl J Med 362:370–372
69. Roman F, Vaman T, Gerlach B, Markendorf A, Gillard P, Devaster JM (2009) Immunogenicity and safety in adults of one dose of influenza A H1N1v 2009 vaccine formulated with and without AS03-adjuvant: preliminary report of an observed-blind, randomized trial. Vaccine 28:1740–1745
70. Del Giudice G, Stittelaar KJ, van Amerongen G, Simon J, Osterhaus ADME, Stohr K, Rappuoli R (2009) Seasonal vaccine provides priming against A/H1N1 influenza. Sci Transl Med 1:12re1
71. Greenbaum JA, Kotturi MF, Kim Y, Oseroff C, Vaughan K, Salimi N, Vita R, Ponomarenko J, Scheuermann RH, Sette A, Peters B (2009) Pre-existing immunity against swine-origin H1N1 influenza viruses in the general human population. Proc Natl Acad Sci USA 106:20365–20370
72. O'Hagan DT, De Gregorio E (2009) The path to a successful vaccine adjuvant – "the long and winding road". Drug Discov Today 14:541–551
73. Mosca F, Tritto E, Muzzi A, Monaci E, Bagnoli F, Iavarone C, O'Hagan D, Rappuoli R, De Gregorio E (2008) Molecular and cellular signatures of human vaccine adjuvants. Proc Natl Acad Sci USA 105:10501–10506
74. Khurana S, Chearwae W, Castellino F, Manischewitz J, King LR, Honorkiewicz A, Rock MT, Edwards KM, Del Giudice G, Rappuoli R, Golding H (2010) MF59-adjuvanted vaccines expand antibody repertoires targeting protective sites of pandemic H5N1 influenza virus. Sci Transl Med 2:15ra5

75. Khurana S, Suguitan A Jr, Rivera Y, Simmons CP, Lanzavecchia A, Sallusto F, Manischewitz J, King LR, Subbarao K, Golding H (2009) Antigenic fingerprinting of H5N1 avian influenza using convalescent sera and monoclonal antibodies reveals potential vaccine and siagnostic tools. PLoS Med 6:e1000049. doi:10.371

76. Wood JM, Robertson JS (2004) From lethal virus to life-saving vaccine: developing inactivated vaccines for pandemic influenza. Nat Rev Microbiol 2:842–847

77. Ulmer JB, Valley U, Rappuoli R (2006) Vaccine manufacturing: challenges and solutions. Nat Biotechnol 24:1377–1383

78. Rappuoli R, Del Giudice G, Nabel GJ, Osterhaus AD, Robinson R, Salisbury D, Stoehr K, Treanor JJ (2009) Rethinking influenza. Science 326:50

79. Black S, Eskola J, Siegrist CA, Halsey N, MacDonald N, Law B, Miller E, Andrews N, Stowe J, Salmon D, Vannice K, Izurieta H, Akhtar A, Gold M, Oselka G, Zuber P, Pfeifer D, Vellozi C (2009) The importance of an understanding of backgrounds rates of diseases in evaluation of vaccine safety during mass immunization with pandemic influenza vaccines. Lancet 374: 2115–2122

80. Klein NP, Ray P, Carpenter D, Hansen J, Lewis E, Fireman B, Black S, Galindo C, Schmidt J, Baxter R (2009) Rates of autoimmune diseases in Kaiser Permanente for use in vaccine adverse event safety studies. Vaccine 28:1062–1068. doi:10.1016

81. Evans D, Cauchemez S, Hayden FG (2009) "Prepandemic" immunization for novel influenza viruses, "swine" flu vaccine, Guillain-Barré syndrome, and the detection of rare severe adverse events. J Infect Dis 200:321–328

82. Libster R, Bugna J, Coviello S, Hijano DR, Dunaiewsky M, Reynoso N, Cavalieri ML, Guglielmo MC, Areso MS, Gilligan T, Santucho F, Cabral G, Gregorio GL, Moreno R, Lutz MI, Panigasi AL, Saligari L, Caballero MT, Egues Almeida RM, Gutierrez Meyer ME, Neder MD, Davenport MC, Del Valle MP, Santidrian VS, Mosca G, Garcia Dominguez M, Alvarez L, Panda P, Pota A, Bolonati N, Dalamon R, Sanchez Mercol VI, Espinoza M, Peuchot JC, Karolinski A, Bruno M, Borsa A, Ferrero F, Bonina A, Ramonet M, Albano LC, Luedicke N, Alterman E, Savy V, Baumeister E, Chappell JD, Edwards KM, Melendi GA, Polack FP (2009) Pediatric hospitalizations associated with 2009 pandemic influenza A (H1N1) in Argentina. N Engl J Med 362:45–55. doi:10.1056/NEJMoa0907673

83. Louie JK, Acosta M, Janieson DJ, Honein MA, California pandemic (H1N1) working group (2010) Severe 2009 H1N1 influenza in pregnant and postpartum women in California. N Engl J Med 362:27–35. doi:10.1056/NEJMoa0910444

84. Pellegrini M, Nicolay U, Lindert K, Groth N, Della Cioppa G (2009) MF59-adjuvanted versus non-adjuvanted influenza vaccines: integrated analysis from a large safety database. Vaccine 27:6959–6965

85. Tsai T, Kyaw MH, Novicki D, Nacci P, Rai S, Clemens R (2009) Exposure to MF59-adjuvanted influenza vaccines during pregnancy – a retrospective analysis. Vaccine 28:1877–1880. doi:10.1016/j.vaccine.2009.11.077

86. Waddington CS, Walker WT, Oeser C, Reiner A, John T, Wilkins S, Casey M, Eccleston PE, Allen RJ, Okike I, Ladhani S, Sheasby E, Hoschler K, Andrews N, Waight P, Collinson AC, Heath PT, Finn A, Faust SN, Snape MD, Miller E, Pollard AJ (2010) Safety and immunogenicity of AS03 adjuvanted split virion versus non-adjuvanted whole virion H1N1 influenza vaccin ein UK children aged 6 months-12 years: open-label, randomised, parallel group, multicentre study. BMJ 340:c2849. doi: 10.1136/bmj.c2649

Economic Implications of Influenza and Influenza Vaccine

Julia A. Walsh and Cyrus Maher

Abstract The objective of this chapter is to review and summarize the current economic estimates of influenza and the cost-effectiveness of its vaccines. We reviewed the published assessments of the economic costs of human seasonal and pandemic influenza internationally. Seasonal influenza costs Germany, France, and the USA between $4 and $87 billion annually. Depending upon the intensity of transmission and severity of disease, pandemic influenza may cause as many as 350 million deaths and result in economic losses topping $1 trillion – an impact great enough to create a worldwide recession. We then reviewed 100 papers primarily from more than a dozen countries which studied the cost-effectiveness of influenza vaccine in children, adults, and the elderly. These studies demonstrate that influenza vaccination is quite cost-effective among children 6 months to 18 years old, in health care workers and pregnant women, and in high-risk individuals. Remarkably, compared with the other recently introduced vaccines for children, such as rotavirus and pneumococcal polysaccharide, vaccinating children and school attendees results in societal *cost savings* because it obviates lost productivity and wages among infected individuals and their caretakers. Vaccination for children is recommended in the USA and in Canada, but public health policy makers in Europe have undervalued this vaccine and not recommended it so widely.

1 Introduction

Annual seasonal outbreaks of influenza result in substantial socioeconomic costs. In addition to the health care costs incurred by sick patients needing outpatient care and hospitalization, the societal costs of lost productivity and ancillary costs are

J.A. Walsh (✉) and C. Maher
School of Public Health, University of California, Berkeley, CA 94720-7360, USA
e-mail: jwalsh@berkeley.edu

G. Del Giudice and R. Rappuoli (eds.), *Influenza Vaccines for the Future*, 2nd edition,   425
Birkhäuser Advances in Infectious Diseases,
DOI 10.1007/978-3-0346-0279-2_19, © Springer Basel AG 2011

substantial. When children become ill, parents take time off from work to care for them and take them for treatment. Ill employees are absent from work, and even when present in the workplace, their productivity can be substantially reduced, while increasing the risk of transmission to coworkers. Retirees suffer some of the most severe complications of influenza and may require more assistance from family members and informal caregivers resulting in societal costs that exceed the usual medical expenditures.

Pandemic influenza that occurs only once every few decades results in much greater socioeconomic costs than annual seasonal influenza. During pandemics, more cases of influenza occur, but more importantly because of widespread fear of infection, people stop working, shopping, going to social and cultural events and venues where large numbers gather. The impact on the economy can be enormous depending upon the severity of the pandemic.

An efficacious influenza vaccine can avert most of the disease, deaths, and socioeconomic consequences, if vaccination programs effect high coverage among those at risk. Among some high-risk populations such as children and elderly, the societal savings from annual vaccination can sometimes outweigh the costs of the vaccination program. However, in other populations such as young adults, the costs of annual vaccination to prevent seasonal disease may substantially exceed any cost savings in health care and societal costs. When the societal cost savings are high, then vaccination becomes a high priority for public health.

This chapter will review three main topics in economics of influenza and vaccination:

1. Macroeconomic estimates of the societal costs of pandemic influenza
2. Estimates of the societal costs of the annual seasonal outbreaks
3. Cost-effectiveness of annual vaccination for the prevention of seasonal outbreaks. The variation in cost-effectiveness values that result from comparisons among different target groups, methods, and locations will be reviewed

2   Methods

2.1   Societal Costs of Pandemic Influenza

Initial search at http://www.pandemicflu.gov provided four references on the economic implications of pandemic flu. Further sources were identified by performing a general internet search, as well as a query of the Web of Science and PubMed databases up to September 5, 2009, utilizing the terms: (flu OR influenza) AND pandemic AND (cost* OR macroeconomic* OR economic* OR death* OR mortality OR impact* OR effect*) AND (worldwide OR global OR "united states").

2.2   Societal Costs of Seasonal Influenza

Sources were identified by performing a general internet search, as well as a query of the Web of Science and PubMed databases up to September 5, 2009, utilizing the terms: (flu OR influenza) AND seasonal AND (cost* OR macroeconomic* OR economic* OR death* OR mortality OR impact* OR effect*) AND (worldwide OR global OR "united states").

2.3   Cost-Effectiveness: Children

We searched Web of Science and PubMed databases up to September 5, 2009, using the following Boolean search: (flu OR influenza) AND (cost* OR economic* OR pharmacoeconomic*) AND (child* OR infant* OR toddler*). The bibliographies of retrieved studies were reviewed to identify studies that may have been missed by these search criteria. This approach produced 417 potentially relevant articles.

These abstracts were assessed and only studies published within the last 15 years that contained an original, quantitative economic comparison of vaccination of children against seasonal influenza (compared with no intervention) were included. This produced a list of nine references that were then analyzed in detail. Of these, the study published by [1] was excluded for failing to include the indirect costs of influenza. Additionally, the paper published by [2] the same year was left out because the cohort study was based on a small number ($N = 303$) of vaccinated and nonvaccinated individuals with wide variation in average costs (e.g., 131.43 ± 1058).

To compare results of the papers, metrics were converted to the prevailing measure in the literature, which was dollars saved per child vaccinated. Schmier et al. [3] presented their results in dollars saved per family with school-age children, which was converted by assuming an average of 1.4 school-age children in families with children between 5 and 18 years. All foreign currencies were converted to US dollars according to the exchange rate during the year of the study (http://www.oanda.com/convert/fxhistory), and all dollar values were adjusted to 2009 dollars using general (as opposed to health care) inflation rates, given that the majority of the costs of influenza were found to be nonmedical.

2.4   Cost-Effectiveness: Elderly

We searched Web of Science and PubMed databases up to September 5, 2009, using the following Boolean search: (flu OR influenza) AND (cost* OR economic* OR pharmacoeconomic*) AND elderly. The bibliographies of retrieved studies were reviewed to identify studies that may have been missed by these search criteria. This approach produced 436 potentially relevant articles, which were filtered using criteria similar to those applied to the articles on influenza in children. This produced a list of 12 references which were then analyzed in greater detail. Of these, the studies published by [4, 5], were excluded as their analyses were based on a small number of people with no significant difference in incidence of influenza-like illness (ILI) and hospitalizations between those vaccinated and those not vaccinated.

For those studies reporting costs per year of life saved, foreign currencies were converted to dollars using exchange rates from the year of the study and then adjusted to 2009 dollars using general inflation rates as described above. For those reporting benefit-to-cost ratios, no such conversions were necessary.

3 Results

3.1 Macroeconomic Costs of Seasonal Influenza

The seasonal influenza epidemics that occur throughout the world each year wreak billions of dollars in economic damage, in addition to killing hundreds of thousands. Figure 1 illustrates the estimated societal costs (in 2009 dollars) for the USA (approximately $240/capita), France ($80/capita), Germany ($50/capita), Thailand ($0.50–$1/capita), and Australia [$5/capita (health care costs only)], the

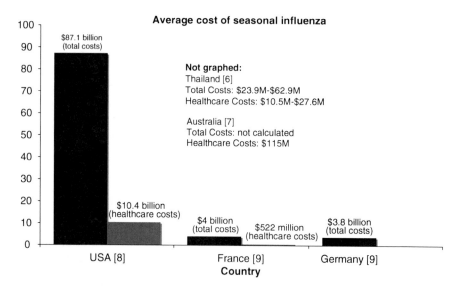

Fig. 1 National estimates of direct and indirect costs of seasonal influenza (2009 US dollars)

only countries for which such an analysis has been undertaken. Even in Thailand, a middle-income country, average costs to society total $40 million annually [6], while the toll is significantly greater in higher income countries ($4–$88 billion) [8, 9]. The differences in these per capita estimates stem from differences in health care costs in value of a day of lost productivity and absenteeism for the sick person and his/her caretaker during illness.

The majority of these costs are not from increased health care expenditures (hospitalizations, ambulatory care visits, drugs, and over-the-counter treatments, etc.) but rather from lost productivity due to illness in jobholders or their dependents. Indeed, these "indirect" costs of illness to be two to seven times greater than the "direct" costs (Fig. 1). Furthermore, because these calculations do not include reduced productivity of employees who come to work despite illness, they are an *underestimate*.

3.2 Macroeconomic Costs of Pandemic Influenza

Figure 2 presents the predicted economic costs of a pandemic (adjusted to 2009 dollars) in the USA performed by several different international agencies. Even a mild pandemic would cost the USA nearly $100 billion and take as many as two million lives (Fig. 3). Worldwide, a pandemic could result in as many as 350 million

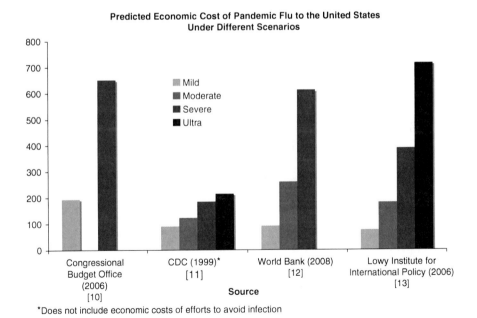

Fig. 2 Predicted economic cost of pandemic flu to the USA under different scenarios (2009 US dollars)

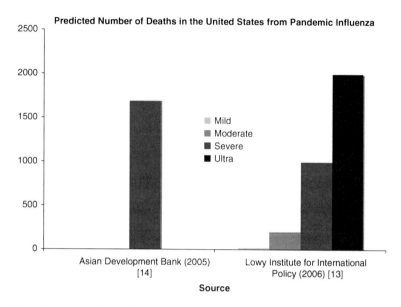

Fig. 3 Deaths from pandemic flu in the USA

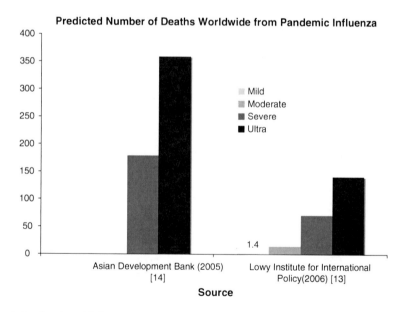

Fig. 4 Predicted worldwide deaths from pandemic flu

lives (Fig. 4), with economic costs predicted to range from several hundred billion (1,000 million) dollars for mild disease to several trillion (million million) dollars for a more severe scenario (Fig. 5). Under the last estimate, the slowdown caused by the disease would be expected to precipitate a worldwide recession [10–14].

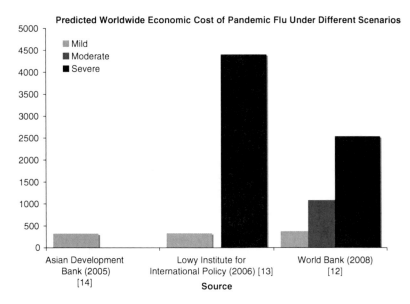

Fig. 5 Predicted worldwide economic cost of pandemic flu (2009 US dollars)

Surprisingly, Fig. 6 demonstrates that 60% of the costs of a pandemic are predicted to result not from infection, but from people's effort to *avoid* infection [12]. These estimates are based on observations of the SARS epidemic in Hong Kong and Canada, where many remained in their homes, avoiding places of work, markets, restaurants, and so on, until the fear of infection subsided.

3.3   Cost-Effectiveness of Vaccination

Given the high societal costs of seasonal influenza and the catastrophic potential of pandemic flu, strategies for mitigating the flow of this disease through populations are of crucial human and economic importance. Although there are several strategies to address the spread of influenza, the low cost and relatively good efficacy (25–85%) of vaccination mean that it is still preferred over other measures such as targeted antiretroviral therapy and, in more extreme cases, quarantine of those who are ill and their susceptible associates through closure of schools and workplaces [15]. This review focuses on vaccination against influenza in the main target groups: children, elderly, and adults.

The studies were conducted in the USA and more than a dozen other countries, and most take the societal perspective for costs and include indirect costs of lost productivity from absenteeism and time off from work.

The societal cost of a vaccination program ideally be calculated as a *net* cost using the following simplified formula:

Fig. 6 Distribution of estimated societal costs during a pandemic

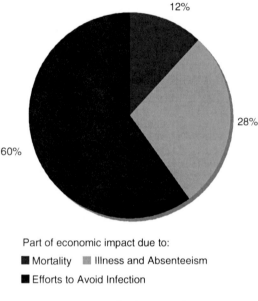

Efforts to avoid infection cause most of the costs during a pandemic

Part of economic impact due to:
■ Mortality   ▨ Illness and Absenteeism
■ Efforts to Avoid Infection

World Bank 2006 [12]

Net Cost of Vaccination Program =
Cost of the Vaccination Program + Cost of Side Effects from the Vaccine
– Cost of Healthcare Averted Resulting from Illnesses Prevented
– Cost of Lost Productivity Averted Resulting from Illnesses Prevented

When estimating cost-effectiveness, the net cost of the program is compared with the illness and disability prevented usually in terms of years of life saved (YOLS), quality adjusted life years (QALYs), or disability adjusted life years (DALYs). When the net costs are less than zero, as is the case in most of the studies of children vaccination, then the program *saves* society money.

Influenza vaccine has extremely few, rare, and usually mild side effects when they occur so that many of the studies do not include any estimates of costs of side effects. Unfortunately, many of the published cost-effectiveness assessments use different methods to estimate costs of health care and lost productivity so that the results are difficult to compare.

3.3.1 Children

Figure 7 reports estimates from nine studies for the cost-effectiveness of annual influenza vaccination in children to prevent seasonal influenza. Nearly all of the

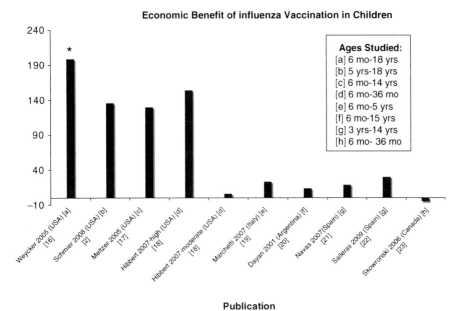

Fig. 7 Economic benefits of influenza immunization in children (cost savings)

analyses demonstrate cost savings to society from vaccination. When health care costs are considered alone, there are no net savings (i.e., the cost of vaccination exceeds the cost of averted health care expenditures in most cases) [2,16–23]. Net savings result only when societal costs of lost productivity are included. Important parameters that affect the results from each of these papers are outlined in Table 1. The inconsistencies in the assumptions and methods are substantial and explain the wide range in results.

Nonetheless, cost savings from childhood vaccination remains robust across a wide range of model input values and assumptions. These include vaccine effectiveness estimates from 25% to 85% (higher efficacy improves cost-effectiveness), attack rates from 13% to 40% (higher attack rates improve cost-effectiveness), and vaccine coverage from 20% to 100% (variable results on cost-effectiveness), and vaccination program costs of $0–$60 (higher vaccine costs decrease cost-effectiveness). The target age group for vaccination varied from day care center attendees, school children, and all children 6 months to 18 years. Some of the studies included only illness in those children in the target age group receiving the vaccine; others included secondary cases in families and general population averted resulting from childhood vaccination. In all studies, indirect benefits such as fewer days off from work for parents to care for sick children greatly exceeded the direct savings from the obviated medical expenses.

Table 1 Assumptions across studies of children

	Weycker et al. [16]	Schmier [2]	Meltzer et al. [17]	Hibbert et al. [18] (season 1)	Hibbert et al. [18] (season 2)	Marchetti et al. [19]	Dayan et al. [20]	Navas et al. [21]	Salleras et al. [22]	Skowronski et al. [23]
Age range	6 months to 18 years	5 to 18 years	6 months to 14 years	6 to 36 months	6 to 36 months	6 months to 5 years	6 months to 15 years	3 to 14 years	3 to 14 years	6 to 23 months
Study design	Population transmission model with Monte Carlo simulation	School-based trial+model	Model	Model based on two season trial	Model based on trial	Markov model	Model	Model based on cohort study	Model based on (the same) cohort study	Model
Vaccine effectiveness (%)	70 (in children), 50 (in adults)	35	Distribution, but 70–80% most likely	83.8	85.3	25–48	70	58.6	58.6	66
Vaccine cost (2009 US dollars)	Not included	51	34–68	72	39	24	25	13	25	14
% Vaccine coverage (intervention)	20 or 80	56	100	100	100	30	100	100	100	100
% Vaccine coverage (control)	5	2	0	0	0	0	0	0	0	0
% Attack rate w/out vaccine	22.2	26	30–40	13.4	32.1	16.8	25	42.9	42.9	25
Change in health care costs post-vaccination (2009 US dollars)	20%: –$101, 80%: –$40	–74	–33	53	–11	6–24 months: 6.2, 6–60 months: 16	–29	0.63	–13	NA
Change in economic costs post-vaccination (2009 US dollars)	20%: –$354, 80%: –$139	–110	–90	–59	–144	6–24 months: –11, 6–60 months: –33	–15	–19	–43	NA
Cohort size	NA	15,000	NA	1,616	1,616	NA	NA	1,951	1,951	NA
Summary estimate	$198 saved/child	$135 saved/child	$129 saved/child	$5.81 saved/child	$154 saved/child	$23 saved/child	$13 saved/child	B:C of 2.5	B:C of 3	$136,000/YOLS
Notes	Includes decline in cases in all age groups	Per household	Costed secondary cases in adults, household				All high risk			

3.3.2 Elderly

Summarized in Fig. 8 are the findings from ten studies of the cost-effectiveness of vaccinating the elderly against seasonal influenza, measured in dollars per year of life saved. Only one study (Scuffham and West, England and Wales) demonstrated cost savings [22]. Others showed that vaccination of the elderly costs $700 to $15,500 per year of life saved (YOLS) [24–31]. Most of the studies only include the direct health care costs as lost economic productivity is assumed to be minimal among the elderly.

A selection of important assumptions included in each of these investigations is presented in Table 2. These models apply varying assumptions, including vaccine efficacies from 18% to 60%, attack rates from 1% to 15%, and vaccination coverages from 37% to 100%. Despite these variations, all studies found the cost/YOLS to be less than $50,000. Interventions less than this threshold are usually considered cost-effective and of societal value for resource allocation.

3.3.3 Working Adults

A number of articles have assessed cost–benefit and cost-effectiveness of influenza vaccination in healthy adults [35–40]. Similar to the analyses of influenza vaccination in children and in the elderly, the studies have involved a variety of population

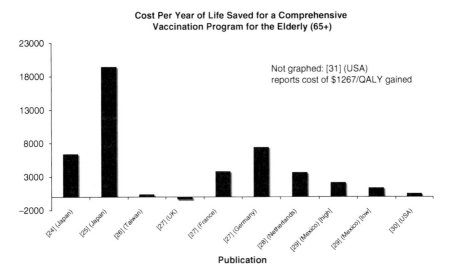

Fig. 8 Cost-effectiveness of influenza immunization in the elderly in terms of years of life saved (YOLS)

Table 2 Assumptions across studies of the elderly

	Hoshi et al. [25]	Ohkusa [32]	Cai et al. [24]	Wang et al. [26]	Fintzer [34]	Scuffham and West [27] (England)	Scuffham and West [27] (France)	Scuffham and West [27] (Germany)	Postma et al. [28]	Gasparini et al. [33]	Gutierrez and Bertozzi [29]	Maciosek et al. [31]	Nichol [30]
Study design	Model	Model based on phone survey	Model	Model based on study	Model	Model	Model	Model	Model	Model based on phone survey	Model	Model	Model based on study
Vaccine effectiveness % (infection)	58 to 29 (over 75)	NA	58	?	60	56	56	56	56	45.7	20	18.9	36
Vaccine effectiveness % (death)	30	72	6.25	29	NA	50	50	50	NA	68	30	42.9	40
Vaccine cost (2009 US dollars)	38	8.60 in subsidy	43	7.60	12.3	11	14	13	NA	18	3	20	4.50–16
% Vaccine coverage (intervention)	54	37	100	100	60	65	65	75	85% (HR), 63% (LR)	63	60–100	100	100
% Vaccine coverage (control)	7.6	30	0	0	0	0	0	0	0	0	0	57.4	0
% Attack rate w/out vaccine	1.4	NA	5	NA	11	10	10	10	10	1.06	NA	15	6 yr Average
% High risk in intervention group	32 (under 75), 49 (over 75)	NA	NA	53	NA	NA	NA	NA	35	NA	NA	NA	20.6
% Probability of death (normal)	1.6	4	4.8	?	0	0.156	0.163	0.1	0.01146	NA	0.125	2.1	NA
Cohort size	NA	NA		227,000		NA	NA	NA	1,000,000	512	4,000,000	NA	66,435
Summary estimate	$19,429/ YOLS	B:C of 22.9	$6,417/ YOLS	$386/ YOLS	B:C of 0.26	−$398/ YOLS	$3,828/ YOLS	$7,449/ YOLS	$3,700/ YOLS	B:C of 8.22	$1,379 to $2,178/ YOLS	$1,267/ QALY	$380/ YOLS
Notes		With partial subsidy											

NA = Information not available in the article

groups (adults 50–64, 18–50, 25–64 years; health care workers, pregnant women, etc.). Similarly, the studies applied a variety of methods, including simulation models, observational models, and randomized double-blind placebo-controlled trials with varied assumptions for some of the key factors.

Despite these differences, the majority of the studies demonstrate that vaccination of high-risk individuals, health care workers, and pregnant women is likely to be cost-effective.

4   Discussion

4.1   Cost-Effectiveness and Cost–Benefit Analysis: Methodological Issues

Although the data, when viewed as a whole, clearly demonstrate influenza vaccination to be quite cost-effective, there were several methodological inconsistencies noted during our review.

First, several studies estimating cost-effectiveness for country #1 applied disease incidence rates from country #2 but cost data from country #1. It is probable that this was done because incidence rates were not available for country #1. However, the authors may draw conclusions based on data that were not necessarily representative of the country for which they were doing the analysis. Several studies did not include one-way, two-way, or multi-way sensitivity analyses to test the robustness of the results to the lack of confidence in the assumptions.

Second, there were several studies that directly compared health outcomes for people who *chose* to be immunized with those who *chose not* to be immunized. It is common knowledge among epidemiological researchers that the groups of individuals who seek medical care are different in many ways from the groups of individuals who do not. Although not every study can be a randomized controlled trial, for those that are not, care should be taken to eliminate confounding by controlling for the differences between intervention and control groups that are unrelated to vaccination. In many of these studies, no such precautions were taken.

Finally, a few studies made an attempt to include the effects of herd immunity. In order to do this, authors must model the spread of infection through the population or make assumptions about transmission rates from infected to susceptibles. Influenza is one of the most transmissible human infections. When herd immunity is included, the estimates of benefits of vaccination rise.

This review includes results from studies using live-attenuated intranasal vaccine and the inactivated injectable vaccine. The estimated vaccine efficacy and community effectiveness of these vaccines overlapped and cost-effectiveness/benefit results were similar; therefore we reviewed them together.

Table 3 The cost-effectiveness of several recently introduced vaccines [2009 US dollars per year of life saved (YOLS) or quality adjusted life year (QALY)]

Vaccine	Target group	Economics
Pneumococcal conjugate	0–15 months	$100,334/YOLS [41]
Meningococcal conjugate	11-year olds	$140,849/YOLS [42]
Rotavirus	2–6 months	$217,848/YOLS [43]
Human papilloma virus	12+ year olds	$4,285/QALY [44]

4.2 Policy Implications

Despite these methodological problems, the consistency of findings across a wide range of methods and study populations demonstrates that among most population groups influenza vaccination is highly cost-effective.

Table 3 lists estimates for the cost-effectiveness of several other recently introduced vaccines that are widely recommended for children and preteens. Influenza vaccination in these age groups is generally cost saving compared with the societal costs of several thousand dollars for saving only one quality adjusted life year (QALY) for human papilloma virus (HPV) vaccine and more than one hundred thousand for only one year of life saved (YOLS) for pneumococcal vaccine. The US Advisory Committee on Immunization Practices recommends that all children (or preteens in the case of HPV) receive the vaccines listed in Table 3 plus an annual influenza vaccination. Canada recommends vaccination of the entire population annually. A few European countries recommend annual vaccination of children, despite the benefits demonstrated in this review, and only recommend vaccination for the elderly and those with chronic illness. Public health policymakers and health providers have undervalued influenza vaccination despite clear benefits.

References

1. Prosser LA, Bridges CB, Uyeki TM, Hinrichsen VL, Meltzer MI, Molinari NA, Schwartz B, Thompson WW, Fukuda K, Lieu TA (2006) Health benefits, risks, and cost-effectiveness of influenza vaccination of children. Emerg Infect Dis 12(10):1548–1558
2. Schmier J, Li S, King JC Jr, Nichol K, Mahadevia PJ (2008) Benefits and costs of immunizing children against influenza at school: an economic analysis based on a large-cluster controlled clinical trial. Health Aff (Millwood) 27(2):w96–104
3. Esposito S, Marchisio P, Bosis S, Lambertini L, Claut L, Faelli N, Bianchi C, Colombo GL, Principi N (2006) Clinical and economic impact of influenza vaccination on healthy children aged 2-5 years. Vaccine 24(5):629–635
4. Allsup S, Gosney M, Haycox A, Regan M (2003) Cost-benefit evaluation of routine influenza immunisation in people 65–74 years of age. Health Technol Assess 7(24):iii-x, 1–65. PMID: 14499051

5. Allsup S, Haycox A, Regan M, Gosney M (2004) Is influenza vaccination cost effective for healthy people between ages 65 and 74 years? A randomised controlled trial. Vaccine 16;23 (5):639–645
6. Simmerman JM, Lertiendumrong J, Dowell SF, Uyeki T, Olsen SJ, Chittaganpitch M et al (2006) The cost of influenza in Thailand. Vaccine 24(20):4417–4426
7. Newall AT, Scuffham PA (2008) Influenza-related disease: the cost to the Australian health-care system. Vaccine. Dec 9;26(52):6818–6823
8. Molinari N, Ortega-Sanchez I, Massonnier M, Thompson W, Wortley P, Weintraub E et al (2007) The annual impact of seasonal influenza in the US: measuring disease burden and costs. Vaccine 25:5086–5096
9. Szucs T (1999) The socio-economic burden of influenza. J Antimicrob Chemother 44:11–15
10. Congressional Budget Office (2006) A potential influenza pandemic: an update on possible macroeconomic effects and policy issues. Congressional Budget Office, Washington
11. Meltzer M, Cox N, Fukuda K (1999) The economic impacts of pandemic influenza in the united states: priorities for intervention. Emerg Infect Dis 5(5):659–671
12. Burns A, van der Mensbrugghe D, Timmer H (2008) Evaluating the economic consequences of avian influenza. The World Bank, Washington
13. McKibbin W, Sidorenko A (2006) Global macroeconomic consequences of pandemic influenza. Lowy Institute for International Policy, Sydney
14. Bloom E, deWit V, Carangal-San Jose MJ (2005) Potential Economic Impact of an Avian Flu Pandemic on Asia. Washington, DC: Asian Development Bank. Economics and Research Department. ERD Policy Brief Series #42
15. Sander B, Nizam A, Garrison LP (2009) Economic evaluation of influenza pandemic mitigation strategies in the united states using a stochastic microsimulation transmission model. Value Health 12(2):226–233
16. Weycker D, Edelsberg J, Halloran ME, Longini IMJ, Nizam A, Ciuryla V et al (2005) Population-wide benefits of routine vaccination of children against influenza. Vaccine 23 (10):1284–1293
17. Meltzer MI, Neuzil KM, Griffin MR, Fukuda K (2005) An economic analysis of annual influenza vaccination of children. Vaccine 23(8):1004–1014
18. Hibbert CL, Piedra PA, McLaurin KK, Vesikari T, Mauskopf J, Mahadevia PJ (2007) Cost-effectiveness of live-attenuated influenza vaccine, trivalent in preventing influenza in young children attending day-care centres. Vaccine 25(47):8010–8020
19. Marchetti M, Kuhnel UM, Colombo GL, Esposito S, Principi N (2007) Cost-effectiveness of adjuvanted influenza vaccination of healthy children 6 to 60 months of age. Hum Vaccin 3 (1):14–22
20. Dayan GH, Nguyen VH, Debbag R, Gomez R, Wood SC (2001) Cost-effectiveness of influenza vaccination in high-risk children in argentina. Vaccine 19(30):4204–4213
21. Navas E, Salleras L, Dominguez A, Ibanez D, Prat A, Sentis J et al (2007) Cost-effectiveness analysis of inactivated virosomal subunit influenza vaccination in children aged 3–14 years from the provider and societal perspectives. Vaccine 25(16):3233–3239
22. Salleras L, Navas E, Dominguez A, Ibanez D, Prat A, Garrido P et al (2009) Economic benefits for the family of inactivated subunit virosomal influenza vaccination of healthy children aged 3–14 years during the annual health examination in private paediatric offices. Vaccine 27(25–26):3454–3458
23. Skowronski DM, Woolcott JC, Tweed SA, Brunham RC, Marra F (2006) Potential cost-effectiveness of annual influenza immunization for infants and toddlers: experience from Canada. Vaccine 24(19):4222–4232
24. Cai L, Uchiyama H, Yanagisawa S, Kamae I (2006) Cost-effectiveness analysis of influenza and pneumococcal vaccinations among elderly people in Japan. Kobe J Med Sci 52 (3–4):97–109
25. Hoshi SL, Kondo M, Honda Y, Okubo I (2007) Cost-effectiveness analysis of influenza vaccination for people aged 65 and over in Japan. Vaccine 25(35):6511–6521

26. Wang ST, Lee LT, Chen LS, Chen TH (2005) Economic evaluation of vaccination against influenza in the elderly: an experience from a population- based influenza vaccination program in Taiwan. Vaccine 23(16):1973–1980

27. Scuffham PA, West PA (2002) Economic evaluation of strategies for the control and management of influenza in Europe. Vaccine 20(19–20):2562–2578

28. Postma MJ, Bos JM, van Gennep M, Jager JC, Baltussen R, Sprenger MJ (1999) Economic evaluation of influenza vaccination. Assessment for the Netherlands. Pharmacoeconomics 16 (Suppl 1):33–40

29. Gutierrez JP, Bertozzi SM (2005) Influenza vaccination in the elderly population in Mexico: economic considerations. Salud Pública Méx 47(3):234–239

30. Nichol KL, Goodman M (1999) The health and economic benefits of influenza vaccination for healthy and at-risk persons aged 65 to 74 years. Pharmacoeconomics 16(Suppl 1):63–71

31. Maciosek MV, Solberg LI, Coffield AB, Edwards NM, Goodman MJ (2006) Influenza vaccination health impact and cost effectiveness among adults aged 50 to 64 and 65 and older. Am J Prev Med 31(1):72–79

32. Ohkusa Y (2005) Policy evaluation for the subsidy for influenza vaccination in elderly. Vaccine 23(17–18):2256–2260

33. Gasparini R, Lucioni C, Lai P, Maggioni P, Sticchi L, Durando P et al (2002) Cost–benefit evaluation of influenza vaccination in the elderly in the Italian region of Liguria. Vaccine 20 (Suppl 5):B50–B54

34. Fitzner KA, Shortridge KF, McGhee SM, Hedley AJ (2001) Cost-effectiveness study on influenza prevention in Hong Kong. Health Policy 56(3):215–234

35. Lee PY, Matchar DB, Clements DA, Huber J, Hamilton JD, Peterson ED (2002) Economic analysis of influenza vaccination and antiviral treatment for healthy working adults. Ann Intern Med 137(4):225–231

36. Burls A, Jordan R, Barton P, Olowokure B, Wake B, Albon E et al (2006) Vaccinating healthcare workers against influenza to protect the vulnerable – is it a good use of healthcare resources? A systematic review of the evidence and an economic evaluation. Vaccine 24 (19):4212–4221

37. Bridges CB, Thompson WW, Meltzer MI, Reeve GR, Talamonti WJ, Cox NJ et al (2000) Effectiveness and cost-benefit of influenza vaccination of healthy working adults: a randomized controlled trial. JAMA 284(13):1655–1663

38. Turner D, Wailoo A, Nicholson K, Cooper N, Sutton A, Abrams K (2003) Systematic review and economic decision modeling for the prevention and treatment of influenza A and B. Health Technol Assess 7(35):iii–iv, xi–xiii, 1–170

39. Postma MJ, Jansema P, van Genugten ML, Heijnen ML, Jager JC, de Jong-van den Berg LT (2002) Pharmacoeconomics of influenza vaccination for healthy working adults: reviewing the available evidence. Drugs 62(7):1013–1024

40. Rothberg MB, Rose DN (2005) Vaccination versus treatment of influenza in working adults: a cost-effectiveness analysis. Am J Med 118(1):68–77

41. Lieu TA, Ray GT, Black SB, Butler JC, Klein JO, Breiman RF, Miller MA, Shinefield HR (2000) Projected cost-effectiveness of pneumococcal conjugate vaccination of healthy infants and young children. JAMA 283(11):1460–1468

42. Shepard CW, Ortega-Sanchez IR, Scott RD 2nd, Rosenstein NE (2005) Cost-effectiveness of conjugate meningococcal vaccination strategies in the United States. Pediatrics 115 (5):1220–1232

43. Widdowson MA, Meltzer MI, Zhang X, Bresee JS, Parashar UD, Glass RI (2007) Cost-effectiveness and potential impact of rotavirus vaccination in the United States. Pediatrics 119 (4):684–697

44. Kim JJ, Goldie SJ (2008) Health and economic implications of HPV vaccination in the United States. N Engl J Med 359(8):821–832

Index